Advanced Geotechnical Engineering

Soil–Structure Interaction Using
Computer and Material Models

Advanced Geotechnical Engineering

Soil–Structure Interaction Using Computer and Material Models

Chandrakant S. Desai
Musharraf Zaman

CRC Press
Taylor & Francis Group
Boca Raton London New York

CRC Press is an imprint of the
Taylor & Francis Group, an **informa** business

CRC Press
Taylor & Francis Group
6000 Broken Sound Parkway NW, Suite 300
Boca Raton, FL 33487-2742

© 2014 by Taylor & Francis Group, LLC
CRC Press is an imprint of Taylor & Francis Group, an Informa business

No claim to original U.S. Government works

Version Date: 20130819

International Standard Book Number-13: 978-1-4665-1560-4 (Hardback)

Library of Congress Cataloging-in-Publication Data

Desai, C. S. (Chandrakant S.), 1936-
    Advanced geotechnical engineering : soil-structure interaction using computer and material models / Chandrakant S. Desai, Musharraf Zaman.
        pages cm.
    Includes bibliographical references and index.
    ISBN 978-1-4665-1560-4 (hbk. : alk paper) 1. Soil-structure interaction. I. Zaman, Musharraf. II. Title. III. Title: Soil-structure interaction using computer and material models.

TA710.A1D43 2013
624.1′5136--dc23   2013032354

Visit the Taylor & Francis Web site at
http://www.taylorandfrancis.com

and the CRC Press Web site at
http://www.crcpress.com

To

Professor Hudson Matlock

and

Professor Lymon Reese

For their pioneering contributions to

computational geotechnical engineering

and to

Our wives and grandchildren

Patricia Lynn Desai, Lois Mira and Vernon Jay Divoll

and

Afroza Khanam Zaman and Tasneem Ayesha Chowdhury

For their love and support

Contents

Preface

Soil–structure interaction is a topic of significant importance in the solution of problems in geotechnical engineering. Conventional and ad hoc techniques are usually not sufficient to understand the mechanism and model the challenging behavior at the interfaces and joints prevalent in most structural and foundation systems. In addition to mechanical loading, the behavior of structures containing interfaces can be affected by environmental factors such as fluids, temperature, and chemicals.

Understanding and defining the behavior of engineering materials and interfaces or joints are vital for realistic and economic analysis and design of engineering problems. Hence, constitutive modeling that defines the behavior of the materials and interfaces, related testing, and validation assume high importance.

Owing to the complexities involved in many geotechnical problems, conventional procedures based on assumption of the linear elastic and isotropic nature of materials, and limit equilibrium procedures are found to be insufficient. Hence, we need to use modern computer-oriented procedures to account for factors, such as *in situ* stress, stress path, volume change, discontinuities and microcracking (initial and induced), strain softening, and liquefaction, which are not accounted for in most conventional methods. Hence, the objective of this book is to present various computer-based methods such as finite element, finite difference, and analytical.

The details of these methods are presented for the solution of one-, two-, and three-dimensional problems. Various constitutive models for geologic media ("solid") and interfaces, from simple to advanced, are included to characterize appropriately the behavior of a wide range of materials and interfaces.

Wherever possible, we have included simple problems that can be solved by hand, which is an essential step to understanding problems requiring the use of computers. A number of examples for one-, two-, and three-dimensional problems solved by using finite element, finite difference, and analytical methods are also presented. As an exercise for students and readers, a number of problems, often with partial solutions, are included at the end many chapters.

The book can be used for courses at the graduate and undergraduate (senior) levels for students who have backgrounds in geotechnical, structural engineering, and basic mechanics courses, including matrix algebra and numerical analysis; a background in numerical methods (finite element, finite difference, etc.) will be valuable to understand and apply the procedures in the book.

Practitioners interested in the analysis and design of geotechnical structures can benefit by using the book. They can use the available codes or acquire them from sources listed in Appendix 2; most of such codes can be used on desktop and laptop computers. The book can also be useful to researchers to get acquainted with the available developments, and with advances beyond the level of topics addressed in the book.

This book presents the contributions of the authors and other persons and covers a wide spectrum of geotechnical problems that extend over the last four decades or so. It emphasizes the application of modern and powerful computer methods and analytical techniques for the solution of such challenging problems, with special attention to the significant issue of material constitutive modeling.

Pioneering applications of numerical methods for solution of challenging problems in geotechnical engineering have taken place from the start of the computer age. Over the last few decades, impressive advances have occurred for constitutive modeling of geomaterials and interfaces/joints. Applications of computer and constitutive models for analysis and design are expected to continue and increase. We believe that this book can provide an impetus to the continuing growth.

A number of our students and coworkers have participated in the development and application of constitutive and computer models presented in this book; their contributions are cited through references in various chapters. We express our sincere thanks for their contributions. We cite only a few of them here: M. Al-Younis, G.C. Appel, S.H. Armaleh, B. Baseghi, B. Barua, E.C. Drumm, M.O. Faruque, K.H. Fuad, G. Frantziskonis, H.M. Galagoda, M. Gong, M.M. Gyi, Q.S.E. Hashmi, K.E. El-Hoseiny, D.R. Katti, D.C. Koutsoftas, T. Kuppusamy, G.C. Li, Y. Ma, S. Nandi, A. Muqtadir, I.J. Park, J.V. Perumpal, S. Pradhan, H.V. Phan, S.M. Rassel, D.B. Rigby, M.R. Salami, N.C. Samtani, S.K. Saxena, F. Scheele, H.J. Siriwardane, K.G. Sharma, C. Shao, S. Somasundaram, J. Toth, V. Toufigh, K. Ugai, A. Varadarajan, L. Vulliet, G.W. Wathugala, and D. Zhang.

We express our deep appreciation to our parents who have been sources of learning and loving support.

Chandrakant S. Desai
Tucson, Arizona

Musharraf Zaman
Norman, Oklahoma

Authors

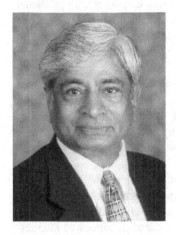

Chandrakant S. Desai is a Regents' Professor (Emeritus), Department of Civil Engineering and Engineering Mechanics, University of Arizona, Tucson, Arizona, USA. From January to April, 2012, he was a visiting professor at the Indian Institute of Technology, Gandhinagar, Gujarat, India (IITGn), and a distinguished visiting professor at the Indian Institute of Technology, Bombay, from January to February, 2013.

Dr. Desai is recognized internationally for his significant and outstanding contributions in research, teaching, applications, and professional work in a wide range of topics in engineering. The topics he has contributed to include material (constitutive) modeling, laboratory and field testing, and computational methods for interdisciplinary problems in civil engineering related to geomechanics/geotechnical engineering, structural mechanics/structural engineering; dynamic soil–structure interaction and earthquake engineering; coupled fluid flow through porous media; and some areas in mechanical engineering (e.g., electronic packaging).

Dr. Desai has authored/coauthored/edited 22 books in the areas of finite element method and constitutive modeling, and 19 book chapters, and has authored/coauthored about 320 technical papers in refereed journals and conferences.

Dr. Desai's research on the development of the innovative *disturbed state concept* (DSC) for constitutive modeling of materials and interfaces/joints has been accepted for research and teaching in many countries. In conjunction with the nonlinear finite element method, it provides an innovative and alternative procedure for analysis, design, and reliability of challenging nonlinear problems of modern technology.

Nowadays, the finite element method using computers has been the premier procedure for research, teaching, analysis and design in engineering. Dr. Desai's book, *Introduction to the Finite Element Method* (coauthored with J.F. Abel), published in 1972, was the first formal textbook on the subject in the United States (second internationally). It has been translated into a number of languages, including an Indian edition. In 1979, he authored the first text (*Elementary Finite Element Method*) for teaching the finite element method to undergraduate students.

Understanding and defining the behavior of materials that compose engineering systems are vital for realistic and economical solutions. His book on *Constitutive Laws for Engineering Materials* (Desai and Siriwardane) in 1984 is considered to be pioneering on the subject and presented various material models based on continuum mechanics. In 2001, he authored the book *Mechanics of Materials and Interfaces: The Disturbed State Concept (DSC)* that presents an innovative approach for modeling materials and interfaces in a unified manner, combining the continuum mechanics models and a novel idea for introducing the important aspect of discontinuities in deforming materials. In 1977, he coedited (Desai and Christian) the first book on *Numerical Methods in Geotechnical Engineering* that deals with problems from geotechnical and structural engineering, which included his own contributed chapters. These books on constitutive modeling and the finite element method have been adopted and used in academia and practiced in a number of engineering disciplines.

He was the founding general editor of the *International Journal for Numerical and Analytical Methods in Geomechanics* from 1977 to 2000, published by John Wiley, UK. He was the founding editor-in-chief of the *International Journal of Geomechanics* (IJOG) from 2001 to 2010, published by the Geo Institute, American Society of Civil Engineers (ASCE); he continues to serve as the advisory editor for this journal.

He was the founding president of the International Association for Computer Methods and Advances in Geomechanics (IACMAG). He is credited with introducing the interdisciplinary definition of *geomechanics* that involves various areas such as geotechnical engineering and rock mechanics, static and dynamics of interacting structures and foundations, fluid flow through porous media, geoenvironmental engineering, natural hazards such as earthquakes, landslides, and subsidence, petroleum engineering, offshore and marine technology, geological modeling, geothermal energy, ice mechanics, and lunar and planetary structural systems. He has served on the editorial boards of 14 journals, and has been the chair/member of a number of committees of various national and international societies and conferences.

Dr. Desai has been involved in consulting work for the solutions of practical problems for a number of private, public, and international agencies. For the latter, he has served as a consultant for UNESCO for (i) computer analysis and design in the Narmada Sardar Sarovar Project; (ii) tunneling projects in the Himalayas; (iii) the development of testing equipments at Central Material Testing Laboratory, New Delhi; and (iv) the development of material testing equipment at the Technological Institute, M.S. University of Baroda, Vadodara, Gujarat.

The body of his research, publications, and professional work has been seminal and original and has changed the direction of research, teaching, and design applications for a number of areas in civil and other engineering disciplines, for which he has received national and international awards and recognitions such as the Distinguished Member Award by the American Society of Civil Engineers (ASCE), which is the second-highest recognition (the first being president of the ASCE); Distinguished Alumni Award by VJTI, Mumbai (VJTI was established in 1887); the Nathan M. Newmark Medal by the Structural Engineering and Engineering Mechanics Institutes, ASCE; the Karl Terzaghi Award by the Geo Institute, ASCE; the Diamond Jubilee Honor by the Indian Geotechnical Society, New Delhi, India; the Suklje Award/Lecture by the Slovenian Geotechnical Society, Slovenia; the HIND Rattan (Jewel of India) Award by the NRI Society, New Delhi; the Meritorious Civilian Service Award by the U.S. Corps of Engineers; the Alexander von Humboldt Stiftung Prize by the German Government; the Outstanding Contributions Medal by the International Association for Computer Methods and Advances in Geomechanics;

the Meritorious Contributions Medal in Mechanics by the Czech Academy of Science, Czechoslovakia; and the Clock Award for outstanding contributions in thermomechanical analysis in electronic packaging by the Electrical and Electronic Packaging Division, American Society of Mechanical Engineers (ASME).

In 1989, the Board of Regents of the Arizona's Universities System conferred upon him the prestigious award Regents' Professor.

For his teaching excellence, he has received the Five Star Faculty Teaching Finalist Award and the El Paso Natural Gas Foundation Faculty Achievement Award at the University of Arizona, Tucson, Arizona. He has received two Certificates of Excellence in Teaching at Virginia Tech, Blacksburg, Virginia.

 **Musharraf Zaman** holds the David Ross Boyd Professorship and Aaron Alexander Professorship in Civil Engineering and Environmental Science at the University of Oklahoma in Norman. He is also an Alumni Chair Professor in the Mewbourne School of Petroleum and Geological Engineering. He has been serving as the associate dean for research and graduate programs in the University of Oklahoma College of Engineering since July 2005. Under his leadership, research and scholarship in the OU College of Engineering (COE) have expanded at an accelerated pace. Under his leadership, the research expenditures in CoE increased steadily, reaching more than $22 million annually. The total funding in force in 2012 was $83 million.

During the past 15 years, he has provided leadership in introducing an interdisciplinary, graduate-level program in the asphalt area and developed two major laboratories: Broce Asphalt Laboratory and Asphalt Binders Laboratory.

He received his baccalaureate degree from the Bangladesh University of Engineering and Technology, Bangladesh, his master of science degree from Carlton University, Canada, and his PhD degree from the University of Arizona, Tucson; all his degrees were in civil engineering. During his 30 plus years of service at the University of Oklahoma, he has introduced six new graduate courses, taught a variety of undergraduate and graduate courses, supervised more than 80 master theses and doctoral dissertations to completion, received more than $15.5 million in external funding, published more than 150 journal and 215 peer-reviewed conference proceedings papers, published 8 book chapters, edited 2 books, and served on the editorial boards of several prestigious journals including the *International Journal of Numerical and Analytical Methods of Geomechanics, Journal of Petroleum Science and Engineering, International Journal of Geotechnical Engineering*, and *International Journal of Pavement Research and Technology*. He has been serving as the editor-in-chief of the *International Journal of Geomechanics, ASCE* for the past three years.

In recognition of his teaching excellence, he has received several regional and national awards from the American Society of Engineering Education and the David

Ross Boyd Professorship—the highest lifetime teaching award given by the University of Oklahoma. He has also received several research awards, including the Regent's Award for Superior Research and Creative Activity and Presidential Professorship. His research papers have received international awards from the International Association for Computer Methods and Advances in Geomechanics (IACMAG) and the Indian Geotechnical Society. In 2011, he received the Outstanding Contributions Award from IACMAG, in recognition of his lifetime achievement in geomechanics.

1 Introduction

Engineering structures are made of materials from matter found in the universe. At every level—atomic, nano, micro, and macro—the components in the matter interact with each other and merge together continuously, assuming various states that, for example, identify initiation (birth) to the end or failure (death). The interaction or coupling plays a significant role in the response of materials and hence, engineering systems. Then, the coupling under external influences or forces at the micro level between particles, fluid, air, temperature, and chemicals in a material element, between structural and geologic materials at interfaces, and between rock masses and joints is of utmost importance. Thus, the understanding and the characterization of the behavior of materials are vital for the solution of geotechnical systems. This book attempts to address these issues.

For a long time, geotechnical engineers have used simplified and computer-oriented schemes to analyze and design problems that involve advanced conditions such as soil–structure interaction and the effect of coupled behavior of fluid–geologic materials on the response of structures and foundations. They have involved analytical and closed-form solutions based on linear elastic Boussinesq's theory, and limit equilibrium for evaluating ultimate or failure loads. However, many geotechnical problems are affected by important factors that cannot be handled by simplified and conventional solution procedures. Such factors may include nonhomogeneity and layering, arbitrary geometries, nonlinear material behavior, special behavior of interfaces and joints, interaction between structures and geologic materials, effects of fluids and other environmental factors, and repetitive and dynamic loadings. The realistic solutions of practical problems involving those factors require the use of numerical techniques such as the finite difference (FD), finite element (FE), and boundary element (BE) methods based on the use of modern computers.

The computer-based methods have been developed and are available for practical analysis and design by the geotechnical engineer. They have been published in a wide range of professional journals, and sometimes briefly described in geotechnical books. However, a systematic and comprehensive treatment is not yet available.

The main emphasis in this book is on soil–structure interaction problems. Since many problems involve some level of interaction between structures and geologic materials (soils and rocks), often influenced by fluids (seepage, consolidation, and coupled fluid–solid effect), they are also covered. In view of the scope of the book, it will be useful to practitioners, students, teachers, and researchers.

The rational analysis of the soil–structure interaction, one of the most challenging topics, can be achieved by using modern computer methods. *Relative motions* between the structure and surrounding geologic materials occur at interfaces (or joints) between them. Figure 1.1 shows the schematic of a building on soil or rock foundations, which can contain interfaces and joints; the latter two are collectively

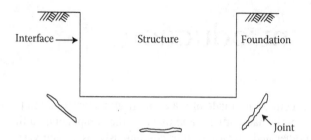

FIGURE 1.1   Schematic of structure–foundation interaction.

referred to as interfaces. The interaction or coupling at interfaces can change the distributions and magnitudes of strains and stresses compared to those evaluated from conventional methods, which often assume that the soil and the structure are "glued" together, and they do not experience any relative motions. The influence of the relative motions can cause, for instance, reduction of stresses leading to arching effects. In dynamic analysis, neglecting the interaction can result in far different stresses and displacements compared to those that actually occur. Also, the liquefaction predictions can be more realistic if the relative motions are taken into account.

Another significant factor is the *nonlinear behavior* for modeling the mechanical response of geologic materials and interfaces. This issue presents a rather formidable task because the behavior of geologic materials is much more challenging to understand and define. The topic of constitutive models of materials and interfaces thus becomes vital for realistic solution of practical geotechnical problems. This book presents theoretical, experimental, and validation aspects for constitutive models. However, more comprehensive treatments for constitutive models are available in other books and papers, for example, Refs. [1,2].

Computer methods for a wide range of geotechnical problems are presented in this book, together with advanced and realistic constitutive models. For the application of computer methods, it is essential to ascertain the validity of computer and constitutive models. Hence, a part of this book is devoted to practical problems solved by hand calculations and by the use of computer methods. Detailed treatments of various computer methods are available in the FE method—[3,4 (revised by Desai and Kundu), 5,6]; the FD method [7–9]; and the BE method [10–13].

1.1   IMPORTANCE OF INTERACTION

Engineering structures very often involve the use of more than one material in contact. Each material possesses specific behavioral characteristics; however, when in contact, the behavior of the composite system is influenced by the response of each material as well as the interaction or coupling between the materials.

Figure 1.2 shows a schematic of two materials, as a composite. The deformation mechanism at the interface or joint is influenced by the relative motions between the materials, which may constitute translations (Figure 1.2b), rotation (Figure 1.2c), and interpenetration (Figure 1.2d). The behavior of a "solid" material is influenced by the mechanisms between the particles at micro or higher levels. The behavior of an

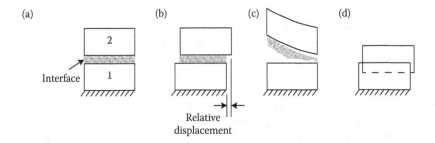

FIGURE 1.2 Modes of deformation in interfaces and joints. (a) Interface (joint); (b) translation; (c) rotation; and (d) interpenetration.

interface is influenced by mechanisms between particles of two different materials, as well as particles of different materials such as gouge (in joints) (Figure 1.2a).

Many times in the past, structure–foundation systems were analyzed and designed by assuming no relative motions between them, that is, by assuming that they are "glued" to each other. However, relative motions do occur at the interfaces causing a significant effect on the overall behavior of the system. Hence, it is imperative to define and include the behavior of interfaces in the analysis and design of structures founded on or in geologic materials.

A main objective of this book is to give comprehensive consideration to the modeling, testing, and calibration of models for the interfaces or joints, and application for the solution of practical problems in geotechnical engineering.

1.2 IMPORTANCE OF MATERIAL BEHAVIOR

The behavior of engineering materials and interfaces is influenced significantly by the nature and composition of the materials; mathematical models to define the behavior based on laboratory and/or field testing are called *constitutive models*. Such a model for appropriate and realistic characterization of the mechanical behavior of the "solid" materials (geologic and structural) and interfaces is vital for the solution of practical problems. To apply a model for practical problems, it is necessary to validate constitutive models at the specimen level and boundary value problem level, toward their safe use for practical problems. The importance of realistic modeling cannot be overemphasized!

1.2.1 LINEAR ELASTIC BEHAVIOR

In simplified methods for computing displacements and stresses in a geologic (soil or rock) mass, an assumption is generally made that the material (geologic) behaves as linear elastic or piecewise linear elastic. Indeed, this is based on a gross assumption that the stress–strain behavior is linear, and when the applied load is removed, the geologic material returns to its original configuration. As most geologic materials are not linear, sometimes, nonlinearity is simulated incrementally by using a higher-order mathematical function to represent the stress–strain behavior, for example, hyperbola, parabola, and splines. In such piecewise elastic models, the theory of elasticity is still assumed in each increment. The linear elastic behavior is valid for very limited analysis and design because the actual behavior of geologic materials exhibits irreversible or inelastic or plastic deformations.

1.2.2 INELASTIC BEHAVIOR

In the case of inelastic or plastic behavior in which a material does not return to its original configuration, but retains certain levels of strains (deformation), various models based on the theory of plasticity can be invoked. In the conventional plasticity models such as von Mises, Drucker–Prager, and Mohr–Coulomb, the behavior is assumed to be elastic until the material reaches a specific yield condition, defined often by the yield stress. Thereafter, the material enters into the plastic range, guided by conditions such as a yield criterion and "flow" rule that defines the plastic flow, like a "liquid." Although these models provide some improvements over the linear (or nonlinear) elastic models, particularly in computing the ultimate or failure strength of the material, they do not provide realistic predictions for the entire stress–strain response, as explained later.

1.2.3 CONTINUOUS YIELD BEHAVIOR

Most geologic materials exhibit inelastic or plastic behavior almost from the beginning of the loading; in other words, every point on the stress–strain curve designates a yield point. For a plasticity formulation, this requires the application of a yield criterion and the flow rule from the beginning of the loading. The critical state—Ref. [14], cap—Ref. [15], and hierarchical single surface (HISS) plasticity models—Refs. [2,16,17] can handle the condition of yield from the start, and can be called "continuous yield" models.

1.2.4 CREEP BEHAVIOR

Also, many geologic materials and interfaces exhibit time-dependent creep behavior, that is, they continue to experience increase (growth) in strain under constant stress or continue to experience change (relaxation) in stress under constant strain [2].

1.2.5 DISCONTINUOUS BEHAVIOR

We cannot miss the fact that a geologic material, made of (billions) of particles, contains "discontinuities" due to microstructural modifications during loading affected by factors such as particle sliding, separation, and riding over each other. The discontinuities can occur from the beginning (initially) and during deformations. Then, the models based on elasticity, plasticity, viscoelasticity, and so on may have only limited validity for geologic materials because they are based on the assumption of "continuity" between particles, that is, particles in a material maintain their neighborhoods during deformations.

It is difficult to develop a theoretical model for discontinuous materials. Hence, almost all available models proposed for accounting discontinuities are based on a combination of continuous and discontinuous behaviors. In other words, they

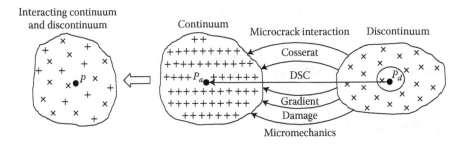

FIGURE 1.3 Interacting continuum and discontinuum and approximate models. (Adapted from Desai, C.S., *Mechanics of Materials and Interfaces: The Disturbed State Concept*, CRC Press, Boca Raton, FL, 2001.)

introduce, or superimpose, models for discontinuities on those for continuous behavior. There are a number of models proposed to account for discontinuities, for example, classical damage, damage with external enrichments, microcrack interaction, micromechanics, gradient, and Cosserat theories. Figure 1.3 shows a schematic of various models that introduce schemes for discontinuities into the procedures based on continuum theories [2].

Under a combination of loading (mechanical and environmental), a deforming material can experience microstructural modifications which may result in microcracking leading to fracture, and failure (softening), or healing leading to stiffening. The major attention is given to microcracking, leading to failure because most of the problems considered in this book involve that aspect.

The disturbed state concept (DSC) is based on the consideration that the observed behavior of a (dry) deforming material can be expressed in terms of the behaviors of the continuum part called relative intact (RI), and the other part generated by the asymptotic state reached by the microcracked part called fully adjusted (FA). The disturbance function connects the two parts, thereby representing the coupling between the RI and FA parts. The DSC allows for the microstructural modifications *intrinsically*; hence, it provides certain advantages such as avoidance of spurious mesh dependence over other models. A major advantage of the DSC is that its mathematical framework can be specialized for interfaces and joints. Finally, the DSC provides a unique unified approach, it contains most available models as special cases, and it is hierarchical. Further details of the DSC are presented in Ref. [2].

Appendix 1 presents a brief description of various constitutive models for solids and interfaces or joints that are used successfully for the solution of soil–structure interaction problems. The applications of some practical problems are presented in this book.

1.2.6 MATERIAL PARAMETERS

The definition of a constitutive model involves a number of parameters that can be functions of certain factors influencing the behavior, or can be constant. For realistic prediction of practical problems, the values of the parameters must be determined from appropriate laboratory and/or field tests. This topic is not within the scope of this book; it has been covered in other publications such as Ref. [2]. A brief description is given in Appendix 1.

Also, when a constitutive model is used for a given application in this book, the values of the parameters with brief background and references are provided with the application.

1.3   RANGES OF APPLICABILITY OF MODELS

Figure 1.4 shows the schematic nonlinear behavior exhibited by a typical geologic material. It shows the linear elastic behavior for a small range, marked 1 in the figure. Hence, under a load P (stress σ), only the strain $\bar{\varepsilon}$ can be predicted. In other words, a linear elastic model can be used safely when the load is limited to the initial elastic region, with a larger factor of safety. However, in reality, the actual strain under the load will be ε^a, and it is necessary to adopt a plasticity model. If the behavior exhibits strain softening after the peak, it is essential to adopt models that account for (induced) discontinuities and occurrences of instabilities such as failure and liquefaction.

In general, for realistic analysis and design of soil–structure interaction problems, nonlinear behavior involving elastic, plastic, and creep, microcracking leading to fracture and failure may have to be considered. Indeed, specialized version(s) accounting for only relevant factors can be used, depending upon the need of specific material in a given problem. For example, a soft clay may need only an associative plasticity model, whereas an overconsolidated clay may need a version that allows microcracking. In this context, the hierarchical property of the DSC model allows the user to choose a model depending upon the specific material behavior.

1.4   COMPUTER METHODS

Before the advent of the computer, the analysis and design of engineering problems were based very often on empirical consideration and the use of closed-form solution

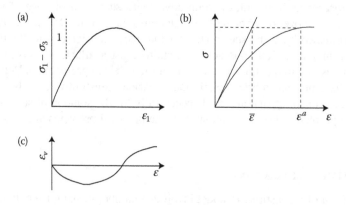

FIGURE 1.4 Schematic of nonlinear behavior of geologic material: (a) stress–strain and elastic range; (b) stress–strain and plastic range; and (c) volumetric.

of simplified mathematical equations that govern the behavior. However, these methods usually were not capable of accounting for behavioral aspects such as nonlinear response of the materials and interfaces, complex geometrics, and loadings. With the use of the computer for methods such as FE, many complex factors can be accounted for, which was not possible by the empirical and mathematical closed-form solutions.

Although computers can be used for closed-form solutions, they can be used much more efficiently for the solution of much more realistic problems involving nonlinear behavior, complex boundary conditions, nonhomogeneous nature of material systems, and realistic loading conditions. FD, FE, and BE methods have been employed for realistic solutions of soil–structure problems. This book provides details of the FE and FD methods in various chapters, relevant to the specific problems considered.

1.5 FLUID FLOW

Many geotechnical and soil–structure interaction problems are affected by the existence of fluid, and fluid or pore water pressure in the foundation soils. Hence, the effect of fluid pressure needs to be considered in the analysis and design of geotechnical and soil–structure interaction problems. We have presented descriptions of confined and unconfined (free surface) seepage, consolidation, and coupled fluid–solid behavior, in various chapters.

1.6 SCOPE AND CONTENTS

Conventional methods are often not capable of handling many significant factors that influence the behavior of geotechnical systems. This book emphasizes the application of modern and powerful computer methods and analytical techniques for the solution of such challenging problems. The mechanical behavior of the materials involved in geotechnical problems plays a vital role in the reliable and economical solutions for analysis and design. Hence, the use of computer methods with appropriate and realistic constitutive models is the main theme of this book.

Chapter 1 introduces the objective and main factors important in the application of computer methods and constitutive models for geologic material and interfaces/joints. Chapter 2 presents a comprehensive treatment of analytical and numerical (FD and FE) methods, for solution of problems that can be idealized as one-dimensional (1-D) (beams, piles, and retaining walls). Simplified constitutive models such as linear elastic, nonlinear elastic, and resistance–displacement curves are used in this chapter.

Chapter 3 comprises the FE method for problems that are idealized as two-dimensional (2-D) and three-dimensional (3-D). 2-D applications included in this chapter are footings (circular, square, and rectangular), piles, dams, embankments, tunnels, retaining structures, reinforced earth, and pavements. Constitutive models based on conventional elasticity and plasticity, elastoviscoplasticity, continuous yield plasticity, HISS plasticity, and DSC, capable of modeling softening and degradation in materials, are used in this and subsequent chapters. The parameters for constitutive models used are presented with applications, while their details are presented in Appendix 1. The application for 3-D FE method and an approximate procedure called *multicomponent* method are covered in Chapter 4.

In Chapter 5, 2-D and 3-D FE and 2-D FD methods are applied to the *seepage* (flow through porous nondeformable media), which is a specialized form of the general flow through porous deformable media. A number of problems involving 2-D and 3-D simulation for dams, embankments, river banks, and wells are described in this chapter.

Chapter 6 contains FD and FE applications for 1-D consolidation and settlement analyses. Chapter 7 contains formulation of the general problem, which involves coupled behavior between deformation and fluid pressure for static, quasi-static, and dynamic analyses. It includes the application to a number of problems such as dams and piles involving field measurements, shake table and centrifuge tests. Such important topics and the effect of interface response on the behavior of geotechnical systems and liquefaction (considered as a microstructural instability) are discussed in Chapter 7.

Appendix 1 gives details of various constitutive models with the parameters used in this book. A major emphasis is given on the models developed and used by the authors of this book; appropriate references are cited for the use of other models. Thus, the reader can consult Appendix 1 for details of the constitutive models used in various chapters. Appendix 2 presents concise descriptions of various computer codes developed by the authors and used for the solution of problems in various chapters. The list also includes some other available computers codes that can be used for the solution of problems covered in this book.

REFERENCES

1. Desai, C.S. and Siriwardane, H.J., *Constitutive Laws of Engineering Materials*, Prentice-Hall, Englewood Cliffs, NJ, 1984.
2. Desai, C.S., *Mechanics of Materials and Interfaces: The Disturbed State Concept*, CRC Press, Boca Raton, FL, 2001.
3. Desai, C.S. and Abel, J.F., *Introduction to the Finite Element Method*, Van Nostrand Reinhold, New York, 1972.
4. Desai, C.S., *Elementary Finite Element Method*, Prentice-Hall, Englewood Cliffs, NJ, 1977. Revised as Desai, C.S. and Kundu, T., *Introductory Finite Element Method*, CRC Press, Boca Raton, Fl, 2001.
5. Zienkiewicz, O.C., *The Finite Element Method*, 3rd Edition, McGraw-Hill, London, UK, 1997.
6. Bathe, K.J., *Finite Element Procedures*, Prentice-Hall, Englewood Cliffs, NJ, 1996.
7. Crandall, S.H., *Engineering Analysis*, McGraw-Hill Book Company, New York, 1956.
8. Forsythe, G.E. and Wasow, W.R., *Finite Difference Methods for Partial Differential Equations*, Dover Publications, UK, 2001.
9. Leveque, R.J., *Finite Difference Methods for Ordinary and Partial Differential Equations: Steady State and Time Dependent Problems*, Society for Industrial and Applied Mathematics (SIAM), Philadelphia, PA, USA 2007.
10. Brebbia, C.A. and Walker, S., *Boundary Element Technique in Engineering*, Newnes-Butterworths, London, 1980.
11. Banerjee, P.K. and Butterfield, R, *Boundary Element Method in Engineering Science*, McGraw-Hill Book Co., UK, 1981.
12. Liggett, J.A. and Liu, P.L.F., *The Boundary Integral Equation Method for Porous Media Flow*, George Allen and Unwin, London, 1983.

13. Aliabadi, F., *The Boundary Element Method Applications, Vol. 2: Solids and Structures*, Wiley, UK, 2002.
14. Roscoe, H.H., Sohofield, A.N., and Wroth, C.P., On yielding of soil, *Geotechnique*, 8, 1, 1958, 22–53.
15. DiMaggio, F.L. and Sandler, I., Material model for granular soils, *Journal of Engineering Mechanics, ASCE*, 97, 3, 1971, 935–950.
16. Desai, C.S., A general basis for yield, failure and potential functions in plasticity, *International Journal for Numerical and Analytical Methods in Geomechanics*, 4, 1980, 361–375.
17. Desai, C.S., Somasundaram, S., and Frantziskonis, G., A hierarchical approach for constitutive modeling of geologic materials, *International Journal for Numerical and Analytical Methods in Geomechanics*, 10, 3, 1986, 225–257.

2 Beam-Columns, Piles, and Walls
One-Dimensional Simulation

2.1 INTRODUCTION

If a 1-D structural member is symmetrical about its axis and the loading is also symmetric, then it can be idealized as a 1-D column, simulated by an equivalent line. If such a symmetrical structure is subjected to a lateral load, it can also be idealized by using an equivalent 1-D line (Figure 2.1). For axial and lateral loads, the beam-column idealization is shown in Figure 2.1b. It should be noted that such a beam-column idealization represents a linear superposition of effects of the column and the beam, and it does not take into account nonlinearity, which may lead to effects such as buckling.

Figure 2.1 shows a pile structure subjected to axial and lateral loads; it can be analyzed as a classical beam bending and column problem. Hence, we will first present closed-form solutions for the beam-bending problem that has been used to analyze laterally loaded single piles embedded in linear elastic soils. We will also present, in a subsequent chapter, a brief description of closed-form and numerical solutions for slab (2-D) on linear elastic soils; such solutions can be used for analysis of foundation slabs, rafts, and similar other structures.

2.2 BEAMS WITH SPRING SOIL MODEL

Deformable beams and slab resting on soils can involve complex coupled behavior between the structure and soil foundation. The simple spring model commonly used to represent the soil behavior was proposed by Winkler [1], which was also used by Euler, Füss, and Zimmerman [2]. It is based on the assumption that the soil behavior can be modeled by using a series of *independent* springs (Figure 2.2), which are often assumed as linear elastic. Since the behavior of soil is often nonlinear and coupled, the Winkler model is considered to be an approximate representation of the soil response. A number of publications [2–9] present applications of the Winkler soil model.

2.2.1 GOVERNING EQUATIONS FOR BEAMS WITH WINKLER MODEL

In the Winkler model, the displacement at a point in the soil, v, along the y-direction is assumed to be a linear function of the soil pressure, p, at the point, and is given by

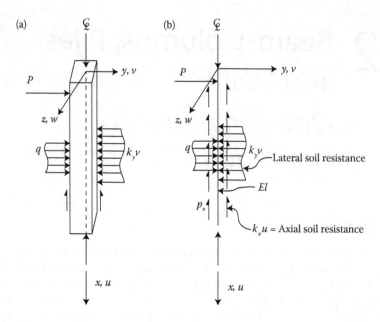

FIGURE 2.1 One-dimensional idealization of pile. (a) Pile (long); (b) idealized pile.

$$p = k_o v \tag{2.1}$$

For linear elastic response, k_o represents the property (stiffness) of soil in the y-direction (Figure 2.2); the units of k_o are F/L^3. It (k_o) is referred to by various names such as *subgrade modulus, coefficient of subgrade reaction*, and *spring modulus*, with dimension, F/L^3. The Winkler model was initially used for computing stresses and deformations under railroad systems.

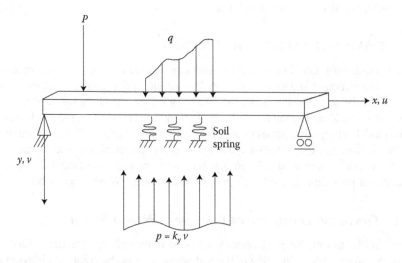

FIGURE 2.2 One-dimensional beam.

Since the soil is considered to be a "solid" body, the definition of continuum model may require more than one constants; even for simple linear elastic and isotropic behavior, two constants such as Young's modulus, E, and Poison's ratio, v, or shear modulus, G, and bulk modulus, K, are required. Then, for a 2-D and 3-D medium, the coupled behavior between the vertical and horizontal responses is included. However, the Winkler model considers only vertical behavior, and the coupled effect in the horizontal direction is ignored. Hence, the Winkler model can provide only an approximate response for the continuum material. Vlaslov and Leontiev [5] modified the Winkler model by adopting the linear elastic medium with E and v, thereby improving the coupling effect. In the following, we consider flexible beams; rigid beams are not included because of their limited application in practice.

2.2.2 GOVERNING EQUATIONS FOR FLEXIBLE BEAMS

For a 2-D beam, the subgrade modulus, k_o (Equation 2.1), is multiplied by the width of the beam, b, to give

$$k = k_o b \qquad (2.2)$$

The units of k, then, are F/L^2.

The beam can be subjected to different forces such as distributed (q) and point (P) loads (Figure 2.2). Then, the effective pressure on the beam is $(p - q)b$, where p is the soil resistance.

The governing differential equation (GDE) for the beam can be derived as [10]

$$\frac{d^2 M}{dx^2} = pb \qquad (2.3)$$

Substituting $p = k_o v$ and $k = k_o b$, Equation 2.3 transforms to

$$\frac{d^2 M}{dx^2} = k_o v \cdot b = kv \qquad (2.4)$$

Now, the relation between the bending moment, M, and displacement, v, can be expressed as

$$EI \frac{d^2 v}{dx^2} = -M \qquad (2.5)$$

where EI is the flexural or bending stiffness and I is the moment of inertia of the beam. Differentiating Equation 2.5 twice with respect to x, we obtain

$$EI \frac{d^4 v}{dx^4} = -\frac{d^2 M}{dx^2} \qquad (2.6)$$

Then, substitution from Equation 2.4 gives

$$EI \frac{d^4 v}{dx^4} + kv = 0 \tag{2.7a}$$

or

$$EI \frac{d^4 v}{dx^4} + (kv - q) = 0 \tag{2.7b}$$

where q is applied distributed lateral load (Figure 2.2).

Alternatively, Equation 2.4 can be differentiated twice to obtain

$$\frac{d^4 M}{dx^4} + \frac{k}{EI} M = 0 \tag{2.8}$$

in which $d^2 v/dx^2$ from Equation 2.5 has been substituted.

2.2.3 SOLUTION

We can express the closed-form solution for v in Equation 2.7 as follows [11–16]:

$$v = (A \cos \lambda x + B \sin \lambda x)e^{\lambda x}$$
$$+ (C \cos \lambda x + D \sin \lambda x)e^{-\lambda x} \tag{2.9}$$

where A, B, C, and D are constant coefficients that can be determined from given boundary conditions, and

$$\lambda = \sqrt[4]{\frac{k}{4EI}} \quad \text{or} \quad \frac{k}{EI} = 4\lambda^4 \tag{2.10}$$

This definition of λ includes effects of the subgrade reaction and the bending stiffness (EI), which is the property of the structure (beam). The dimension of λ is 1/L, and its inverse is the characteristic length of the beam–soil system.

If the beam is much stiffer compared to the soil, the characteristic length is large, resulting in greater beam displacement for a significant distance from the point to where the load is applied. Inversely, when the characteristic length is smaller and the beam is softer compared to the soil, the deflection can be localized near the zone where the load is applied.

The solution for v (Equation 2.9) contains two terms that indicate exponential growth, whereas the other two terms show exponential decay with distance. The coefficients A, B, C, and D in Equation 2.9 can be determined from the available boundary conditions. We will discuss the derivation of A, B, C, and D when the beam equation is applied for piles with various boundary conditions.

2.3 LATERALLY LOADED (ONE-DIMENSIONAL) PILE

As discussed before, a laterally loaded pile can be analyzed as a beam on soil foundation, which is often idealized as linear elastic, represented by the Winkler model.

Figure 2.3b shows a schematic of a 1-D pile subjected to a moment M_t and load P_t at the top of the pile or the mudline. Assuming that the soil can be represented by a linear elastic spring, a series of such springs are shown along the length of the pile in Figure 2.3a. We now consider the sign convention for the problem, which, for various quantities such as lateral pressure, p, shear force, V, bending moment, M, gradient or slope, S, and deflection or displacement, v, are shown in Figure 2.4.

The analytical or closed-form solution for the displacement, v, for the pile can be expressed as for the beam (Equation 2.9).

2.3.1 COEFFICIENTS A, B, C, D: BASED ON BOUNDARY CONDITIONS

The boundary conditions are usually given in terms of displacement, v, and/or derivatives of v. Hence, the constants A, B, C, and D are determined from the boundary conditions using Equation 2.9. We consider below solutions based on boundary

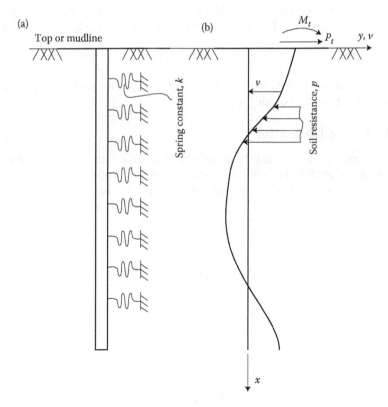

FIGURE 2.3 Laterally loaded pile and deformed shape. (a) Soil resistance by springs; (b) schematic of deformed shape of pile: P_t = applied load at top, M_t = applied moment at top.

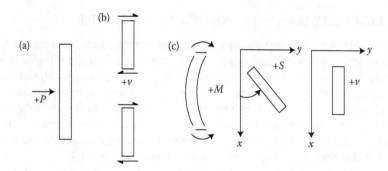

FIGURE 2.4 Sign convention. (a) Positive load; (b) positive shear, V; and (c) positive moment, M, positive slope, S, positive displacement, v.

conditions, first for piles that are very long, and can be considered as having an "infinite" length.

2.3.2 PILE OF INFINITE LENGTH

For infinite length (Figure 2.5), $e^{\lambda x} \to \infty$, that is, the terms $A \cos \lambda x$ and $B \sin \lambda x$ (Equation 2.9) will be very small. Hence, coefficients A and B are approximately zero. Then, the solution will depend only on the last two terms in Equation 2.9, that is

$$v = e^{-\lambda x}(C \cos \lambda x + D \sin \lambda x) \tag{2.11}$$

Now, we can derive solutions for piles involving specific loads and/or moments and particular boundary conditions.

2.3.3 LATERAL LOAD AT TOP

For a concentrated lateral load, P_t, at top, that is, at the ground level or the mudline (Figure 2.6), where the pile is free to move (horizontally) at the top, and the moment, $M_t = 0$, that is, the second derivative, $(d^2v/dx^2) = 0$ at $x = 0$, or

$$EI \frac{d^2v}{dx^2} = 0 \quad \text{at } x = 0 \tag{2.12}$$

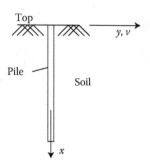

FIGURE 2.5 Pile of "infinite" length.

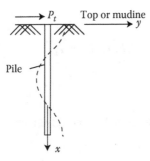

FIGURE 2.6 Pile with lateral load at top.

Now, the first derivative of v with respect to x is given by

$$\frac{dv}{dx} = \lambda e^{\lambda x} \left(A \cos \lambda x - A \sin \lambda x \right) + B \sin \lambda x + B \cos \lambda x)$$
$$+ \lambda e^{-\lambda x} (-C \cos \lambda x + C \sin \lambda x - D \cos \lambda x - D \sin \lambda x) \quad (2.13)$$

The second derivative of v can be obtained as

$$\frac{d^2v}{dx^2} = 2\lambda^2 e^{\lambda x}(-A \sin \lambda x + B \cos \lambda x)$$
$$+ 2\lambda^2 e^{-\lambda x}(C \sin \lambda x - D \cos \lambda x) \quad (2.14a)$$

Since the first two terms are zero, d^2v/dx^2 can be written as

$$\frac{d^2v}{dx^2} = 2\lambda^2 e^{-\lambda x}(C \sin \lambda x - D \cos \lambda x) \quad (2.14b)$$

At $x = 0$, $\sin \lambda x = 0$, $\cos \lambda x = 1$, $e^{-\lambda x} = 1$; therefore, using Equation 2.12, we can write

$$2\ EI\ \lambda^2(-D) = 0, \quad \text{hence, } D = 0$$

because EI and λ^2 are not zero.

Now, at $x = 0$, the shear force $V = P_t$; hence, the third derivative of v is given by

$$\frac{d^3v}{dx^3} = 2\lambda^3 e^{\lambda x}(-A \sin \lambda x + B \cos \lambda x - A \cos \lambda x - B \sin \lambda x)$$
$$+ 2\lambda^3 e^{-\lambda x}(-C \sin \lambda x + D \cos \lambda x + C \cos \lambda x + D \sin \lambda x) \quad (2.14c)$$

Hence, since $A = B = D = 0$, and at $x = 0$, $\sin \lambda x = 0$, $\cos \lambda x = 1$, and $e^{-\lambda x} = 1$.

Therefore

$$2EI\lambda^3(C) = P_t$$

$$C = \frac{P_t}{2EI\lambda^3} \tag{2.15}$$

Then, the specific solution for deflection, v, is given by

$$v = \frac{P_t e^{-\lambda x}}{2EI\lambda^3}\cos\lambda x \tag{2.16}$$

By substituting k from Equation 2.10 into Equation 2.16, we obtain

$$v = \frac{2P_t\lambda e^{-\lambda x}}{k}\cos\lambda x \tag{2.17a}$$

Various quantities for analysis and design can be derived from Equation 2.17a as

$$\text{Moment} \quad M = \frac{4P_t\lambda^3 EI\, e^{-\lambda x}}{k}\sin\lambda x \tag{2.17b}$$

$$\text{Shear force} \quad V = \frac{4P_t\lambda^4 EI\, e^{-\lambda x}}{k}(\cos\lambda x - \sin\lambda x)$$

$$= P_t e^{-\lambda x}(\cos\lambda x - \sin\lambda x) \tag{2.17c}$$

$$\text{Soil resistance} \quad p = 2P_t\lambda e^{-\lambda x}(-\cos\lambda x) \tag{2.17d}$$

To simplify the above equations, let

$$A_1 = e^{-\lambda x}(\cos\lambda x + \sin\lambda x) \tag{2.18a}$$

$$B_1 = e^{-\lambda x}(\sin\lambda x - \cos\lambda x) \tag{2.18b}$$

$$C_1 = e^{-\lambda x}\cos\lambda x \tag{2.18c}$$

$$D_1 = e^{-\lambda x}\sin\lambda x \tag{2.18d}$$

Then, various quantities of interest can be expressed in a simplified form as

$$v = \frac{2P_t\lambda}{k}C_1 \tag{2.19a}$$

$$\frac{dv}{dx} = S = -\frac{2P_t\lambda^2}{k}A_1 \tag{2.19b}$$

$$M = \frac{P_t}{\lambda}D_1 \tag{2.19c}$$

$$V = P_tB_1 \tag{2.19d}$$

$$p = -2P_t\lambda C_1 \tag{2.19e}$$

2.3.4 MOMENT AT TOP

Consider moment M_t applied at the top or the mudline (Figure 2.7). Let the boundary conditions at the top or the mudline be expressed as follows:

1. $M = M_t$ at $x = 0$, which implies that $(d^2v/dx^2) = M_t$ at $x = 0$.
2. $P_t = 0$ at $x = 0$, which implies that $EI\,(d^3v/dx^3) = 0$ at $x = 0$.

Since $A = B = 0$, and at $x = 0$, $\sin \lambda x = 0$, and $\cos \lambda x = 1$, substitution in Equation 2.14c gives

$$2EI\lambda^3 e^{-\lambda x}(D + C) = 0$$

Therefore, $D = -C$. Now, Equation 2.11 can be expressed as

$$v = Ce^{-\lambda x}(\cos \lambda x - \sin \lambda x) \tag{2.20}$$

Boundary condition $M = M_t$ at $x = 0$ leads to

$$EI\frac{d^2v}{dx^2} = M_t \quad \text{at } x = 0$$

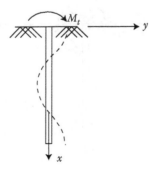

FIGURE 2.7 Pile with moment at top.

Substitution in Equation 2.14a gives $2\ EI\ \lambda^2\ (-D) = M_t$. Therefore, $D = M_t/(2\ EI\ \lambda^2)$ and $C = -M_t/(2\ EI\ \lambda^2)$.

Substitution of D and C in Equation 2.11 gives

$$v = \frac{M_t e^{-\lambda x}}{2EI\lambda^2}\left(\cos\lambda x - \sin\lambda x\right) \quad \text{or} \quad v = \frac{M_t}{2EI\lambda^2}B_1 \qquad (2.21a)$$

For this case, the other quantities of interest can be expressed as

$$S = -\frac{M_t}{EI\,\lambda}C_1 \qquad (2.21b)$$

$$M = M_t A_1 \qquad (2.21c)$$

$$V = -2M_t\lambda D_1 \qquad (2.21d)$$

$$p = -2M_t\lambda^2 B_1 \qquad (2.21e)$$

2.3.5 PILE FIXED AGAINST ROTATION AT TOP

Figure 2.8 shows a schematic of the above condition, that is, pile fixed at the top. For this case, the boundary conditions at the top are given by the following equations:

1. $dv/dx = 0$ at $x = 0$
2. $V = P_t$ at $x = 0$

Substitution of dv/dx in Equation 2.13 gives

$$-C + D = 0$$

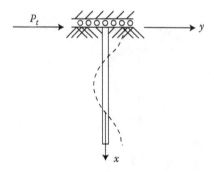

FIGURE 2.8   Pile fixed against rotation at top.

and

$$v = Ce^{-\lambda x}(\cos \lambda x + \sin \lambda x) \tag{2.22}$$

Now, the boundary condition $V = P_t$ at $x = 0$, or

$$EI\frac{d^3v}{dx^3} = P_t \quad \text{at } x = 0$$

Substitution in Equation 2.14c leads to

$$2\,EI\,\lambda^3(2C) = P_t$$

Therefore,

$$C = \frac{P_t}{4EI\,\lambda^3}$$

Substitution of C in Equation 2.11 gives

$$v = \frac{P_t e^{-\lambda x}}{4EI\lambda^3}(\cos \lambda x + \sin \lambda x) \tag{2.23a}$$

or

$$v = \frac{P_t \lambda e^{-\lambda x}}{k}(\cos \lambda x + \sin \lambda x)$$

or

$$v = \frac{P_t \lambda}{k}A_1$$

The other quantities of interest for this case can be expressed as follows:

$$S = -\frac{P_t}{2EI\lambda^2}D_1 \tag{2.23b}$$

$$M = -\frac{P_t}{2\lambda}B_1 \tag{2.23c}$$

$$V = P_t C_1 \tag{2.23d}$$

$$p = -P_t \lambda A_1 \tag{2.23e}$$

TABLE 2.1
Values of Various Parameters in Solution for One-Dimensional Laterally Loaded Pile

λx	A_1	B_1	C_1	D_1	λx	A_1	B_1	C_1	D_1
0	1.0000	1.0000	1.0000	0	2.4	-0.0056	-0.1282	-0.0669	0.0613
0.1	0.9907	0.8100	0.9003	0.0903	2.6	-0.0254	-0.1019	-0.0636	0.0383
0.2	0.9651	0.6398	0.8024	0.1627	2.8	-0.0369	-0.0777	-0.0573	0.0204
0.3	0.9267	0.4888	0.7077	0.2189	3.2	-0.0431	-0.0383	-0.0407	-0.0024
0.4	0.8784	0.3564	0.6174	0.2610	3.6	-0.0366	-0.0124	-0.0245	-0.0121
0.5	0.8231	0.2415	0.5323	0.2908	4.0	-0.0258	0.0019	-0.0120	-0.0139
0.6	0.7628	0.1413	0.4530	0.3099	4.4	-0.0155	0.0079	-0.0038	-0.0117
0.7	0.6997	0.0599	0.3798	0.3199	4.8	-0.0075	0.0089	0.0007	-0.0082
0.8	0.6354	-0.0093	0.3131	0.3223	5.2	-0.0023	0.0075	0.0026	-0.0049
0.9	0.5712	-0.0657	0.2527	0.3185	5.6	0.0005	0.0052	0.0029	-0.0023
1.0	0.5083	-0.1108	0.1988	0.3096	6.0	0.0017	0.0031	0.0024	-0.0007
1.1	0.4476	-0.1457	0.1510	0.2967	6.4	0.0018	0.0015	0.0017	0.0003
1.2	0.3899	-0.1716	0.1091	0.2807	6.8	0.0015	0.0004	0.0010	0.0006
1.3	0.3355	-0.1897	0.0729	0.2626	7.2	0.0015	-0.00014	0.00045	0.0006
1.4	0.2849	-0.2011	0.0419	0.2430	7.6	0.00061	-0.00036	0.00012	0.00049
1.5	0.2384	-0.2068	0.0158	0.2226	8.0	0.00028	-0.00038	-0.0005	0.00033
1.6	0.1959	-0.2077	-0.0059	0.2018	8.4	0.00007	-0.00031	-0.00012	0.00019
1.7	0.1576	-0.2047	-0.0235	0.1812	8.8	-0.00003	-0.00021	-0.00012	0.00009
1.8	0.1234	-0.1985	-0.0376	0.1610	9.2	-0.00008	-0.00012	-0.00010	0.00002
1.9	0.0932	-0.1899	-0.0484	0.1415	9.6	-0.00008	-0.00005	-0.00007	-0.00001
2.0	0.0667	-0.1794	-0.0563	0.1230	10.0	-0.00006	-0.00001	-0.00004	-0.00002
2.2	0.0244	-0.1548	-0.0652	0.0895					

Table 2.1 shows values of various parameters for the analytical solution for 1-D laterally loaded pile.

2.3.6 EXAMPLE 2.1: ANALYTICAL SOLUTION FOR LOAD AT TOP OF PILE WITH OVERHANG

Figure 2.9 shows a pile of infinite length with the following properties:

Elastic modulus of pile $E = 30 \times 16^6$ psi (207×10^6 kPa)
Elastic modulus of soil $E_s = 50$ psi $= 0.05$ ksi (345 kPa)
Moment of inertia of pile $I = 8000$ in⁴ (333×10^3 cm⁴)
and lateral load at 10 ft (3.048 m) from the top.

Find (a) the maximum (positive) moment, M_{max}, and (b) the maximum deflection, v_{max}.

Solution: The problem can be converted to an equivalent system with a load applied at the top (P_t), and an equal and opposite load, P_t, at the mudline (Figure 2.9b). The solution for the equivalent system in Figure 2.9b can be expressed as a

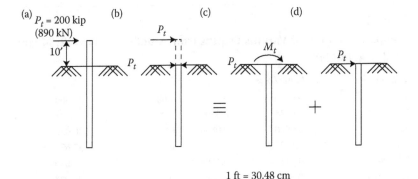

1 ft = 30.48 cm

FIGURE 2.9 Pile with overhang: Example 2.1. (a) Pile with overhang; (b) equivalent load; (c) moment at top; and (d) load at top.

summation of moment, M_t, at the mudline (Figure 2.9c) and lateral P_t at the mudline (Figure 2.9d).

Then, the overall solution can be found by adding responses due to two loads as follows:

$$\text{Moment} \quad M = \frac{P_t}{\lambda} D_1 + M_t A_1$$

$$\text{Deflection} \quad v = \frac{2 P_t \lambda}{k} C_1 + \frac{M_t}{2 E I \lambda^2} B_1$$

Now, $P_t = 200$ kip (889.6 kN)

$$M_t = 10 \times 12 \, (200) = 24{,}000 \text{ in-kip } (271{,}200 \text{ cm-N})$$

$$\lambda = \sqrt[4]{\frac{k}{4EI}} = \sqrt[4]{\frac{50}{4 \, (30 \times 10^6 \times 8000)}}$$

$$= 0.0027$$

Therefore,

$$M = \frac{200}{0.0027} D_1 + 24{,}000 \, A_1$$

$$= 74{,}074 \, D_1 + 24{,}000 \, A_1$$

Values of D_1 and A_1 can be obtained from Table 2.1 by interpolation, and the maximum moment can be found by using a trial-and-error procedure (Table 2.2) for various values of depth and corresponding values of λx. Accordingly, the maximum value will occur at $x \approx 15.0$ ft (4.57 m), that is, $\lambda x = 0.486$. Hence

TABLE 2.2

Computation of M at Various Depths from Mudline (1.0 in = 2.54 cm; 1.0 lb = 4.448 N)

Depth ft (in)	$\lambda x \; 0.0027 \times x$	D_1	A_1	M
0	0.0	0.0	1.000	24,000
1′ (12″)	0.0324	0.0297	0.997	26,128
2.5′ (30″)	0.0810	0.0850	0.990	30,056
5′ (60″)	0.162	0.1352	0.944	32,671
7.5′ (90″)	0.243	0.1869	0.949	36,620
10′ (120″)	0.324	0.229	0.9152	38,928
12.5′ (150″)	0.405	0.2670	0.876	40,473
15′ (180″)	0.486	0.2866	0.831	41,169
				(approximate maximum)
20′ (240″)	0.648	0.3147	0.7300	40,623

$$M_{max} = \frac{200}{0.0027} \times (0.2866) + 2400 \times 0.8308$$
$$= 21{,}230 + 19{,}939$$
$$= 41{,}169 \text{ kip-in at } 15.0' \text{ or } 180 \text{ in} \left(47 \times 10^7 \text{cm-N at } 457 \text{ cm}\right)$$

Deflection:

$$v = \frac{2 \times 200 \times 0.0027}{0.050} C_1$$
$$+ \frac{24{,}000 \times 1000 \times B_1}{2 \times 30 \times 10^6 \times 8000 (0.0027)^2}$$

The maximum deflection occurs at the mudline, that is, $x = 0$; hence, $C_1 = 1.00$ and $B_1 = 1.00$.

Therefore,

$$v_{max} = \frac{0.54 \times 1.0}{0.05} + \frac{2.4 \times 1.0}{3.5}$$
$$= 10.8 + 6.86$$
$$= 17.66 \text{ in } (45 \text{ cm})$$

Deflection at 15 ft (4.57 m) (at maximum moment) is given by

$$v = 10.8 \times C_1 + 6.86 \times B_1$$
$$= 10.8 \times 0.5442 + 6.86 \times 0.2576$$
$$= 5.88 + 1.77 = 7.65 \text{ in } (19 \text{ cm}).$$

2.3.7   EXAMPLE 2.2: LONG PILE LOADED AT TOP WITH NO ROTATION

The long pile (Figure 2.8) is fixed against rotation at the top, while it can experience movements at the top. The properties are given as follows:

$E = 30 \times 10^6$ psi $(207 \times 10^6$ kPa$)$; $I = 12,000$ in^4 $(500 \times 10^3$ cm^4)
$E_s = k = 100$ psi (690 kPa)
P_t, Load at top $= 150$ K (6,670,200 N)

Find (i) maximum moment, M_{max}, and (ii) maximum displacement, v_{max}.

Solution:

$$\lambda = \sqrt[4]{\frac{100}{4 \times 30 \times 10^6 \times 12000}} = \frac{3.1623}{1.414 \times 2.34 \times 31.623 \times 10.466}$$

$$= \frac{3.1623}{1095} = 0.0029$$

Moment

$$M = -\frac{P_t}{2\lambda} B_1 \qquad\qquad (2.23c)$$

The moment will be maximum when B_1 is maximum, that is, at the top. From Table 2.1, $B_1 = 1.0$. Therefore,

$$M_{max} = -\frac{150 \times 1.0}{2 \times 0.0029} = 25,862 \text{ kip-in } (292 \times 10^6 \text{ N-cm})$$

Deflection: The maximum deflection will occur at the top, for which A_1 in Equation 2.23a is 1.0. Therefore

$$v_{max} = \frac{150 \times 0.0029 \times 1.0}{0.10} = \frac{0.435}{0.10}$$

$$= 4.35 \text{ in } (11.0 \text{ cm})$$

2.4   NUMERICAL SOLUTIONS

The closed-form solutions used above are possible, but require a number of simplifying assumptions, for example, the pile has uniform geometry and is long, and the soil resistance is constant. However, in many practical situations, the pile can have variable geometry, the boundary conditions may be different and complex, and the soil resistance can be nonlinear and may vary with depth and displacement. Hence, to solve realistic problems, it is often necessary to resort to the use of numerical or computer methods. Chief among those are the FD and FE methods. Now, we will describe the formulation of FD and FE methods for the 1-D pile problems.

2.4.1 FINITE DIFFERENCE METHOD

The FD method is based on expressing the continuous derivatives in the governing differential equations (GDEs) by their finite or discreet values [17,18]. Let us consider the GDE for the pile problem (Equation 2.7). The derivatives in that equation can be obtained from Figure 2.10.

2.4.1.1 First-Order Derivative: Central Difference

Using central difference scheme, dv/dx can be expressed as

$$\frac{dv}{dx} \approx \frac{v_{m-1} - v_{m+1}}{2\Delta x} \tag{2.24a}$$

where v_m represents the value of displacement at point m and Δx represents the (equal) interval between the consecutive points; the intervals need not be equal. Note that dv/dx represents a continuous derivative while the right-hand side in Equation 2.24a represents the slope between two consecutive points x_{m-1} and x_{m+1}. Equation 2.24a represents approximate derivative using the central difference method. Although the central difference method is used commonly, other kinds of approximate derivatives such as forward difference and backward difference can be formed, as shown below.

$$\text{Forward difference: } \frac{dv}{dx} \approx \frac{v_m - v_{m+1}}{\Delta x} \tag{2.24b}$$

$$\text{Backward difference: } \frac{dv}{dx} \approx \frac{v_{m-1} - v_m}{\Delta x} \tag{2.24c}$$

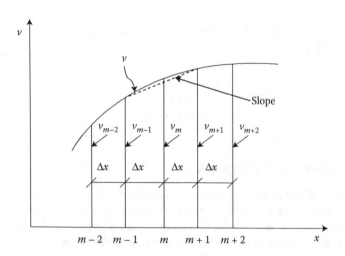

FIGURE 2.10 Finite difference approximation.

2.4.1.2   Second Derivative

The second derivative can be expressed in terms of the first derivatives as follows:

$$\left(\frac{d^2v}{dx^2}\right)_{x=m} \cong \left[\frac{v_{m-1} - v_m}{\Delta x} - \frac{v_m - v_{m+1}}{\Delta x}\right]\frac{1}{\Delta x}$$

$$\approx \frac{v_{m-1} - 2v_m + v_{m+1}}{\Delta x^2} \tag{2.25}$$

Likewise, the third and fourth derivatives can be expressed as

$$\left(\frac{d^3v}{d\Delta^3}\right)_{x=m} = \frac{v_{m-2} - 2v_{m-1} + 2v_{m+1} - v_{m+2}}{2(\Delta x)^3} \tag{2.26}$$

$$\left(\frac{d^4v}{dx^4}\right)_{x=m} = \frac{v_{m-2} - 4v_{m-1} + 6v_m - 4v_{m+1} + v_{m+2}}{(\Delta x)^4} \tag{2.27}$$

Substitution of the above FD approximations in Equation 2.7 leads to

$$v_{m-2} - 4v_{m-1} + 6v_m - 4v_{m+1} + v_{m+2}$$

$$\left(+\frac{k_m}{EI}v_m - \frac{q_m}{EI}\right)\Delta x^4 = 0 \tag{2.28}$$

If $q = 0$, Equation 2.28 becomes

$$v_{m-2} - 4v_{m-1} + 6v_m - 4v_{m+1} + v_{m+2}$$

$$= \left(-\frac{k_m}{EI}v_m\right)\Delta x^4 = -A_m v_m \tag{2.29}$$

Here, we define $(k_m/EI)\,\Delta x^4 = A_m$.

When Equation 2.29 is applied to all points from $m = 0, 1, ..., M$ (Figure 2.11), it results into a set of simultaneous equations in which the displacements at various nodal points are unknown. However, for the FD equation to be applicable at the top and bottom points, we need to use the boundary conditions first. For this purpose, we introduce two phantom or hypothetical nodal points at the top, M_{t+1} M_{t+2}, and at the bottom, -1 and -2 (Figure 2.11).

2.4.1.3   Boundary Conditions

2.4.1.3.1   At the End of a Long Pile

For a long pile, let us assume that the moment and shear force at the bottom are zero. Then, $m = 0$ that is, $x = L$

$$EI\frac{d^2v}{dx^2} = 0 \quad \text{at point } m = 0 \tag{2.30a}$$

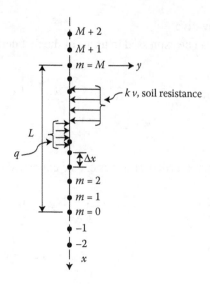

FIGURE 2.11   Finite difference discretization of pile of length L.

Therefore, as per Equation 2.25:

$$\frac{v_{-1} - 2v_0 + v_1}{\Delta x^2} = 0 \tag{2.30b}$$

Hence

$$v_{-1} = 2v_0 - v_1 \tag{2.30c}$$

When shear force $V = 0$ at $x = L$ or $m = 0$, we have

$$EI\frac{d^3v}{dx^3} = 0 \quad \text{at } m = 0 \tag{2.31a}$$

Hence, from Equation 2.26, we get

$$v_{-2} - 2v_{-1} + 2v_1 - v_2 = 0 \tag{2.31b}$$

Substitution of v_{-1} from Equation 2.30c gives

$$
\begin{aligned}
v_{-2} &= -2v_1 + 2(2v_0 - v_1) + v_2 \\
&= -2v_1 + 4v_0 - 2v_1 + v_2 \\
&= v_2 - 4v_1 + 4v_0
\end{aligned} \tag{2.31c}
$$

As noted earlier, the FD equations (Equation 2.29) without q give

$$v_{m-2} - 4v_{m-1} + 6v_m - 4v_{m+1} + v_{m+2} = -A_m v_m \qquad (2.29)$$

For $m = 0$, Equation 2.29 becomes

$$v_{-2} - 4v_{-1} + 6v_0 - 4v_1 + v_2 = -A_o v_0 \qquad (2.32a)$$

By substitution of v_{-1} and v_{-2} from Equations 2.30c and 2.31c leads to

$$(v_2 - 4v_1 + 4v_0) - 4(2v_0 - v_1) + 6v_0 - 4v_1 + v_2 = -A_o v_0$$

Therefore

$$2v_0 + A_0 v_0 - 4v_1 + 2v_2 = 0 \qquad (2.32b)$$

Solving for v_0, we get

$$v_0 = \frac{4v_1 - 2v_2}{A_0 + 2} = \frac{2(2v_1 - v_2)}{A_0 + 2} \qquad (2.32c)$$

Thus, Equation 2.32c is expressed only in terms of the physical points on the pile and does not contain any phantom or hypothetical points, v_{-1}, v_{-2}. Now, let $B_0 = 2/(A_0 + 2)$ and $B_1 = 2B_0$. Then, Equation 2.32c reduces to

$$v_0 = -B_0 v_2 + B_1 v_1 \qquad (2.32d)$$

Now, the FD equations can be expressed from bottom to top, that is, from $m = 1$ to M as follows:
At $m = 1$:

$$v_{-1} - 4v_0 + 6v_1 - 4v_2 + v_3 = -A_1 v_1 \qquad (2.33a)$$

The use of Equation 2.30c for v_{-1} in Equation 2.33a gives

$$(2v_0 - v_1) - 4v_0 + 6v_1 - 4v_2 + v_3 = -A_1 v_1 \qquad (2.33b)$$

Now, substituting Equation 2.32d for v_0 in Equation 2.33b leads to the following equation:

$$-2(-B_0 v_2 + B_1 v_1) + 5v_1 - 4v_2 + v_3 = -A_1 v_1$$

Therefore,

$$v_1 = \frac{-v_3 + (4 - 2B_0)v_2}{5 + A_1 - 2B_1} \qquad (2.33c)$$

Now, let

$$B_2 = \frac{1}{5 + A_1 - 2B_1}$$

and

$$B_3 = B_2 (4 - B_1)$$

Therefore,

$$v_1 = -B_2 v_3 + B_3 v_2 \qquad (2.33d)$$

At m = 2:
Using Equation 2.29 for $m = 2$, with v_0 and v_1 from Equations 2.32d and 2.33c, respectively, gives

$$v_4 + (-4 + 4B_2 - B_1 B_2)v_3 + (6 - 4B_3 - B_0 + A_2 + B_1 B_3)v_2 = 0$$

Therefore,

$$v_2 = \frac{-v_4 + [4 - B_2(4 - B_1)]v_3}{6 + A_2 - B_0 - B_3(4 - B_1)} \qquad (2.34a)$$

or

$$v_2 = \frac{v_4 + (4 - B_3)v_3}{6 + A_2 - B_0 - B_3(4 - B_1)} \qquad (2.34b)$$

or

$$v_2 = -B_4 v_4 + B_5 v_3 \qquad (2.34c)$$

where

$$B_4 = \frac{1}{6 + A_2 - B_0 - B_3(4 - B_1)}$$
$$B_5 = B_4(4 - B_3)$$

At m = 3:
The expression for v_3 can be derived as before

$$v_3 = -B_6 v_5 + B_7 v_4 \qquad (2.35a)$$

in which

$$B_6 = \frac{1}{6 + A_3 - B_2 - B_5(4 - B_3)} \qquad (2.35b)$$

and

$$B_7 = B_6(4 - B_5) \qquad (2.35c)$$

2.4.1.3.2 General Expression for Displacement, v

The general expression for v can be written as

$$v_m = -B_{2m}v_{m+2} + B_{2m+1}v_{m+1} \qquad (2.36)$$

and the general expression for B, except for B_0, B_1, and B_2, is given as follows:

$$B_{2m} = \frac{1}{6 + A_m - B_{2m-4} - B_{2m-1}(4 - B_{2m-3})} \qquad (2.37a)$$

and

$$B_{2m+1} = B_{2m}(4 - B_{2m-1}) \qquad (2.37b)$$

For $m = 2$ and onwards until $m = M - 2$, the application of the FD equation (Equation 2.36) is straightforward. However, for $m = M - 1$ and $m = M$, we need expressions for v with v at $M + 1$ and $M + 2$. They can be obtained by using the boundary conditions at the top of the pile, which are described below.

2.4.1.3.3 Lateral Load at Top

Let us consider that the lateral load P_t is applied at the top of the pile, and the bending moment at the top is zero (Figure 2.6). Then, we have

1. $M = 0$ at $m = M$
2. $V = P_t$ at $m = M$

The boundary condition (1) can be expressed in the FD form as follows:

$$v_{M+1} - 2v_M + v_{M-1} = 0$$

or

$$v_{M+1} = 2v_M - v_{M-1} \qquad (2.38)$$

Likewise, for boundary condition (2), using Equation 2.26, we get

$$-v_{M+2} + 2v_{M+1} - 2v_{M-1} + v_{M-2} = C_1 \qquad (2.39)$$

where $C_1 = (2P_t/EI)(\Delta x^3)$.

Now, we can use Equation 2.36 to express displacements at M, $M-1$, and $M-2$ as follows:

$$v_M = -B_{2M}v_{M+2} + B_{2M+1}v_{M+1} \qquad (2.40a)$$

$$v_{M-1} = -B_{2M-2}v_{M+1} + B_{2M-1}v_M \qquad (2.40b)$$

$$v_{M-2} = -B_{M-4}v_M + B_{2M-3}v_{M-1} \qquad (2.40c)$$

From Equation 2.40a, the expression for v_{M+2} can be written as

$$v_{M+2} = D_1 B_{2M+1}v_{M+1} - D_1 v_M \qquad (2.41)$$

where $D_1 = 1/B_{2M}$.

Substituting v_{M-1} from Equation 2.40b into Equation 2.38 leads to

$$v_{M+1} = \frac{(2 - B_{2M-1})v_M}{1 - B_{2M-2}} \qquad (2.42)$$

Substitution of Equation 2.40c into Equation 2.39 gives

$$-v_{M+2} + 2v_{M+1} - 2v_{M-1} - B_{2M-4}v_M + B_{2M-3}v_{M-1} = C_1 \qquad (2.43)$$

Now, we can substitute Equation 2.40b into Equation 2.43 to obtain

$$-v_{M+2} + 2v_{M+1} + 2B_{2M-2}v_{M+1} - 2B_{2M-1}v_M - B_{2M-3}B_{2M-2}v_{M+1} \\ + B_{2M-3}\cdot B_{2M-1}v_M - B_{2M-4}v_M = C_1 \qquad (2.44)$$

Similarly, substituting Equation 2.41 into Equation 2.44 leads to

$$(-D_1 B_{2M+1} + 2B_{2M-2} - B_{2M-3}B_{2M-2} + 2)v_{M+1} \\ + (D_1 - 2B_{2M-1} + B_{2M-3}B_{2M-1} - B_{2M-4})v_M = C_1 \qquad (2.45a)$$

Rearranging these equations, they can be written in terms of coefficients C_1, D_2, and D_3 as follows:

$$D_3 v_M - D_2 v_{M+1} = C_1 \qquad (2.45b)$$

where

$$D_2 = D_1 B_{2M+1} - B_{2M-2}(2 - B_{2M-3}) - 2 \qquad (2.46\text{a})$$

and

$$D_3 = D_1 - B_{2M-4} - B_{2M-1}(2 - B_{2M-3}) \qquad (2.46\text{b})$$

Substitution of Equation 2.42 into Equation 2.45b gives

$$v_M = \frac{(1 - B_{2M-2})C_1}{D_3(1 - B_{2M-2}) - D_2(2' - B_{2M-1})} \qquad (2.47)$$

2.4.1.3.4 Pile Fixed against Rotation at Top

For this case (Figure 2.8), the boundary conditions at the top with four segments for the pile (i.e., $m = M = 5$) (Figure 2.12) can be expressed as

1. Shear force, $V = P_t$
2. Slope, $S = 0$

For the boundary condition (1), we have, from Equation 2.26

$$-v_{M+2} + 2v_{M+1} - 2v_{m-1} + v_{M-2} = C_1 \qquad (2.48)$$

and for the boundary condition (2), we have, from Equation 2.24a

$$-v_{M+1} + v_{M-1} = 0 \qquad (2.49)$$

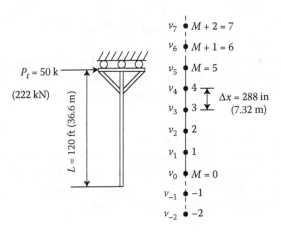

FIGURE 2.12 Example: Long pile restrained against rotation at top.

The solutions for v_M, v_{M+2}, and v_{M+2} can be obtained by following a similar procedure as for load at the top of the pile. By using Equations 2.40 through 2.40c, the expression for v_{M+2} can be written as in Equation 2.41, that is

$$v_{M+2} = D_1 B_{2M+1} v_{M+1} - D_1 v_M \tag{2.41}$$

Now, using Equation 2.40b in Equation 2.49, we have

$$-v_{M+1} - B_{2M-2} v_{M+1} + B_{2M-1} v_M = 0$$

Solving for v_{M+1} gives

$$v_{M+1} = \frac{B_{2M-1} v_M}{1 + B_{2M-2}} \tag{2.50}$$

Substitution of Equation 2.40c into Equation 2.48 gives

$$-v_{M+2} + 2v_{M+1} - 2v_{M-1} - B_{2M-4} v_M + B_{2M-3} v_{M-1} = C_1 \tag{2.51}$$

Now, substitution of Equation 2.41 into Equation 2.51, and the use of Equation 2.40b leads to

$$D_3 v_M - D_2 v_{M+2} = C_1 \tag{2.52}$$

where D_2 and D_3 are the same as in Equations 2.46a and 2.46b, respectively.
Substitution of Equation 2.42 into Equation 2.44 leads to

$$v_M = \frac{(1 + B_{2M-2}) C_1}{(1 + B_{2M-2}) D_3 - B_{2M-1} \cdot D_2} \tag{2.53}$$

Similarly, equations can be derived for other loading cases such as moment at the top, and so on, and the procedures can also be used to address other types of boundary conditions. The solution procedures can be used with hand calculations, and also computerized.

2.4.1.3.5 General Computer Procedure with FD Method

The FD method (Equation 2.29) can be used to develop a general computer procedure resulting in a set of simultaneous equations, which can be solved for displacements at all nodes ($m = 0$ to M) for various boundary conditions. In such a procedure, the final equations can be written in matrix rotation as

$$K\underset{\sim}{v} = Q \tag{2.54}$$

where K is the stiffness matrix dependent on the material properties, including subgrade modulus k, v is the vector of nodal displacements from 0 to M, and Q is the vector of applied lateral loads from subgrade or soil resistance and applied distributed loads.

Once the displacements, v, at the node points are obtained, we can solve for approximate values of slope (S), moment (M), shear forces (V), and soil pressure (p) at any point m by using the FD equations for those quantities:

$$S_m = \frac{dv}{dx}$$

$$M_m = EI_m \frac{d^2v}{dx^2}$$

$$(2.55)$$

$$V_m = EI_m \frac{d^3v}{dx^3}$$

$$p_m = -kv_m$$

The above recursive FD solution can be used for hand calculations. However, it is often easier to develop a computer program for Equation 2.29, which can be used for solving most problems by substituting values for v at $m = -1$, $m = -2$, $m = M + 1$ and $m = M + 2$ directly in the equations corresponding to specific boundary conditions. Such a procedure is presented subsequently.

2.4.2 EXAMPLE 2.3: FINITE DIFFERENCE METHOD: LONG PILE RESTRAINED AGAINST ROTATION AT TOP

Figure 2.12 shows a long pile restrained against rotation at the top. The properties are given below:

$E = 30 \times 10^6$ psi (207×10^6 kPa); $I = 5000$ in⁴ (208116 cm⁴), $k_o = 10$ lb/in³ (2.71 N/cm³), $E_s = k_o x$, $L = 120$ ft or 1440 in (3658 cm); $P_t = 50,000$ lbs (222400 N); number of divisions with the FD method, $M = 5$, that is, $\Delta x = 1440/5 = 288$ in (731.5 cm). Here, we have used $E_s = k$ to be linear with depth x. However, in general, $k = k_o b$, where b is the diameter or width (in case of square or rectangular cross section) of the pile.

Required: Displacements at points 0–5, verification of boundary conditions, and identification of displacement and moment at the top of the pile.

Solution:

Values of A_m:

$$A_m = \frac{k_o x}{EI}(\Delta x)^4$$

$$= \frac{10 \times (288)^4}{30 \times 10^6 \times 5000} x = \frac{68.8}{150} = 0.460x$$

Hence, the values of A_m from 0–5 are computed as follows:

Node Point (m)	Depth, x in (cm)	A_m
0	1440 (3658)	$A_0 = 662$
1	1152 (2926)	$A_1 = 530$
2	864 (2195)	$A_2 = 397$
3	576 (1463)	$A_3 = 265$
4	288 (732)	$A_4 = 132$
5	0 (0)	$A_5 = 0$

also

$$C_1 = \frac{2P_t}{EI} (\Delta x)^3 = \frac{2 \times 50{,}000}{80 \times 10^6 \ (5000)} \cdot (288)^3 = 16$$

To calculate displacements, we need to find B_m and D_m. The expressions for B_{2m}, and B_{2m+1} are given below:

$$B_{2m} = \frac{1}{6 + A_m - B_{2m} - B_{2m-1} \ (4 - B_{2m-3})}$$

and

$$B_{2m+1} = B_{2m} \ (4 - B_{2m-1})$$

These expressions are used for values of B_m greater than B_3. The expressions B_0, B_1, B_2, and B_3 are given before. Hence, the computed values of B_m are given below:

$$B_0 = \frac{2}{A_0 + 2} = \frac{2}{662 + 2} = 0.0030$$

$$B_1 = 2B_0 = 2 \times 0.0030 = 0.0060$$

$$B_2 = \frac{1}{5 + A_1 - 2B_1} = \frac{1}{5 + 530.0 - 2(0.006)} = 0.00190$$

$$B_3 = B_2(4 - B_1) = 0.00190(4 - 0.006) = 0.00760$$

$$B_4 = \frac{1}{6 + A_2 - B_0 - B_3(4 - B_1)}$$

$$= \frac{1}{6 + 397 - 0.0030 - 0.0076(4 - 0.0060)} = 0.00248$$

$$B_5 = B_4(4 - B_3) = 0.00248 \ (4 - 0.0076) = 0.0099$$

$$B_6 = \frac{1}{6 + A_3 - B_2 - B_5 (4 - B_3)}$$

$$= \frac{1}{6 + 265 - 0.00190 - 0.0099 (4 - 0.00760)} = 0.00369$$

$$B_7 = B_6(4 - B_5) = 0.00369(4 - 0.0099) = 0.01470$$

$$B_8 = \frac{1}{6 + A_4 - B_4 - B_7 (4 - B_5)}$$

$$= \frac{1}{6 + 132 - 0.00248 - 0.0147 (4 - 0.0099)} = 0.00746$$

$$B_9 = B_8(4 - B_7) = 0.00746(4 - 0.0147) = 0.0297$$

$$B_{10} = \frac{1}{6 + A_5 - B_6 - B_9(4 - B_7)}$$

$$= \frac{1}{6 + 0 - 0.00369 - 0.0297(4 - 0.0147)} = 0.1701$$

$$B_{11} = B_{10}(4 - B_9) = 0.1701(4 - 0.0297) = 0.6753$$

The computations for D_m are shown below:

$$D_1 = \frac{1}{B_{2m}} = \frac{1}{B_{10}} = \frac{1}{0.1701} = 5.879$$

$$D_2 = D_1 B_{2m+1} - B_{2m-2}(2 - B_{2m-3}) - 2$$

$$= D_1 B_{11} - B_8 (2 - B_7) - 2$$

$$= 5.879(0.6753) - 0.00746(2 - 0.0147) - 2 = 1.9552$$

$$D_3 = D_1 - B_{2M-4} - B_{2M-1}(2 - B_{2M-3})$$

$$= D_1 - B_6 - B_9(2 - B_7)$$

$$= 5.879 - 0.0369 - 0.0297(2 - 0.0147) = 5.816$$

Now, the displacements can be computed using the values of C_1, B_m, and D_m.

The displacement, v_5, at the top of the pile, that is, $m = M = 5$, is calculated using Equation 2.53:

$$v_5 = v_M = \frac{C_1(1 + B_{2M-2})}{D_3(1 + B_{2M-2}) - D_2 \cdot B_{2M-1}}$$

$$= \frac{C_1(1 - B_8)}{D_3(1 + B_8) - D_2(B_9)}$$

$$= \frac{16.00 (1 - 0.00746)}{5.816(1 + 0.00746) - 1.9552 (0.0297)}$$

$$= 2.77825 \text{ in } (7.05739 \text{ cm})$$

Now, using Equation 2.50, we have

$$v_{M+1} = v_6 = \frac{B_{2M-1} \cdot v_M}{1 + B_{2M-2}} = \frac{B_9 \cdot v_5}{1 + B_8}$$

$$= \frac{0.0297\,(2.77825)}{1 + 0.00746}$$

$$= 0.0819\,\text{in}\,(0.208\,\text{cm})$$

Similarly, using Equation 2.41, we get

$$v_{M+2} = v_7 = D_1 B_{2M+1} v_{M+1} - D_1 v_M$$

$$= D_1 B_{11} v_6 - D_1 v_5$$

$$= 5.879 \times 0.6753 \times 0.0819 - 5.879\,(2.77825)$$

$$= -16.008\,\text{in}\,(40.66\,\text{cm})$$

Now, we can use Equation 2.36 to find displacements at nodes 4, 3, 2, 1, and 0 as follows:

$$v_m = -B_{2m} v_{m+2} + B_{2m+1} \cdot v_{m+1} \tag{2.36b}$$

$$v_4 = -B_8 v_6 + B_9 v_5$$

$$= -0.00746 \times 0.0819 + 0.00297 \times 2.77825$$

$$= -0.00611 + 0.08234$$

$$= 0.07623\,\text{in}\,(0.1936\,\text{cm})$$

$$v_3 = -B_6 v_5 + B_7 v_4$$

$$= -0.00369 \times 2.77825 \times 0.0147 \times 0.07623$$

$$= -0.00904\,\text{in}\,(0.0296\,\text{cm})$$

$$v_2 = -B_4 v_4 + B_5 v_3$$

$$= -0.00248 \times 0.07623 + 0.0099 \times (-0.00904)$$

$$= -0.000279\,\text{in}\,(0.000709\,\text{cm})$$

$$v_1 = -B_2 v_3 + B_3 v_2$$

$$= -0.00190 \times (-0.00904) + 0.00760(-0.000279)$$

$$= 0.0000151\,\text{in}\,(0.0000384\,\text{cm})$$

$$v_0 = -B_0 v_2 + B_1 v_1$$

$$= -0.0030(-0.000279) + 0.0060(0.0000151)$$

$$= +0.0000008 + 0.00000009$$

$$= +0.0000008\,\text{in}\,(0.00000203\,\text{cm})$$

Verification

1. *Shear force at the top (Equation 2.26):*

$$V_5 = \frac{EI}{2(\Delta x^3)}(v_3 - 2v_4 + 2v_6 - v_7)$$

$$= \frac{30 \times 10^6 \times 5000}{2(288^3)}[-0.00904 - 2 \times (0.07623) + 2 \times 0.0819 - (-16.008)]$$

$$= 3125 \times 16.1738$$

$$= 50,038 \text{ lbs. } (222,570\,\text{N})$$

The computed shear force is almost equal to the applied load $P_t = 50,000$ lbs, as expected.

2. *Moment and shear at the bottom:*
   The values of v_{-1} and v_{-2} can be found by using Equations 2.30c and 2.31c, respectively:

$$v_{-1} = 2v_0 - v_1$$
$$= 2 \times 0.0000008 - 0.0000151$$
$$= 0.0000016 - 0.0000151$$
$$= -0.0000135\,\text{in}$$

$$v_{-2} = v_2 - 4v_1 + 4v_0$$
$$= -0.000279 - 4 \times (0.0000151)$$
$$+ 4 \times (0.0000008)$$
$$= -0.0003394 + 0.0000032$$
$$= -0.000336\,\text{in}$$

Therefore

$$M_0 = \frac{30 \times 10^6 \times 5000}{(288)^2}(v_{-1} - 2v_0 + v_1)$$

$$= 18.1 \times 10^5(-0.0000135 - 2 \times (0.0000008 + 0.0000151)$$

$$= 18.1 \times 10^5 (-0.0000151 + 0.0000151)$$

$$= 0 \text{ lb-in} (0\,\text{N-cm})$$

$$V_0 = \frac{30 \times 10^6 \times 5000}{2 \times (288)^3}(v_{-2} - 2v_{-1} + 2v_1 - v_2)$$

$$= 3125[(-0.000336 - 2 \times (-0.0000135) + 2(0.0000151) - (-0.000279)]$$

$$= 3125 [-0.000363 + 0.0003092]$$

$$= 3125 \times (0.0000538)$$

$$= 0.168\,\text{lb} (\sim \text{zero}) (0.747\,\text{N})$$

Thus, the moment and shear force at the bottom of the long pile satisfy the boundary conditions that $M = V = 0$.

Moment at the top

$$
\begin{aligned}
M_5 &= \frac{EI}{(\Delta x)^2}(v_4 - 2v_5 + v_6) \\
&= \frac{30 \times 10^6 \times 5000}{(288)^2}(0.07626 - 2 \times 2.77825 + 0.0819) \\
&= -18.1 \times 10^5 \, (5.400) \\
&= -97.75 \times 10^5 \text{ lb-in } (-1104\,\text{N-cm})
\end{aligned}
$$

2.5 FINITE ELEMENT METHOD: ONE-DIMENSIONAL SIMULATION

The literature on the finite element method (FEM) is wide and available in many publications, including textbooks, for example, Refs. [19–23]. It has been applied successfully to many problems such as in civil, mechanical, aerospace, mining, geological, and electrical engineering and applied physics. The FEM possesses a number of advantages such as consideration of nonlinear behavior: material and geometric nonlinearity; nonhomogeneous materials; arbitrary geometries; different loadings: static, repetitive, cyclic, dynamic, thermomechanical, including environmental factors such as moisture and chemical effects, and arbitrary boundary conditions. Also, the FEM possesses a number of advantages compared to the FD method, for example, easier implementation of the boundary conditions, arbitrary geometries, and loading conditions.

In this book, we first describe the 1-D FEM simulation applicable to piles, retaining walls, and so on. Then, in the subsequent chapters, descriptions for 2-D and 3-D problems will be presented.

2.5.1 ONE-DIMENSIONAL FINITE ELEMENT METHOD

The following descriptions are adopted from various publications, for example, Refs. [16,19–21,24]. They are related to piles loaded axially and laterally, and retaining walls with tie-backs and reinforcements.

In the FEM, an engineering problem is divided into a number of elements. For 1-D idealization of an axially and laterally loaded pile as a beam-column (Figure 2.13a), the 3-D loadings can involve axial and lateral loads and moments. The beam-column is divided into 1-D elements (Figure 2.13b). A generic FE with possible six degrees of freedom (u, v, w, θ_x, θ_y, and θ_z) is shown in Figure 2.13c. The rotation about the x-axis (θ_x) represents torsion, which is described briefly below.

Soil behavior in the 1-D idealization is represented by linear or nonlinear springs for translations (u, v, w), and rotations (θ_x, θ_y, θ_z) (Figure 2.13d). These springs are assumed to be uncoupled. Figure 2.14 shows the combined bending and axial loadings and the 1-D FE.

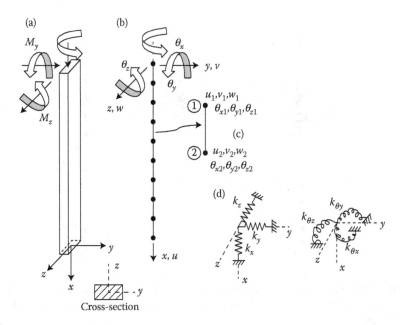

Cross-section

FIGURE 2.13 One-dimensional idealization of axially and laterally loaded pile. (a) Structure; (b) idealization; (c) generic element; and (d) translational and rotational springs.

Here, the effects of lateral and axial loadings in respective axes can be coupled. However, for simplicity, they are evaluated separately and superimposed for overall effects on the behavior of the beam-column. The FE equations are first derived for the simplified 1-D simulation, for the bending cases in y- and z-directions, in Figures 2.15a and b, respectively. Then, the axial behavior is considered.

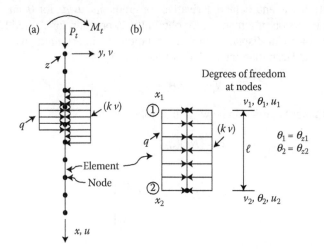

FIGURE 2.14 Finite element discretization for 1-D pile. (a) 1-D discretization in x–y space; (b) generic 1-D element of length, ℓ.

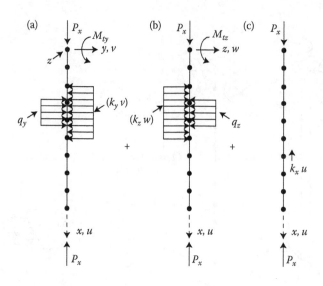

FIGURE 2.15 Superposition for three-dimensional behavior with one-dimensional ideal-izations. (a) Bending y-direction and axial load; (b) bending z-direction and axial load; and (c) axial load.

2.5.2 Details of Finite Element Method

2.5.2.1 Bending Behavior

The displacement, v, in the y-direction due to bending can be expressed as a cubic polynomial in x as follows:

$$v = a_1 + a_2 x + a_3 x^2 + a_4 x^3 \tag{2.56a}$$

By substituting v_1 and v_2 (and derivatives or gradients, dv/dx for θ_1 and θ_2, which involves the inversion of a matrix), we obtain the expression for v directly in terms of the values of nodal displacements and rotations (primary unknowns in the FE formulation) and Hermitian interpolation functions N_i as

$$v = N_1 v_1 + N_2 \theta_1 + N_3 v_2 + N_4 \theta_2 \tag{2.56b}$$

In matrix notation, this can be written as follows:

$$v = [N_b]\{q_b\}$$

$$= [N_1 N_2 N_3 N_4] \begin{Bmatrix} v_1 \\ \theta_1 \\ v_2 \\ \theta_2 \end{Bmatrix} \tag{2.56c}$$

where $N_1 = 1 - 3s^2 + 2s^3$; $N_2 = \ell s(1 - 2s + s^2)$; $N_3 = s^2(3 - 2s)$; $N_4 = \ell s^2(s - 1)$, s is the local coordinate, $(x - x_1)/\ell$, which varies from 0 to 1, ℓ is the length of the element, x_1

and x_2 are the coordinates of the element (Figure 2.14b), and $[N_b]$ is the interpolation matrix for the bending behavior.

2.5.2.2 Axial Behavior

For linear approximation of axial displacement, there are two nodal degrees of freedom for each element, u_1 and u_2 (Figure 2.14). The approximation function for linear displacement in the element can be expressed as

$$u = (1 - s)u_1 + su_2 \tag{2.57a}$$

$$= [N_a]\{q_a\} \tag{2.57b}$$

where $[N_a]$ is the interpolation matrix for axial behavior and $\{q_a\}$ is the vector of the nodal displacements.

The rotational or torsional variable about the x-axis can also be expressed by using the following linear function:

$$\theta_x = (1 - s)\theta_{x1} + s\theta_{x2} \tag{2.58a}$$

$$= [N_\theta]\{q_\theta\} \tag{2.58b}$$

where $[N_\theta]$ is the matrix for interpolation function for rotational response and $\{q_\theta\}$ is the vector of rotational nodal variables. Torsional condition can be derived and superimposed on the stiffness matrices for bending and axial behaviors. However, at this time, the torsional part is not included in the derivations and computer solutions. However, a brief description is given later. Now, we derive stiffness matrices for bending and axial behavior. The combined and superimposed bending and axial approximation functions can be expressed as follows:

$$\begin{Bmatrix} v \\ w \\ u \end{Bmatrix} = \begin{bmatrix} N_b & 0 & 0 \\ 0 & N_b & 0 \\ 0 & 0 & N_a \end{bmatrix} \begin{Bmatrix} q_{by} \\ q_{bz} \\ q_a \end{Bmatrix} \tag{2.59}$$

where u, v, and w are displacements in the x-, y-, and z-directions, respectively (Figure 2.15)

$$\{q_{by}\}^T = \begin{bmatrix} v_1 & \theta_{z1} & v_2 & \theta_{z2} \end{bmatrix}$$

$$\{q_{bw}\}^T = \begin{bmatrix} w_1 & \theta_{y1} & w_2 & \theta_{y2} \end{bmatrix}$$

and

$$\{q_a\} = \begin{bmatrix} u_1 & u_2 \end{bmatrix}$$

where θ_{y1} is the slope for w at node 1 and so on. Here, we have extended the expressions in the z-direction, similar to those for bending in the y-direction. Now, assuming that the cross-sectional area of the beam-column is constant, the potential energy, π_p, can be expressed as

$$\pi_p = A\ell \int_0^1 \left\{ \frac{1}{2}(EI_z(v'')^2 + EI_y(w'')^2) \right\} ds$$

$$+ A\ell \int_0^1 E(u')^2 ds - A\ell \int_0^1 (\bar{X}u + \bar{Y}v + \bar{Z}w)\, ds$$

$$- \ell \int_0^1 \left[\bar{T}_x u + \bar{T}_y v + \bar{T}_z w \right] ds \qquad (2.60)$$

where $\{\bar{X}\}^T = \{\bar{X}\ \bar{Y}\ \bar{Z}\}$ is the vector of body forces (weight), which usually occurs only in the vertical direction (here x), $\{T\} = \{\bar{T}_x \bar{T}_y \bar{T}_z\}$ is the vector of traction or uniformly distributed loads (here q), and v'' and w'' are second derivatives of v and w, respectively; for example, v'' can be derived as follows:

$$v' = \frac{dv}{dx}$$

$$= \frac{1}{\ell} \frac{d}{dx} \left[N_1 v_1 + N_2 \theta_{z1} + N_3 v_2 + N_4 \theta_{z2} \right] \qquad (2.61a)$$

$$= \frac{1}{\ell} \left[-6s + 6s^2 \quad \ell(1 - 4s + 3s^2) \quad 6s - s^2 \quad \ell(3s^2 - 2s) \right] \begin{Bmatrix} v_1 \\ \theta_{z1} \\ v_2 \\ \theta_{z2} \end{Bmatrix} \qquad (2.61b)$$

Then,

$$v'' = \frac{d^2 v}{dx^2} = \frac{1}{\ell} \frac{d}{ds}\left(\frac{dv}{dx} \right)$$

$$= \frac{1}{\ell^2} \left[-6 + 12s \quad -4\ell + 6\ell s \quad 6 - 12s \quad 6\ell s - 2\ell) \right] \begin{Bmatrix} v_1 \\ \theta_{z1} \\ v_2 \\ \theta_{z2} \end{Bmatrix}$$

$$= [B_y]\{q_{by}\} \qquad (2.61c)$$

Similarly, w'' can be derived as

$$w'' = \frac{d^2w}{dx^2}$$

$$= \frac{1}{\ell^2}[-6+12s \quad -4\ell+6\ell s \quad 6-12s \quad 6\ell s-2\ell]\begin{Bmatrix} w_1 \\ \theta_{y1} \\ w_2 \\ \theta_{y2} \end{Bmatrix}$$

$$= [B_z]\{q_{bz}\} \tag{2.62}$$

where $[B_y]$ and $[B_z]$ are transformation matrices.

For axial behavior, u' can be found as

$$u' = \frac{du}{dx} = \frac{d}{dx}[N_1u_1 + N_2u_2]$$
$$= [B_a]\{q_a\} \tag{2.63}$$

where $[B_a] = (1/\ell)[-1 \quad 1]$ is the transformation matrix. Substitution of v, w, u, v'', w'', and u' in Equation 2.60 and using the minimum potential energy principle, the element equations are derived as follows [19,20]:

$$[k]\{q\} = \{Q\} \tag{2.64}$$

where $[k]$ is the stiffness matrix for the element, $\{q\}$ is the vector of nodal unknowns (displacements and rotations or gradients), and $\{Q\}$ is the vector of nodal forces. In the following equations, $\bar{X}$, $\bar{Y}$, and $\bar{Z}$ indicate the body force or weight of the pile in the x-, y-, and z-directions, respectively, and $\bar{T}_x$ involves distributed axial load, including the axial soil resistance ($= k_xu$) and $\bar{T}_y$ and $\bar{T}_z$ represent distributed lateral loads and soil resistance in the y-direction ($= k_yv$) and the z-direction ($= k_zw$), respectively, and k_x, k_y, and k_z are subgrade moduli in the x-, y-, and z-directions, respectively. Detailed forms of the matrices for a uniform pile are given as follows:

$$[k] = \begin{bmatrix} \alpha_y[k_y] & o & o \\ & \alpha_z[k_z] & o \\ \text{symmetrical} & & \alpha_x[k_x] \end{bmatrix} \tag{2.65a}$$

where $\alpha_y = (EI_z/\ell^3)$, $\alpha_z = (EI_y/\ell^3)$, $\alpha_x = (EA/\ell)$, E is the elastic modulus, A is the cross-sectional area, and I_z and I_y are moment of inertias about the z- and y-axes, respectively, of the element, and $[k_y]$ and $[k_z]$ are given by

$$[k_y] = [k_z] = \begin{bmatrix} 12 & 6\ell & -12 & 6\ell \\ & 4\ell^2 & -6\ell & 2\ell^2 \\ & & 12 & -6\ell \\ \text{symmetrical} & & & 4\ell^2 \end{bmatrix} \tag{2.65b}$$

and

$$[k_x] = \begin{bmatrix} 1 & -1 \\ -1 & 1 \end{bmatrix}$$

(2.65c)

The nodal load vector is given by

$$\{Q\} = \{Q_B\} + \{Q_T\}$$

(2.66)

where

$$\{Q_B\} = A\int [N]^T \{\bar{X}\} d\ell$$

$$\{Q_T\} = \int_{y_1}^{y_z} [N]^T \{\bar{T}\} d\ell$$

The element equations are assembled to yield the global or assemblage equations by enforcing the compatibility of displacements and rotations at element nodal points as

$$[K]\{r\} = \{R\}$$

(2.67)

where $[K]$ is the global stiffness matrix, $\{r\}$ is the global vector of nodal displacements and gradients, involving $\{q_b\}$ and $\{q_a\}$, and $\{R\}$ is the global vector of nodal forces.

2.5.3 BOUNDARY CONDITIONS

The boundary conditions are introduced in the global Equation 2.67. In contrast to the FD method, in the FEM, it is not necessary to introduce hypothetical nodes near the boundaries. The boundary conditions on displacement and slopes are introduced directly in Equation 2.67. For instance, in Figure 2.6, for the long pile with load P_t at the top, the boundary conditions are needed to be introduced at the bottom node:

$$\text{Displacement } v_0 = 0$$

(2.68a)

$$\text{Slope}\left(\frac{dv}{dx}\right)_0 = \theta_0 = 0$$

(2.68b)

For a (long) pile fixed against the rotation at the top (Figure 2.8), the value of $\theta_M = 0$ is introduced at the top node, M, similar to Equation 2.68b.

2.5.3.1 Applied Forces

The pile can be subjected to distributed loads ($\bar{T} = \bar{q}$) and/or soil resistance ($p = kv$), concentrated nodal loads ($\bar{P}$), and the weight of the pile ($\bar{X}$).

2.6 SOIL BEHAVIOR: RESISTANCE–DISPLACEMENT (p_y–v OR p–y) REPRESENTATION

Almost all geologic materials (soils, rocks, concrete, etc.) exhibit nonlinear behavior when subjected to loading. Hence, the analysis and design by assuming linear (elastic) behavior can yield only approximate results, often for preliminary investigation. In general, however, it is essential that we understand and define the nonlinear behavior affected by significant practical factors such as elastic, plastic and creep strains, stress paths, *in situ* stress, volume change, anisotropy, loading types, microstructural adjustments leading to degradation or softening, and healing or strengthening.

The nonlinear behavior, in general, should consider 3-D responses (Figure 2.16), which can be different in different directions due to (inherent) anisotropy. Often, it is assumed that the material is isotropic so that the behavior in three dimensions is assumed to be the same.

The subject of characterization of multidimensional material behavior is wide in scope. Recently, significant developments have occurred for defining the material behavior, which is called stress–strain response or constitutive response. There are many publications including books on this topic [25–27].

To simplify the analysis, we often resort to approximate simulations of the 3-D behavior as 2-D or 1-D. The 1-D idealization has been commonly used for structure–foundation systems that can be idealized as 1-D because of simple loading and geometrical conditions. For instance, an axially and/or laterally loaded single beam or pile can be idealized as 1-D for the symmetric (square, circular, etc.) geometry and symmetric applied loading (axial) about the main axis and the lateral loading on the structure (Figure 2.16).

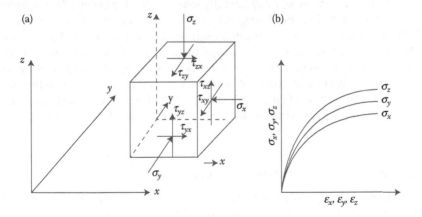

FIGURE 2.16 Schematic of three-dimensional behavior. (a) Element with six stresses; (b) schematic of stress-strain behavior.

Hence, for such 1-D idealization of a pile, the resistance of the soil can be replaced by 1-D springs (e.g., Winker springs) for three translations and three rotations, with spring moduli, k_x, k_y, and k_z, and $k_{\theta x}$, $k_{\theta y}$, and $k_{\theta z}$, respectively (Figure 2.13d). Often, each of the six stiffnesses representing the soil resistance is assumed to be independent; this may simplify the problem. However, it may be noted that the actual responses are indeed coupled.

2.6.1 ONE-DIMENSIONAL RESPONSE

Consider the lateral resistance simulated by independent springs in the y-direction. Then, the resistance–displacement relation can be expressed as

$$p_y = k_y(x)v \tag{2.69}$$

where p_y is the soil resistance (F/L), k_y is the spring stiffness (F/L^2), and v is the lateral displacement (L).

Similarly, the lateral resistance, p_z, in the z-direction can be expressed as

$$p_z = k_z(x)w \tag{2.70}$$

where k_z is the lateral spring stiffness, and w is the lateral displacement. The axial behavior can also be nonlinear. Then

$$p_x = k_x(x)u \tag{2.71}$$

where p_x is the axial soil resistance, k_x is the axial stiffness, and u is the axial displacement.

Now, the spring stiffness (k) can be considered to be constant if it is possible to assume that the resistance does not vary with depth (x) and/or displacements v, w, or u. If the resistance is not dependent on the displacement, the behavior can be assumed to be variable, often linear with depth.

In the literature for axially and laterally loaded piles, Equation 2.71 is referred to as the t–z curve [28], and Equations 2.69 and 2.70 are called the p–y curves (12–16). For the t–z curves, t is the axial soil resistance and z is the displacement in the axial direction. For p–y curves, p is the lateral soil resistance and y is the displacement in the lateral direction. In this book, we adopt the notation p_y–v and p_z–w to be consistent with the coordinate axes (y and z) and corresponding displacements (v and w). We first provide a description for p_y–v representation using nonlinear curves. As is evident, such curves represent equivalent 1-D simulation of the 3-D response of the soil.

2.6.2 P_Y–V (P–Y) REPRESENTATION AND CURVES

The concept of p–y curves has been developed, usually, in relation to research and applications for offshore structures (platforms) for oil and gas explorations. Significant lateral (and axial) loads arise because of the actions of waves, drilling and resulting dynamic loads, and often earthquake loads. Hence, significant effort and resources have been spent by various researchers such as H. Matlock and L.C. Reese to define behavior involving soil–structure interaction, often using 1-D idealizations for economical analysis and design. The idea of p–y curves has been developed and is available in various publications [12–16].

Figure 2.17 shows the various possibilities for simulating the soil resistance. If the stiffness (resistance) relating p_y and v is not dependent on the depth (x) and the displacement v, the relation represents constant (linear) variation of the resistance, k_y (Figure 2.17b). If the (stiffness) resistance varies with depth, but does not vary with v, its value can vary with depth (Figure 2.17c). If the stiffness is dependent on both x and v, it represents a nonlinear variation with v (Figure 2.17d).

The differential equation (Equation 2.7) for a pile with displacement, v, in the y-direction due to bending can be expressed as

$$EI \frac{d^4 v}{dx^4} = -p(x, v)$$

$$= -k_y(x, v)v \qquad (2.72a)$$

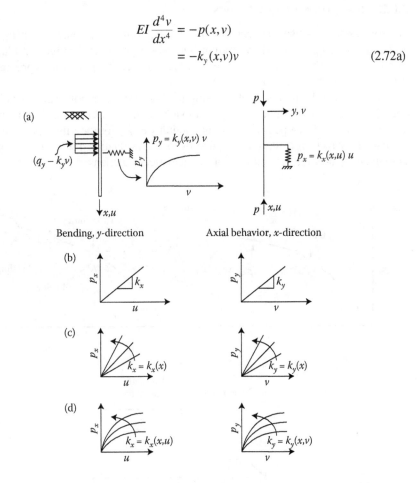

FIGURE 2.17 Schematic of soil resistance and displacement responses in x, y, and (z) directions; x and y shown here. (a) Behavior in x- and y-directions; (b) stiffness constant with x and linear with u, v or (w); (c) stiffness linear with x and linear with u, v or (w); and (d) stiffness nonlinear with x, and with u, v or (w).

For constant stiffness, $\bar{k}_y$, Equation 2.72a reduces to the following form:

$$EI\frac{d^4v}{dx^4} = -\bar{k}_y v \qquad (2.72b)$$

The closed-form solution for constant stiffness was derived before (Equation 2.9). Now, we consider the solution of Equation 2.72a, when the stiffness is nonlinear and a function of x and v. For such a nonlinear behavior, it is usually necessary to resort to numerical solutions such as the FD and FE methods.

2.6.3 SIMULATION OF p_y–v CURVES

The soil resistance–displacement (p_y–v) curves are often represented by using a set of data points (p_{yi}, v_i), joined by straight lines, for a given depth (Figure 2.18a) and then computing the slope of the line between consecutive points to represent the stiffness, k_y, for the y-direction.

The curve can be represented by using mathematical functions such as polynomial, hyperbola, and Ramberg–Osgood models (Figure 2.18b). As a simplification, the curve can be developed by using the modulus $E_s(k_y)$ for the initial part, and the ultimate resistance, p_u (Figure 2.18c); details for the evaluation of the latter are given subsequently, after the presentation of mathematical functions.

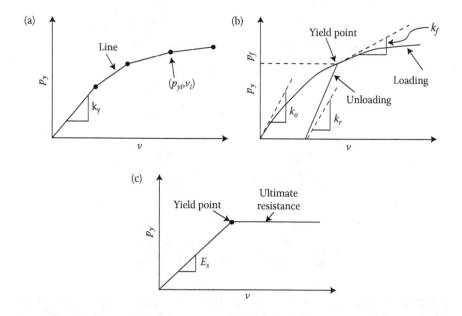

FIGURE 2.18 Representation of p_y–v curves. (a) Piecewise linear by data points; (b) Ramberg–Osgood representation; and (c) simplified representation.

In a functional form, the p_y–v relation can be expressed as

$$p_y(x,v) = \alpha_0 + \alpha_1 x + \alpha_2 v + \alpha_3 x^2 + \alpha_4 v^2 + \cdots \tag{2.73}$$

The first derivative of p_y in Equation 2.73, if it is dependent on v only, can be expressed as the stiffness or soil modulus as follows:

$$k_y(v) = \frac{\partial p_y(v)}{\partial v} \tag{2.74}$$

The Ramberg–Osgood model [25,29–31] is used in this book (Figure 2.18b). For this case, the p_y–v relationship can be expressed as

$$p_y = \frac{(k_o - k_f)v}{\left[1 + \left\{((k_o - k_f)/p_f)v\right\}^m\right]^{1/m}} + k_f v \tag{2.75}$$

where k_o is the initial spring stiffness, k_f is the final spring stiffness, p_f is the resistance corresponding to the "yield" point, and m is the order of the curve. For $m = 1$ and $k_f = 0$ and $p_f = p_u$ (ultimate/asymptotic value), Equation 2.75 reduces to the hyperbolic form:

$$p_y = \frac{v}{a + bv} \tag{2.76}$$

where $(1/a) = k_o$ and $1/b = p_u$. For linear variation, that is, constant modulus, set $k_f = 0$, $p_f = 1$, and $m = 0$. In the above expression, the initial stiffness k_o can be expressed as the function of depth x. Thus, p_y–v in Equation 2.75 can be functions of both x and v.

Using the FD or FE equations, the tangent modulus after an increment is evaluated based on Equation 2.75. In the nonlinear incremental analysis, those values are computed often at the end of the previous increment.

2.6.4 DETERMINATION OF P_Y–V (P–Y) CURVES

The p_y–v curves and soil stiffness at various points are required for the solution of problems using the FD and FE methods. Laboratory and/or field test data are required to develop the p_y–v curves.

The response (Figure 2.18) includes the yielding of the soil, which may occur almost from the start of loading. However, very often, yielding is defined at a specific point and the behavior is assumed linearly or nonlinearly elastic before the yield or ultimate resistance (Figure 2.18c). After the yield point, the behavior is assumed to be "elastic" with a much reduced fraction of the initial stiffness, k_o. Alternatively, as will be discussed in Appendix 1, the behavior can be considered to be elastoplastic based on the plasticity theory, in which case the yield and postyield behavior are defined by using various rules regarding initiation and growth of plastic flow.

2.6.4.1   Ultimate Soil Resistance

The ultimate soil resistance will be different for cohesive (clayey) and cohesionless (sandy) soils. The behavior of the soil will be different near the surface and at some depth from the surface. Near the surface, the pile, under a lateral load, may push up a soil wedge under lateral motion. At some depth, however, the overburden pressure may prevent the formation of the wedge, but the soil may experience motions around the pile; such behavior can occur for both clays and sands. However, the methods for computations of the ultimate resistance may be different.

2.6.4.2   Ultimate Soil Resistance for Clays

2.6.4.2.1   Soil Resistance near Surface

Figure 2.19a shows the wedge proposed by Reese [15] and Reese and Matlock [32], which participates in developing the ultimate soil resistance causing a push out of the wedge along the plane abfe. The forces acting on various faces of the wedge to resist the motion are shown in Figure 2.19a.

The (shear) forces on various faces of the wedge are mobilized to resist the motion of the wedge. In Figure 2.19a, F_1 is the weight of the soil, F_2 is the force acting on surface abfe, F_3 is the force acting on surface bcf, F_4 is the force acting on surface ade, F_5 is the force acting on surface cdef, and $F_6 = F_p$ is the total force.

By assuming the clay to be saturated and undrained and $\beta = 45°$, the summation of the forces (F_1 through F_5) leads to the total force F_6. Differentiation of F_6 with respect to x leads to the ultimate resistance per unit length of the pile, p_u, as

$$p_u = \gamma bx + 2cb + 2.83cx \qquad (2.77)$$

where c is the average undrained shear strength of soil over the wedge depth. For fissured clay, the side resistance may be reduced; hence, the coefficient in the third term in Equation 2.77 may be reduced to around 0.50.

2.6.4.2.2   Soil Resistance at Depth

Beyond a depth of about 9–10 times the pile diameter (b), the clay may not be mobilized as a wedge, but may flow around the pile [15,32]. Figure 2.20a shows such soil motion with stresses acting on various blocks around the pile.

It is assumed that the movement of the soil would cause failure by shearing of block nos. 1, 2, 4, and 5, and by sliding for block no. 3. The ultimate resistance can be found by using the difference between σ_6 and σ_1 and the Mohr–Coulomb diagram (Figure 2.20b).

$$p_u = (\sigma_6 - \sigma_1) \cdot b = (8 \text{ to } 11)cb \qquad (2.78)$$

Skempton [33] showed that p_u can vary from 7.6 cb to 9.4 cb. Very often, p_u is adopted as 11 cb.

To find the critical depth, x_c, at which Equation 2.78 becomes operational, we equate Equations 2.77 and 2.78 to find

$$x_c = \frac{9\,cb}{\gamma b + 2.83c} \qquad (2.79)$$

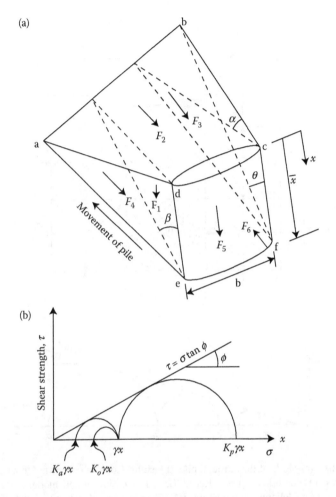

FIGURE 2.19 Ultimate soil resistance at shallow depths. (a) Passive wedge failure; (b) Mohr–Coulomb diagram for state of stress for earth pressure conditions. (From Reese, L.C., Discussion of soil modulus for laterally loaded piles, by McClelland, B. and Focht, J.A., *Transactions, ASCE*, 123, 1958, 1071–1074; Reese, L.C. (1) and Matlock, H. (2), *(1) Soil-Structure Interaction; (2) Mechanics of Laterally Loaded Piles, Lecture Notes,* Courses Taught at the University of Texas, Austin, TX, 1967–1968. With permission.)

If the shear strength and the unit weight vary with depth, x_c can be obtained by plotting p_u versus x as in Figure 2.21.

2.6.4.2.3 Ultimate Soil Resistance for Cohesionless Soils

Just like in cohesive soil, we need to use two mechanisms for cohesionless soils: the ultimate resistance near surface and at depth. The procedures for the ultimate resistance have been developed by Reese [21,32,34,35]. The assumed wedge for near surface is similar to that shown in Figure 2.19a. The pile is assumed to be rigid and it experiences

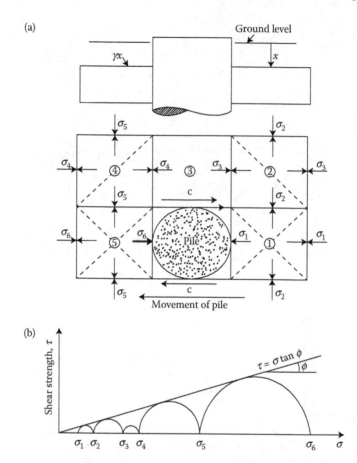

FIGURE 2.20 (a) Mode of flow around pile; (b) Mohr–Coulomb diagram for state of stress for earth pressure conditions. (From Reese, L.C., Discussion of soil modulus for laterally loaded piles, by McClelland, B. and Focht, J.A., *Transactions, ASCE*, 123, 1958, 1071–1074; Reese, L.C. (1) and Matlock, H. (2), *(1) Soil-Structure Interaction; (2) Mechanics of Laterally Loaded Piles, Lecture Notes*, Courses Taught at the University of Texas, Austin, TX, 1967–68. With permission.)

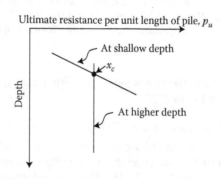

FIGURE 2.21 Schematic for transition depth (x_c).

motions such that the wedge moves under the passive condition. By considering equilibrium of forces on the wedge, the following expression for p_u was derived [34,35]:

$$p_u = \gamma x \left[\frac{K_o x \tan \phi \sin \beta}{\tan(\beta - \phi) \cos \alpha} + \frac{\tan \beta}{\tan(\beta - \phi)} (b + x \tan \beta \tan \alpha) \right.$$
$$\left. + K_o x \tan \beta (\tan \phi \sin \beta - \tan \alpha) - K_a b \right] \qquad (2.80)$$

where K_o is the coefficient of earth pressure at rest and K_a is the minimum coefficient of active earth pressure, and β and α are shown in Figure 2.19a. Based on laboratory tests, Bowman [36] suggested that α for cohesionless soils be modified as $\phi/2$ to $\phi/3$ for loose sands, and up to ϕ for dense sands. If $\alpha = 0$, the solution given by Reese [32,34] would apply, which will also be equal to that presented by Terzaghi [37]. The value of β can be found approximately by using the following equation:

$$\beta = 45 + \phi/2 \qquad (2.81)$$

2.6.4.2.4   Ultimate Resistance at Depth

As for cohesive soils, the mechanism of soil motions will take place around the pile at some distance from the ground surface. The states of stresses on the blocks are shown in Figure 2.20a with the Mohr–Coulomb diagram in Figure 2.20b. Then, the ultimate resistance can be derived as [34,35]:

$$p_u = K_a b \gamma x \left(\tan^8 \beta - 1 \right) + K_o b \gamma x \tan^4 \beta \tan \phi \qquad (2.82)$$

The critical depth, x_c, at which Equation 2.82 will be applicable can be computed by equating Equations 2.80 and 2.82, and by developing the plot as shown in Figure 2.21. Since the term $x (= x_c)$ appears in both equations, it may be necessary to use an iterative procedure to solve for x_c.

2.6.4.3 p_y–v Curves for Yielding Behavior

In the previous section, we derived equations for ultimate resistance and critical depths for applications involving near-surface and finite depths. In general, such equations may be used to develop the p_y–v curves. However, sometimes the ultimate resistance, p_u, formulas can be further simplified; such simplifications are used in the following descriptions. As noted before, the procedures are different for clays and sands.

2.6.4.3.1   For Soft Clays

Schematics of the p_y–v curves for soil are shown in Figure 2.17. The procedure for soft clay involves the following steps [15,32]:

1. For a practical problem under consideration, obtain the undrained shear strength of the soil, say, from triaxial tests on specimens of soft clay at various depths. Also, obtain unit weight of the soil at various depths. Find

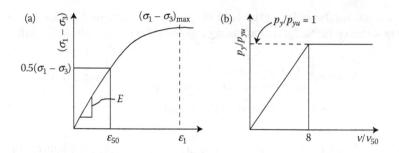

FIGURE 2.22 p_y–v Curves for clay. (a) From triaxial test; (b) simplified nondimentional representation.

the strain, ε_{50}, related to 50% of maximum principal stress difference, $(\sigma_1 - \sigma_3)_{max}$ (Figure 2.22a). If the values of ε_{50} are not available from test data, an approximate value of $\varepsilon_{50} = 0.020$ can be assumed.

2. The ultimate soil resistance, p_u (or p_{yu}) (Figure 2.22b) is computed on the basis of the cohesive strength (c), average unit weight (γ), and the width or diameter (b) of the pile; the following simplified forms are often used for soft clays:

$$p_u = \left(3 + \frac{\gamma}{c} x + \frac{0.5}{b} x \right) cb \qquad (2.83a)$$

or

$$p_u = (8 \text{ to } 11)cb \qquad (2.83b)$$

where x is the depth at which the p_y–v curve is considered. The smaller value of p_u is adopted. For medium soft clays, the third term in Equation 2.83a is reduced by using 0.25 instead of 0.50.

3. Compute the displacement v_{50} by using the following formula:

$$v_{50} = 2.5 \, \varepsilon_{50} b \qquad (2.84)$$

where $\varepsilon_{50} = ((\sigma_1 - \sigma_3)_{max})/(2E)$ and E is the secant modulus from the origin to point A (Figure 2.22a). Now, the p_y–v curve for a soft soil can be expressed as a relation between p_y and v in the following form [34]:

$$\frac{p_y}{p_u} = 0.50 \left(\frac{v}{v_{50}} \right)^{1/3} \qquad (2.85a)$$

The p_y–v curve (Equation 2.85a) is often assumed to intersect the constant-level yield line p/p_u (or p/p_{yu}) = 1 at v/v_{50} = 8.0 (Figure 2.22b).

2.6.4.4 p_y–v Curves for Stiff Clay

The procedure for p_y–v curves for stiff clay is similar to that for soft clay. However, the expression for p/p_u is slightly different, as shown below:

$$\frac{p}{p_u} = 0.50\left(\frac{v}{v_{50}}\right)^{1/4} \tag{2.85b}$$

2.6.4.5 p_y–v Curves for Sands

A schematic of the p_y–v curve for sand is shown in Figure 2.23 [15,32,35]. The steps for constructing p_y–v curves for sands are given below:

1. Obtain the soil properties and the following parameters from laboratory and/or geotechnical field tests, for evaluating the soil resistance:

Angle of frictional resistance	ϕ
Parameter	$\alpha = \phi/2$
Coefficient of earth pressure at rest	$K_o = 0.4$
Coefficient of active earth pressure	$K_a = \tan^2(45 - \alpha)$
Parameter	$\beta = 45° + \phi/2$

2. Compute the ultimate resistance of soil near the ground surface by using Equation 2.80. The ultimate resistance well below the ground surface is obtained by using Equation 2.82. The critical depth x_c is found by equating Equations 2.80 and 2.82. Then, Equation 2.80 is used for depths above x_c, and Equation 2.82 is used for depths below x_c.
3. Now, find the ultimate value of soil resistance, p_{yu}, and the corresponding displacement using the following formulas [38]:

$$p_{yu} = Ap_u \tag{2.86a}$$

and

$$v_u = 3b/80 \tag{2.86b}$$

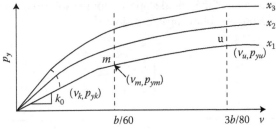

FIGURE 2.23 Representation of p_y–v curves for sand. (From Reese, L.C., Discussion of soil modulus for laterally loaded piles, by McClelland, B. and Focht, J.A., *Transactions, ASCE*, 123, 1958, 1071–1074; Reese, L.C. (1) and Matlock, H. (2), *(1) Soil-Structure Interaction; (2) Mechanics of Laterally Loaded Piles, Lecture Notes*, Courses Taught at the University of Texas, Austin, TX, 1967–68; Reese, L.C. and Desai, C.S., Chapter 9 in *Numerical Methods in Geotechnical Engineering*, C.S. Desai and J.T. Christian (Editors), McGraw-Hill Book Co., New York, USA, 1977. With permission.)

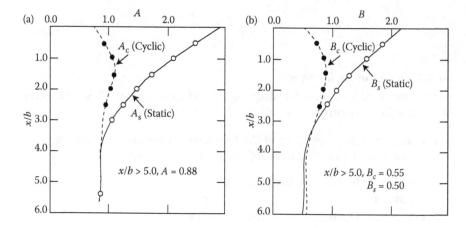

FIGURE 2.24 Nondimensional coefficients A and B for ultimate soil resistance. (From Reese, L.C. and Desai, C.S., Chapter 9 in *Numerical Methods in Geotechnical Engineering*, C.S. Desai and J.T. Christian (Editors), McGraw-Hill Book Co., New York, USA, 1977. With permission.)

where p_u is obtained from Equation 2.80 or 2.82, and A is a parameter that varies with depth and is found from Figure 2.24a, which shows the curves for both static and cyclic loading; the latter is described subsequently.

4. Find v_m and p_{ym} as follows:

$$v_m = b/60 \qquad (2.87a)$$

$$p_{ym} = Bp_u \qquad (2.87b)$$

The parameter B is obtained from Figure 2.24b.

5. Now, select a value of the initial slope, k_o, for the soil from Table 2.3.
6. Compute v_k from the following expression:

$$v_k = \left(\frac{C}{kx} \right)^{n/(n-1)} \qquad (2.88a)$$

TABLE 2.3

Recommended Value of k

State of Soil	Loose	Medium	Dense
K_o (lb/in²)	20	60	125

1 psi = 6.895 kPa.

where $C = p_{ym}/v_m^{1/n}$, $n = p_{ym}/mv_m$, and $m = (p_{yu} - p_{ym})/(v_u - v_m)$. The portion of the p_y–v curve between points k and m is represented by a parabola as follows:

$$p_y = Cv^{1/n} \qquad\qquad (2.88b)$$

Further details are given later in Example 2.7.

2.6.5 P_y–v CURVES FOR CYCLIC BEHAVIOR

Cyclic loading due to wave forces is common for offshore piles in which the number of loading cycles is usually very large. Figure 2.25 shows the p_y–v curve for soft clays proposed by Matlock [39].

Cyclic loading can cause degradation or softening in the soil strength, usually after certain number of loading cycles, N_c. Matlock [39] reported that such degradation does not occur significantly before point D (Figure 2.25) at the resistance ratio of about 0.72; thus, until point D, the cyclic response is about the same as the static response. The peak cyclic resistance (point D) occurs at v/v_c of about 3 and p_y/p_{yu} of about 0.72. At greater displacements, the soil resistance diminishes or degrades and may approach zero at v/v_c of about 15. According to Matlock [39], the value of v_c can be obtained as follows:

$$v_c = \frac{2.5c}{E}b \qquad\qquad (2.89a)$$

where E is the secant modulus and b the pile diameter. The p_y–v curve from the origin to point D′ can be obtained from [39]:

$$\frac{p_y}{p_{yu}} = 0.5\left(\frac{v}{v_c}\right)^{1/3} \qquad\qquad (2.89b)$$

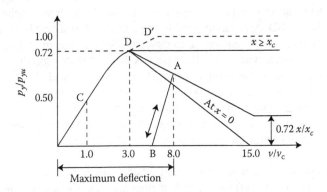

FIGURE 2.25 p–v curve for cyclic loading: clay. (From Matlock, H., Correlations for the design of laterally loaded pikes in soft clays, *Proceedings of the 11th Offshore Technology Conference*, Paper No. 1204, Houston, TX, 577–594, 1970. With permission.)

It intersects the constant yield line (D') at $p_y/p_{yu} = 1.0$ at $v/v_c = 8$.

By equating Equations 2.83a and 2.83b, with 9 as the coefficient, the critical depth x_c is given by

$$x_c = \frac{6cb}{\gamma b + 0.5c} \tag{2.89c}$$

where γ is the buoyant unit weight of the soil. If the depth to the p_y–v curve is greater than or equal to x_c, p_y is taken equal to $0.72\,p_{yu}$, for all values of v greater than $3v_c$.

If the depth to the p_y–v curve is less than x_c, the value of p_y decreases linearly from $0.72\,p_{yu}$ at $v = 3v_c$ to the value of p_y expressed as follows (below $v = 15v_c$):

$$p_y = 0.72 p_{yu} \frac{x}{x_c} \tag{2.89d}$$

The value of p_y remains constant after $v = 15v_c$ (Figure 2.25).

2.6.6 RAMBERG–OSGOOD MODEL (R–O) FOR REPRESENTATION OF P_y–V CURVES

Once a nonlinear p_y–v curve is developed, we can also use the Ramberg–Osgood (R–O) model to simulate the curve. Then, the soil modulus (k) can be derived by taking a derivative of p_y with respect to v for the function in Equation 2.75. Note that the required R–O parameters are defined based on the p_y–v curve (Figure 2.18b).

2.7 ONE-DIMENSIONAL SIMULATION OF RETAINING STRUCTURES

The behavior of unit length of a long retaining structure (including anchors and reinforcements) can be assumed to be the same over the entire length. Then, the beam-column simulation can be used, with certain modifications, for the approximate analysis of the behavior of retaining structures. Figure 2.26a shows a schematic of a retaining wall; the 1-D idealization considering unit length is shown in Figure 2.26b.

For a retaining structure embedded in clay, the secant elastic modulus, E_s, can be assumed to be constant with depth. Because it is difficult to develop the variation of E_s with depth in sand, it is also often assumed to be constant with depth.

Walls with anchor bulkheads sometimes experience inward displacements, which may cause compression in the soil above the anchor (Figure 2.26); inward deflections are also possible below the point of contraflexure (Figures 2.27b and 2.27d). For a trench with multiple braces, inward deflections can occur between pairs of braces.

The depth of embedment of the wall may be determined by making the following assumptions: (1) select the effective depth of embedment as the depth of the soil mass in front of the toe of the bulkhead (Figure 2.27); (2) select the depth of embedment as the distance between the bulkhead anchors and the surface of the retained soils (Figure 2.27), in which inward deflections may occur; and (3) for braced trenches,

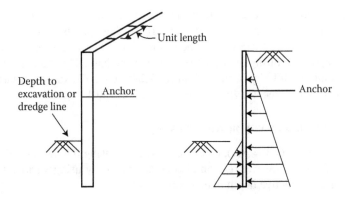

FIGURE 2.26 Retaining wall and one-dimensional idealization. (a) Retaining structure and (b) one-dimensional idealization (unit length).

select it as the largest vertical distance between braces or the distance between the top of the trench and the first brace, whichever is greater [40,41].

Two types of anchored bulkheads are shown in Figures 2.27a and 2.27b. Figure 2.27a shows a free support near the lower end and Figure 2.27b shows a fixed support near the lower end. The mechanisms of deformation are shown differently for the

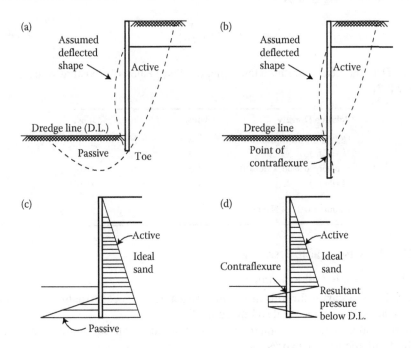

FIGURE 2.27 Anchor bulkheads: (a) free at soil support; (b) fixed at soil support; (c) conventional earth pressure for free support; and (d) conventional earth pressure for fixed support. (From Haliburtan, T.A., *Journal of the Soil Mechanics and Foundations Division, ASCE*, 44(SM6), 1968, 1233–1251. With permission.)

two conditions. For the free support, the displacements are considered to be similar to those of a simply supported vertical beam. If the sheet pile is driven to a significant depth, the lower end can be assumed to be fixed. The free support structures fail by bending of the piling or sliding of soil near the base. The fixed support case involves a failure mainly by bending. Figures 2.27c and 2.27d show the earth pressure diagrams for the free and fixed cases, respectively [40].

2.7.1 CALCULATIONS FOR SOIL MODULUS, E_s

The soil modulus, E_s (= $k = p_y/v$), can be computed by using the methods described previously for piles. However, E_s can also be computed by using various methods depending on the effective depth of embedment, D.

2.7.1.1 Terzaghi Method

The value of E_s can be evaluated for clays as [42]

$$E_s = 0.67 E_{s1}\left(\frac{1}{D}\right)$$ (2.90a)

where E_{s1} is the soil modulus for a 1 ft^2 plate on clays. For the bulkhead in sand, the expression for E_s is given by

$$E_s = k\left(\frac{x}{D}\right)$$ (2.90b)

where D is the effective embedment depth. The values of constant k are given below [37]:

Relative Density	Loose	Medium	Dense
Dry or moist sand, k (tons/ft^3)	2.5	8.0	20.0
k (lb/in^3)	2.9	9.0	23.0
Submerged sand, k (tons/ft^3)	1.6	5.0	13.0
k (lb/in^3)	1.9	6.0	15.0

1 lb/in^3 = 0.2714 N/cm^3.

2.7.2 NONLINEAR SOIL RESPONSE

2.7.2.1 Ultimate Soil Resistance

To develop p_y–v curves for soils in retaining structures, we need to define terms such as ultimate soil resistance and soil modulus, E_s.

If the soil experiences compression, the resistance can be obtained from the Mohr–Coulomb passive expressions:

$$\sigma_p = 2c\tan\left(45 + \frac{\phi}{2}\right) + \sigma_v \tan^2\left(45 + \frac{\phi}{2}\right)$$ (2.91a)

If the soil expands, the (active) pressure can be obtained from

$$\sigma_a = -2c \tan\left(45 - \frac{\phi}{2}\right) + \sigma_v \tan^2\left(45 - \frac{\phi}{2}\right) \qquad (2.91b)$$

where σ_v is the total vertical stress or pressure. It can be computed as γx (density and depth), including the effect of layered deposits and surcharge, if any.

2.7.2.2 p_y–v Curves

Figure 2.28 shows a p_y–v curve for retaining walls, shown by the dashed line [40,41]. A simplified form can be used by joining the active pressure to the passive resistance, where E_s and E_s' are soil moduli in the passive and active zones, respectively. In the passive zone, a line is drawn with a slope of E_s from the at-rest condition to the point where the asymptotic response initiates. In the active zone, the line is drawn with E_s' from the at-rest state to the asymptote in the active region. Figures 2.29a and 2.29b show soil resistance–displacement curves for soil mass on the right and left sides of the wall, respectively [40]. The combined curve is also shown in Figure 2.29d [40]. Here, the existing at-rest pressure on the structure is equal to the algebraic sum of at-rest pressures acting on the right and left sides; usually the at-rest pressure in the right zone is greater than in the left zone. The sign convention is also shown in Figure 2.29c [40]. The values of E_s and the active at-rest and passive pressures, which increase linearly with depth, are affected by factors such as overburden, surcharge, saturation, and submergence.

The soil resistance will increase to a maximum value if the structure displaces to the right, and the ultimate resistance can be defined as the passive resistance of the soil on the right. If the toe of the wall displaces to the left, the limiting soil resistance can be expressed as the maximum passive pressure for the left soil minus the active pressure exerted by the right soil. Figure 2.29d shows the combined curve as average of right and left curves.

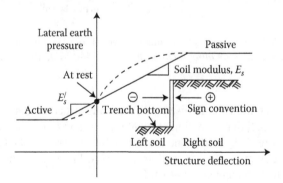

FIGURE 2.28 Simplified nonlinear soil resistance for retaining structure. (From Haliburtan, T.A., *Journal of the Soil Mechanics and Foundations Division, ASCE*, 44(SM6), 1968, 1233–1251; Halliburton, T.A., Soil-structure interaction: Numerical analysis of beams and beam-columns, Technical Publication No. 14, School of Civil Engineering, Oklahoma State Univ., Stillwater, Oklahoma, USA, 1971. With permission.)

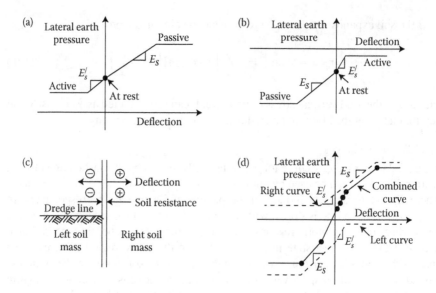

FIGURE 2.29 Nonlinear soil resistance below the bulkhead dredge line (a) soil resistance for right soil mass; (b) soil resistance for left soil mass; (c) sign conventions; and (d) combined soil resistance. (From Haliburtan, T.A., *Journal of the Soil Mechanics and Foundations Division, ASCE*, 44(SM6), 1968, 1233–1251. With permission.)

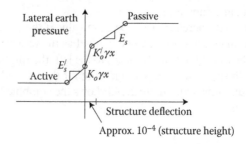

FIGURE 2.30 Modification of soil resistance curve. (From Haliburtan, T.A., *Journal of the Soil Mechanics and Foundations Division, ASCE*, 44(SM6), 1968, 1233–1251. With permission.)

When anchored bulkheads are used with retaining structure in sands, it was proposed by Terzaghi [42] that a small movement into the soil results into an increase in the coefficient "at rest" K_o to K_o'. The magnitude of increase can be 0.40, 0.80, and 1.20 for loose, medium, and dense sand, respectively. Figure 2.30 shows a modified soil resistance–displacement curve [40,42].

2.8 AXIALLY LOADED PILES

Pile foundations, with symmetrical geometry, can involve symmetric axial loads in the direction of the centerline of the pile. We consider the displacement in

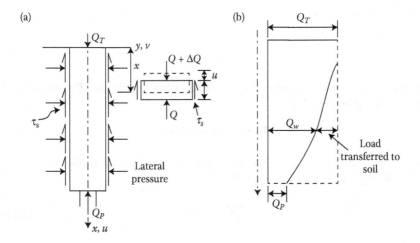

FIGURE 2.31 Axially loaded pile and load distribution along pile. (a) Applied load and pressure; (b) load distribution along pile.

the vertical x-direction to be u (Figure 2.31a). Figure 2.32a shows a mechanical analog of the pile. The axial strain, du/dx, is denoted as ε_x. Then, the force $Q = EA\varepsilon_x$, where A is the cross-sectional area of the pile, and Q is (increment of) the axial load.

Figure 2.31a shows an element of pile at a distance x from the origin (top). The difference of forces Q and $Q + (dQ/dx)dx$ is $(dQ/dx)dx$, which is equal to the net

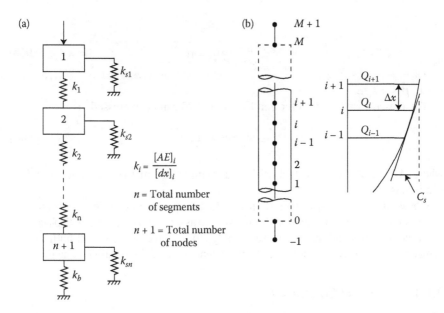

FIGURE 2.32 Mechanical analog and finite difference model. (a) Mechanical analog; (b) finite difference representation.

(shear) force carried by the soil. Now, the shear stress, τ_s, acting over the circumferential section of the pile dx is given by

$$\tau_s = k_s u \qquad (2.92)$$

where k_s is the spring constant of the soil in the vicinity of the pile. The total force Q due to shear stress is given by

$$Q_s = \tau_s C dx = C k_s u dx \qquad (2.93)$$

where C is the perimeter of the pile. We equate the vertical force in the pile with that in the soil at the section (Figure 2.31a) as follows:

$$\frac{dQ}{dx} = C k_s u \qquad (2.94a)$$

This equation represents the change in Q with depth (Figure 2.32b). Now, we can write for a pile of uniform cross section

$$\frac{dQ}{dx} = \frac{d}{dx}\left(EA \frac{du}{dx} \right) = EA \frac{d^2u}{dx^2} \qquad (2.94b)$$

By equating Equations 2.94a and 2.94b, we obtain

$$EA \frac{d^2u}{dx^2} = C k_s u \qquad (2.95a)$$

or

$$\frac{d^2u}{dx^2} - \beta^2 u = 0 \qquad (2.95b)$$

where $\beta^2 = C k_s / EA$.

The closed-form solution for u in Equation 2.95 can be expressed as

$$u = A_1 e^{-\beta x} + A_2 e^{\beta x} \qquad (2.96)$$

where A_1 and A_2 are constants to be determined from the available boundary conditions.

Coyle and Reese [28] proposed an incremental iterative numerical procedure for the solution of the above equation. It involves the assumption of a (small) displacement at the bottom (tip) and moving incrementally from segment to segment until the top is reached. The procedure provides displacement and load distribution along the pile.

2.8.1 BOUNDARY CONDITIONS

The applied force at the top of the pile is P_t ($= Q_T$). Hence, from Equation 2.96, we can write

$$P_t = -EA\frac{du}{dx}$$

$$= \beta EA(A_1 e^{-\beta x} - A_2 e^{\beta x}) \tag{2.97a}$$

Now, substituting $x = 0$ at the top to yield

$$P_t = \beta EA(A_1 - A_2) \tag{2.97b}$$

We can assume approximately that the displacement is zero at the tip, where $x = L$. This boundary condition can be expressed as

$$0 = A_1 e^{-\beta L} + A_2 e^{\beta L} \tag{2.98a}$$

Solution of Equations 2.97b and 2.98a leads to expression for the constants as

$$A_1 = \frac{P_t}{\beta EA(1 + e^{-2\beta L})}$$

$$\tag{2.98b}$$

$$A_2 = \frac{P_t}{\beta EA(1 + e^{2\beta L})}$$

We can also use the following boundary conditions.
At the top, the applied load is P_t, that is, $Q = P_t$; hence

$$\left(\frac{du}{dx}\right)_{x=0} = \frac{-P_t}{EA}$$

At the bottom, $x = L$ and assuming no load at the tip, we have

$$\left(\frac{du}{dx}\right)_{x=L} = 0$$

We can derive the expressions for A_1 and A_2 by using the above boundary conditions.

2.8.2 Tip Behavior

The applied load at top, P_t, is carried by the side friction discussed above and the tip resistance at the base of the pile. Let us represent the Winkler spring at the tip as k_b and displacement at tip as u_L (Figure 2.32a). Then, the (uniform) resistance or pressure, p, over the area of the base can be expressed as

$$p_{xL} = k_b u_L \tag{2.99a}$$

Here, k_b is the axial spring stiffness at the base and p_{xL} is the pressure applied over the area of the base. The total load taken by the tip is given by

$$P = pA = Ak_b(A_1 e^{-\beta L} + A_2 e^{\beta L}) \tag{2.99b}$$

Here, we used the expression for u given in Equation 2.96 and A is the area of the base, which can be different from that of the pile. Figure 2.32a shows a mechanical analog for the side shear and normal stress at the base and the variation of Q along the pile is shown in Figure 2.32b.

2.8.3 Soil Resistance Curves at Tip

The soil resistance–displacement p_t–u (τ_t–u) curves for the tip can be derived in a similar way as for p_y–v curves described before and used in the numerical (FD and FE) procedures, except that in this case, the axial behavior of soil needs to be considered.

2.8.4 Finite Difference Method for Axially Loaded Piles

We can express the FD forms of Equation 2.95b at point i (Figure 2.32b) as

$$\frac{u_{i-1} - 2u_i + u_{i+1}}{\Delta x^2} - \beta^2 u_i = 0 \tag{2.100a}$$

or

$$u_{i-1} - (2 + \beta^2 \Delta x^2)u_i + u_{i+1} = 0 \tag{2.100b}$$

In the FD method, we need hypothetical points -1 and $M + 1$. From the boundary conditions on the slope, we obtain

$$\frac{u_{M+1} - u_{M-1}}{2\Delta x} = \frac{-P_t}{EA} \tag{2.101a}$$

Therefore

$$u_{M+1} = \frac{-2P_t \Delta x}{EA} + u_{M-1} \tag{2.101b}$$

Assuming a linear and symmetric variation from u_{-1} to u_1, we have

$$u_{-1} = u_1 \tag{2.102}$$

Now, we write the FD equations using Equation 2.100b at all points and then substitute the above values of u_{M+1} and u_{-1}. Solutions of the resulting algebraic simultaneous equations will lead to the displacements u_o to u_M.

Once the displacements u_o to u_M are known, we can find other quantities as follows:

$$\tau_{si} = (k_s u)_i \tag{2.103}$$

$$Q_i = \tau_{si} C_i dx \tag{2.104}$$

where dx is the increment along the x-axis, and i denotes a node.

2.8.5 Nonlinear Axial Response

In the literature for piles, the axial soil resistance–displacement response is termed as t–z curves [28,32,43], where t denotes (shear) stress in the axial direction and z denotes axial displacement. To be consistent with the nomenclature in this book, we call it τ_s–u curves. The procedure for finding τ_s–u curves is discussed next.

2.8.6 Procedure for Developing τ_s–u (t–z) Curves

The axial displacement due to shear on the pile element is given by [28,32]

$$u = \frac{\tau_s r_o}{G_i} \ell n \left(\frac{(r_m / r_o) - a}{1 - a} \right) \tag{2.105a}$$

where u is the displacement of the pile element, r_o is the pile radius, τ_s is the shear the pile–soil interface, G_i is the initial shear modulus for small strain, and

$$a = \frac{\tau_s R_f}{\tau_{max}} \tag{2.105b}$$

where R_f is the stress–strain curve-fitting constant ($R_f \cong 0.9$ for sand), τ_{max} = shear stress at failure [$\tau_{max} \cong 0.45$ ksf (47.90 kN/m²) for sand], and r_m = radial distance. The latter is expressed as

$$r_m = 2.5 \, L\rho(1 - v) \tag{2.105c}$$

where L is the length of the pile, v is the Poisson's ratio of soil, and ρ is the ratio of the soil shear modulus at depth $L/2$ and the pile tip given by

$$\rho = \frac{G_{L/2}}{G_L} \tag{2.105d}$$

2.8.6.1 Steps for Construction of τ_s–u (t–z) Curves

Step 1. First, we find the initial shear modulus:

$$G_i = \frac{E_i}{2(1 + v)} \tag{2.105e}$$

where v is the Poisson's ratio often assumed as 0.30 for sand and 0.45 for clay, and E_i is the initial Young's modulus of soil, which is expressed in the functional form as

$$E_i(\sigma_3) = K' p_a \left(\frac{\sigma_3}{p_a} \right)^n \tag{2.105f}$$

where n and K' are parameters in the formulation, p_a is the atmospheric pressure (= 14.7 psi or = 101.36 kPa), σ_3 is the minor principal stress ($\sigma_3 = K\gamma x$ at desired x), K is the lateral earth pressure coefficient ($K = 1.23$ for some sands), and γ is the equivalent density of soil.

Step 2. Use Equation 2.105e to find G_i at $x = L$ and G_i at $x = L/2$ by adopting E_i as a function of σ_3 (i.e., x) and then determine ρ using Equation 2.105d.

Step 3. Determine r_m using Equation 2.105c.

Step 4. Determine a as a function of τ_s using Equation 2.105b.

Step 5. Use Equation 2.105a as follows:
 – At a specific depth, we find the appropriate G_i, for example, at $x = 0.10\ L$, use Equations 2.105e and 2.105f to obtain G_i at 0.1 L
 – then substitute G_i in Equation 2.105a and find u as a function of τ_s.

Step 6. Develop τ_s–u curves at various depths. Use Equation 2.105a to find u for different values of τ_s for a given curve. Choose values of τ_s based on the τ_{max} defined above.

In the solution for axial loading, when the FD or FE method is used, we introduce the τ_s–u curves in the x-direction, along the pile, in the same way as we introduce the p_y–v curves in the y- (and z-) directions. Also, we introduce a special spring, k_b, at the base to develop p_z–u curves.

2.9   TORSIONAL LOAD ON PILES

Figure 2.33 shows a schematic of a pile subjected to torsion or torque. The soil resistance in this case will be in the circumferential direction (θ) along the wall, and at the tip. We consider first the 1-D idealization; then, we will consider 2-D idealization using the FEM.

The GDE for torsion for 1-D simulation can be obtained by following the procedure similar to axially loaded pile. Here, we adopt the central (vertical) axis as z. We can express shear stress τ as [9]

$$\tau = k_\theta r \theta \tag{2.106}$$

where r is the radius of the pile, θ is the angular deformation, and k_θ is the soil (shear) stiffness in the circumferential direction. The torque on the pile, T, over the area r θdx can be expressed as

$$T = 2\pi r^3 k_\theta \theta dx \tag{2.107}$$

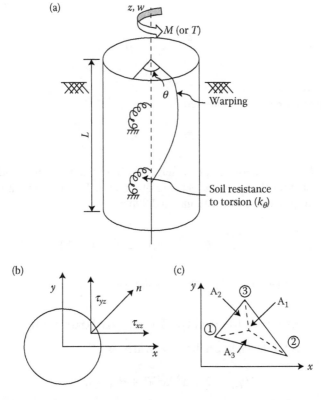

FIGURE 2.33 Torsional loading on pile. (a) Pile subjected to torsional load; (b) cross-section of pile; and (c) triangular element.

Therefore

$$\frac{dT}{dx} = 2\pi r^3 k_\theta \cdot \theta \qquad (2.108)$$

Now

$$\frac{dT}{dx} = \frac{d}{dx}\left(GJ\frac{d\theta}{dx}\right) \quad \text{or} \quad GJ\frac{d^2\theta}{dx^2} \qquad (2.109)$$

where G is the shear modulus and J is the polar moment of inertia of the cross section of the pile.

Equating Equations 2.108 and 2.109, we obtain

$$GJ\frac{d^2\theta}{dx^2} - 2\pi r^3 k_\theta \theta = 0 \qquad (2.110)$$

or

$$\frac{d^2\theta}{dx^2} - \alpha^2\theta = 0$$

where $\alpha^2 = (2\pi r^3 k_\theta)/(GJ)$.

The solution for θ can be expressed as

$$\theta = D_1 e^{-\alpha x} + D_2 e^{\alpha x} \tag{2.111}$$

Constants D_1 and D_2 above can be determined from the boundary conditions.

The expression for the torque at the base of the pile, T_p, can be expressed by following a similar procedure as for Equation 2.99b as

$$T_p = \frac{\pi r^4}{2} k_b \theta = \frac{\pi r^4}{2} k_b \left(D_1 e^{-\alpha L} + D_2 e^{\alpha L} \right) \tag{2.112}$$

where k_b is the stiffness for soil resistance at the base.

2.9.1 Finite Difference Method for Torsionally Loaded Pile

The FD equations for the GDE can be now written as

$$\theta_{i-1} - (2 + \alpha^2 \Delta x^2)\theta_i + \theta_{i+1} = 0 \tag{2.113}$$

which can be written for all nodes resulting into a set of simultaneous equations for the unknown, θ. Here, i denotes a node point (Figure 2.32b). The boundary conditions can be approximated, as discussed below.

1. The applied torque T at the top gives (Equation 2.107)

$$T_M = 2\pi r^3 L k_\theta \frac{\theta_{M-1} + \theta_{M+1}}{2} \tag{2.114a}$$

Therefore

$$\theta_{M+1} = \frac{T_M}{a} - \theta_{M-1} \tag{2.114b}$$

where $a = \pi r^3 L k_\theta$ and M denotes the top node (Figure 2.32b).
2 The value of torque T_p at the tip may be assumed as zero. Then,

$$0 = 2\pi r^3 L k_p \left(\frac{\theta_{-1} + \theta_1}{2} \right) \tag{2.115a}$$

Therefore,

$$\theta_{-1} = -\theta_1 \qquad (2.115b)$$

Now, we derive the FEM for pile subjected to torsional load.

2.9.2 Finite Element Method for Torsionally Loaded Pile

The FE procedure presented before includes the derivation of the equilibrium equations for axial loading. Now, since a 1-D formulation for the torsion problem may be unrealistic, we present a 2-D idealization and procedure. Then, the potential energy function can be written as follows [20,44–46]:

$$\pi_p = \frac{LG}{2} \iint_A \left[\theta^2 \left(\frac{\partial \psi}{\partial x} - y \right)^2 + \theta^2 \left(\frac{\partial \psi}{\partial y} + x \right)^2 \right] dxdy \qquad (2.116)$$

Here, we have considered the pile as a 2-D structure with coordinates x and y (Figure 2.33c). Also, L is the length of the pile and θ is the angle of twist under applied torsional moment or torque T, and ψ is the warping function. Here, the x–y plane denotes the cross section and z denotes the longitudinal axis of the pile (Figure 2.33). We assume linear function for ψ over the triangular element as

$$\psi = [N]\{q_\psi\} \qquad (2.117)$$

where $[N]$ is the matrix of interpolation functions, $L_i = A_i/A$, and A_i $(i = 1, 2, 3)$ is indicated in Figure 2.33c. Then, the relation between the gradient of ψ and nodal ψ is derived as

$$\begin{Bmatrix} \partial\psi/\partial x \\ \partial\psi/\partial y \end{Bmatrix} = \frac{1}{2A} \begin{bmatrix} b_1 & b_2 & b_3 \\ a_1 & a_2 & a_3 \end{bmatrix} \begin{Bmatrix} \psi_1 \\ \psi_2 \\ \psi_3 \end{Bmatrix}$$

$$= [B]\{q_\psi\} \qquad (2.118)$$

where b_i and a_i $(i = 1, 2, 3)$ are differences between coordinates of nodal points for the triangular element, ψ_i $(i = 1, 2, 3)$ is the nodal warping function and $[B]$ is the transformation matrix. The strain (gradient) vector $\{\varepsilon\}$ is given by

$$\{\varepsilon\} = [B]\{q_\psi\} + \begin{Bmatrix} -\theta_y \\ \theta_x \end{Bmatrix} \qquad (2.119)$$

Now, we can substitute the foregoing derivations in π_p. Then, by variation of π_p with respect to ψ_i and equating to zero, the equations for an element are derived as [20]

$$[k_\psi]\{q_\psi\} = \{Q_\psi\} \qquad (2.120)$$

where

$$[k_\psi] = GL\theta^2 \iint_A [B]^T [B] dx dy$$

$$\{Q_\psi\} = GL\theta^2 \iint_A [B]^T \begin{Bmatrix} y_m \\ -x_m \end{Bmatrix} dx dy$$

where x_m and y_m are mean coordinates values evaluated as $x_m = (x_1 + x_2 + x_3)/3$ and $y_m = y_1 + y_2 = y_3)/3$.

The element equations now are assembled by maintaining the compatibility for ψ over the adjoining elements. Since the values of the warping functions are relative, the boundary condition can be introduced by adopting a reference value for ψ to one of the nodes of the triangle. For example, $\psi_1 = 0$ can be assumed. Then, the computed values are relative, but are appropriate for required formulation and computations.

2.9.3 DESIGN QUANTITIES

When a pile is subjected to only lateral loads, the design may be based on critical displacements, moments, shear forces, and stresses induced. Usually, the maximum moment (M_{max}) and shear force (V_{max}) are considered. The stress (axial), σ_m, due to moment is computed as

$$\sigma_m = \frac{M_{max}c}{I} \tag{2.121a}$$

where c is the distance from the neutral axis of the pile. If the pile is also subjected to axial load (Q), the total design stress, σ, is given by

$$\sigma = \frac{M_{max}c}{I} \pm \frac{Q}{A} \tag{2.121b}$$

where A is the cross-sectional area of the beam-column.

2.10 EXAMPLES

Now, we present a number of example problems for the piles and retaining wall problems considered in this chapter.

2.10.1 EXAMPLE 2.4: p_y–v CURVES FOR NORMALLY CONSOLIDATED CLAY

We develop p_y–v response curves for normally consolidated clay by using the details given in McCammon and Ascherman [47]. The soil properties considered in this example are given below:

TABLE 2.4
Soil Properties

Depth, x (ft)	Moisture Content (%)	Dry Density (lb/ft³)	Liquid Density lb/ft³ Limit	Plasticity Index	Void Ratio
12	194	71	100	57	6.40
17	144	80	83	47	4.20
25	187	76	96	47	5.80
33	148	78	78	41	4.50
41	122	81	75	41	3.80
48	183	75	95	51	5.40
57	102	82	70	40	3.40
65	33	116	35	16	0.92
73	6	118	–	–	0.51

1 ft = 0.3048 m, 1 lb/ft³ = 157.06 N/m³.

Pile data: The pile is a concrete-encased hollow steel cylinder with diameter ≈54 in (137 cm) and $EI = 10 \times 10^{12}$ lb-in² (69×10^{12} kPa). The total length of the pile is ≈170 ft (51.8 m). We perform calculations for two cases of strain: $\varepsilon_{50} = 0.01$ and 0.02.

No shear strength test data were available. Hence, based on the data in Table 2.4, we can obtain cohesive strength (c) values by using Figure 2.34 proposed by Skempton and Bishop [48], which shows the ratio c/p versus plasticity index (PI), where p is the overburden pressure and c is the cohesive strength.

Table 2.5 shows the computation of cohesive strength, c, with depth using Figure 2.34. Figure 2.35 shows the plot of the shear strength (c) with depth based on Table 2.5 below. Now, the peak stress difference, σ_D, is adopted as 2c, which is shown at various depths in Table 2.6.

Figure 2.36 shows a log–log plot of σ_D (lb/in²) versus strain for $\varepsilon_{50} = 0.01$ and 0.02. The curves are obtained by plotting a point for 0.50 σ_D versus $\varepsilon_{50} = 0.01$ (or 0.02), and

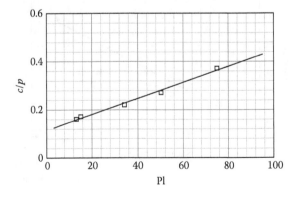

FIGURE 2.34 *c/p* versus PI.

TABLE 2.5

Values of Various Quantities with Depth

Density (Submerged) (lb/ft³)	Depth (ft)	PI	c/p	$p = \gamma h$ (lb/ft²)	$c = (c/p)p$ (lb/ft²)
71 − 62 = 9	12	57	0.30	9 × 12 = 108	32.50
80 − 62 = 18	17	47	0.265	18 × 5 + 108 = 198	52.50
76 − 62 = 14	25	47	0.265	14 × 8 + 198 = 310	82.00
78 − 62 = 16	33	41	0.245	16 × 8 + 310 = 438	107.00
81 − 62 = 19	41	41	0.245	19 × 8 + 438 = 590	145.00
75 − 62 = 13	49	51	0.280	13 × 8 + 530 = 634	178.00
82 − 62 = 20	57	40	0.242	20 × 8 + 694 = 854	207.00
116 − 62 = 54	65	16	0.170	54 × 8 + 794 = 1226	208.00

1 ft = 0.3048 m, 1 lb/ft³ = 157.06 N/m³, 1 lb/ft² = 47.88 N/m².

then drawing a line with the slope of 1:2 through that point. Next, a horizontal line is drawn at the ultimate value of σ_D (Table 2.6). Thus, we developed curves for σ_D versus ε, for $\varepsilon_{50} = 0.01$, and also for $\varepsilon_{50} = 0.02$.

Development of p_y–v curves: The ultimate resistance (p_u) of clay soil can be expressed by Equations 2.77 and 2.78 by assuming the coefficient as 11 in the latter equation:

At shallow depths:

$$p_u = \gamma bx + 2\, cb + 2.83\, cx \tag{2.122a}$$

At higher depths:

$$p_u = 11\, cb \tag{2.122b}$$

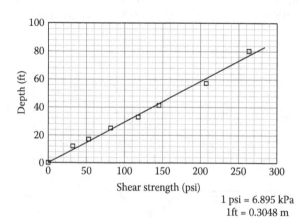

1 psi = 6.895 kPa
1 ft = 0.3048 m

FIGURE 2.35 Shear strength versus depth.

TABLE 2.6
Peak Stress Difference versus Depth

Depth (ft)	c (lb/ft²)	$\sigma_D = 2 \times c$ (lb/ft²)	σ_D (lb/in²)
0	2.0	4	0.028
5	18	36	0.25
10	34	68	0.472
15	55	110	0.765
20	66	132	0.916
25	84	168	1.17
30	100	200	1.39
35	116	232	1.61
40	132	264	1.84
50	166	332	2.30
60	188	376	2.61
70	220	440	3.05
80	264	528	3.67

1 ft = 0.3048 m, 1 lb/ft² = 47.88 N/m², 1 lb/in² = 6.895 N/m² (kPa).

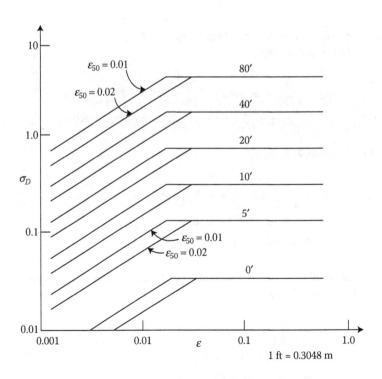

FIGURE 2.36 Plots of σ_D versus ε for $\varepsilon_{50} = 0.01$ and 0.02.

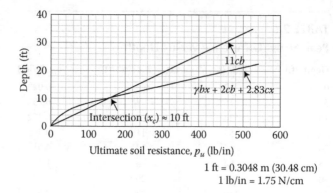

FIGURE 2.37   Plot for critical or transition depth.

Equating the above two equations or plotting p_u versus depth (Figure 2.37), we find the depth at which these curves intersect, which is about 10 ft (3.048 m). Now, Equation 2.122a is applied from depths 0–10 ft (3.048 m):

At x = 0:

$$P_u = 2\ cb = 2 \times \frac{2}{144} \times 54 = 1.5\ \text{lb/in} (262.7\ \text{N/m})$$

Note: The value of c is obtained from Figure 2.35.
x = 1 ft (12 in) (0.3048 m):

$$P_u = \frac{9}{1728} \times 54 \times 12 + 2\frac{5}{144} \times 54 + 2.83\frac{5}{144} \times 12$$
$$= 3.37 + 3.75 + 1.18 = 8.30\ \text{lb/in} (1453.48\ \text{N/m})$$

x = 3 ft (36 in) (0.9144 m):

$$P_u = \frac{9}{1728} \times 54 \times 36 + 2\frac{12}{144} \times 54 + 2.83\frac{12}{144} \times 36$$
$$= 10.10 + 9.00 + 8.50$$
$$= 27.60\ \text{lb/in} (4833.3\,\text{N/m})$$

Note: Value of submerged γ_s is obtained from Table 2.5; it is used up to depth = 12 ft.
x = 5 ft (60 in) (1.524 m):

$$P_u = \frac{9}{1728} \times 54 \times 60 + 2\frac{18}{144} \times 54 + 2.83\frac{18}{144} \times 60$$
$$= 16.90 + 13.50 + 4.25$$
$$= 34.65\ \text{lb/in} (6067.\ 9\ \text{N/m})$$

x = 10 ft (120 in) (3.048 m):

$$p_u = \frac{9}{1728} \times 54 \times 120 + 2 \frac{34}{144} \times 54 + 2.83 \frac{34}{144} \times 120$$
$$= 33.70 + 25.50 + 80.20$$
$$= 139.40 \text{ lb/in } (24411.7 \text{ N/m})$$

For x > 10 ft (3.048 m)
x = 20 ft (240 in) (6.096 m)

$$p_u = 11 \, cb = 11 \times \frac{66}{144} \times 54$$
$$= 272 \text{ lb/in } (47632.6 \text{ N/m})$$

x = 40 ft (480 in)(12.19 m)

$$p_u = 11 \times \frac{132}{144} \times 54$$
$$= 545.00 \text{ lb/in } (95440.3 \text{ N/m})$$

x = 60 ft (720 in)

$$p_u = 11 \times \frac{188}{144} \times 54$$
$$= 776.00 \text{ lb/in } (135893.1 \text{ N/m})$$

x = 80 ft (960 in)

$$p_u = 11 \times \frac{264}{144} \times 54$$
$$= 1089.00 \text{ lb/in } (190705.7 \text{ N/m})$$

Now, to develop the p_y–v curves at various depths, we use the following expressions:

$$p_y = 5.5 \, \sigma_D b \quad \text{and} \quad v_1 = \frac{b}{2} \varepsilon \quad [49]: \text{Method I}$$

or

$$p_y = 5.5 \sigma_D b \quad \text{and} \quad v_2 = 2b\varepsilon \quad [33,48]: \text{Method II}$$

For example, we select a number of values of ε and locate the corresponding values of σ_D from Figure 2.36, for ε_{50} either equal to 0.01 or 0.02.

For $\varepsilon_{50} = 0.01$

Typical points on the p_y–v curve are obtained as

At $x = 0.0$ ft (0.0 m)

$\sigma_D = 0.01$, $\varepsilon = 0.005$

$p_y = 5.5 \times 0.01 \times 54 = 2.97$ lb/in (520.08 N/m)

$v_1 = \dfrac{54}{2} \times .005 = 0.135$ in (0.343 cm)

$v_2 = 2b\varepsilon = 2 \times 54 \times 0.005 = 0.54$ in. (1.37 cm)

and at $x = 5$ ft (1.524 m)

$\sigma_D = 0.11$, $\varepsilon = 0.01$

$p = 5.5 \times 0.11 \times 54 = 32.70$ lb/in (5726 N/m)

$v_1 = \dfrac{54}{2} \times 0.01 = 0.27$ in (0.69 cm)

$v_2 = 2 \times 54 \times 0.01 = 1.08$ in. (2.74 cm)

The following data give the p_y–v curves at different depths for $\varepsilon_{50} = 0.01$ and $\varepsilon_{50} = 0.02$ for selected values of ε.

	For $\varepsilon_{50} = 0.01$				For $\varepsilon_{50} = 0.02$				
σ_D	p_y (lb/in)	ε	$v_1{}^a$ (in)	$v_2{}^a$ (in)	σ_D	p_y (lb/in)	ε	v_1 (in)	v_2 (in)
				Depth $x = 0$ ft (0.0 m)					
0.010	2.97	0.005	0.135	0.54	0.01	2.97	0.01	0.27	1.08
0.014	4.16	0.010	0.270	1.08	0.014	3.40	0.02	0.54	2.16
0.020	5.94	0.020	0.54	2.16	0.020	5.94	0.04	1.08	4.32
0.028	8.32	0.040	1.08	4.32	0.028	8.32	0.075	2.02	8.08
				Depth = 5 ft (1.524 m)					
0.08	23.80	0.005	0.135	0.54	0.055	16.35	0.005	0.135	0.54
0.11	32.70	0.100	0.270	1.08	0.078	23.2	0.010	0.27	1.08
0.16	47.50	0.020	0.54	2.16	0.11	32.70	0.02	0.54	2.16
0.25	74.00	0.050	1.35	5.40	0.18	53.2	0.05	1.35	5.40
					0.25	74.00	0.10	2.70	9.80
				Depth = 10 ft (3.048 m)					
0.168	50.0	0.005	0.135	0.54	0.116	34.4	0.005	0.135	0.54
0.240	71.50	0.010	0.27	1.08	0.118	50.0	0.01	0.27	1.08
0.318	94.5	0.020	0.54	2.16	0.238	70.8	0.02	0.54	2.16
0.472	140.0	0.050	1.35	5.40	0.380	113.0	0.05	1.35	5.40
					0.472	140.0	0.075	2.70	9.80
				Depth = 20 ft (6.10 m)					
0.32	98.0	0.005	0.135	0.54	0.24	71.4	0.005	0.135	0.54
0.45	134.0	0.010	0.27	1.08	0.32	98.0	0.010	0.27	1.08

0.65	193.0	0.020	0.54	2.16	0.45	134.0	0.020	0.54	2.16
0.916	272.0	0.050	1.35	5.40	0.64	190.0	0.05	1.35	5.40
					0.916	272.0	0.08	2.16	8.64

Depth = 40 ft (12.2 m)

0.65	193.0	0.005	0.135	0.54	0.44	131.0	0.005	0.135	0.54
0.92	274.0	0.01	0.27	1.08	0.63	188.0	0.01	0.27	1.08
1.30	388.0	0.02	0.54	2.16	0.90	268.0	0.02	0.54	2.16
1.84	545.0	0.05	1.35	5.40	1.42	422.0	0.05	1.35	5.40
					1.84	545.0	0.0805	2.18	8.70

Depth = 80 ft (24.38 m)

1.28	370.0	0.005	0.135	0.54	0.90	267.0	0.005	0.135	0.54
1.80	535.0	0.010	0.27	1.08	1.30	386.0	0.010	0.27	1.08
2.62	780.0	0.020	0.54	2.16	1.80	535.0	0.020	0.54	2.16
3.67	1090.0	0.050	1.35	5.40	2.90	885.0	0.050	1.35	5.40
					3.67	1090.0	0.08	2.16	8.64

[a] v_1 and v_2 refer to the two methods used. (1 lb/in = 175.11 N/m, 1.1 in = 2.54 cm).
1 in = 2.54 cm, 1 lb = 4.448 N.

Figure 2.38 shows the plots of p_y–v curves at various depths for $\varepsilon_{50} = 0.01$ for both v_1 and v_2 using the methods of McClelland and Focht [49] and Skempton [48]. It can be seen that the two methods yield significantly different p_y–v curves below the ultimate resistance.

The p_y–v curves can be expressed by using various functions including the Ramberg–Osgood model (Equation 2.75). Such forms can be implemented in the FE and FD procedures, which can be used for computer predictions of the behavior of piles.

Tables 2.7a and 2.7b show a comparison of the observed [47] and computed displacements for three different values of load at top by using the methods of McClelland and Focht (Method I) and Skempton (Method II). It can be seen that Skempton's method gives much higher predictions for displacements than McClelland and Focht's method. This can be attributed to the higher values of displacements ($v = 2b\varepsilon$), that is, lower stiffnesses in the Skempton method.

2.10.2 EXAMPLE 2.5: LATERALLY LOADED PILE IN STIFF CLAY

Reese et al. [50] have presented static and dynamic analyses of laterally loaded piles in stiff clay, where predicted and measured behaviors were compared. We consider here one of the piles with diameter 24 in (61 cm) and the total length 60 ft (18.3 m); the embedded length of the pile was 49 ft (15 m) (Figure 2.39). The upper 32 ft (9.75 m) length of the pile was instrumented. Both static and dynamic loadings were applied to the test pile; we consider here only static loading. Reese et al.

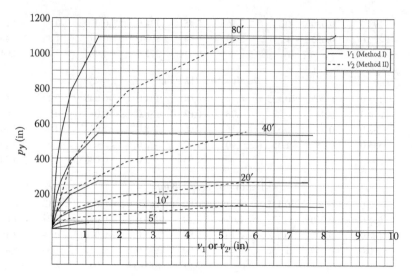

FIGURE 2.38 p_y–v curves for clay, $\varepsilon_{50} = 0.01$. (1 inch = 2.54 cm, 1 ft = 0.305 m, 1 lb/in = 1.75 N/cm)

TABLE 2.7
Comparison of Observed and Computed Results

(a) Method I

Observed [47]		Computed p_y–v Curves by Using $\varepsilon_{0.01}$ at $p_{50\%}$ Ultimate, and $p_y = 5.5\ \sigma_D b$; $v = (b/2)\ \varepsilon$		
Load at Top (lb)	Deflection at Top (in)	Load at Top (lb)	Deflection at Top (in)	% Difference
5000	3.85	5000	3.986	−3.5%
10,000	8.500	10,000	8.587	−1.021
15,000	14.00	15,000	13.75	+1.75

Observed moment$_{\text{max}}$ = $1410 \times 10^3 \times 12 = 1.692 \times 10^7$ lb in (0.19 × 10⁷ N.m).
Computed moment$_{\text{max}}$ = 1.628×10^7 lb in (0.184 × 10⁷ N.m).
% Diff = 3.75%.

(b) Method II

Observed [47]		Computed p_y–v Curves by Using $\varepsilon_{0.001}$ at $p_{50\%}$ Ultimate, and $p = 5.5\ \sigma_D\ v = 2\ b\ \varepsilon$		
Load at Top (lb)	Deflection at Top (in)	Load at Top (lb)	Deflection at Top (in)	% Difference
5000	3.85	5000	4.88	26.7
10,000	8.50	10,000	10.22	20.1
15,000	14.00	15,000	16.46	17.5

1 lb = 4.448 N, 1 in = 2.54 cm.
Observed moment$_{\text{max}}$ = 1.692×10^7 lb in (0.19 × 10⁷ N.m).
Computed moment$_{\text{max}}$ = 1.692×10^7 lb in (0.19 × 10⁷ N.m).

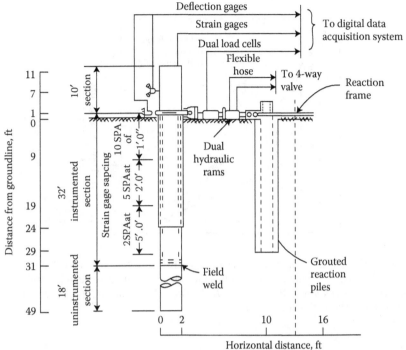

FIGURE 2.39 Field pile load set-up (left part) for 24 inch diameter pile (1 ft = 0.305 m, 1 inch = 2.54 cm). (From Reese, L.C., Cox, W.R., and Koop, F.D., Field testing and analysis of laterally loaded piles in stiff clay, *Proceedings Offshore Technology Conference (OTC)*, Paper No. 2312, Houston, TX, 1975. With permission.)

[50] obtained the p_y–v curves by using the procedure in Ref. [50] on the basis of the experimental (field) behavior of the pile.

However, for use in this analysis, we derive p_y–v curves by using a different procedure presented in Ref. [49]. Such curves were simulated by using the Ramberg–Osgood model. Then, the predictions were obtained by using an FE code SSTIN-IDFE [24], and were compared with typical field data.

2.10.2.1 Development of p_y–v Curves

The following procedure is adopted from Ref. [49]; it allows the computation of p_y–v curves at various depths, x. The undrained shear strengths, c, are obtained from Figure 2.40. The submerged unit weight of soil is about 53 lb/cu ft (8324 N/m³). The initial part of the curve can be assumed to be a straight line from the origin and expressed as (Figure 2.41)

$$E_{si} = \frac{p_{yi}}{v_i} \tag{2.123}$$

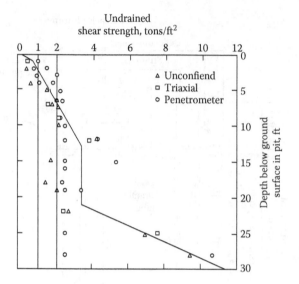

FIGURE 2.40 Shear strength versus depth (1 ton/ft² = 9.58 N/cm², 1 ft = 0.305 m). (From Reese, L.C., Cox, W.R., and Koop, F.D., Field testing and analysis of laterally loaded piles in stiff clay, *Proceedings of the Offshore Technology Conference (OTC)*, Paper No. 2312, Houston, TX, 1975. With permission.)

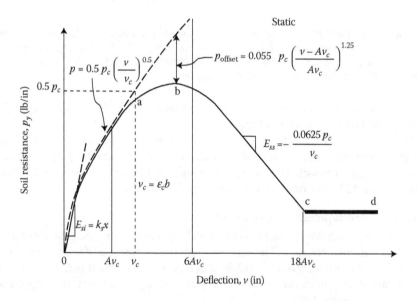

FIGURE 2.41 Proposed *p*–*v* curve for static loading: stiff clay (1 lb/inch = 0.69 N/cm, 1 inch = 2.54 cm). (From Reese, L.C., Cox, W.R., and Koop, F.D., Field testing and analysis of laterally loaded piles in stiff clay, *Proceedings of the Offshore Technology Conference (OTC)*, Paper No. 2312, Houston, TX, 1975. With permission.)

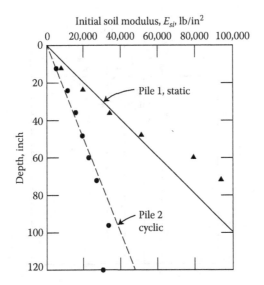

FIGURE 2.42 Initial soil modulus versus depth (1 inch = 2.54 cm, 1 lb/in^2 = 28.71 N/cm^2). (From Reese, L.C., Cox, W.R., and Koop, F.D., Field testing and analysis of laterally loaded piles in stiff clay, *Proceedings of the Offshore Technology Conference (OTC)*, Paper No. 2312, Houston, TX, 1975. With permission.)

where E_{si} is the initial modulus and p_{yi} and v_i are the coordinates of a point on the initial portion. The values of E_{si} at different depths are plotted in Figure 2.42. Then, the value of the modulus can be expressed as

$$E_{si} = kx \tag{2.124}$$

The value of k for static analysis was found to be about k_s = 1000 lb/cu in (271.35 N/cm^3) (Figure 2.42). As described later, the initial part can also be represented by a parabola.

Now the ultimate soil resistance was found from

$$p_{yu} = 2cb + \gamma'bx + 2.83c_a x \tag{2.125}$$

where p_{yu} is the ultimate soil resistance at depth, x, c is the average undrained shear strength of clay over depth (Figure 2.40), b is the diameter of the pile, and γ' is the submerged unit weight of soil [=55 lb/cu ft (8638 N/m^3)]. Equation 2.125 is considered to be valid in the upper zone of the soil. For higher depths, the ultimate resistance is given by

$$p_{yu} = 11\, cb \tag{2.126}$$

where c is the undrained shear strength of the clay at higher depths. It was found in Ref. [50] that the computed values of ultimate resistance (Equations 2.125 and 2.126)

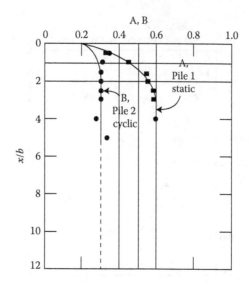

FIGURE 2.43 Coefficients A or B for static and cyclic loadings. (From Reese, L.C., Cox, W.R., and Koop, F.D., Field testing and analysis of laterally loaded piles in stiff clay, *Proceedings of the Offshore Technology Conference (OTC)*, Paper No. 2312, Houston, TX, 1975. With permission.)

were much higher than those obtained from experiments. Hence, the computed values of p_{yu} were adjusted by multiplying by factor A:

$$(p_{yu})_s = A p_{yu} \tag{2.127}$$

where A is the empirical factor (Figure 2.43) for static loading, p_{yu} is computed from Equations 2.125 or 2.126 and $(p_{yu})_s$ is a corrected (static) value guided by experiments.

The initial part (o–a) (Figure 2.41) can be simulated by a parabolic function. We first define the critical displacement as

$$v_c = \varepsilon_c b \tag{2.128}$$

where v_c is the displacement corresponding to the critical strain, ε_c, at 50% of the ultimate stress. It can be obtained from plots such as in Figure 2.44. Then, the initial part (o–a) (Figure 2.41) can be represented by the following parabola:

$$p_y = 0.5 \, p_{yu} \left(\frac{v}{v_c} \right)^{0.50} \tag{2.129}$$

The next parabolic part (a–b) (Figure 2.41) is expressed as

$$p_y = 0.5 \, p_{yu} \left(\frac{v}{v_c} \right)^{0.50} - 0.055 \, p_{yu} \left(\frac{v - A v_c}{A v_c} \right)^{1.25} \tag{2.130}$$

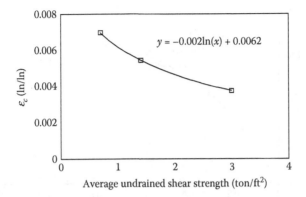

FIGURE 2.44   Variation of ε_c versus undrained shear strength.

Here, $Av_c \leq v \leq 6\,Av_c$.
The subsequent straight line part (b–c) is given by

$$p_y = 0.5\,p_{yu}(6A)^{0.50} - 0.411\,p_{yu} - \frac{0.0625}{v_c}\,p_{yu}(v - 6\,Av_c) \qquad (2.131)$$

for $6\,Av_c \leq v \leq 18\,Av_c$.
The final straight line part (c–d) of the p_y–v curve is given by

$$p_y = 0.5\,p_{yu}(6A)^{0.5} - 0.411\,p_c - 0.75\,p_{yu}A \qquad (2.132)$$

for $18\,Av_c \leq v$.
The p_y–v curves were obtained for various depths by using the above procedure. To determine the parameters for the R–O simulation (Figure 2.18b) the curves were smoothened. Typical parameters at the ground level and at the bottom of the pile are given below:

	At Ground Level	At End (Butt)
k_o	566 lb/in² (3902.6 kPa)	2246 lb/in² (15486 kPa)
k_r	566 lb/in² (3902.6 kPa)	2246 lb/in² (3902.6 kPa)
k_f	0.00	0.00
p_f	1167 lb/in (204.4 kN/m)	15840 lb/in (2774 kN/m)
m	1.0	1.0

The FE code SSTIN-1 DFE [24] was used to compute displacements, moments, shear forces, and so on for the 24 in (61 cm) diameter pile (Figure 2.39). The pile was divided into 20 elements, and the total lateral load of 160 kip (711.68 kN)was applied at the top in increments of $\Delta P = 20$ kip (89 kN). The predictions are labeled as R–O model.

Figure 2.45 shows the computed load–displacement values at the ground line in comparison with the field measurements. The computed and observed bending moments versus depth for two load levels of 71.43 and 136.28 kip (318 and 606 kN),

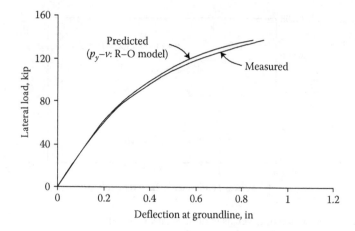

FIGURE 2.45 Comparison of load-displacement curves at ground line (1 inch = 2.54 cm, 1 kip = 4450 N).

the latter being the maximum load, are shown in Figure 2.46. It can be seen from these figures that the FE (1-D) procedure with the R–O model for the p_y–v curves provides very good correlations with the field measurements.

2.10.3 EXAMPLE 2.6: P_y–v CURVES FOR COHESIONLESS SOIL

Determine p_y–v curves for various depths for three sands (Figure 2.47), which gives the dimensions of three layers, the density (γ), and the angle of friction (ϕ) for pile no. 2 [51]. The measured field curves for wall and tip resistances, load distribution along the pile, and pile movements are shown in Figure 2.48 [51]. We assume the sand to be medium to dense. Other properties are given below:

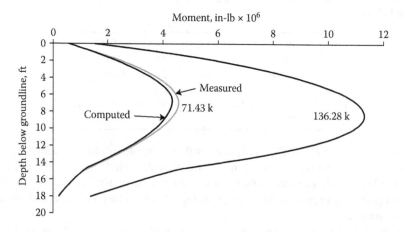

FIGURE 2.46 Bending moment curves for two loads: comparison between computed and observed results (1 inch = 2.54 cm, 1 lb = 4.45 N, 1 kip = 1000 lbs, 1 ft = 0.305 m).

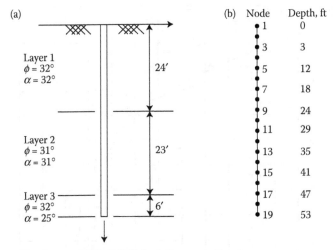

Density, $\gamma = 100$ lb/ft^3(1600 kg/m^3), same in three layers

FIGURE 2.47   Pile and soil properties. (a) Pile; (b) idealization.

Length of the pile = 53 ft (16.1 m)
Diameter of the pile = 16 in = 1.333 ft (0.406 m)
Unit weight of soil = 100 lb/ft^3 (15,706 N/m^3)
Subgrade modulus $k = 100$ lb/in^3 = 172.80 kip/ft^3 (27.14 N/cm^3)
Coefficient of earth pressure at rest $K_0 = 0.45$
Coefficient of active earth pressure:

$$K_a = \tan^2\left(45° - \frac{\phi}{2}\right)$$

Coefficient of passive earth pressure:

$$K_p = \tan^2\left(45° + \frac{\phi}{2}\right) = \tan^2\beta$$

Computed values from various quantities are given in Table 2.8.
To determine the critical depth, x_c, we equate Equations 2.80 and 2.82. Hence, x_c was found to be 10.95 ft (3.34 m).
Various parameters to develop the p_y–v curves (Figure 2.23) are given below:

$$p_{yu} = A_s p_u$$

$$v_u = \frac{3b}{80} = \frac{3 \times 1.33}{80} = 0.05 \text{ ft } (1.524\,\text{cm})$$

$$p_{ym} = B_s p_u$$

$$v_m = \frac{b}{60} = \frac{1.333}{60} = 0.022 \text{ ft } (0.67\,\text{cm})$$

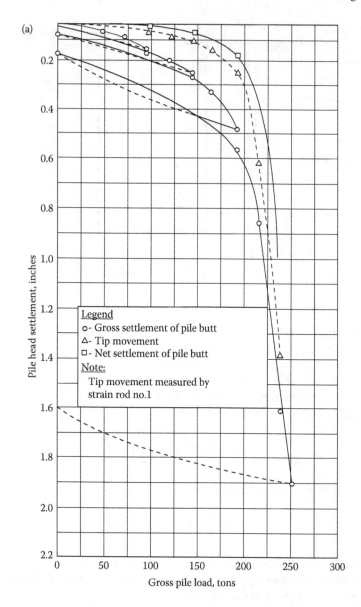

FIGURE 2.48 Field measurements for Pile no. 2, Arkansas Lock and Dam 4 (1 inch = 2.54 cm, 1 ton = 8.9 kN). (a) Pile movement versus load; (b) load distribution in pile; and (c) gross load versus tip and wall load. (Adapted from Fruco and Associates, Results of Tests on Foundation Materials, Lock and Dam No. 4, Reports 7920 and 7923, U.S. Army Engr. Div. Lab., Southwestern, Corps of Engineers, SWDGL, Dallas, TX, 1962.)

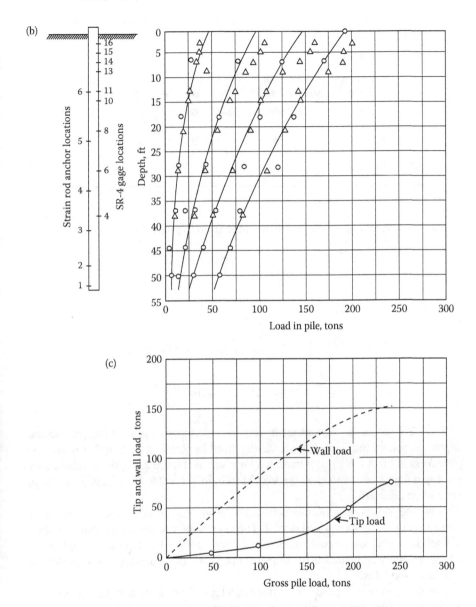

FIGURE 2.48   (continued) Field measurements for Pile No. 2, Arkansas Lock and Dam 4 (51). (1 inch = 2.54 cm, 1 ton = 8.9 kN). (a) Pile movement versus load; (b) load distribution in pile; and (c) gross load versus tip and wall load. (Adapted from Fruco and Associates, Results of Tests on Foundation Materials, Lock and Dam No. 4, Reports 7920 and 7923, U.S. Army Engr. Div. Lab., Southwestern, Corps of Engineers, SWDGL, Dallas, TX, 1962.)

TABLE 2.8

Various Quantities

Layer	γ (Uniform) (lb/ft³)	k Subgrade Modulus (lb/ft³)	K_0	φ^0	α^0	β^0	K_a	K_p
1 and 3	100	172,800	0.45	32	25°	61°	0.31	3.25
2	100	172,800	0.45	31	27°	60.5°	0.32	3.124

1 lb/ft³ = 157.06 N/m³.

The static values of A_s and B_s are obtained from Figure 2.24:

$$v_k = \left(\frac{C}{kx}\right)^{\frac{n}{n-1}}$$

where $C = p_{ym}/v_m^{1/n}$ lb/ft²

$$m = \frac{p_{yu} - p_{ym}}{v_u - v_m} = \frac{p_{yu} - p_{ym}}{0.05 - 0.022} = 36(p_{yu} - p_{ym})$$

$$n = \frac{p_{ym}}{mv_m} = \frac{p_{ym}}{m(0.022)} = 45\left(\frac{p_{ym}}{m}\right)$$

Table 2.9 shows the various terms required to construct p_y–v curves at selected depths, as shown in Figure 2.49. It was reported by Reese et al. [50] that if the displacement at point k is greater than that for point m (Figure 2.23), the parabola connecting points k and m may be ignored. Hence, in Table 2.9, the values of v_k and p_k are chosen arbitrarily.

Now, the values of coordinates (v, p_y) of points are used to construct the p_y–v curve (Figure 2.49) for various depths. The curves for various depths can be implemented in a computer-based solution procedure. Here, the slope at a given point, which is usually computed as the slope between two consecutive points, provides the resistance modulus.

Since the curve in Figure 2.23 is discontinuous, it would be preferable to develop a continuous function to facilitate the computation of the subgrade modulus as the derivative, dp_y/du, at a point. Such a function is presented below.

2.10.4 SIMULATION OF p_y–v CURVE BY USING RAMBERG–OSGOOD MODEL

The following parameters are needed to model the smooth p_y–v curve using the R–O model (Figure 2.18b):

Initial modulus $k_0 = \dfrac{p_k}{v_k}$

TABLE 2.9
Parameters for Computation of p_y–v Curves

Node No.	Layer	Depth x (ft)	p_u (lb/ft²)	Depth Ratio x/b	A_s	B_s	$p_{yu} = A_s P_u$ (lb/ft)	$p_{ym} = B_s P_u$ (lb/ft)	m	n	C (kip/ft²)	v_k (ft)	P_{yk} (kip/ft)
1	I	0 ≈ 0.1	44	0.075	2.8	2.15	122	94	61,380	3.572	14.15	0.00056	0.10
2		3	4872	2.25	1.35	1.0	6577	4872	502,812	1.645	186	0.00673	3.490
4		9	36,756	6.75	0.88	0.5	32,345	18,378	1,001,376	1.645	370.23	0.00445	6.915
6		15	73,200	11.25	0.88	0.5	64,416	36,600	1,602,180	1.645	592.4	0.007	18.134
9		24	117,120	18	0.88	0.5	103,065	58,560				0.007	29.016
10	II	26	111,150	19.5	0.88	0.5	97,812	55,575	1,520,532	1.645	562.2	0.005	22.443
13		35	149,625	26.5	0.88	0.5	131,670	74,812	2,046,870	1.645	756.8	0.005	30.212
15		41	175,275	30.75	0.88	0.5	154,242	87,638	2,397,762	1.645	886.5	0.005	35.390
17		47	200,925	35.25	0.88	0.5	176,814	100,462	2,748,654	1.645	1,016.2	0.005	40.567
18	II	50	244,000	37.5	0.88	0.5	214,720	122,000	3,337,920	1.645	1234.1	0.007	60.448
		53	258,640	39.75	0.88	0.5	227,603	129,320	3,538,195	1.645	1308.7	0.007	64.077

1 ft = 0.3048 m, 1 lb/ft = 14.6 N/m, 1 lb/ft² = 47.88 N/m², 1 kip = 1000 lb (4448 N).

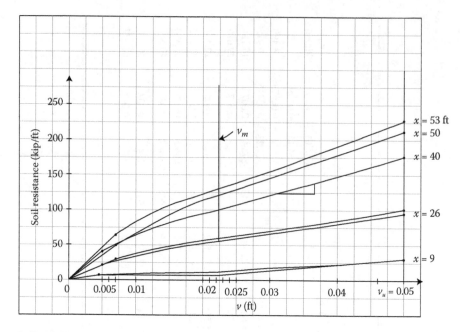

FIGURE 2.49 p_y–v curves for sands. (1 ft = 0.305 m, 1 kip = 4.45 kN)

Final modulus $k_f = \dfrac{p_{yu} - p_{ym}}{v_u - v_m}$

Modulus $k_i - k = k_r$

The Ramberg–Osgood (R–O) model is expressed as (Figure 2.18b)

$$p_y = \frac{(k_0 - k_f)v}{[1 + \{((k_0 - k_f)v)/(p_{yu})\}^m]^{(1/m)}} + k_f v \qquad (2.75)$$

where $k_0 (= k_i)$ is the initial modulus, k_f is the modulus at the yield point, p_{yu} is the soil resistance at the yield point, and m is the order of the curve.

If $m = 1$ and $k_f = 0$, the above equation reduces to the following hyperbola:

$$p = \frac{v}{a + bv} \qquad (2.76)$$

where $a = 1/k_0$, $b = 1/p_{yu}$, and p_{yu} denotes the asymptotic values for the hyperbola.

Procedures for finding the parameters in the R–O model are described below:

Initial modulus, k_0, is found as the slope of the p_y–v curve at the origin, or it can be found as the ratio, p_{yk}/v_k in Figure 2.23. The final modulus, k_f, can be found as the value corresponding to the asymptote to the curve, or it can be computed as the ratio $(p_{yu} - p_{ym})/(v_u - v_m)$ (Figure 2.23).

The value m can be found by an iterative procedure based on Equation 2.75, which is rewritten as

$$\frac{(k_0 - k_f)v}{p_y - k_f v} - \left[1 + \left\{\frac{(k_0 - k_f)v}{p_{yu}}\right\}^m\right]^{\frac{1}{m}} = 0 \qquad (2.133)$$

Now, Equation 2.133 is solved for a specific point, for example, (v_m, p_{ym}) by selecting various values of m until the equation is satisfied. One can select more than one point and then find the average as the value of m for a given depth. Thus, the R–O curve can be defined at various depths. Since some of the parameters, for example, k_0, k_f, and p_{yu}, vary with depth, they can be expressed as a function of depth.

2.10.5 Example 2.7: Axially Loaded Pile: τ_s–u (τ–z), Q_p–U_p Curves

2.10.5.1 τ_s–u Behavior

Develop τ_s–u or t–z curves for the pile (No. 2) in Example 2.6; the term t–z curve is used in the literature [28]. However, the term τ_s–u is used in this book to be consistent with the coordinates and displacement components. As noted before, the soil resistance in the x (vertical) direction is represented by the shear stress (τ_s) on the skin of the pile and axial displacement is represented by u. The resistance on the tip is given by normal stress (q_p) on the cross section of the pile tip and the tip displacement (u_p). Further details are given below:

Pile outer diameter = 16 in (40.60 cm)
Pile area = 201 in^2 (1296.8 cm^2)
Modulus of elasticity, $E = 3.44 \times 10^6$ psi (27.72×10^6 kPa)
Assume $v = 0.20$
K_o based on field observation = 1.29
Density $\gamma = 100$ pcf (1.58 g/cm^3)

The following terms relate to Equations 2.105b and 2.105f:

Layer	K′	N	R$_f$	φ Degrees
I	1500	0.60	0.90	32
II	1200	0.50	0.90	31
III	1500	0.60	0.90	32

We follow the foregoing steps described under Section 2.8.6, in a nondimensional form. The axial displacement is given by

$$u = \frac{\tau_s r_o}{G_i} \ell n \left(\frac{(r_m/r_o) - (\tau_s R_f/\tau_{max})}{1 - (\tau_s R_f/\tau_{max})}\right) \qquad (2.134)$$

For maximum shear, τ_{max}, u will tend to u_{max}; therefore, from Equation 2.134

$$u_{max} = \frac{\tau_{max} r_o}{G_i} \cdot \ell n \left(\frac{(r_m/r_o) - R_f(\tau_{max}/\tau_{max})}{1 - R_f \cdot (\tau_{max}/\tau_{max})} \right)$$

$$= \frac{\tau_{max} r_o}{G_i} \ell n \left(\frac{(r_m/r_o) - 0.9}{0.10} \right) \tag{2.135}$$

Hence

$$\frac{u}{u_{max}} = \frac{\tau_s}{\tau_{max}} \left[\frac{\ell n \{[(r_m/r_o) - 0.9(\tau_s/\tau_{max})]/[1 - 0.9(\tau_s/\tau_{max})]\}}{\ell n [\{(r_m/r_o) - 0.9\}/\{0.10\}]} \right] \tag{2.136}$$

We compute various quantities in the above equation as follows:

$$r_o = 16/2 = 8.0 \text{ in } (20.32 \text{ cm})$$

$$E_i = K' p_a \left(\frac{\sigma_3}{p_a} \right)^n$$

where p_a is the atmospheric pressure constant = 14.7 psi = 2116.80 psf (101.35 kPa)

$$\sigma_3 = K_o \sigma_1 = K_o \gamma x = 1.29 \times 100 \times x$$
$$= 129 \ x \text{ psf}$$

Layers I and III

$$E_i = 1500 \times 2116.8 \left(\frac{129 \ x}{2116.8} \right)^{0.6} = 5.93 \times 10^5 \times x^{0.6} \text{ psf}$$

$$G_i = \frac{E_i}{2(1+v)} = \frac{5.93 \times 10^5}{2 \ (1+0.20)} \times 0.6 = 2.47 \times 10^5 \times x^{0.6} \text{ psf}$$

Now G_i at the bottom of the pile, $L = 53$ ft (16.15 m)

$$G_L = 2.47 \times 10^5 \ (53)^{0.6} = 26.7 \times 10^5 \text{ psf (128 MPa)}$$

Layer II

$$E_i = 1200 \times 2116.8 \left(\frac{129x}{2116.8} \right)^{0.50}$$

$$G_i = \frac{E_i}{2(1+v)} = \frac{E_i}{2.40} = 1.975 \times 10^5 \times x^{0.6} \text{ psf}$$

$$G_{L/2} = 1.975 \times 10^5 \left(\frac{53}{2}\right)^{0.60} = 14.10 \times 10^5 \, \text{psf} (68 \, \text{MPa})$$

$$\rho = \frac{G_{L/2}}{G_L} = \frac{14.10 \times 10^5}{26.7 \times 10^5} = 0.528$$

$$r_m = 2.5 \times 53 \times 0.528 \, (1 - 0.20) = 55.8 \, \text{ft} \, (17.0 \, \text{m})$$

$$\frac{r_m}{r_o} = \frac{55.8}{0.6667} = 83.70$$

Now, we assume $\tau_{max} = 0.45$ ksf $= 450$ psf (21,546 kPa), which is usually obtained for sands from laboratory tests. Then

$$u_{max} = \frac{450 \times 0.667}{G_i} \, \ell n \left(\frac{83.70 - 0.90}{0.10}\right)$$

$$= \frac{2016}{G_i} \, \text{ft}$$

Then

$$\frac{u}{u_{max}} = \frac{\tau_s}{\tau_{max}} \left[\frac{\ell n \left[\{83.70 - 0.90(\tau_s/\tau_{max})\} / \{1 - 0.90(\tau_s/\tau_{max})\} \right]}{6.72} \right] \qquad (2.137)$$

Table 2.10 shows the computations for u/u_{max} and τ_s/τ_{max}. The plotted curve is shown in Figure 2.50, together with the predictions by using the R–O model; they correlate very well.

In the nonlinear incremental analysis by the FEM, Equation 2.67, the axial displacements at increment i are computed by using the soil stiffness, k_s, as computed in the previous step $(i - 1)$. At the current increment i, we can compute τ_{si} at any node point (i) by using the displacement u_i and Equation 2.136. The solution of Equation 2.136 will require an iterative approach because τ_s appears in the right side of the function repeatedly. Such an iterative procedure is described below:

- Let the computed average displacement at the middle of an element be

$$\bar{u} = \frac{u_1 + u_2}{2}$$

  where u_1 and u_2 are the displacements at the nodes of an element; here, we have assumed a linear variation of u.
- Assume an initial value of the shear stress τ_s; here, we can adopt the shear stress at the end of the last increment $(i - 1)$, that is, $\tau_{s(i-1)}$.

TABLE 2.10

Computation of u/u_{max} and τ_s/τ_{max}

Layer Node 1	Node	Depth (ft)	τ_s/τ_{max} Adopted	u_{max}	u/u_{max}
	1	0.05[a]	0.00	4.92×10^{-2}	0
	2	3	0.050	4.22×10^{-3}	0.0333
	3	6	0.100	2.79×10^{-3}	0.0673
	4	9	0.150	2.18×10^{-3}	0.1020
	5	12	0.20	1.84×10^{-3}	0.1376
I	6	15	0.25	1.61×10^{-3}	0.1741
	7	18	0.30	1.44×10^{-3}	0.2116
	8	21	0.35	1.31×10^{-3}	0.2501
	9	24	0.40	1.21×10^{-3}	0.2900
	10	26	0.45	1.16×10^{-3}	0.3300
	11	29	0.50	1.08×10^{-3}	0.3735
II	12	32	0.55	1.02×10^{-3}	0.4178
	13	35	0.60	9.67×10^{-4}	0.4641
	14	38	0.65	9.20×10^{-4}	0.5127
	15	41	0.70	8.80×10^{-4}	0.5640
	16	44	0.75	8.43×10^{-4}	0.6187
	17	47	0.80	8.10×10^{-4}	0.6776
III	18	50	0.85	7.81×10^{-4}	0.7421
	19	53	0.90	7.54×10^{-4}	0.8141
			0.95		0.8975
			1.000	–	1.000

[a] This value is assumed as near zero.

1 ft = 0.3048 m

- Let j be the iteration for the increment n. Then, the initial shear stress will be $\tau_{s(i-1)}^0$; the superscript o denotes iteration $j = 0$.
- Compute a temporary value of displacement $u_i^{(o)}$ by using Equation 2.136 with $\tau_{s(i-1)}$ in which u_{max} is computed at the middle of the element at corresponding depth.
- Compare $\bar{u}$ and $u_i^{(o)}$ by computing the difference as

$$\Delta u_i^{(o)} = \left| \bar{u} - u_i^{(o)} \right|$$

- If $\Delta u_i^{(o)} \leq \varepsilon$ a small value, $\tau_{s(i-1)}$ can be accepted as the shear stress. Otherwise, the value of the shear stress is revised for iteration $j = 1$ as follows:

$$\tau_{s(i)}^1 = \tau_{s(i-1)}^0 \pm \Delta \tau_{i-1}^1$$

where $\Delta \tau_{s(i-1)}^1 = \tau_{s(i-1)}^0 / M$ and M denotes the number of divisions.

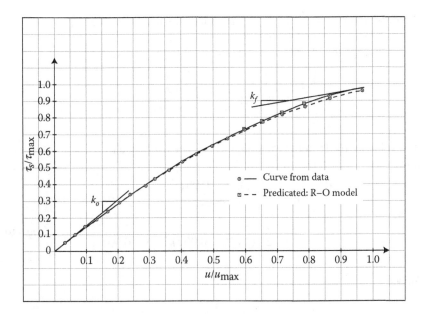

FIGURE 2.50   Plot of normalized τ and u curves.

- Now, we compute the new value of u_i^1 by using Equation 2.137 and the value $\tau_{s(i)}^1$ just computed. The process is continued till $\Delta u_i^1 \leq \varepsilon$. Then, the final values for the increment are $u_i^{(j)}$ and $\tau_{s(i)}^j$, where j denotes the last iteration when the difference ε is satisfied.

Thus, such iterated value can be found for middle points of all elements in the pile. The values of τ_s along the pile can be used to compute the side or wall friction for an element as follows:

$$Q = Ck_s u\ell$$
$$= C\tau_s \ell$$

where ℓ is the length of the element. These values can be used to plot the load distribution curve (Figure 2.32b). Moreover, the value of k_s to be used in the subsequent load increment can be found as

$$k_{si} = \frac{\tau_{s(i)}^j}{u_i^j}$$

Simulation of τ_s–u (t–z) response using Ramberg–Osgood (R–O) model: We consider Equation 2.137 as

$$U = \frac{\tau}{6.72}\left[\frac{\ell n(83.70 - 0.9\tau)}{1 - 0.90\,\tau}\right] \tag{2.138}$$

where $U = u/u_{max}$ and $\tau = \tau_s/\tau_{max}$.

According to Figure 2.18b, we require initial slope, k_o, ultimate slope, k_f, $k_r = k_o - k_f$ and parameter m to define the R–O model.

Based on Equation 2.138, we can define k_o and k_f as follows:

$$k_o = \frac{1}{(dU/d\tau)|_{\tau=o}} = \frac{d\tau}{dU}\Big|_{U=o}$$

The value of $dU/d\tau$ from Equation 2.138 can be expressed as

$$\frac{dU}{d\tau} = \frac{1}{6.72} \ell n\left[\frac{83.70 - 0.90\tau}{1 - 0.9\tau} \right]$$

$$- \frac{0.9\tau}{6.72} \left[\frac{1}{(83.70 - 0.90\tau)} \right] - \frac{1}{1 - 0.90\tau} \tag{2.139}$$

Hence

$$\frac{dU}{d\tau}\Big|_{\tau=o} = \frac{1}{6.72} \ell n(83.70) - 0 = 0.6588$$

Therefore, k_o is found as

$$k_o = \frac{1}{0.6588} = 1.518$$

The final slope, k_f, can be computed from Equation 2.139 by substituting $\tau = 1.0$, that is, in the final zone of the curve as

$$\frac{dU}{d\tau}\Big|_{\tau=1} = 1 - \frac{0.9}{6.72}\left[\frac{1}{83.70 - 0.9} - \frac{1}{0.1} \right]$$

$$= 1 - 0.134[0.0121 - 10]$$

$$= 1 - 0.134(-9.988)$$

$$= 1 - (-1.338) = 2.338$$

Therefore,

$$k_f = \frac{d\tau}{dU}\Big|_{U=1} = \frac{1}{(dU/d\tau)|_{\tau=1}} = \frac{1}{2.338} = 0.4300$$

Hence, $k_r = k_o - k_f = 1.518 - 0.430 = 1.088$.

2.10.5.2 Parameter, m

The R–O model (Equation 2.75) can be expressed as

$$\tau = \frac{\tau_s}{\tau_{max}} = \frac{k_r(u/u_{max})}{[1+((k_r(u/u_{max}))/\tau_f)^m]^{(1/m)}} + k_f \frac{u}{u_{max}} \tag{2.140a}$$

Hence

$$\tau = \frac{1.088\,U}{[1+(1.088\,U/1)^m]^{(1/m)}} + 0.430\,U \tag{2.140b}$$

Here, we have assumed the final value of τ, that is, $\tau_f = 1$. We can express Equation 2.140b in a residual form, R as

$$R = \frac{1.088\ U}{\tau - 0.430\ U} - \left[1 + (1.088\ U)^m\right]^{\frac{1}{m}} \tag{2.141}$$

The solution for m using Equation 2.141 needs an iterative procedure. We can choose various points from Figure 2.50 and then choose a number of values of m. The lowest value of R (near zero) gives the final value m, which can be found as the average for points chosen. Let us adopt $\tau = 0.50$ and $U = 0.374$ from Figure 2.51. From Equation 2.141, we have

$$|R| = \frac{1.088 \times 0.374}{0.50 - 0.43 \times 0.374} - \left[1 + (1.088 \times 0.374)^m\right]^{\frac{1}{m}}$$

$$= \frac{0.407}{0.340} - \left[1 + (0.407)^m\right]^{\frac{1}{m}}$$

$$= 1.197 - \left[1 + (0.407)^m\right]^{\frac{1}{m}}$$

Let us choose a number of values m and find |R| for each value, as shown below:

Assumed m	R
4	0.1900
2	0.117
1.5	0.0628
1.4	0.00145
1.3	0.04345
1.0	0.210

Therefore, $m = 1.4$ can be accepted. Hence, the R–O parameters for the nondimensional response (Figure 2.50) are

$$k_o = 1.518,\ k_f = 0.430,\ k_r = 1.088,\ \text{and}\ m = 1.4$$

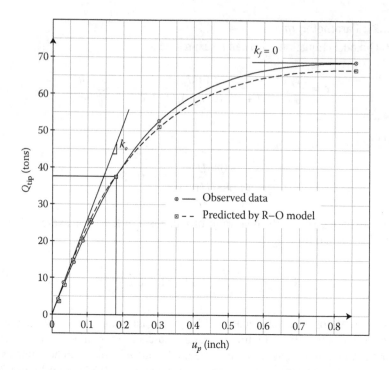

FIGURE 2.51 Comparison between observed and predicted tip load versus tip movement curves. (1 inch = 2.54 cm; 1 ton = 8.90 kN)

2.10.5.3 Back Prediction for τ_s–u Curve

We can use the foregoing parameters to predict the τ_s/τ_{max} (τ) versus $u/u_{max}(U)$ curve. We adopt the values of u/u_{max} and τ_s/τ_{max} from Table 2.10. Then, we adopt various values of u/u_{max} from Table 2.10 to compute the corresponding τ_s/τ_{max} predicted by the R–O model (Equation 2.140). Table 2.11 shows the values of τ_s/τ_{max} from the developed curve with those from the R–O model. The developed predictions and those by the R–O model show a very good agreement (Table 2.11 and Figure 2.50).

2.10.5.4 Tip Resistance

Figures 2.48a through 2.48c show measured gross pile load versus pile head deflections, load distribution along the pile versus depth, and gross pile load versus tip and wall load, respectively. To find the relation between tip load and displacements, the following steps are followed: first, a gross load (Q_T) is chosen (Table 2.12) and by using Figure 2.48c, we determine the tip pile load (Q_p). Then, we use Figure 2.48a to find the tip movement, u_p, which is the displacement at the bottom of the pile, corresponding to the selected gross load.

Figure 2.51 shows the plot of Q_p versus u_p based on Table 2.12. We can determine the initial and final slopes by using Figure 2.51; here, we have used the pressure $q_p = Q_p/A$, where $A = 201$ in^2 (1296 cm^2), for example, at 40 T (Figure 2.51):

TABLE 2.11

Developed τ–U Curves and R–O Predictions

Developed Curve $\tau = \tau_s/\tau_{max}$	$U = u/u_{max}$	R–O Predictions τ_s/τ_{max}
0.05	0.0333	0.051
0.15	0.1020	0.1550
0.25	0.1741	0.253
0.35	0.250	0.3600
0.45	0.330	0.455
0.55	0.4178	0.550
0.65	0.5127	0.660
0.75	0.6187	0.755
0.85	0.7421	0.865
0.95	0.8975	0.988
1.00	1.00	1.070

$$k_o = \frac{40}{201 \times 0.151} = \frac{40}{30.35} = 1.32\,\text{T/in}^3\ (716.6\,\text{N/cm}^3)$$

$k_f = 0$, as the curve appears to be horizontal in the ultimate region. Therefore, $k_r = 1.32$ T/in^3 (716.6 N/cm^2). Now, the R–O model for tip behavior can be expressed as

$$\frac{Q_p}{A} = q_p = \frac{k_r u_p}{[1 + (k_r u_p/q_f)^m]^{(1/m)}} + k_p u_p \tag{2.142}$$

where q_f is the final or ultimate pressure from Figure 2.51, as 70/201 = 0.348 T/in^2 (480 N/cm^2).

TABLE 2.12

Developed Tip Behavior and R–O Predictions

Gross Load Q_T (Ton)	Tip Load Q_p (Ton)	Tip Displacement u_p (in)	Q_p from R–O Model q_p, Ton (T/in^2)
25	2.00	0.008	2.10 (0.0104)
50	4.50	0.015	4.00 (0.020)
75	8.00	0.030	8.10 (0.040)
100	13.50	0.055	14.00 (0.070)
125	19.00	0.080	20.00 (0.100)
150	25.00	0.105	25.50 (0.127)
175	38.00	0.175	38.20 (0.190)
200	53.00	0.300	52.50 (0.261)
225	69.00	0.85	68.00 (0.338)

1 ton = 2000 lb = 8896 N, 1 in = 2.54 cm, 1 ton/in^2 = 1379 N/cm^2.

Let us write Equation 2.142 at Q_p/A, for $Q_p = 54.5$ tons [0.271 T/in² (373.7 N/cm²)] and 0.30 in (0.76 cm) for finding m by the iterative procedure. Then, the residual R is given by

$$R = \frac{1.32\,(0.30)}{0.271} - \left[1 + \left\{ \frac{1.32\,(0.30)}{0.348} \right\}^m \right]^{1/m}$$

The above equation is satisfied for the approximate value of $m = 1.99$.

Hence, the R–O parameters for the tip resistance q_p versus tip displacement u_p are found as

$$k_o = 1.32 \text{ T/in}^2 \text{ (1820 N/cm}^2\text{)}; \quad k_f = 0$$

$$k_r = 1.32 \text{ T/in}^2 \text{ (1820 N/cm}^{2)}; \quad m = 1.99$$

We can back predict q_p–u_p by using the R–O model with the above parameters. The predicted values are shown in Table 2.12 and plotted in Figure 2.51.

2.10.6   EXAMPLE 2.8: LATERALLY LOADED PILE—A FIELD PROBLEM

A wooden pile was driven in dense sand in the field at the site of Arkansas Lock and Dam No. 4 [51]. The properties of the pile and soil are given below:

Diameter = 14 in (35.6 cm)
Length = 40 ft (12.2 m)
E (Pile) = 1.6×10^6 psi (11.03×10^4 kPa)
I (Pile) = 1980 in⁴ (82.4×10^3 cm⁴)
$\Phi = 40°$
$K_o = 0.40$
$K_a = 0.22$

Table 2.13 shows values of v and p_y at various depths. They were obtained by using procedures presented in Refs. [12,14,49–51]. The data in Table 2.13 were used to determine the R–O model parameters; values of these parameters at various depths are given in Table 2.14.

2.10.6.1   Linear Analysis

We first analyze the pile as a simple problem assuming an average value of $k = 15,000$ lb/in² (10,3425 kPa) and applied load $P_t = 10,000$ lbs (10 kip) (44.48 kN). A number of FE and FD meshes, with segments (elements) $N = 20$, 40, and 120 were used.

We present typical results using FE and FD methods for $N = 120$ and a load of 10,000 lbs (44.48 kN). Figures 2.52a through 2.52e show predictions for displacements, slopes, bending moments, shear forces, and soil reactions, respectively.

TABLE 2.13
Points on p_y–v Curves at Various Depths

Depth = 0		Depth = 8 ft		Depth = 10 ft		Depth = 24 ft		Depth = 32 ft	
v (in)	p_y (lbf/in)	v (in)	p_y (lbf/in)	v (in)	p_y (lbf/in)	u (in)	p_y (lbf/in)	u (in)	p_y (lbf/in)
0.000	0.00	0.000	0	0.000	0	0.000	0	0.000	0
0.045	29.46	0.240	1292	0.330	4051	0.330	6076	0.330	8102
0.106	58.93	0.606	2585	0.819	8102	0.819	12,153	0.819	16,204
0.195	88.39	1.090	3877	1.470	12,153	1.470	18,229	1.470	26,306
0.352	117.85	1.930	5170	2.620	16,204	2.620	24,306	2.620	32,408
0.700	117.85	3.860	5170	5.240	16,204	5.240	24,306	5.240	32,408

1 ft = 30.48 cm, 1 in = 2.54 cm, 1 lb/in = 1.75 N/cm.

Because we have used an average value of $k = 15,000$ lb/in^2 (10,3425 kPa) (Table 2.14) for a one-step linear analysis for the total load $P_t = 10,000$ lbs (44.48 kN), the computed displacement at the top (Figure 2.52a) of about 0.042 in (0.107 cm) is much smaller than that in the field (Figure 2.53) of about 0.25 in (0.635 cm) at 10,000 lbs (44.48 kN).

2.10.6.2 Incremental Nonlinear Analysis

In the incremental nonlinear (FE) analysis using the R–O model, increments of load equal to 2 kip (8.9 kN) were applied. Figure 2.53 shows comparisons between the FE predictions from the SSTIN-1DFE code and field observations. It can be seen that the FE predictions compare very well with the field measurements [51]. Also, the predictions show good improvement with increasing number of elements, from $N = 20$ to 120. Hence, nonlinear behavior, including varying properties with depth, is essential for realistic predictions.

TABLE 2.14
Parameters for Ramberg–Osgood Model

Depth	k_o (lb/in^2)	p_f (lb/in)	k_f (lb/in^2)	m
0	654.7	117.8	0.0	1.0
8	5383.0	5170.0	0.0	1.0
16	12,275.0	16,204.0	0.0	1.0
24	18,412.0	24,306.0	0.0	1.0
32	24,551.0	22,408.0	0.0	1.0
40	24,551.0	22,408.0	0.0	1.0

1 lb/in = 1.75 N/cm, 1 lb/in^2 = 6.895 kPa.

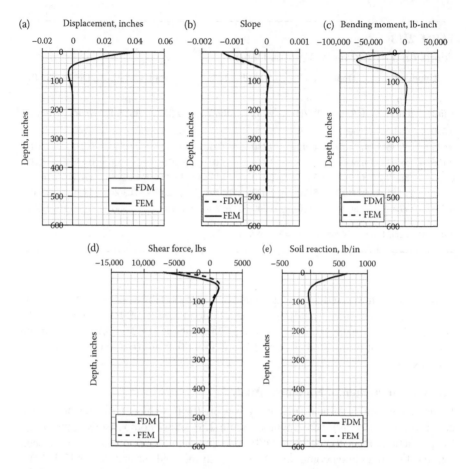

FIGURE 2.52 Predictions for various quantities: (a) displacement; (b) slope; (c) bending moment; (d) shear force; and (e) soil reaction. ($p = 10,000$ lbs (44.48 kN), $k = 15,000$ lb/in² (103.4 MPa), number of element = 120)

2.10.7 EXAMPLE 2.9: ONE-DIMENSIONAL SIMULATION OF THREE-DIMENSIONAL LOADING ON PILES

The computer procedure, SSTIN-1DFE, was used for a (hypothetical) pile subjected to general loading in the three-directions, for example, loads F_x, F_y, and F_z and moments M_x and M_y at the top (Figure 2.54a) [16]; note that the coordinate axes are labeled different from the previous cases. We did not consider the moment or torque about the x-axis. The applied load involves a linearly varying distributed load (traction) in the y- and z-directions over the top 40 cm (0.40 m) of the pile above the ground level (Figure 2.54b); the values of the load are 980 and 4900 kN/m² at the top and bottom, respectively. This type of loading may occur due to various factors such as external pulls by mooring forces in offshore piles.

The loadings (concentrated load, surface tractions, and moments) applied to the pile are divided into three increments as follows:

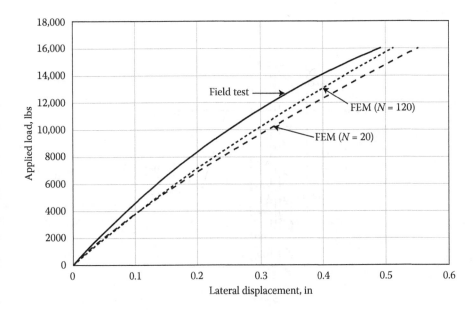

FIGURE 2.53 Comparison between FEM predictions and field test data at top of wooden pile-Arkansas Lock & Dam no. 4. *Note*: The predictions by using the code FDM-COM52 [32,50] were essentially the same as by the code FEM-SSTIN-IDFE [24] (1 inch = 2.54 cm, 1 lb = 4.448 N). (Field test data adapted from Fruco and Associates, Results of Tests on Foundation Materials, Lock and Dam No. 4, Reports 7920 and 7923, U.S. Army Engr. Div. Lab., Southwestern, Corps of Engineers, SWDGL, Dallas, TX, 1962.)

Axial:	$\Delta F_x = 19.6$ kN
Lateral:	$\Delta F_y = 4.9$ kN
	$\Delta F_z = 4.9$ kN
Moments:	$\Delta M_x = \Delta M_y = 0.10$ kN-m

The soil is represented by nonlinear springs in x-, y-, and z-directions, with the assumptions that they are the same in all directions. The following R–O model parameters are used for the simulation of the p_y–v curves:

k_o at the top = 0.0
k_o at the bottom = 980.7 kN/m^2
p_f at the top = 0.0
p_f at the bottom = 1961.4 kN/m^2
k_f for all depths = 0.0
$m = 1.0$

The following properties of the pile are used:

Length of pile = 2.80 m
$E = 19.60 \times 10^6$ kN/m^2

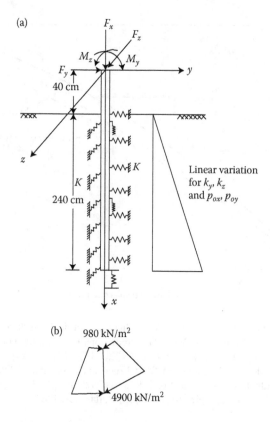

FIGURE 2.54 Three-dimensional pile: (a) pile; (b) surface loading.

Area = 0.10 m^2
$I_y = I_z = 1.0 \times 10^{-4}$ m^4

The pile was divided into 20 elements, starting with No. 1 at the top. The code, SSTIN-1DFE, was used to solve this 3-D pile problem approximately. Some of the computer predictions, for example, for moments with depth (Figure 2.54a), and displacements (in plan) for three selected top nodes (1, 3, and 5) are shown in Figure 2.55b, in the x- and y-directions. The computer results show reasonable trends and magnitudes.

2.10.8   EXAMPLE 2.10: TIE-BACK SHEET PILE WALL BY ONE-DIMENSIONAL SIMULATION

Figure 2.56 shows a steel pile wall analyzed using a 2-D FE idealization by Clough et al. [52]; it was used to support a deep and open excavation for a building in Seattle, Washington. However, the 1-D idealization was adopted here by assuming that the behavior of a unit of length is the same for other units along the wall [16]. The code, SSTIN-1DFE was used for this approximate analysis. The 1-D code also includes

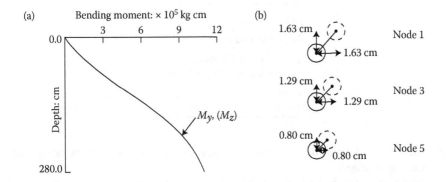

FIGURE 2.55 Computer results: (a) bending moment; (b) displacements for selected nodes (in plan).

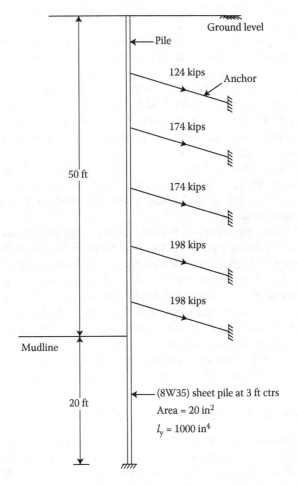

FIGURE 2.56 Sheet pile wall with tie-backs (1 kip = 4.448 kN, 1 inch = 2.54 cm, 1 ft = 30.50 cm).

TABLE 2.15
R–O Parameters for Soils

k_o = 8315 lbf/in^2 (57,332.00 kPa)
p_f = 1764.0 lbf/in^2 (12,163.00 kPa)
k_f = 0.0
m = 1.0

the simulation of construction sequences such as excavations, embankments, and installation of support.

The tie-backs (anchors) were installed in stages with loading capacity in the range of 120–200 kips (533.76–889.60 kN). The foundation and back-fill soil were cohesive, with an average undrained shear strength of 2.10 tonf/ft^2 (20 N/cm^2). The p_y–v curves were derived using this shear strength; then the R–O model parameters were determined and are shown in Table 2.15.

The excavation process was simulated approximately by applying an (excavation) surface loading equal to the lateral earth pressure due to the initial (*in situ*) stresses caused by the overburden. A total of 11 stages of construction sequences including six excavations and five installations of anchors were simulated. The surface force during a stage of excavation was found based on the shear strength equal to 2.10 tonf/ft^2 (201.10 kPa). Such a force equals to about 32.7 lbf/in^2 (225.5 kPa) was applied in increments up to the depth excavated at each stage. Five load increments were applied for each excavation stage.

The anchors (tie-backs) were installed at 3.0 ft (0.91 m) intervals in the direction along the length of the wall; hence, the total anchor load was obtained by dividing the anchor load in Figure 2.56 by the interval of 3.0 ft (0.91 m). Then the load was applied in five increments.

Figure 2.57 shows a comparison between displacement predictions by the code SSTIN-1DFE and field measurements for excavation up to the mudline and after installation of the five tie-backs. The predictions by the 2-D computations [52] are also shown in Figure 2.57. It can be seen that the 1-D analysis gives satisfactory predictions, which sometimes improved compared to the predictions by the 2-D analysis. This may be indeed fortuitous; however, the trends of the results are considered to be very good. Also, the cost of the 1-D analysis would be lower than that for the 2-D analysis.

Sometimes, such problems are solved approximately by using the strength of materials procedures [45,53]. However, because of the nonlinearity involved, they may not yield realistic results. Then the FE procedure can provide improved and realistic results.

2.10.9 EXAMPLE 2.11: HYPERBOLIC SIMULATION FOR P_y–v CURVES

Different mathematical forms such as bilinear, hyperbolic, and exponential functions have been used by researchers to simulate experimental p_y–v curves from the lateral load tests; see above and Refs. [53–56]. A hyperbolic type p_y–v curve is illustrated in this example for cohesive soil. The general form of the p_y–v curve is expressed in the following form:

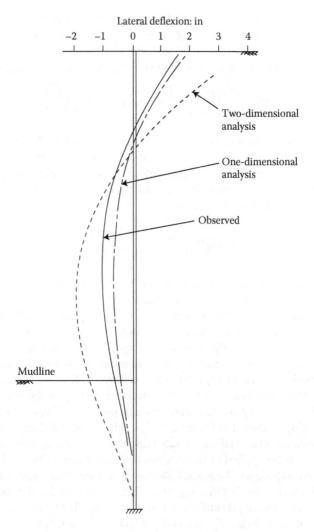

FIGURE 2.57 Comparisons between predicted and observed behavior of sheet pile wall (1 inch = 2.54 cm).

$$p_y = \frac{v}{(1/K) + (v/p_u)} \tag{2.143}$$

where K is the initial tangent slope to the p_y–v curve, also called the modulus of subgrade reaction. Experimental studies show that the modulus of subgrade reaction increases with an increase in pile diameter. Carter [57] suggested a linear relationship between the modulus of subgrade reaction, k, and pile diameter of the following form:

$$K = i \frac{E_s}{1 - \mu_s^2} \frac{D}{D_{\text{ref}}} \left[\frac{E_s D^4}{E_p I_p} \right]^j \tag{2.144}$$

where μ_s is the Poisson's ratio of soil; D is the pile diameter; D_{ref} is the reference pile diameter (assumed as 1.0 m); $E_p I_p$ is the flexural rigidity of pile (kN·m²); E_s is the modulus of elasticity of soil (kPa); and i and j represent fitting parameters. Rearranging terms in Equation 2.144 and using logarithm, it can be expressed as

$$\log\left[\frac{K(1-\mu_s^2)D_{ref}}{E_s\,D}\right] = \log i + j\log\left[\frac{E_s D^4}{E_p I_p}\right] \qquad (2.145)$$

To evaluate the fitting parameters i and j, the values of log $[E_s\,D^4/E_p I_p]$ and log $[K(1-\mu_s^2)\,D_{ref}/E_s D]$ can be plotted in the x- and y-axes, respectively, as a straight line. The slope and intercept of this line are equal to j and log i, respectively. From the linear regression analysis of measurement and predicted K, Kim et al. [56] reported the following expression for K:

$$K = 16.01\frac{E_s}{(1-\mu_s^2)}\frac{D}{D_{ref}}\left[\frac{E_s D^4}{E_p I_p}\right]^j \qquad (2.146)$$

The ultimate soil resistance, p_u, in Equation 2.143 can be determined by assuming a 3-D passive wedge (defined by angles α and θ in Figure 2.58) of soil in front of the pile (see, e.g., Refs. 50,58). The bottom angle, θ, was approximated by Reese et al. [50] as 45° for wedge-type failure in clay. For Mohr–Coulomb-type failure, however, θ will be associated with the effective friction angle, ϕ, of the soil (i.e., $\theta = 45° + \phi'/2$). The fanning angle, α, was assumed as zero by Reese et al. [58] for piles in clay. Based on detailed considerations of force equilibrium conditions, Kim et al. [56] have shown that $\alpha \approx \phi'/5$. This value is used in the numerical example below. From the results of nonlinear FE analyses, a possible failure mode of the pile–soil system was reported by Kim et al. [56], as shown in Figure 2.58. For this assumed failure mode, the ultimate soil resistance was obtained by satisfying the force and moment equilibrium conditions. The forces applied on the side and bottom surfaces of the wedge are classified into two components: the normal force and the shear force. The normal forces on the side, F_n, and on the bottom, F_{nb}, of the wedge are considered as the side friction force, F_f, between the pile and the soil. The shear applied to the side and bottom surfaces are defined as F_s and F_{sb}, respectively. The total resistance F_{tot} in the loading direction is determined by the horizontal force equilibrium as follows:

$$F_{tot} = 2F_s \cos\alpha\sin\theta + F_{sb}\sin\theta + F_{nb}\cos\theta \qquad (2.147)$$

The ultimate soil resistance, p_u, of the soil is obtained by differentiating F_{tot} with respect to the height of the pile, H, as

$$p_u = \frac{dF_{tot}}{dH} = \begin{bmatrix} J c_u D\left(\tan\theta + \dfrac{1}{\sin\theta} + \dfrac{\pi\omega}{2\tan\theta}\right) + J c_u H \\[2mm] \times\left(2\tan\theta\sin\theta + 2\tan^2\theta\tan\alpha + \dfrac{2\tan\alpha}{\cos\theta} + 2\cos\theta\right) \\[2mm] + H\tan\alpha\tan\theta\left(2\sigma'_{vo} + \gamma'H\right) + \sigma'_{vo}D + \gamma'HD \end{bmatrix} \qquad (2.148)$$

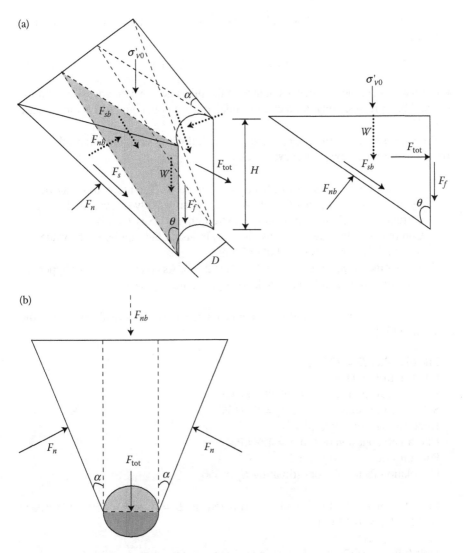

FIGURE 2.58 Three-dimensional wedge failure mode: (a) soil-pile system; (b) sectional view. (From Kim, Y., Jeong, S., and Lee, S., *Journal of Geotechnical and Geoenvironmental Engineering, ASCE*, 137(7), 2011, 678–694. With permission.)

where c_u is the undrained shear strength, J is the empirical soil constant (assumed as 0.5, as per Kim et al. [56]); H is the wedge height; ω is the adhesion factor; γ' is the effective unit weight of soil; and σ'_{vo} is the effective normal stress applied to the top wedge. For a short pile, the total length, L, can be taken as the wedge height, H. For a long pile, however, the smaller value ($1/\beta$ and 7D) should be taken as the wedge height, where β is the characteristic pile length [59,60].

A possible mode for ultimate soil resistance may be plane failure instead of a wedge failure. As discussed by Kim et al. [56], for this case, p_u can be expressed as a sum of the ultimate frontal normal soil resistance, Q, and the ultimate lateral shear resistance, F, as follows:

$$p_u = Q + F = [\eta 10\, c_u + \varepsilon \cdot 2\, c_u]D \qquad (2.149)$$

where η and ε are the coefficients related to the pile shape, which are assumed to be 0.75 and 0.50, respectively, for a circular pile, and are both equal to 1.0 for a square pile [61].

Procedure for constructing p_y–v curve: The following steps are followed to construct a hyperbolic type p_y–v curve for clay:

1. Compute the initial slope, K, from Equation 2.146. The modulus of elasticity of soil, E_s, can either be determined experimentally or evaluated using correlations such as undrained shear strength, c_u (USACE) [61].
2. Compute the ultimate soil resistance, p_u, by using the smaller of the values obtained from Equations 2.148 or 2.149.
3. Determine the p_y–v curve from Equation 2.143. As noted earlier, this hyperbolic type is generally applicable to large diameter piles in clay.

Numerical example: The following data are used in the numerical example of the p_y–v curve [56]:

Pile diameter $D = 0.76$ m
Pile length $L = 9.1$ m
Pile flexural rigidity $E_p I_p = 460,000$ kNm^2
Saturated unit weight of soil $\gamma_{sat} = 17.9$ kPa
Effective friction angle $\phi' = 30°$
Undrained cohesion of soil $c_u = 128$ kPa
Poisson's ratio of soil $\mu_s = 0.3$
Correlation factor K_c (for estimating E_s) $= 550$

Modulus of elasticity of soil is estimated as $E_s = K_c \times c_u = 550 \times 128$ kPa $= 70,400$ kPa $= 70.4$ MPa

Step 1: For $D = 0.76$ m, compute initial slope, K, from Equation 2.146

$$K = 16.01 \times \frac{70,400}{1-0.3^2}\frac{0.76}{1}\left[\frac{(70,400)(0.76)^4}{460,000}\right]^{0.8} = 87,134.8.$$

Step 2: Compute the ultimate soil resistance, p_u, from Equation 2.146:

$$\theta = 45° + \frac{\phi'}{2} = 45° + \frac{30}{2} = 60°$$

$$\alpha = \frac{\phi'}{5} = \frac{30}{5} = 6°$$

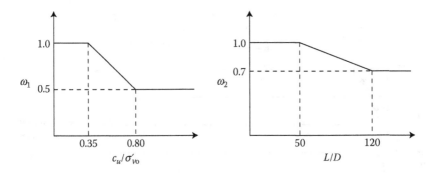

FIGURE 2.59 Adhesion factors in clay soil. (From Kim, Y., Jeong, S., and Lee, S., *Journal of Geotechnical and Geoenvironmental Engineering, ASCE*, 137(7), 2011, 678–694. With permission.)

Characteristic length $\beta = \sqrt[4]{\dfrac{4E_pI_p}{K}} = \sqrt[4]{\dfrac{4 \times 460,000}{87,134.8}} = 2.14$ m

Pile length, 9.1 m > 3β. So, it is a long pile.

Estimate wedge height, H (smaller of $1/\beta = 0.466$ and $7D = 5.32$) = 0.466 m.

Assuming empirical soil constant, J, as 0.5 and the adhesion factor, ω, as 1 (since both ω_1 and ω_2 are 1; see Figure 2.59), compute p_u from Equation 2.148, $p_u = 307.2$ kPa.

Assuming $\eta = 0.75$ and $\varepsilon = 0.5$, compute p_u from Equation 2.149

$$p_u = [0.75 \times 10 \times 128 + 0.5 \times 2 \times 128] \times 0.76 = 826.88 \text{ kPa}$$

Here, p_u obtained from Equation 2.148 is smaller, and is used in constructing the p_y–v curve. Knowing the K and p_u values, the soil reaction, p_y, can be obtained from Equation 2.143 for any given displacement, v.

A plot of the p_y–v curve is shown in Figure 2.60. To examine the effect of the pile diameter, two different diameters ($D = 0.76$ and 0.38 m) are used, keeping other parameters unchanged. By reducing the pile diameter from 0.76 to 0.38 m, the characteristic length, β, reduces from 2.14 m to 1.27 m. For both cases, the pile length $L = 9.1$ m is larger than 3β, and the pile is considered a long pile. The wedge height for the smaller diameter pile is larger and the pile head displacement, v, of the smaller diameter pile is larger, as expected. As noted by Kim et al. [56], the pile head deflections obtained from most conventional methods (e.g., Ref. [50]) are generally larger than those predicted by the hyperbolic p_y–v curve.

2.10.10 EXAMPLE 2.12: p_y–v CURVES FROM 3-D FINITE ELEMENT MODEL

3-D finite element model (FEM) can be used to construct p_y–v curves for laterally loaded piles. FEM-based p_y–v curves are able to capture the behavior of laterally loaded piles where significant and measurable deflections occur at large depths due to causes such as liquefaction and landslide. Also, the behavior of pile–soil interfaces can be adequately characterized in a FEM-based analysis using interface or contact

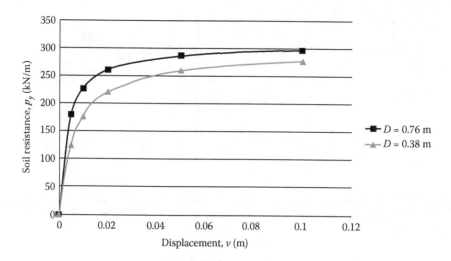

FIGURE 2.60 p_y-v curve for two selected diameters (0.76 m and 0.38 m).

elements. Conventional p_y-v curves based on simplified analytical models (e.g., beam-column on nonlinear Winkler foundation) do not account for these situations.

McGann et al. [62] have used 3-D FEM to construct p_y-v curves for piles of different diameters (small, medium, and large), using the OpenSees FE platform—an object-oriented, open source software framework created at the Pacific Earthquake Engineering Center (see http://peer.berkeley.edu for details). It allows users to create FE applications for simulating response of structural and geotechnical systems subjected to earthquake and other loading. Displacement-based beam-column elements are used to model the pile, while eight-noded brick elements are used to model the soil. Pile–soil contact is modeled using an interface element that serves as a link between the line elements (pile) and the brick elements (soil), enabling the use of standard beam-column elements to model the pile. The contact elements in OpenSees are capable of undergoing separation, rebonding, and frictional slip in accordance with the Coulomb law [63]. Material nonlinearity in the soil is represented using the Drucker–Prager model, which includes pressure-dependent strength, tension cutoff, and nonassociative plasticity (see Appendix 1 for details on Drucker–Prager model). Piles are treated either as a linear elastic material or as an elastoplastic material in which reinforcing bars are represented by a bundle of nonlinear fibers. Liquefaction or lateral spreading is also included using a simplified approach and undrained condition. The elastic modulus of liquefied elements is considered about one-tenth of the modulus of unliquefied elements, and the Poisson's ratio is set to 0.485. Also, a small amount of cohesion ($c = 3.5$ kPa) is introduced for numerical stability, and the internal friction (φ) is set to zero (see Ref. [62] for details).

Figure 2.61 shows the typical FE mesh used in the analysis. The mesh has a height of $21D$ (D = pile diameter) and lateral dimensions of $13D$ and $11D$, and is refined around the pile to adequately capture the pile–soil interaction. For cases involving liquefaction, the liquefied layer is considered to have a thickness of $1D$ and is located at the center of the mesh. The soil and pile nodes on the base of the model are not allowed

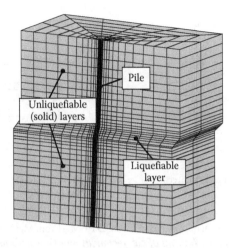

FIGURE 2.61 Typical 3D finite element mesh used. (From McGann, C.R., Arduino, P., and Mackenzie-Heinwein, P., *Journal of Geotechnical and Geoenvironmental Engineering, ASCE*, 137(6), 2011, 557–567. With permission.)

to undergo any vertical displacement. Also, torsional and out-of-plane rotations are assumed to be zero to enforce symmetry. All soil nodes on the outer boundary are assumed to have zero horizontal displacements, and no translational movements are allowed to ensure stability and to allow for the kinematic loading for cases involving liquefaction.

2.10.10.1 Construction of p_y–v Curves

Compared to back-calculating pile forces from bending moment diagrams as done conventionally, the interface elements used in the FEM directly provide the forces exerted on the pile. As noted by McGann et al. [62], the lateral component of the interface force at each interface node is distributed over the tributary length of the pile associated with the node to obtain the force densities (related to p_y) of the corresponding pile segment (Figure 2.62). The corresponding lateral displacement of the pile node is considered as v in constructing the p_y–v curves. The p_y–v values thus obtained pertain to discrete points along the length of the pile. For comparison with conventional p_y–v curves, smooth curves were fitted (using a hyperbolic function) by McGann et al. [62] through these points, and two characteristic parameters were evaluated: initial tangent stiffness, k_o, and ultimate lateral resistance, p_u, where k_o represents the initial slope of the p_y–v curve.

Application of the 3-D FEM: The geometric and material properties used in the FE simulation of the laterally loaded piles are summarized in Table 2.16, while pertinent soil properties are summarized in Table 2.17 [62]. An example of the p_y–v curve constructed from the FEM is shown in Figure 2.63 for a pile having a diameter of 1.3716 m. The p_y and v values obtained from the 3-D FEM analysis are represented by open circles. The fitted initial tangent stiffness (k_o) and the hyperbolic curve are also shown for comparison. Figure 2.64 shows representative computed p_y–v data and fitted hyperbolic tangent curves for the same pile. Conventional p_y–v curves

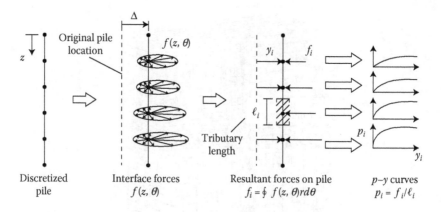

FIGURE 2.62 Construction of p_y–v curve from 3D FEM. (From McGann, C.R., Arduino, P., and Mackenzie-Heinwein, P., *Journal of Geotechnical and Geoenvironmental Engineering, ASCE*, 137(6), 2011, 557–567. With permission.)

(recommended by API [64]) are also shown on the same figure for comparison. It is seen that the two sets of curves are fairly close at shallow depths, but they show significant difference at higher depths, indicating the limitation of conventional p_y–v curves for the analysis of piles exhibiting measurable deformations at large depths due to causes such as liquefaction and landslide.

TABLE 2.16
Material and Section Properties Used in 3-D FEM

Pile Diameter (m)	Area (m²)	E (GPa)	G (GPa)	I_y (m⁴)	I_z (m⁴)
0.6069	0.154	31.3	12.52	0.0038	0.0038
1.3716	0.739	28.7	11.48	0.0869	0.0869
2.5	2.454	102.4	40.96	0.9587	0.9587

Source: From McGann, C.R., Arduino, P., and Mackenzie-Heinwein, P., *Journal of Geotechnical and Geoenvironmental Engineering, ASCE*, 137(6), 2011, 557–567. With permission.

TABLE 2.17
Soil Properties Used in 3-D FEM

Soil Layer	E_s (kPa)	v_s	φ (°)	c (kPa)	γ_{sat} (kN/m³)
Sand	25,000	0.35	36	3.5	17
Liquefied sand	2500	0.485	0	3.5	17

Source: From McGann, C.R., Arduino, P., and Mackenzie-Heinwein, P., *Journal of Geotechnical and Geoenvironmental Engineering, ASCE*, 137(6), 2011, 557–567. With permission.

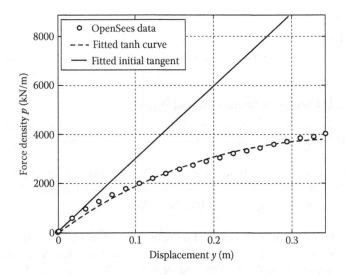

FIGURE 2.63 Computed p_y–v curve with fitted tangent stiffness and hyperbolic curve. (From McGann, C.R., Arduino, P., and Mackenzie-Heinwein, P., *Journal of Geotechnical and Geoenvironmental Engineering, ASCE*, 137(6), 2011, 557–567. With permission.)

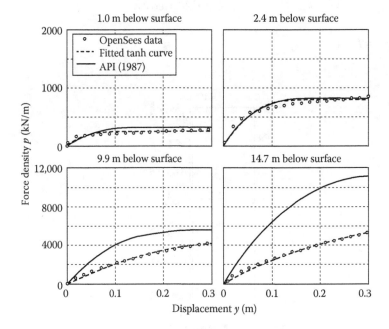

FIGURE 2.64 Comparison of 3D FEM-based and conventional p_y–v curve. (From McGann, C.R., Arduino, P., and Mackenzie-Heinwein, P., *Journal of Geotechnical and Geoenvironmental Engineering, ASCE*, 137(6), 2011, 557–567. With permission.)

PROBLEMS

Problem 2.1

Derive the differential equation for a beam on Winkler foundation subjected to bending.

Problem 2.2: Cantilever Beam with Soil Support

Figure P2.1 shows a beam fixed at one end and free at the other. It is subjected to a load of 10^5 lbs (4.448×10^2 kN) at the end. The properties of beam, soil, and loading are

Length $L = 120$ ft (36.60 m), area $A = 12 \times 12 = 144$ in^2 (929.00) cm^2),
$I = 1728$ in^4 (719,124.8 cm^4), $E = 30 \times 10^6$ psi (201×10^6 kPa), $k_o = 10.8$ lb/in^3
(2.93 N/cm^3), $k = bk_o = 129.60$ lb/in^2 (894 kPa)

Obtain and compare analytical and numerical solutions for displacement, slope, bending moment, shear force, and soil reaction. The numerical solutions can be obtained by using FDM and/or FEM, and by adopting $N = 10$, 20, and 120 segments (elements). The analytical solutions can be obtained by using the following equations given by Hetenyi [4]:

Displacement:

$$v = \frac{2P\lambda}{k} \frac{\sinh \lambda x \cos \lambda x' \cosh \lambda L - \sin \lambda x \cosh \lambda x' \cos \lambda L}{\cosh^2 \lambda \ell + \cos^2 \lambda \ell}$$

where $x' = L - x$.

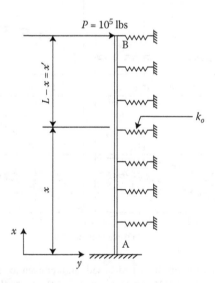

FIGURE P2.1   Beam with soil support—closed form solution (Problem 2.2) (1 lb = 4.45 N).

Displacement at B:

$$v_B = \frac{P\lambda \sin 2\lambda L - \sin 2\lambda L}{k \cosh^2 \lambda L + \cos^2 \lambda L}$$

Slope:

$$\theta = \frac{2P\lambda^2}{k} \frac{1}{\cosh^2 \lambda L + \cos^2 \lambda L}$$
$$\times [\cosh \lambda L(\cosh \times \cos \lambda x' + \sinh \lambda x \sin \lambda x')$$
$$- \cos \lambda L(\cos \lambda x \cosh \lambda x' - \sin \lambda x \sinh \lambda x')]$$

Slope at B:

$$\theta_B = \frac{2P\lambda^2}{k} \frac{\cosh^2 \lambda L - \cos^2 \lambda L}{\cosh^2 \lambda L + \cos^2 \lambda L}$$

Bending moment:

$$M = \frac{P}{\lambda} \cdot \frac{\cosh \lambda x \sin \lambda x' \cosh \lambda L + \cos \lambda x \sinh \lambda x' \cos \lambda L}{\cosh^2 \lambda L + \cos^2 \lambda L}$$

$$M_A = -\frac{P}{\lambda} \frac{\sinh \lambda L \cos \lambda L + \cosh \lambda L \sin \lambda L}{\cosh^2 \lambda L + \cos^2 \lambda L}$$

Shear force:

$$V = -\frac{P}{\cosh^2 \lambda L + \cos^2 \lambda L}$$
$$\times [\cosh \lambda L (\sinh \lambda x \sin \lambda x' - \cosh \lambda x \cos \lambda x')$$
$$- \cos \lambda L (\sin \lambda x \sinh \lambda x' + \cos \lambda x \cosh \lambda x')]$$

$$V_A = \frac{P2 \cosh \lambda L \cos \lambda L}{\cosh^2 \lambda L + \cos^2 \lambda L}$$

Soil resistance:

$$p = kv$$

where

$$\lambda = \sqrt[4]{\frac{k}{4EI}} = \sqrt[4]{\frac{129.6}{4 \times 30 \times 10^6 \times 1728}}$$
$$= 0.005$$

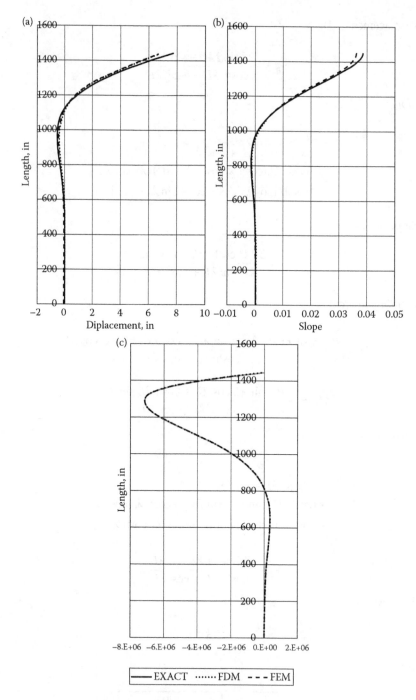

FIGURE P2.2 (a) Displacement versus length of beam, $N = 10$; (b) slope versus length of beam, $N = 20$; (c) bending moment versus length of beam, $N = 120$; (d) shear force versus length of beam, $N = 20$; and (e) soil resistance versus length of beam, $N = 10$.

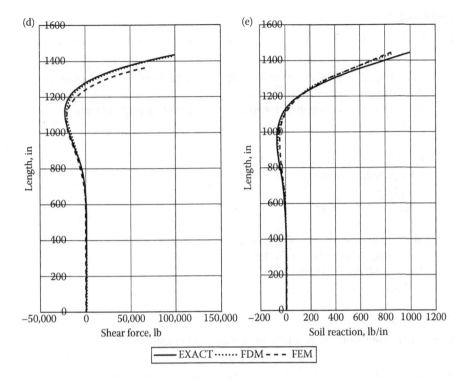

FIGURE P2.2   (continued) (a) Displacement versus length of beam, $N = 10$; (b) slope versus length of beam, $N = 20$; (c) bending moment versus length of beam, $N = 120$; (d) shear force versus length of beam, $N = 20$; and (e) soil resistance versus length of beam, $N = 10$.

Partial solutions: Typical solutions obtained by using the code SSTIN-1DFE, and showing comparisons between analytical (exact) and numerical methods (FE and FE) are given in the above figures for various values of segment (N):

Problem 2.3

Derive the expression for deflection of a column fixed at base and subjected to a concentrated load P_t at the top. Assume properties and boundary conditions.

Problem 2.4

Tabulate and plot C_1, A_1, D_1, and B_1: Use following expressions versus λx:

$$v = \frac{2P_t\lambda}{k}C_1 \qquad \text{where } C_1 = e^{-\lambda x}\cos\lambda x$$

$$S = -\frac{2\,P_t\,\lambda^2}{k}A_1 \qquad \text{where } A_1 = e^{-\lambda x}(\cos\lambda x + \sin\lambda x)$$

$$M = \frac{P_t}{\lambda}D_1 \qquad \text{where } D_1 = e^{-\lambda x}(\sin\lambda x)$$

$$V = P_t B_1 \qquad \text{where } B_1 = e^{-\lambda x}(\cos\lambda x - \sin\lambda x)(\sin\lambda x - \cos\lambda x)$$

$$p = -2P_t\lambda C_1$$

Problem 2.5

The measured pile deflections for a segment of a pile are shown in Figure P2.3:

1. Set up equations and compute bending stress and shear at a depth of 60 cm. Also compute stress due to bending.
2. Write down the equation you would employ for computing soil resistance at 60 cm.

$E = 20 \times 10^6$ N/cm²
$EI = 26 \times 10^{10}$ N-cm²
$I = 13{,}000$ cm⁴
Diameter $= 43.00$ cm
Answer:
Moment at 60 cm $= 115{,}560$ kN-cm $= 1156.0$ kN-m
Stress due to bending $= 180.0$ kN/cm²
Shear at 60 cm $= 385$ kN
Soil resistance, $p = -k_{60}\, v_{60}$

Problem 2.6

Write a FD expression for the following differential equation:

$$\frac{d}{dx}\left[R(x,y)\frac{dv}{dx} \right] - (x)\frac{dv}{dx} = Q$$

where $R(x,y)$ is dependent on x and y. Provide critical comments on the effect of this dependence on the resulting difference equations.

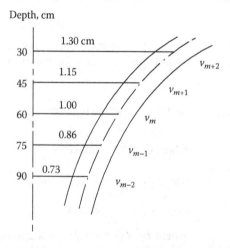

FIGURE P2.3 Measured displacements for a pile segment (Problem 2.5).

Problem 2.7

Develop p_y–v curves for the stiff clay presented in Bhusan et al. [65].
  Make any required assumptions but state them clearly with justification.

- Diameter of piles = 4.0 ft (1.22 m) and embedded length = 15 ft (4.58 m)
- Derive p_y–v curves at every 6 in (15.24 cm)
  Note: The p_y–v curves may be used (in a later problem) to predict the pile behavior using the FD and/or FE procedures.
- The computer analysis can be performed by using 10, 15, and 30 elements

Problem 2.8

1. Divide the beam (Figure P2.4) into 20 elements and 21 nodes. Assuming the following geometric or displacement boundary conditions:

$$v(0) = v(L) = 0$$

for the simply supported beam, compute and plot displacement, moment, shear force, and soil resistance diagrams by using available FD and/or FE code. Compare computer results with closed-form solutions. Assume required properties and cross-section.
2. Assume the moment of inertia of the beam varies linearly from 6000 cm^4 at the left end to 4000 cm^4 at the right end. Assume average of moment of inertia within an element. Compare results with those in item (A) above.

Problem 2.9

Solve the beam-bending problem (Figure P2.5) by using FE and/or FD code.
Load

1. $q(x) = 100$ kg/cm constant (Figure P2.5a)
2. $q(x) =$ linear variation with $q_A = 150$ kg/cm and $q_B = 50$ kg/cm (Figure P2.5b)
   Divide the beam in three sets of elements (segments): $N = 2$, 4, and 16
   Analyze convergence with respect to bending moment M using closed-form (v^*) solution, that is, $\bar{v} = v^* - v$, where v is computed displacement and v^* for the constant q is given by

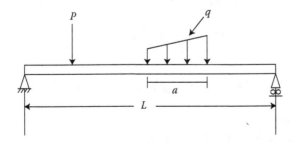

FIGURE P2.4 Simply supported beam (Problem 2.8).

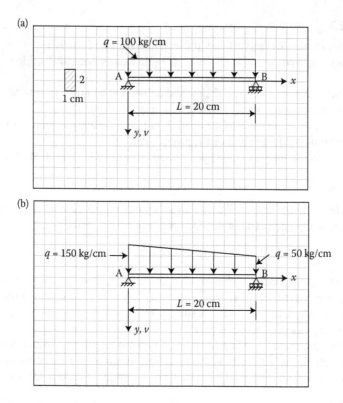

FIGURE P2.5 Simply supported beam (Problem 2.5). (a) Uniform load; (b) linearly varying load.

$$v^* = \frac{qx}{24\,EI}(L^3 - 2Lx^2 + x^3)$$

where x is the coordinate, L is the length of the beam, and EI is the flexural rigidity.

For linearly varying load, v^* is given by [53]

$$v^* = \frac{1}{12\,EI}\left\{\frac{145}{3}L^3(L-x) - \frac{250}{3}(L-x)^3\right.$$
$$\left. + 25(L-x)^4 + \frac{10}{L}(L-x)^5\right\}$$

$$E = 1 \times 10^6 \text{ kg/cm}^2, \quad I = \frac{2}{3} \text{ cm}^4$$

Partial solution:
   Constant load: $N = 4$ (see Figure P2.6)
   Linearly varying load: $N = 16$ (see Figure P2.7).

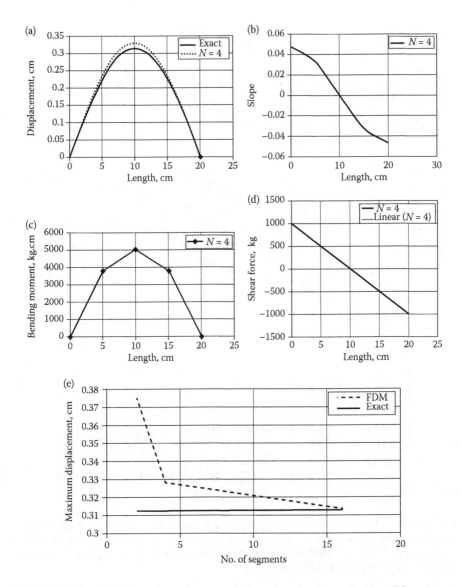

FIGURE P2.6 Predictions for various quantities and comparisons: Problem 2.9, constant load. (a) Displacements versus length; (b) slope versus length; (c) bending moment versus length; (d) shear force versus length; and (e) convergence: maximum displacement versus number of segments or elements.

Problem 2.10

Review the two papers: Refs. [66,67]

1. Write an essay on the procedures presented in these papers, and compare them with the other procedures, for example, with those described in this chapter,
2. Derive p_y–v curves for the test data as available.

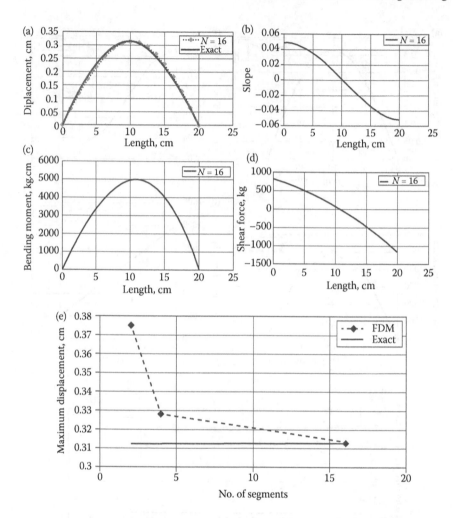

FIGURE P2.7   Predictions for various quantities and comparisons: Problem 2.9, linear load. (a) Displacement versus length, $N = 16$; (b) slope versus length, $N = 16$; (c) bending moments versus length, $N = 16$; (d) shear force versus length, $N = 16$; and (e) convergence: maximum displacement versus number of segments or elements.

Problem 2.11

Consider a long pile with load and fixity at the top. Figure P2.8 shows the long pile with the following properties:

$I = 5000$ in⁴ (208,116 cm⁴)

$k_o = 10$ lb/in³ (2.764 N/cm³)

$k = E_s = k_o d = 10$ d, where $d =$ diameter of the pile $= 17.90$ in (45.47 cm (see below)

$P_t = 100$ kip (4.448 × 10⁵ N)

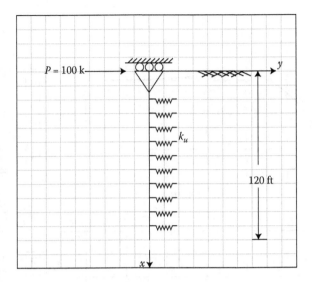

FIGURE P2.8   Pile with load and fixity at top: Problem 2.11 (1 kip = 4.448 kN, 1 ft: 0.3048 m).

Boundary conditions:

At top:

1. $\dfrac{dv}{dx} = 0$
2. $V = P_t$

At bottom:

1. $V = 0$
2. $M = 0$

Divide the pile into $N = 5$, 10, 20, and 120 elements (segments) and find

1. Moment at the ground surface
2. Deflection at the ground surface
3. Plot displacements (v), slopes (S), moment (M), shear force (V), and soil reaction (p) along the pile
4. Find maximum values of above quantities
5. Plot convergence for v with respect to the number of elements

Find diameter, d:

$$I = \frac{\pi d^4}{64} = 5000 \, \text{in}^4 (208{,}116 \, \text{cm}^4)$$

$$\therefore d = 17.9 \text{ in (45.47 cm)}$$

Therefore, $k = 10 \times 17.9 = 179$ lb/in² (1234.2 kPa)
Partial solution: See Figure P2.9.

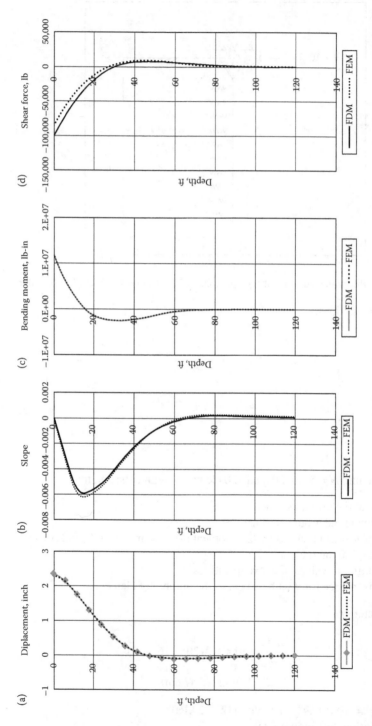

FIGURE P2.9 Computer predictions for pile with fixity at top: Problem 2.11, (1 ft = 0.3048 m, 1 inch = 2.54 cm, 1 lb-in = 11.3 N-cm, 1 lb = 4.448 N, 1 lb/in = 175 N/m). (a) Displacement versus depth, $N = 20$; (b) slope versus depth, $N = 20$; (c) bending moment versus depth, $N = 10$; (d) shear force versus depth, $N = 20$; (e) soil reaction versus depth, $N = 20$. Convergence for maximum displacement versus number of elements: pile fixed at top; (f) finite difference method; and (g) finite element method.

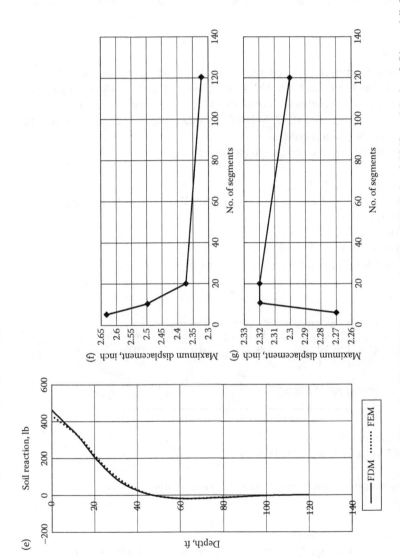

FIGURE P2.9 (continued) Computer predictions for pile with fixity at top: Problem 2.11, (1 ft = 0.3048 m, 1 inch = 2.54 cm, 1 lb-in = 11.3 N-cm, 1 lb = 4.448 N, 1 lb/in = 175 N/m). (a) Displacement versus depth, $N = 20$; (b) slope versus depth, $N = 10$; (c) bending moment versus depth, $N = 120$; (d) shear force versus depth, $N = 20$; (e) soil reaction versus depth, $N = 10$. Convergence for maximum displacement versus number of elements: pile fixed at top: (f) finite difference method; and (g) finite element method.

Note: The convergence behavior appears to be reasonable. However, convergence from FEM shows some initial irregularity.

Problem 2.12

Derive the FD equation for variable cross section of the beam (pile) for bending and axial loads.

Problem 2.13

The observed load–displacement curves for butt (top) and tip (bottom) for pile No. 2 tested at Arkansas Lock and Dam. No. 4 are shown in Figures 2.48a through 2.48c [51]. Develop (p_x-u) $t-z$ curves for one of the piles and compute load–displacement curves by using the FE code (SSTIN-1DFE) or other available code. Compare the predictions with the observations.

Properties of the pile and soil are given below:

Length of pile = 53.0 ft (16.15 m)
Three soil layers: layer I of 24 ft (7.30 m), layer II of 23 ft (7.01 m), and layer
    III of 6 ft (1.83 m)
Pile diameter = 16 in (40.64 cm)
Pile area = 201.0 in² (1297 cm²)
$E = 29 \times 10^6$ lb/in² (200×10^6)
$K = 1.23$
Values for E_i:

Layer	K′	n	γ (pcf)
I and III	1500	0.60	100
II	1200	0.50	100

1 pcf = 1.6 kg/m³

Poison's ratio: $v = 0.30$
$\tau_{max} = 0.45$ ksf = 450 lb/ft² (21546 N/m²).

Make any necessary assumptions but state them clearly.
Required:

1. Develop p_x-u curves at various depths.
2. Develop nonlinear spring at tip by using the field observations for tip load and displacement.
3. By using the computer code SSTTN-IDFE or any other suitable code, find
   a. Load–displacement curves at butt and tip and compare with field observations.
   b. Plot total load versus wall friction and tip loads.
   c. Plot distributions of load in pile and wall friction.

Problem 2.14

Construct a set of p_y–v curves for stiff clay at 5, 10, and 15 ft (1.52, 3.04, and 4.57 m) depths. Assume that the pile is circular in cross section having a diameter of 12 in (30.48 cm). Also, assume that the unit weight of clay increases linearly with depth from $\gamma = 100$ pcf (160 kg/m³) at the ground surface to $\gamma = 115$ pcf (184 kg/m³) at a depth of 15 ft (4.57 m). Given: cohesion $c = 500$ psf (23.94 kN/m²), and friction angle $\varphi = 5°$.

Problem 2.15

Construct a set of p_y–v curves for dense sand at 5, 10, and 15 ft depths (1.52, 3.04, and 4.57 m). Assume that the pile is circular in cross section having a diameter of 12 in (30.48 cm). Also, assume that the unit weight of clay increases linearly with depth from $\gamma = 110$ pcf (184 kg/m³) at the ground surface to $\gamma = 120$ pcf (192 kg/m³) at a depth of 20 ft (6.1 m). The friction angle φ for sand is 35°, and coefficient of earth pressure at rest, $K_o = 0.5$. Assume any other data, if necessary, and state them clearly.

Problem 2.16

Write a computer program to evaluate lateral deflections of pile using the FD method discussed in this chapter. Use a computer program to determine the lateral deflections of the pile shown in Figure P2.10.

Given: pile length = 30 ft (9.14 m); pile cross-section = 18 × 18 in (45.7 × 45.7 cm); pile $E = 3000$ ksi (21 × 10⁶ kPa); lateral load $P_t = 40$ kip (178 kN), lateral moment $M_t = 20$ kip-ft (1.36 kN-m); and axial load $P_x = 5$ kip (22.24 kN). The modulus of subgrade reaction is assumed to vary linearly from $k = 750$ lb/in³ (203.6 kN/m³) at the ground surface to 1250-lb/in³ (339.25 kN/m³) at 30 ft (9.14 m) depth.

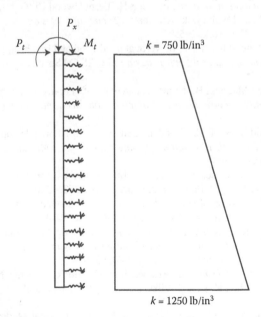

FIGURE P2.10 Pile loading and modulus of subgrade reaction.

Partial answer:

Depth (ft)	Deflection (in)	Depth (ft)	Deflection (in)
0	0.0959	6	0.0159
1	0.07635	7	0.0077
3	0.0575	8	0.0019
4	0.0409	9	−0.0018
5	0.0269	10	−0.0039

Note: 1 in = 2.54 cm.

REFERENCES

1. Winkler, E., *Theory of Elasticity and Strength*, H. Dominicas, Prague (in German), 1867.
2. Zimmerman, H., *Calculation of the Upper Surface Construction of Railway Tracks*, Ernest and Korn Verlag, Berlin (in German), 1888.
3. Hayashi, K., *Theory of Beams on Elastic Foundation*, Springer-Verlag (in German), 1921.
4. Hetenyi, M., *Beams on Elastic Foundation*, University of Michigan Press, Ann Arbor, 1946.
5. Vlasov, V.Z. and Leotiev, N.N., Beams, Plates, and Shells on Elastic Foundations, NTIS No. 67-14238 (translated from Russian by Israel Program for Scientific Translations), 1966.
6. Wolfer, K.H., *Elastically Supported Beams*, Baurerlag GMBH, Berlin, 1971.
7. Sherif, G., *Elastically Fixed Structures*, Ernest and Sohn Verlag, Berlin, 1974.
8. Vesic, A.S., Slabs on elastic subgrade and winkler hypothesis, *Proceedings of the 8th International Conference on Soil Mechanics and Foundation Engineering*, Moscow, 1973.
9. Scott, R.F., *Foundation Analysis*, Prentice-Hall, Englewood Cliffs, NJ, 1981.
10. Timoshenko, S. and Woinowski-Krieger, S., *Theory of Plates and Shells*, McGraw-Hill, Second Edition, New York, 1959.
11. Gleser, S.M., Lateral loads tests on vertical fixed-head and free-head piles, *Proceedings of the Symposium on Lateral Load Tests on Piles*, ASTM Special Tech. Publications No. 154, 1953, 75–101.
12. Reese, L.C. and Matlock, H., Numerical analysis of laterally loaded piles, *Proceedings of the ASCE 2nd Structural Division Conference on Electronic Computation*, Pittsburgh, 1960.
13. Matlock, H. and Reese, L.C., Foundation analysis of offshore pile-supported structures, *Proceedings of the 5th International Conference on Soil Mechanics and Foundation Engineering*, Paris, 1961.
14. Matlock, H. and Reese, L.C., Generalized solutions for laterally loaded piles, *Transactions, ASCE*, 127, Part I, Proc. Paper 3770, 1962, 1220–1249.
15. Reese, L.C., Discussion of soil modulus for laterally loaded piles, by McClelland, B. and Focht, J.A., *Transactions, ASCE*, 123, 1958, 1071–1074.
16. Desai, C.S. and Kuppusamy, T., Application of a numerical procedure for offshore piling, *Proceedings of the International Conference on Numerical Methods in Offshore Piling*, Inst. of Civil Engineers, London, 1980, 93–99.
17. Crandall, S.H., *Engineering Analysis*, McGraw-Hill Book Company, New York, 1956.
18. Carnahan, B., Luther, H.A., and Wilkes, J.O., *Applied Numerical Methods*, John Wiley, Newark, 1969.
19. Desai, C.S. and Abel, J.F., *Introduction to the Finite Element Method*, Van Nostrand Reinhold Company, New York, 1972.

20. Desai, C.S., *Elementary Finite Element Method*, Prentice-Hall, Englewood Cliffs, NJ, USA, 1979; revised as Desai C.S. and Kundu, T., *Introductory Finite Element Method*, CRC Press, Boca Raton, 2001.
21. Desai, C.S. and Christian, J.T. (Editors), *Numerical Methods in Geotechnical Engineering*, McGraw-Hill Book Company, New York, 1977.
22. Zienkiewicz, O.C. and Taylor, R.L., *The Finite Element Method*, 4th Edition, McGraw-Hill, London, UK, 1989.
23. Bathe, K.J., *Finite Element Procedures in Engineering Analysis*, Prentice-Hall, Englewood Cliffs, NJ, 1996.
24. Desai, C.S., *User's Manual for SSTIN-1DFE: Code for Axially and Laterally Loaded Piles and Walls*, Tucson, AZ, 2001.
25. Desai, C.S. and Siriwardane, H.J., *Constitutive Laws for Engineering Materials*, Prentice Hall, Englewood Cliffs, NJ, 1984.
26. Desai, C.S., *Mechanics of Materials and Interfaces: The Disturbed State Concept*, CRC Press, Boca Raton, 2001.
27. Chen, W.F. and Han, D.J., *Plasticity for Structural Engineers*, Springer-Verlag, New York, 1988.
28. Coyle, H.M. and Reese, L.C., Load transfer for axially loaded piles in clay, *Journal of the Soil Mechanics and Foundations Division, ASCE*, 2(SM2), 1966.
29. Ramberg, W. and Osgood, W.R., Description of Stress-Strain Curves by Three Parameters, Tech. Note 902, National Advisory Comm. Aeronaut., Washington, DC, 1943.
30. Desai, C.S. and Wu, T.H., A General function for stress-strain curves, *2nd International Conference Numerical Methods in Geomechanics*, C.S. Desai (Ed.), Blacksburg, VA, ASCE, 1976.
31. Richard, R.M. and Abott, B.J., Versatile elastic-plastic strain-strain curves by three parameters, Tech Note, *Journal of Engineering Mechanics Division, ASCE*, 101(EM4), 1975, 511–515.
32. Reese, L.C. (1) and Matlock, H. (2), *(1) Soil-Structure Interaction; (2) Mechanics of Laterally Loaded Piles, Lecture Notes*, Courses Taught at the University of Texas, Austin, TX, 1967–1968.
33. Skempton, A.W., The bearing capacity of clays, *Proceedings of the Building Research Congress*, Inst. of Civil Engineers, London, 180–189, 1951.
34. Reese, L.C., Ultimate resistance against a rigid cylinder moving laterally in cohesionless soil, *Journal of the Society of Petroleum Engineering*, 2(4), 1962, 355–359.
35. Reese, L.C. and Desai, C.S., Laterally loaded piles, Chapter 9 in *Numerical Methods in Geotechnical Engineering*, C.S. Desai and J.T. Christian (Editors), McGraw-Hill Book Co., New York, USA, 1977.
36. Bowman, E.R., Investigation of Lateral Resistance to Movement of a Plate in Cohesionless Soil, Doctoral Dissertation, University of Texas, Austin, TX, USA, 1958.
37. Terzaghi, K., *Theoretical Soil Mechanics*, John Wiley and Sons, New York, 1943.
38. Reese, L.C., Cox, W.R., and Koop, F.D., Analysis of Laterally Loaded Piles in Sand, Offshore Technology Conference Paper No. 2080, Houston, TX, 1974.
39. Matlock, H., Correlations for the design of laterally loaded pikes in soft clays, *Proceedings of the 11th Offshore Technology Conference*, Paper No. 1204, Houston, TX, 577–594, 1970.
40. Haliburtan, T.A., Numerical analysis of flexible retaining structures, *Journal of the Soil Mechanics and Foundations Division, ASCE*, 94(SM6), 1968, 1233–1251.
41. Halliburton, T.A., Soil-Structure Interaction: Numerical Analysis of Beams and Beam-Columns, Technical Publication No. 14, School of Civil Engineering, Oklahoma State Univ., Stillwater, Oklahoma, USA, 1971.
42. Terzaghi, K., Anchored bulkheads, *Transactions, ASCE*, 119, 1243–1280, 1954, and Discussion, 1281–1324.

43. Bogard, D. and Matlock, H., A Model Study of Axially Loaded Pile Segments Including Pore Pressure Measurements, Report to American Petroleum Institute, Austin, TX, 1979.

44. Love, A.E.H., *The Mathematical Theory of Elasticity*, Dover Publishing Co., New York, 1944.

45. Timoshenko, S. and Goodier, J.N., *Theory of Elasticity*, McGraw-Hill, New York, 1951.

46. Herrmann, L.R., Elastic torsional analysis of irregular shapes, *Journal of Engineering Mechanics Division, ASCE*, 91(EM6), 1965, 11–19.

47. McCammon, G.A. and Ascherman, J.C., Resistance of long hollow piles to applied lateral loads, *ASTM, STP* 154, 1960, 3–11.

48. Skempton, A.W. and Bishop, A.W., Building materials, their elasticity and plasticity, Chapter 10 in *Reiner* 461.

49. McClelland, B. and Focht, J.A., Soil modulus for laterally loaded piles, *Transactions, ASCE*, 123, 1958, 1049–1086.

50. Reese, L.C., Cox, W.R., and Koop, F.D., Field testing and analysis of laterally loaded piles in stiff clay, *Proceedings of the Offshore Technology Conference (OTC)*, Paper No. 2312, Houston, TX, 1975.

51. Fruco and Associates, Results of Tests on Foundation Materials, Lock and Dam No. 4, Reports 7920 and 7923, U.S. Army Engr. Div. Lab., Southwestern, Corps of Engineers, SWDGL, Dallas, TX, 1962.

52. Clough, G.W., Weber, P. R., and Lamunt Jr. J., Design and observations of tied back wall, *Proceedings of the Special Conference on Performance of Earth and Earth Supported Structures*, Purdue Univ., W. Lafayette, IN, 1972, 1367–1389.

53. Pytel, A. and Singer, F.L., *Strength of Materials*, Harper Collins Publ., New York, 1987.

54. Yang, K. and Liang, R., Lateral responses of large diameter drilled shafts in clay, *Proceedings of the 30th Annual Conference on Deep Foundations*, Deep Foundation Institute, NJ, 115–126, 2005.

55. Liang, R., Shantnawi, E.S., and Nusairat, J., Hyperbolic p-y criterion for cohesive soils, *Jordan Journal of Civil Engineering*, 1(1), 2007, 38–58.

56. Kim, Y., Jeong, S., and Lee, S., Wedge failure analysis of soil resistance on laterally loaded piles in clay, *Journal of Geotechnical and Geoenvironmental Engineering, ASCE*, 137(7), 2011, 678–694.

57. Carter, D.P., A Nonlinear Soil Model for Predicting Lateral Pile Response, Rep. No. 359, Civil Engineering Dept., Auckland, New Zealand, 1984.

58. Ashour, M., Pilling, P., and Norris, G. Lateral behavior of pile group in layered soils, *Journal of Geotechnical and Geoenvironmental Engineering, ASCE*, 130(6), 2004, 580–592.

59. Briaud, J.L., SALLOP: Simple approach for lateral loads on piles, *Journal of Geotechnical and Geoenvironmental Engineering, ASCE*, 123(10), 1997, 958–964.

60. Briand, J.L., Smith, T., and Mayer, B., Laterally loaded piles and the pressure meter: Comparison of existing methods, *Laterally Loaded Deep Foundation: Analysis and Performance, STP835, ASTM*, West Conshohocken, PA, 97–111, 1984.

61. United States Army Corps of Engineers (USACE), *Engineering and Design—Settlement Analysis*, Washington, DC, D1-D12, 1990.

62. McGann, C.R., Arduino, P., and Mackenzie-Helnwein, P., Applicability of conventional p-y relations to the analysis of piles in laterally spreading soil, *Journal of Geotechnical and Geoenvironmental Engineering*, ASCE, 137(6), 2011, 557–567.

63. Wriggers, P., *Computational Contact Mechanics*, Wiley, West Sussex, United Kingdom, 2002.

64. American Petroleum Institute (API), Recommended practice for planning, designing and constructing fixed offshore platforms, *API Recommended Practice 2A (RP-2A)*, 17th Edition, Washington, D.C., 1987.

65. Bhusan, K., Haley, S.C., and Tong, P.T., Lateral load tests on drilled piers in stiff clays, *Journal of the Geotechnical Engineering Division, ASCE*, 105(GT8), 1979, 969–985.

66. Broms, B.B., Lateral resistance of piles in cohesionless soils, *Journal of the Soil Mechanics and Foundations Division, ASCE*, 90(SM3), 1964, 123–156.

67. Broms, B.B., Lateral resistance of piles in cohesive soils, *Journal of the Soil Mechanics and Foundations Division, ASCE*, 90(2), 1964, 27–63.

3 Two- and Three-Dimensional Finite Element Static Formulations and Two-Dimensional Applications

3.1 INTRODUCTION

It is possible to develop analytical solutions for engineering problems based on the equations of equilibrium and compatibility; however, it is necessary to make certain simplifying assumptions regarding material behavior, geometry, and boundary conditions. Numerical procedures such as FD, FE, BE, and energy methods can be developed for the solution of such problems by reducing the simplifying assumptions. It is believed that the FEM possesses certain advantages and generality compared to other methods. Hence, in this chapter, we primarily focus on the FEM for problems involving plane deformations in solids. Limited applications of some other methods (stiffness and energy) are also included.

We first present 2-D and 3-D FE formulations. Then, applications for 2-D idealizations are presented in this chapter. In Chapter 4, we will present applications for 3-D idealizations.

3.2 FINITE ELEMENT FORMULATIONS

Comprehensive details for the FE formulations and applications are given in various publications, for example, Refs. [1–4]. Here, we present rather brief descriptions of the formulations for plane problems.

The discretization phase in the FEM involves dividing the domain of interest of a problem by using 2-D and 3-D elements. Figure 3.1 shows typical isoparametric brick (3-D) and quadrilateral (2-D) elements.

Approximation Functions: The approximation functions for displacements at a point in an element depend on the degrees of freedom at the point. For 3-D and 2-D elements shown in Figure 3.1, the expressions for displacements at a point are given by

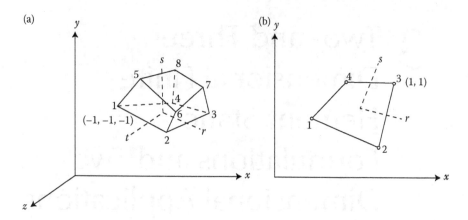

FIGURE 3.1   Three- and two-dimensional elements isoparametric elements. (a) 3-D: brick element; (b) 2-D: quadrilateral element.

$$\{u\} = \begin{Bmatrix} u \\ v \\ w \end{Bmatrix} = [N]\{q\} = \begin{bmatrix} [N_b] & o & o \\ o & [N_b] & o \\ o & o & [N_b] \end{bmatrix} \{q\} \qquad (3.1a)$$

where $[N_b] = [N_1 \ N_2 \ N_3 \ N_4 \ N_5 \ N_6 \ N_7 \ N_8]$ is the matrix of interpolation (shape or basis) functions, N_i $(i = 1, 2, \ldots, 8)$, $\{q\}^T = [u_1 \ u_2 \ldots u_8; \ v_1 \ v_2 \ldots v_8; \ w_1 \ w_2 \ldots w_8]$ is the vector of nodal displacements for eight-noded isoparametric element, and $N_i = (1/8)(1 + rr_i)(1 + ss_i)(1 + tt_i)$, where r, s, t are local coordinates (Figure 3.1). The interpolation functions, N_i, can be obtained by substituting appropriate values of (r_i, s_i, t_i) at the nodes. For example

$$N_1 = \frac{1}{8}(1 - r)(1 - s)(1 - t) \qquad (3.1b)$$

and so on.

For the 2-D plane stress, plane strain, and axisymmetric idealizations, the approximation function for the quadrilateral element (Figure 3.1b) can be expressed as a vector of displacements at a point (u, v) as

$$\{u\} = \begin{Bmatrix} u \\ v \end{Bmatrix} = [N]\{q\} \qquad (3.2a)$$

where $[N]$ is the matrix of interpolation functions

$$N_i = \frac{1}{4}(1 + rr_i)(1 + ss_i) \qquad (3.2b)$$

r and s are local coordinates (Figure 3.1b) $\{q\}^T = [u_1 \ v_1 \ u_2 \ v_2 \ u_3 \ v_3 \ u_4 \ v_4]$, and u_i, v_i ($i = 1, 2, 3, 4$) are nodal displacements. The interpolation functions can be derived by substituting the local coordinates of the nodal points. For example

$$N_3 = \frac{1}{4}(1+r)(1+s) \qquad (3.2c)$$

Stress–Strain Relations: Now, we define stress–strain relations. For the 3-D case, involving small strains, the strain vector is given by

$$\{\varepsilon\} = \begin{Bmatrix} \varepsilon_x \\ \varepsilon_y \\ \varepsilon_z \\ \gamma_{xy} \\ \gamma_{yz} \\ \gamma_{zx} \end{Bmatrix} = \begin{Bmatrix} \partial u/\partial x \\ \partial v/\partial y \\ \partial w/\partial z \\ \partial u/\partial y + \partial v/\partial x \\ \partial v/\partial z + \partial w/\partial y \\ \partial w/\partial x + \partial u/\partial z \end{Bmatrix} = [[B_1][B_2]\ldots[B_8]]\{q\} = [B]\{q\} \quad (3.3a)$$

where

$$[B_i] = \begin{bmatrix} \partial N_i/\partial x & 0 & 0 \\ 0 & \partial N_i/\partial y & 0 \\ 0 & 0 & \partial N_i/\partial z \\ \partial N_i/\partial y & \partial N_i/\partial x & 0 \\ 0 & \partial N_i/\partial z & \partial N_i/\partial y \\ \partial N_i/\partial x & 0 & \partial N_i/\partial z \end{bmatrix} \qquad (3.3b)$$

The derivatives in Equation 3.3b are found as

$$\begin{Bmatrix} \partial N_i/\partial x \\ \partial N_i/\partial y \\ \partial N_i/\partial z \end{Bmatrix} = [J]^{-1} \begin{Bmatrix} \partial N_i/\partial r \\ \partial N_i/\partial s \\ \partial N_i/\partial t \end{Bmatrix} \qquad (3.3c)$$

where the 3×3 Jacobian matrix $[J]$ is obtained from the following equation:

$$\begin{matrix} [J] \\ (3 \times 3) \end{matrix} = \begin{bmatrix} \partial\{N\}^T/\partial r \\ \partial\{N\}^T/\partial s \\ \partial\{N\}^T/\partial t \end{bmatrix} \begin{matrix} [\{x_n\} \ \{y_n\} \ \{z_n\}] \\ (8 \times 3) \end{matrix} \qquad (3.3d)$$

$$(3 \times 8)$$

For 2-D problems, the strain vectors for different idealizations are given by
Plane stress (Figure 3.2a)

$$\{\varepsilon\} = \begin{Bmatrix} \varepsilon_x \\ \varepsilon_y \\ \gamma_{xy} \end{Bmatrix} \tag{3.4a}$$

Plain strain (Figure 3.2b)

$$\{\varepsilon\} = \begin{Bmatrix} \varepsilon_x \\ \varepsilon_y \\ \gamma_{xy} \end{Bmatrix} \tag{3.4b}$$

Axisymmetric (Figure 3.2c)

$$\{\varepsilon\} = \begin{Bmatrix} \varepsilon_r \\ \varepsilon_\theta \\ \varepsilon_z \\ \gamma_{rz} \end{Bmatrix} \tag{3.4c}$$

The stress–strain relations for linear elastic isotropic material behavior, according to the theory of elasticity [5], are given by
Three-dimensional

$$\{\sigma\} = \begin{Bmatrix} \sigma_x \\ \sigma_y \\ \sigma_z \\ \tau_{xy} \\ \tau_{yz} \\ \tau_{zx} \end{Bmatrix} = \frac{E}{(1+v)(1-2v)} \begin{bmatrix} 1-v & v & v & 0 & 0 & 0 \\ & 1-v & v & 0 & 0 & 0 \\ & & 1-v & 0 & 0 & 0 \\ & \text{symmetrical} & & \frac{1-2v}{2} & 0 & 0 \\ & & & & \frac{1-2v}{2} & 0 \\ & & & & & \frac{1-2v}{2} \end{bmatrix} \{\varepsilon\}$$

$$= [C]\{\varepsilon\} \tag{3.5}$$

The above 3-D expression can be specialized for the 2-D idealizations such as plane stress, plane strain, and axisymmetric as follows [1,5]:

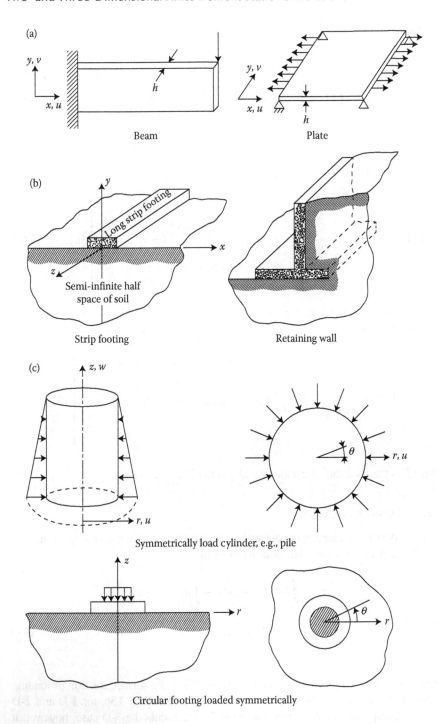

FIGURE 3.2   2-D idealizations. (a) Plane stress idealization; (b) plane strain idealization; and (c) axisymmetric idealization.

Plane stress

$$\{\sigma\} = [C]\{\varepsilon\} = \frac{E}{(1 - v^2)} \begin{bmatrix} 1 & v & 0 \\ v & 1 & 0 \\ 0 & 0 & \dfrac{1 - v}{2} \end{bmatrix} \{\varepsilon\} \qquad (3.6a)$$

Plane strain

$$\{\sigma\} = [C]\{\varepsilon\} = \frac{E}{(1 + v)(1 - 2v)} \begin{bmatrix} 1 - v & v & 0 \\ & 1 - v & 0 \\ \text{symmetrical} & & \dfrac{1 - 2v}{2} \end{bmatrix} \{\varepsilon\} \qquad (3.6b)$$

In the plane strain case, $\sigma_z = v(\sigma_x + \sigma_y)$.

Axisymmetric

$$\{\sigma\} = [C]\{\varepsilon\} = \frac{E}{(1 - v)(1 - 2v)} \begin{bmatrix} 1 - v & v & v & 0 \\ & 1 - v & v & 0 \\ \text{symmetrical} & & 1 - v & 0 \\ & & & \dfrac{1 - 2v}{2} \end{bmatrix} \{\varepsilon\} \qquad (3.6c)$$

where E is the modulus of elasticity and v is the Poisson's ratio.

3.2.1 ELEMENT EQUATIONS

Now, we derive the element equations by minimizing the potential energy function, π_p, with respect to the nodal displacement in $\{q\}$:

$$\pi_p = \int_V \{\varepsilon\}^T [C]\{\varepsilon\}\, dV - \int_V \{u\}^T \{\bar{X}\}\, dV$$
$$- \int_{S_1} \{u\}^T \{\bar{T}\}\, dS \qquad (3.7)$$

where $\{\bar{X}\}$ is the body force (or weight) vector, $\{\bar{T}\}$ is the surface traction or loading vector, and V is the volume of the element (Figures 3.3a and 3.3b, for 3-D and 2-D problems, respectively). Note that Equation 3.7 represents the 3-D case; however, it can be specialized for 2-D problems (plane stress, plane strain, and axisymmetric) by substituting appropriate expressions for $\{u\}$ and $\{\varepsilon\}$, which will be of different orders.

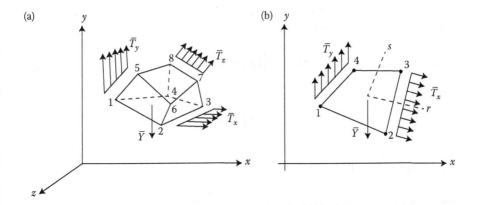

FIGURE 3.3   Body force and surface. (a) 3-D: brick element; (b) 2-D: quadrilateral element.

By substituting the expressions for $\{u\}$ (Equation 3.2a) and for $\{\varepsilon\}$ (Equation 3.3a) into Equation 3.7 and by minimizing π_p with respect to u_i, v_i, w_i ($i = 1, 2, ..., 8$) for the 3-D case, and u_i, v_i ($i = 1, 2, 3, 4$) for the 2-D case, we obtain the element equations as

$$[k]\{q\} = \{Q\} = \{Q_B\} + \{Q_T\} \tag{3.8a}$$

where $[k]$ is the element stiffness matrix, $\{Q_B\}$ is the nodal body force vector, and $\{Q_T\}$ is the nodal surface load vector:

$$[k] = \int_V [B]^T [C] [B] \, dV \tag{3.8b}$$

$$
\begin{aligned}
\{Q\} &= \{Q_B\} + \{Q_T\} \\
&= \int_V [N]^T \{\bar{X}\} \, dV + \int_{S_1} [N]^T \{\bar{T}\} \, dS
\end{aligned} \tag{3.8c}
$$

where S_1 denotes the part of the surface where $\bar{T}$ is applied. The sizes of the matrices will be different for 3-D and 2-D cases, as stated below:

Three-dimensional case (Figure 3.1a)

$[k]$ is a 24×24 stiffness matrix
$\{Q\}$ is a 24×1 nodal force vector

Two-dimensional case (Figure 3.1b)

$[k]$ is an 8×8 stiffness matrix
$\{Q\}$ is an 8×1 nodal force vector

If the thickness for the 2-D element is denoted by h, the stiffness matrix and load vectors due to body force can be expressed as

$$[k] = h \int_A [B]^T [C] [B] \, dA \tag{3.9a}$$

$$\{Q_B\} = h \int_A [N]^T \{\bar{X}\} \tag{3.9b}$$

where A is the element area.

3.2.2 NUMERICAL INTEGRATION

For certain approximation functions, the integration in Equations 3.8b and 3.8c may not be easily obtained in the closed form. Hence, numerical integration is often used. Accordingly, $[k]$ and $\{Q_B\}$ can be evaluated by using Gauss–Legendre schemes [1,3,4,6]:

Three-dimensional

$$[k] = \sum_{i=1}^{N} [B(r_i, s_i, t_i)] [C] [B(r_i, s_i, t_i)] |J(r_i, s_i, t_i)| W_i \tag{3.10a}$$

$$\{Q_B\} = \sum_{i=1}^{N} [N(r_i, s_i, t_i)]^T \{\bar{X}\} |J(r_i, s_i, t_i)| W_i \tag{3.10b}$$

Two-dimensional

$$[k] = h \sum_{i=1}^{N} [B(r_i, s_i)]^T [C][B(r_i, s_i)] |J(r_i, s_i)| W_i \tag{3.11a}$$

$$\{Q_B\} = h \sum_{i=1}^{N} [N(r_i, s_i)]^T \{\bar{X}\} |J(r_i, s_i)| W_i \tag{3.11b}$$

where N denotes the number of integration points, for example, 8 for 3-D and 4 for 2-D elements (Figures 3.4a and 3.4b) and W_i are weights.

3.2.3 ASSEMBLAGE OR GLOBAL EQUATION

Once the equations for a generic element are derived, the equations for all elements are obtained within the idealized domain. We can assemble equations for all elements in a structure by enforcing the condition that the unknowns (displacements) at common node points and sides between elements are compatible, that is, they are the

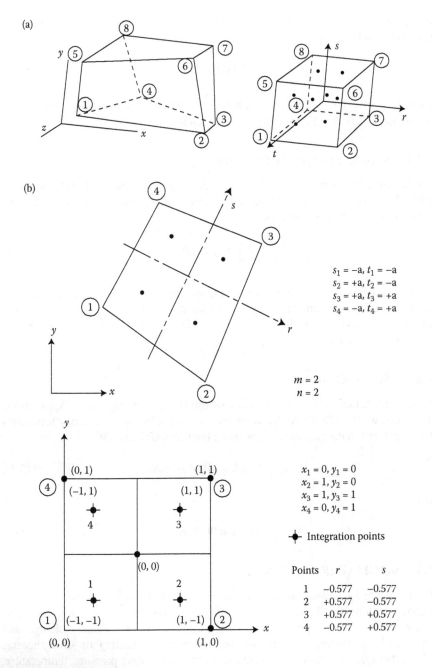

FIGURE 3.4 Numerical integration. (a) 3-D: brick element, projection: 8 Gauss points; (b) 2-D: quadrilateral element, projection: 4 Gauss points.

same. The "direct stiffness method" can be used for such an assembly. The global equations can be written as

$$[K]\{r\} = \{R\}$$ (3.12a)

where $[K]$ is the global stiffness matrix, $\{r\}$ is the global vector of nodal displacements, and $\{R\}$ is the global vector of nodal loads.

3.2.4 SOLUTION OF GLOBAL EQUATIONS

The global stiffness matrix in Equation 3.12a is singular, that is, its determinant is zero. The stiffness matrix, modified for given boundary conditions, will be nonsingular, which can be expressed as

$$\left[\bar{K}\right]\{\bar{r}\} = \{\bar{R}\}$$ (3.12b)

The overbar denotes modified items. For plane problems, the boundary conditions involve (nodal) displacements. However, for bending problems, slopes or gradient of displacement can also constitute boundary conditions; a beam bending case for piles is given in Chapter 2.

3.2.5 SOLVED QUANTITIES

A method like Gaussian elimination can be used for computing nodal displacements $\{\bar{r}\}$ in Equation 3.12b. Once displacements are available, we can derive secondary items such as strains and stresses, by using the previous equations

$$\{\varepsilon\} = [B]\{q\}$$ (3.3a)

and

$$\{\sigma\} = [C][B]\{q\}$$ (3.5)

3.3 NONLINEAR BEHAVIOR

The foregoing formulation uses linear elastic model for isotropic materials. However, geologic materials and interfaces or joints exhibit nonlinear behavior, affected by factors, such as elastic, plastic, and creep strains, stress path, volume change, and microcracking, leading to softening and fracture, and healing or strengthening. Also, other types of loads such as environmental (pore water pressure, temperature, chemicals, etc.) factors can lead to nonlinear behavior.

For linear problems, the total load can be applied and the displacements can be computed in one step (Equation 3.12). However, for nonlinear problems, it is required to apply the total load in several increments, and the stiffness matrix $[k]$ and the constitutive relation matrix $[C]$ for the elements are computed (revised) after each increment, and used for the next increment. Very often, a number of iterations are needed after each increment to converge to the equilibrium state. There are a number of schemes available for the nonlinear analysis. The details of the incremental–iterative methods are given in many publications, for example, Refs. [1–4].

Nonlinear elasticity, conventional plasticity, continuous yield plasticity, HISS plasticity, fracture and damage models, and the DSC can be used to allow for the nonlinear behavior of geologic materials and interfaces and joints. Comprehensive details of these models are given in various publications [1–4,7–13].

Since geologic materials and interfaces are influenced by many of the foregoing factors, it is essential to use realistic and advanced models for solutions of geotechnical problems. Descriptions of a number of constitutive models are given in Appendix 1.

Now, we present a number of practical 2-D problems by using the FEM. Applications for 3-D problems are presented in Chapter 4. We also present some examples based on other methods (stiffness and energy). Before presenting these examples, we first give the details of an important item, *sequential construction*.

3.4 SEQUENTIAL CONSTRUCTION

Engineering structures are usually constructed in operational sequences, which involve changes in stress, deformation, and subsurface water regime. They also cause additional disturbance in the foundation [1,14–18]. Hence, the effects of construction sequences should be taken into account for evaluating the behavior of the structure–foundation systems. Since the behavior of soils (or rocks) is nonlinear, it becomes necessary to take into account the effect of nonlinearity on incremental sequences and response of the geotechnical system. For instance, for computations of the behavior of a geotechnical structure such as building foundations and dams, the use of linear elastic soil model for the total load will not be precise. In other words, if we take into account the effect of construction sequences with the nonlinear behavior, the resulting solution will be more realistic.

Examples of construction sequences are considered to be (1) *in situ* stresses and strains, (2) dewatering of foundation soil, (3) excavation, (4) embankment, (5) installation of support such as anchors, and (6) installation of superstructure. We will now describe each sequence.

In situ condition: The initial state of stress occurs due to geostatic conditions such as overburden and tectonic effects. It is the first state before construction sequences begin. Often, the initial stress vector is defined as $\{\sigma_0\}$. A 2-D (plane strain or plane stress) condition (Figure 3.5a) contains three components σ_x, σ_y, and τ_{xy}. The vertical stress, σ_y, at a point is often computed as

$$\sigma_y = \gamma\, y \qquad\qquad (3.13a)$$

where γ is the (submerged) density of soil and y is the vertical distance to the point from the ground surface. Then, the horizontal stress, σ_x, is computed as

$$\sigma_x = K_0 \sigma_y \qquad\qquad (3.13b)$$

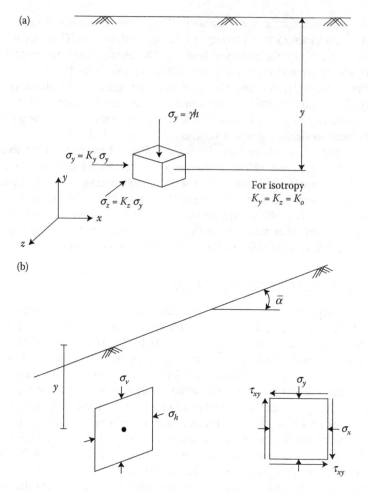

FIGURE 3.5 *In situ* stresses. (a) Horizontal ground surface; (b) inclined ground surface.

where K_o is the coefficient of lateral earth pressure at rest. Its value for an elastic material is found as

$$K_o = \frac{v}{1-v} \tag{3.13c}$$

For general and for the nonlinear behavior, it can be found from laboratory and/or field tests that the value of K_o can be greater than 1 under certain conditions. For a horizontal ground surface (Figure 3.5a), the shear stress, τ_{xy}, is set equal to zero.

For an inclined surface (Figure 3.5b), the state of *in situ* stress can be computed from the following expressions [19]:

$$\sigma_y = \gamma y (1 + K_o \sin^2 \bar{\alpha}) \qquad (3.14a)$$

$$\sigma_x = K_o \sigma_y \qquad (3.14b)$$

$$\tau_{xy} = K_o \gamma y \sin \bar{\alpha} \cos \bar{\alpha} \qquad (3.14c)$$

Equations 3.13a and 3.13b, with $\tau_{xy} = 0$, for horizontal surface, and Equations 3.14a through 3.14c for inclined surface are used to compute initial stresses, which are introduced in the FE procedure.

We can use the FE analysis to evaluate the initial stresses under the gravity load. In such an analysis, the body force or weight of the material (soils or rocks) is applied to evaluate the initial stress vector $\{\sigma_o\} = [\sigma_x, \sigma_y, \tau_{xy}]$. Then, depending on the value of K_o, σ_x is computed using σ_y found from the FE analysis, and τ_{xy} is set equal to zero for horizontal ground surface or computed using Equation 3.14c. Very often, initial strains (deformations) are set equal to zero.

3.4.1 DEWATERING

Excavation is the first sequence for which it is usually necessary to dewater the region of interest for the construction. Let the initial groundwater level be y_o and the water level after dewatering be y_f. The change in pressure Δp_o due to dewatering can be expressed as follows (Figure 3.6):

$$\Delta p_o = \gamma_s (y_f - y_o) \qquad (3.15)$$

where γ_s is the submerged unit weight of soil. We could include the effect of layering with different γ_s in Equation 3.15.

Now, the effect of dewatering on the foundation can be found by converting Δp_o into a force vector, $\{Q_d\}$, for an element:

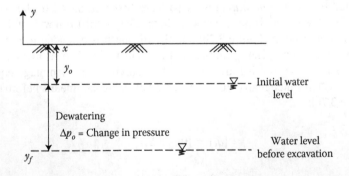

FIGURE 3.6   Schematic of dewatering.

$$\{Q_d\} = \int_V [B]^T \{\Delta p_o\} \, dV \tag{3.16}$$

where V is the volume of the element and $[B]$ is defined before. The force $\{Q_d\}$ for all elements undergoing dewatering is applied in the FE analysis after the initial stress condition. The resulting displacements, strains, and stresses are computed by solving Equation 3.12b:

$$\{q_d\} = \{0\} + \{\Delta q_d\} \tag{3.17a}$$

$$\{\varepsilon_d\} = \{0\} + \{\Delta \varepsilon_d\} \tag{3.17b}$$

$$\{\sigma_d\} = \{\sigma_o\} + \{\Delta \sigma_d\} \tag{3.17c}$$

where $\{q_d\}$, $\{\varepsilon_d\}$, and $\{\sigma_d\}$ are displacements, strains, and stresses due to dewatering.

3.4.2 EMBANKMENT

A mechanistic procedure for embankment and excavation was proposed by Goodman and Brown [14] and has been commonly used in FE analysis [15–17]. The conventional analysis by considering the total forces (in one step) at the end of the construction cannot yield a realistic behavior as found from measurements. Clough and Woodward [18] have identified the effect of sequential embankment for the field behavior of Otter Brook Dam.

3.4.2.1 Simulation of Embankment

Dams, earth banks, and so on are installed almost always in finite-sized layers, or in increments. Figure 3.7 shows the schematic of a mechanistic procedure for simulating embankment. The soil or rock mass is divided into an FE mesh with side and bottom boundaries located at appropriate distances for approximate modeling of the "infinite" domain.

As stated before, the *in situ* stresses $\{\sigma_o\}$ are first introduced in the soil mass (Figure 3.7a); dewatering is not considered here. The first lift is now installed after proper field compaction (Figure 3.7b) and the FE mesh is extended to include the first lift with an appropriate number of FEs. Then, the displacements, strains, and stresses after the first lift are computed by solving the assemblage equations (Equation 3.12) in which matrix $[C]$ is evaluated by using parameters before the first lift (Figure 3.7b):

$$\{q_1\} = \{0\} + \{\Delta q_1\} \tag{3.18a}$$

$$\{\varepsilon_1\} = \{0\} + \{\Delta \varepsilon_1\} \tag{3.18b}$$

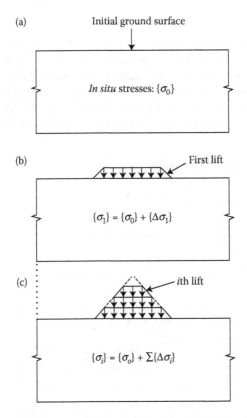

FIGURE 3.7   Simulation of embankment. (a) Initial state; (b) lift 1; and (c) lift *i*.

$$\{\sigma_1\} = \{\sigma_0\} + \{\Delta\sigma_1\} \tag{3.18c}$$

The stress–strain or constitutive relation matrix [C] is modified for the new stresses $\{\sigma_1\}$, and it is used for the next lift. For the *i*th lift (Figure 3.7c), these expressions can be written as

$$\{q_i\} = \Sigma\{\Delta q_i\} \tag{3.19a}$$

$$\{\varepsilon_i\} = \Sigma\{\Delta\varepsilon_i\} \tag{3.19b}$$

$$\{\sigma_i\} = \{\sigma_o\} + \Sigma\{\Delta\sigma_i\} \tag{3.19c}$$

Note that at the end of each lift, the modified constitutive relation matrix [C] is found, and then the stiffness matrix is modified for use in the next lift. Thus, the incremental embankment procedure is based on the nonlinear behavior in which the matrix [C] is modified after each increment. Such a procedure can lead to the prediction of realistic behavior of the problem.

3.4.3 EXCAVATION

Excavations are almost always performed in stages or incrementally. Figure 3.8 shows a schematic of the procedure to simulate excavation in increments. Before excavation, the *in situ* stress in the soil mass $\{\sigma_o\}$ is computed (Figure 3.8a). If dewatering is required, the *in situ* quantities are modified by using Equation 3.17.

After the first lift, the excavated surface (Figure 3.8b) carries no stress or it is *stress free* because it is open to the atmosphere. To create the stress-free surface,

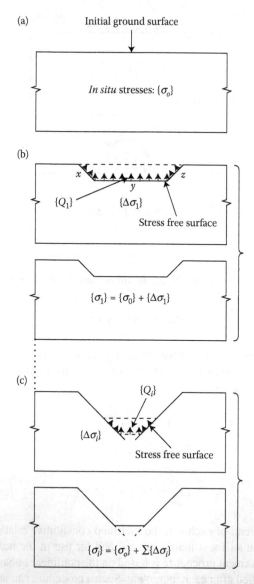

FIGURE 3.8 Simulation of excavation. (a) Initial state; (b) lift 1; and (c) lift *i*.

we apply a load $\{Q_1\}$ which is equal to the (gravity) force at the surface before the excavation lift. It is applied in the opposite direction to the force that existed before the excavation lift. Now, we perform the FE analysis under load $\{Q_1\}$ and compute increments $\{\Delta q_1\}$, $\{\Delta \varepsilon_1\}$, and $\{\Delta \sigma_1\}$, and compute current values as

$$\{q_1\} = \{0\} + \{\Delta q_1\} \tag{3.20a}$$

$$\{\varepsilon_1\} = \{0\} + \{\Delta \varepsilon_1\} \tag{3.20b}$$

$$\{\sigma_1\} = \{\sigma_o\} + \{\Delta \sigma_1\} \tag{3.20c}$$

If dewatering is involved, Equations 3.20 can be modified by using Equations 3.17. Then, we consider lift i (Figure 3.8c) and compute various quantities as

$$\{q_i\} = \Sigma \Delta q_i \tag{3.21a}$$

$$\{\varepsilon_i\} = \Sigma \Delta \varepsilon_i \tag{3.21b}$$

$$\{\sigma_i\} = \{\sigma_o\} + \Sigma \Delta \sigma_i \tag{3.21c}$$

As before, we modify the stress–strain relation matrix, $[C]$, after each lift and use it to compute the modified stiffness matrix for the subsequent lift.

3.4.3.1 Installation of Support Systems

Many times, we need to introduce special supports during construction, which often remain in place after construction. For instance, a system of anchors is installed along the depth of the retaining wall (Figure 3.9). We can simulate the existence of an anchor by applying equal and opposite force in the anchors, at the point at which the anchor crosses the retaining wall. This is done when the excavation reaches (somewhat) below an anchor.

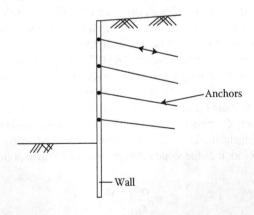

FIGURE 3.9 Anchored wall.

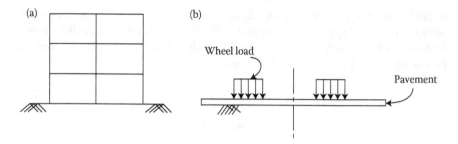

FIGURE 3.10 Superstructure loads to foundation. (a) Building; (b) pavement.

Alternatively, we can provide 1-D elements along the anchors and provide axial stiffness (AE) to anchors. Sometimes, it would be appropriate to use interface elements between the anchor surface and soil to provide for realistic relative motions (e.g., slip) between the soil and anchor.

3.4.3.2 Superstructure

Equivalent loads, arising from superstructures such as building and pavement (Figure 3.10) can be applied to the foundation after the completion of construction.

3.5 EXAMPLES

A number of examples involving 2-D idealizations are presented in this chapter. 3-D problems are included in Chapter 4.

3.5.1 EXAMPLE 3.1: FOOTINGS ON CLAY

The analysis of footing foundations has been one of the initial applications of the FEM for geotechnical problems. In this example, we present comparisons of FE predictions with test data from laboratory models for a circular footing on clays. The clay was classified as the Upper Wilcox of the Calvert Blough formation and was collected from the vicinity of Elgin, Texas; the other clay used was referred to as Taylor Marl I [20,21]. In Test 1, only a single layer of the Wilcox clay was used, while in Test 2, two layers consisting of Upper Wilcox Cay and lower Taylor Marl I were used.

Constitutive model: The clay specimens, 1.4 in (3.6 cm) diameter and 2.8 in (7.1 cm) high, for both clays were tested in a triaxial device under unconsolidated undrained condition and stress control mode. Figure 3.11a shows typical triaxial data for Wilcox Clay and Taylor Marl under different initial confining pressures. For the FE incremental analysis, a number of points $(\sigma_i, \varepsilon_i)$ were input for each stress–strain curve (Figure 3.11b). Then, the tangent modulus, E_t, for a given computed stress (and strain) was computed by interpolation as the slope between two consecutive points as follows:

$$E_t = \left. \frac{\sigma_{i+1} - \sigma_i}{\varepsilon_{i+1} - \varepsilon_i} \right|_{\sigma_3} \tag{3.22}$$

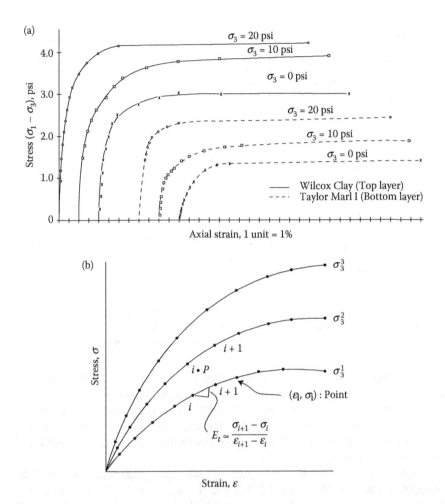

FIGURE 3.11 Stress-strain response and nonlinear elastic simulation by points. (a) Stress-strain curves at various σ_3 : Two layer system; (b) schematic of simulation of stress-strain curves by points.

The tangent modulus can also be interpolated between two curves at different confining pressures (Figure 3.11b). Thus, the constitutive model used is nonlinear elastic or piecewise linear elastic. The Poisson's ratio for the clay was assumed to be a constant equal to about 0.485.

Finite element analysis: The incremental–iterative analysis with axisymmetric idealization was used to solve Equation 3.12. The FE mesh is shown in Figure 3.12. The incremental loadings on the steel footing, 3.0 in (7.6 cm) diameter and 0.50 in (1.27 cm) thick, were applied in two ways: (1) uniform downward pressure and (2) uniform downward displacements to simulate flexible and rigid conditions, respectively. Since the footing is assumed to be rigid, the rigid condition was given the main attention. For the displacement loading, the load (pressure) on the footing was obtained by integrating the computed normal stresses in the elements under the ground surface.

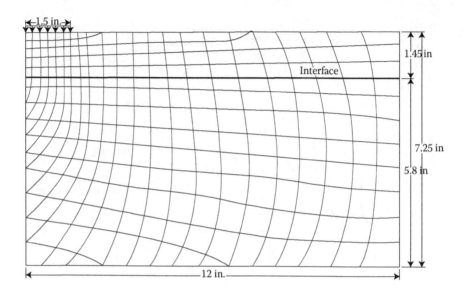

FIGURE 3.12   FE Mesh for two-layer system(1 in = 2.54 cm). (Adapted from Desai, C.S., Solution of Stress-Deformation Problems in Soil and Rock Mechanics Using Finite Element Methods, PhD Dissertation, Dept. of Civil Eng., Univ. of Texas, Austin, TX, 1968.)

The computed load–displacement curves are compared with the laboratory observations for the single-layer and two-layer systems for flexible and rigid conditions: Figures 3.13a and 3.13b and Figures 3.14a and 3.14b, respectively. The correlations between predictions by the FEM and observations are considered to be excellent. It was found that both loading conditions (flexible and rigid) yielded similar correlations [20,21].

Contact pressure distribution: Figure 3.15 shows a typical contact pressure distribution at the junction of the footing and the soil. For the flexible conditions, the pressures are almost uniform for lower applied pressure (Figure 3.15a), while at higher (near ultimate) load, the pressure is nonuniform, concentrated near the edge, and a small part near the edge shows tensile stress condition. For the rigid case, the contact pressures appear to be nonuniform almost from the start of the loading, and the stress concentrates near the edge but without tensile condition (Figure 3.15b).

Ultimate capacity: We consider here only the two-layer system. A critical displacement of 15% of the radius of the footing equal to 0.225 in (0.57 cm) was assumed for ultimate capacity. Accordingly, the ultimate capacity from measurements (Figure 3.14) is about 7.15 psi (49.3 kPa). The ultimate capacity for critical displacement of 0.225 in (0.57 cm) from computed load–displacement curves (Figures 3.14a (flexible) and 3.14b (rigid)) are about 7.15 psi (49.3 kPa) and 6.9 psi (47.5 kPa), respectively. The average of these values is 7.03 psi (48.4 kPa), which compares well with the observed value of 7.15 psi (49.3 kPa).

The ultimate capacities of the two clays from limit equilibrium analysis are computed as follows:

	Wilcox Clay		Taylor Marl	
	psi	(kPa)	psi	(kPa)
Terzaghi (22): $q = 1.3\,c\,N_c \times 2/3$	7.95	(55.0)	4.05	(27.9)
Skempton (23): $q = c\,N_c$	9.92	(68.3)	4.96	(34.0)
Meyerhof (24): $q = c\,N_c$	9.90	(68.2)	4.95	(34.1)

where q is the bearing capacity, $N_c = 5.73$, 6.20, and 6.18 for Terzaghi, Skempton, and Meyerhof methods, respectively [22–24], and cohesion $c = 1.6$ psi (11.0 Pa) and 0.809 psi (5.5 Pa) for Wilcox Clay and Taylor Marl, respectively, found from the test data (Figure 3.11). Here, the angle of friction was neglected. The ultimate bearing

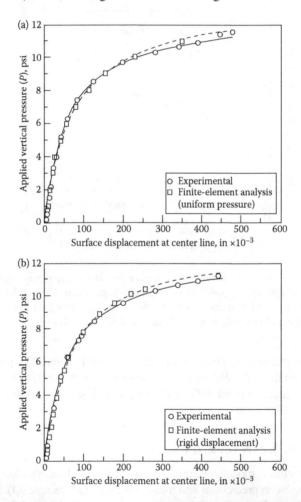

FIGURE 3.13 Comparisons between FE predictions and experimental data: Single layer system (1 in = 2.54 cm, 1 psi = 6.895 kPa). (Adapted from Desai, C.S., Solution of Stress-Deformation Problems in Soil and Rock Mechanics Using Finite Element Methods, PhD Dissertation, Dept. of Civil Eng., Univ. of Texas, Austin, TX, 1968.)

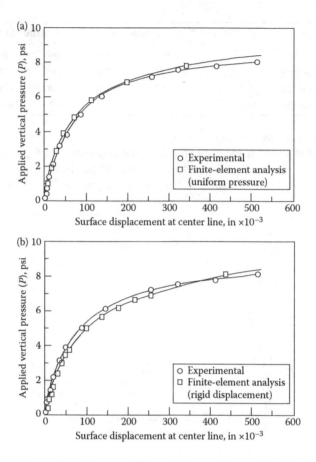

FIGURE 3.14 Comparisons between FE predictions and experimental data: Two-layer system (in = 2.54 cm, 1psi = 6.895 kPa). (Adapted from Desai, C.S., Solution of Stress-Deformation Problems in Soil and Rock Mechanics Using Finite Element Methods, PhD Dissertation, Dept. of Civil Eng., Univ. of Texas, Austin, TX, 1968.)

capacity of the two-layer system for the flexible case is about 7.15 psi (49.3 kPa), which is between those for Wilcox Clay equal to about 9.90 psi (68.2 kPa) and those for Taylor Marl equal to about 5.00 psi (34.5 kPa). This comparison is considered to be reasonable and realistic.

3.5.2 EXAMPLE 3.2: FOOTING ON SAND

The nonlinear elastic model for soft, almost saturated clays, yielded satisfactory comparisons between predictions and measurements (see Example 3.1). However, in the case of cohesionless materials, such a model may not yield satisfactory predictions. In this example, we present comparisons between predictions and test data for a footing on Leighton Buzzard (LB) sand, for which an advanced, plasticity model was warranted [25,26].

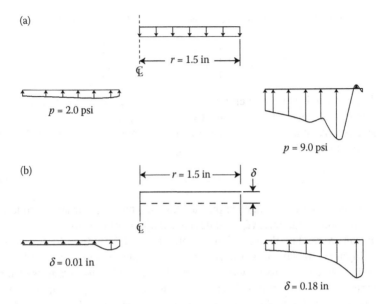

FIGURE 3.15 Predicted pressures distributions below footing: Two-layer system (1 in = 2.54 cm, 1 psi = 6.895 kPa). (a) Uniform pressure; (b) rigid displacement. (Adapted from Desai, C.S., Solution of Stress-Deformation Problems in Soil and Rock Mechanics Using Finite Element Methods, PhD Dissertation, Dept. of Civil Eng., Univ. of Texas, Austin, TX, 1968.)

Constitutive model: The HISS plasticity model (Appendix 1) was used to characterize the LB sand. A series of multiaxial (3-D) tests were performed on $4 \times 4 \times 4$ in ($10 \times 10 \times 10$ cm) specimens of (dry) sand under various confining pressures and stress paths such as conventional triaxial compression (CTC), triaxial extension (TE), and proportional loading (PL). The LB sand was a subrounded, closely graded (U.S. Sieve 20–30) material with a specific gravity of 2.66, and maximum and minimum void ratios equal to 0.81 and 0.53, respectively. The initial dry density of the sand was $\gamma_d = 1.74$ g/cm³, which gives the relative density $D_r = 95\%$.

On the basis of the laboratory multiaxial tests, the parameters in the HISS plasticity model (see Appendix 1 for details) were derived as follows:

Elastic parameters	$E = 11{,}500$ psi (79.2 MPa)
	$v = 0.29$
Plasticity parameters	
Ultimate	$\gamma = 0.1021$, $\beta = 0.362$
Phase change	$n = 2.5$
Hardening/yielding (*)	$\beta_a = 0.0351$
	$\eta_1 = 450.0$
	$\beta_b = 0.0047$
	$\eta_2 = 1.02$
Nonassociative	$\kappa = 0.29$

The following hardening function, α, (*) were adopted for the LB sand:

$$\alpha = \beta_a \exp\left[\eta_1 \xi \left(1 - \frac{\xi_D}{\beta_b + \xi_D \eta_2}\right)\right] \tag{3.23}$$

where β_a, η_1, β_b, and η_2 are hardening parameters, and ξ and ξ_D are total and deviatoric accumulated plastic strains, respectively.

For an isotropic material under the hydrostatic compression (HC) test, ξ_D is zero; hence, Equation 3.23 reduces to the following form:

$$\alpha = \beta_a e^{\eta_1 \xi_v} \tag{3.24}$$

where ξ_v is accumulated volumetric plastic strains. Thus, the hardening function is dependent on both volumetric (ξ_v) and deviatoric (ξ_D) plastic strains.

Laboratory tests: The details of the laboratory model used for footing tests are shown in Figure 3.16. The dimensions of the rectangular box were $54.0 \times 4.0 \times 18.0$ in ($137 \times 10 \times 46$ cm). Glasses with 0.25 in (0.16 cm) thickness were used for the walls of the box. The LB sand with a relative density of about 95%, similar to that of specimen tests using the multiaxial device, was used in the model test. A $4 \times 4 \times 0.75$ ($15.3 \times 10.8 \times 1.2$ cm) steel footing was installed in the box. Three circular cells having 1.5 in (3.8 cm) diameter and 0.25 in (0.64 cm) thickness were embedded in the vicinity of the bottom of the footing, to measure contact pressures [27]. Also, three stress cells were installed at a distance of 2.5 in (6.5 cm) from the center along the center line and at the corner along the diagonal. A stress cell was placed at the bottom of the soil box, about 18.0 in (46.0 cm), from the surface of the sand. A stress cell was installed on the side of the inner wall at about 2.0 in (5.00 cm) below the sand surface.

The displacements of the footing were measured using SRLPSM (Spring Return Linear Position Sensor Modules) at the top of the footing. Displacements at distances of 5.00 in (13.0 cm), 8.0 in (20.0 cm), and 12 in (30.5 cm) from the center of the footing were also measured by using the SRLPSM sensors. A total loading of 25.0 psi (172 kPa) was applied on the footing in increments of 2.50 psi (17 kPa) by using the MTS test frame.

Finite element analysis: Because of the symmetry, only half of the domain was discretized by the FE mesh (Figure 3.17). The mesh contained 58 eight-noded isoparametric elements and 207 nodes. The incremental iterative analysis with the drift correction scheme was used to compute displacements, strains, and stresses by using both HISS associative and nonassociative models.

Figures 3.18a and 3.18b show predicted and measured load–displacement responses at the center of the footing (node No. 15, Figure 3.17) and at 12.0 in (30.50 cm) from the center of the footing (Node 138), respectively. It can be seen from this figure that the nonassociative model provides improved predictions.

Comparisons of predicted and measured data were obtained for normal (vertical) stress under the footing, for both associative and nonassociative models (Figure 3.19a). Similar comparisons for normal (horizontal) stress against the side of the box are shown in Figure 3.19b. The vertical stress predictions by the associative and nonassociative models show similar comparisons with measurement (Figure 3.19a).

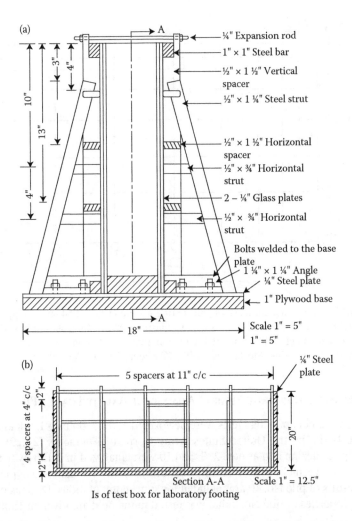

FIGURE 3.16   Test box for laboratory footing (1 in = 2.54 cm). (Adapted from Hashmi, Q.S.E. and Desai, C.S., Nonassociative Plasticity Model for Cohesionless Materials and Its Implementation in Soil-Structure Interaction, Report to National Science Foundation, Dept. of Civil Eng. and Eng. Mechanics, Univ. of Arizona, Tucson, AZ, 1987.)

However, the nonassociative model exhibits improved predictions for the stress (Figure 3.19b).

The above results show that the sand exhibits nonassociative behavior and such models should be used to obtain realistic predictions using the FEM. The load–displacement results indicate that the ultimate region is not reached for the applied total load of 25.0 psi (172 kPa). From the comparisons (Figure 3.18), it can be concluded that the ultimate region with flattening of the load–displacement curve can be reached with the application of loads greater than 25.0 psi (172 kPa). Also, the provision of continuous yielding, and nonassociative character caused by friction in the sand is required through an advanced model such as the HISS plasticity.

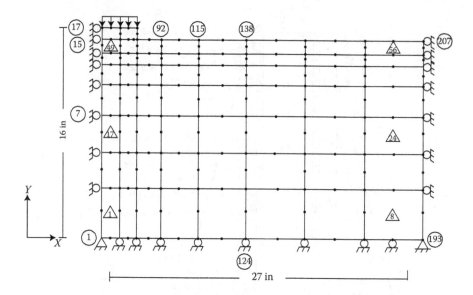

FIGURE 3.17 Finite element mesh of soil-footing system (1 in = 2.54 cm). (Adapted from Hashmi, Q.S.E. and Desai, C.S., Nonassociative Plasticity Model for Cohesionless Materials and Its Implementation in Soil-Structure Interaction, Report to National Science Foundation, Dept. of Civil Eng. and Eng. Mechanics, Univ. of Arizona, Tucson, AZ, 1987.)

3.5.3   EXAMPLE 3.3: FINITE ELEMENT ANALYSIS OF AXIALLY LOADED PILES

Comprehensive field pile load tests were performed at the Arkansas Lock and Dam No. 4 (LD4) site by the United States Army Corps of Engineers [28,29]. Three steel pipe piles designated as nos. 2, 3, and 10 were analyzed in detail by using the FEM [30,31]. We include here mainly the analyses and results for a typical pile (No. 2); results of pile No. 10 are included for some computations. Dimensions and material properties of pile No. 2 that was tested in the field are given in Tables 3.1a and 3.1b.

The pile was assumed to be a solid cylinder for the FE analysis. Hence, equivalent quantities given below were computed by keeping the outer diameter and axial stiffness the same:

Pile: Figure 3.20 shows the pile dimensions with sand layers. It was instrumented with two kinds of strain gages: (1) steel strain rods and (2) electrical resistance strain gages. The pile head movements were measured by three dial gages with an accuracy of 0.001 in (0.025 mm). The pile was driven by using a steam hammer.

Loading: The total load was applied in about 10 increments. Each load increment was maintained for a minimum period of 1 h. The next load increment was applied only after the pile head movement was less than 0.01 in/h (0.25 mm/h). After each load increment, it was released at a rate of 2 tons/min (17,640 N/min) [28,29].

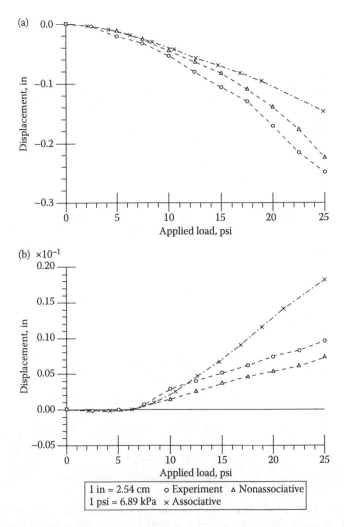

FIGURE 3.18 Comparisons between FE predictions and measurements for displacements. (a) At center of footing (Node 15); (b) at 12 in (30.48 cm) from center of footing (Node 138). (Adapted from Hashmi, Q.S.E. and Desai, C.S., Nonassociative Plasticity Model for Cohesionless Materials and Its Implementation in Soil-Structure Interaction, Report to National Science Foundation, Dept. of Civil Eng. and Eng. Mechanics, Univ. of Arizona, Tucson, AZ, 1987.)

In situ stress: The *in situ* stresses were computed before the FE analysis, based on the coefficient of lateral pressure, $K_0 = 1.17$, which was obtained based on field observations [28].

3.5.3.1 Finite Element Analysis

Axisymmetric idealization, with quadrilateral isoparametric elements, was used in the FE mesh (Figure 3.21). Relatively finer mesh was used in the vicinity of the tip of the pile.

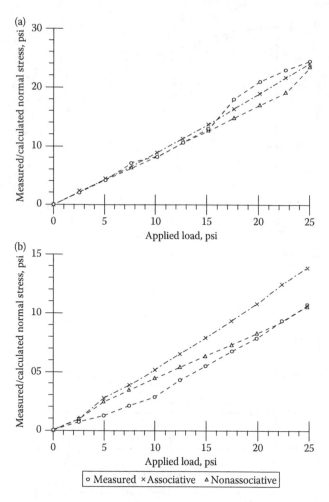

FIGURE 3.19 Comparisons between FE predictions and measurements for normal stress (1 in = 2.54 cm, 1psi = 6.895 kPa). (a) Under footing; (b) against side of box, 2.0 in (5.08 cm) from top. (Adapted from Hashmi, Q.S.E. and Desai, C.S., Nonassociative Plasticity Model for Cohesionless Materials and Its Implementation in Soil-Structure Interaction, Report to National Science Foundation, Dept. of Civil Eng. and Eng. Mechanics, Univ. of Arizona, Tucson, AZ, 1987.)

The nonlinear elastic hyperbolic model was used to characterize the behavior of the sands; the details are given in Appendix 1. Since the required stress–strain data were not available for the sands at the LD4 site, the parameters were adopted from those in similar alluvial sands at other locations such as Jonesville Lock (JL) site [32]. Triaxial tests under various confining pressures and relative densities, $D_r = 60\%$, 80%, and 100%, were conducted for the sands from the JL site [30]. Then, the hyperbolic parameters were derived from the triaxial tests and are given in Table 3.2.

TABLE 3.1a

Properties of Pile No. 2

Diameter (outer)	= 16.0 in (41.0 cm)
Wall thickness	= 0.312 in (0.79 cm)
Cross-sectional area	= 23.86 in² (154 cm²)
Elastic modulus, E	= 29 × 10⁶ psi (200 × 10⁶ kN/m²)
Axial stiffness, AE	= 692 × 10⁶ lbs (3.079 × 10⁶ kN)
Length of pile, L	= 52.8 ft (16.1 m)

TABLE 3.1b

Equivalent Properties of Pile No. 2

Diameter	= 16.0 in (41.0 cm)
Equivalent area	= 201.0 in² (1296.0 cm²)
Equivalent, E	= 3.44 × 10⁶ psi (23.7 kN/m²)
Equivalent density	= 57.0 pci (1.58 kg/cm³)

The load was applied in 12 increments, each increment consisting of 20 tons (178 kN), that is, 14.3 tons/ft² (1370 kN/m²). In the incremental loading, the values of the tangent moduli, E_t, and Poisson's ratio, v_t, were computed by using Equations A1.4 and A1.6 in Appendix 1. The limiting value of v was adopted to be ≤0.495 because for the elastic model, $v \geq 0.5$ is not applicable. If tensile stress is developed during loading, the modulus E was set equal to 10^{-3} tons/ft² (96 N/m²). The parameters for the soil and interface are given in Tables 3.3 and 3.4.

The shear stiffness, k_{st}, of the interface element was evaluated by using Figure A1.27 in Appendix 1. The normal stiffness, k_{nt}, was adopted as a high value, for example, 10^8 tons/ft³ (313 × 10¹¹ N/m³). If the shear stress exceeded the maximum strength given by the Mohr–Coulomb criterion based on the angle of interface friction, δ, the shear stiffness was set equal to a small value of 0.01 tons/ft³ (3130 N/m³). When tensile stress was developed in an interface, both shear and normal stiffnesses were assigned a low value of 10^{-4} tons/ft³ (3130.0 N/m³).

3.5.3.2 Results

As mentioned before, this example includes results for pile No. 2, although other piles, Nos. 3 and 10, were also analyzed [30].

Figure 3.22a shows the computed load–displacement curves for pile No. 2. The distributions of tip loads and wall friction, and the distribution of load along the pile are shown in Figures 3.22b and 3.22c, respectively. The computed results show very good correlation with measurements, particularly in the earlier region.

In Figure 3.22b, the tip load was computed as

$$Q_{tip} = \left[\sigma_y - \sigma_{yo}\right]_{tip} \times A_{eq} = q_p \cdot A_{eq} \qquad (3.25a)$$

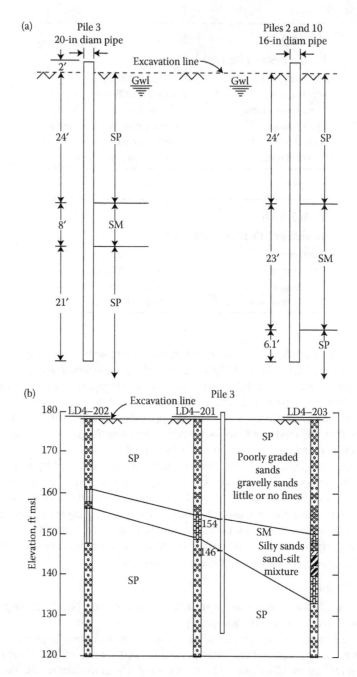

FIGURE 3.20 Details of piles and foundation soils. (a) Dimensions of piles; (b) details of soil layers. (1 in = 2.54 cm, 1ft = 0.305 m). (Adapted from Fruco and Associates, Pile Driving and Load Tests: Lock and Dam No. 4, Arkansas River and Tributaries, Arkansas and Oklahoma, United States Army Engineer District, Corps of Engineers, Little Rock, AR, Sept. 1964.)

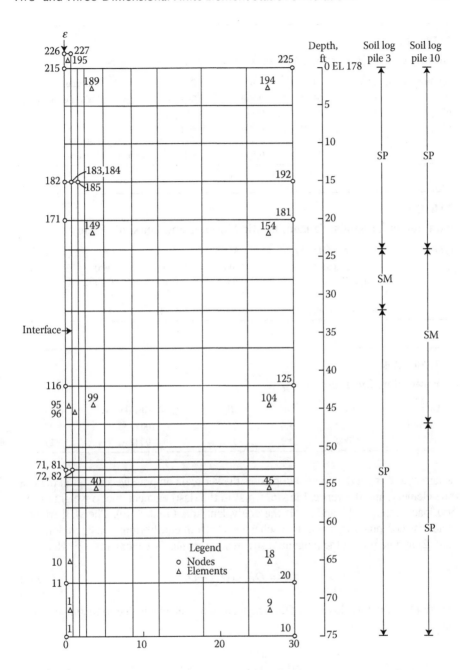

FIGURE 3.21   FE Mesh and details of soils. (Adapted from Desai, C.S., Finite Element Method for Design Analysis of Deep Pile Foundations, Technical Report I, U.S. Army Engineer Waterways Expt. Stn., Vicksburg, MS, 1974.)

TABLE 3.2

Parameters for Hyperbolic Model for Sands at Different Sites

Relative Density (%)	K'	n	R_f	Source
60	909	0.490	0.895	Jonesville Lock, Desai [30,32]
80	1080	0.516	0.844	
100	1530	0.600	0.889	
	1200	0.500	0.800	Kulhawy et al. [33]
	1160	0.500	0.850	Clough and Duncan [34]

TABLE 3.3

Parameters for Sands: Tangent Elastic Modulus and Poisson's Ratio

E_t or v_t	Layer (1)	K'	n	R_f	φ (deg)
E_t	Layers I and III	1500	0.60	0.90	32
	Layer II	1200	0.50	0.80	31
		G	F	D	
v_t	Layers I through III	0.54	0.24	4.0	

TABLE 3.4

Parameters for Interface: k_{st}

Layer	K_i^j	n	R_f	γ_a (lbs/ft³) (kg/cm³)	δ (deg)
I and III	25,000	1.0	0.87	100 (2.8)	25
II	20,000	1.2	0.88	100 (2.8)	27

where σ_y and σ_{yo} at the tip are adopted as the vertical compressive stresses at the last pile element, and the vertical *in situ* stress in the (last) element due to weight (gravity) load, respectively, and A_{eq} is the equivalent area (Table 3.1b). The wall friction load was computed by using two schemes: (1) from equilibrium consideration, Q_w^1, and (2) by integrating the tangential stresses in the interface elements, Q_w^2. Hence

$$Q_w^1 = Q_T - (Q_{\text{tip}} + Q_o) \tag{3.25b}$$

where Q_T is the total load and Q_o is the load in pile elements due to *in situ* stresses, and

$$Q_w^2 = \sum_{i=1}^{M} \sigma_{ti} \cdot \Delta h_i \times \pi D_i \tag{3.25c}$$

where σ_{ti} is the tangential stress in the interface element, Δh_i is the length of an element, D_i is the mean diameter, i denotes an element, and M denotes the total number of interface elements.

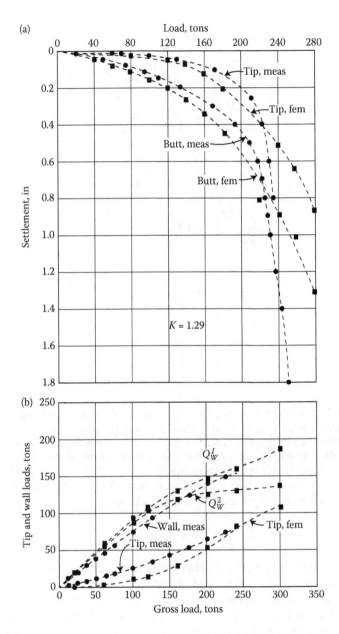

FIGURE 3.22 Comparison of settlements, tip and wall loads and distribution of load in pile for pile No. 2. (a) Load-settlement curves; (b) gross load versus tip and wall friction loads; and (c) distribution of load in pile. (Adapted from Desai, C.S., Finite Element Method for Design Analysis of Deep Pile Foundations, Technical Report I, U.S. Army Engineer Waterways Expt. Stn., Vicksburg, MS, 1974.)

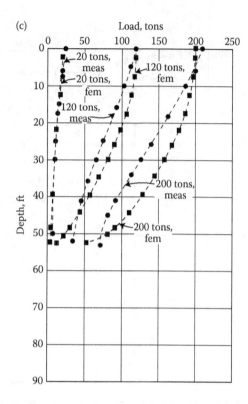

FIGURE 3.22 (continued) Comparison of settlements, tip and wall loads and distribution of load in pile for pile No. 2. (a) Load-settlement curves; (b) gross load versus tip and wall friction loads; and (c) distribution of load in pile. (Adapted from Desai, C.S., Finite Element Method for Design Analysis of Deep Pile Foundations, Technical Report I, U.S. Army Engineer Waterways Expt. Stn., Vicksburg, MS, 1974.)

The load in the pile, Q_{pi} (Figure 3.22b) was computed as

$$Q_{pi} = A_{eqi} \times \sigma_{yi} - Q_{oi} \qquad (3.25d)$$

where σ_{yi} is the vertical stress in the pile, and i denotes an element of the pile.

Figure 3.23 shows the computed distributions of stress along the wall (σ_t) and normal (σ_n) stress in the interfaces for pile No. 10. The distribution of shear stress shows that it is nonlinear; near the tip, the shear stress (Figure 3.23a) experiences a reduction in the stress, indicating stress relief. The distribution of normal stress (Figure 3.23b) shows that it is approximately linear for the major part of the pile length. However, near the tip, it shows significant reduction, and sometimes tensile conditions [30,35]. The reduction in stresses and the stress relief phenomenon can lead to arching effects [36–40].

Design consideration: Desai [30,31] has proposed modifications of conventional formulas for loads (tip and wall friction), based on the FE analysis. Also, the bearing capacity of a pile can be derived from the FE results. For example, it can be obtained

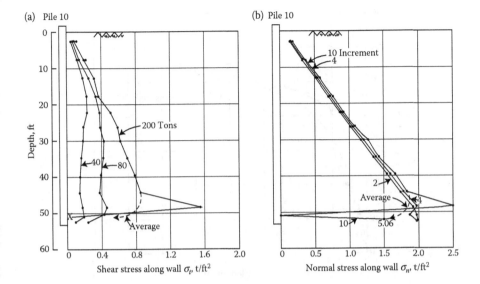

(a) Pile 10

(b) Pile 10

FIGURE 3.23 Computed distributions of shear and normal stresses in interfaces for pile No.10 (1 ft = 0.305 m, 1t/ft² = 95.8 kN/m²). (a) Distribution of σ_t in interface elements; (b) Distribution of σ_n in interface elements. (Adapted from Desai, C.S., Finite Element Method for Design Analysis of Deep Pile Foundations, Technical Report I, U.S. Army Engineer Waterways Expt. Stn., Vicksburg, MS, 1974.)

by adopting a critical displacement [e.g., 0.50 in (1.27 cm)] or at the intersection of tangent to the initial and later part of the load–displacement curve, as shown in Figure 3.24a for pile No. 10.

3.5.4 EXAMPLE 3.4: TWO-DIMENSIONAL ANALYSIS OF PILES USING HRENNIKOFF METHOD

In this example, we use the Hrennikoff method [40] to analyze a pile group. The analysis involves the determination of pile cap response (i.e., horizontal displacement (Δx), vertical displacement (Δy), and rotation ($\Delta \alpha$)) as well as pile forces (i.e., axial force (P), transverse force (Q), and moment (S)). Before presenting a numerical example, we provide a brief description of the method. Only 2-D cases are considered here. 3-D cases are considered in Chapter 4.

The Hrennikoff method is an approximate method, based on stiffness approach, which assumes that the pile cap is rigid and that the load carried by each pile in a group is proportional to the pile head response, namely axial displacement, δ, transverse displacement, δ_t, and rotation, α. With these notations, the equilibrium equations for a pile group, Figure 3.25, can be expressed in the following form:

$$X_x' \, \Delta x' + X_y' \, \Delta y' + M_x' \, \Delta \alpha' + X = 0 \qquad (3.26a)$$

$$X_y' \, \Delta x' + Y_y' \, \Delta y' + M_y' \, \Delta \alpha' + Y = 0 \qquad (3.26b)$$

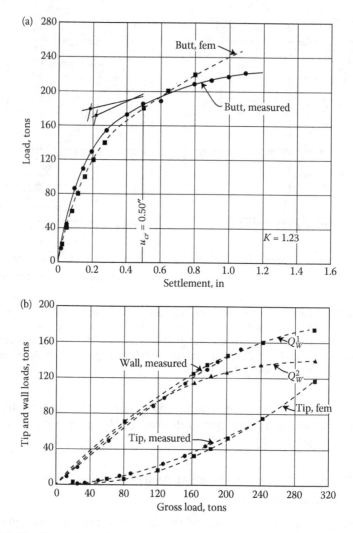

FIGURE 3.24 Comparisons for Load vs. Settlements: Tip and wall loads for pile No. 10 and computation of Bearing Capacity (1 in = 2.54 cm, 1 Ton = 8.896 kN). (a) Load-settlement curve; (b) gross load versus tip and wall friction loads. (Adapted from Desai, C.S., Finite Element Method for Design Analysis of Deep Pile Foundations, Technical Report I, U.S. Army Engineer Waterways Expt. Stn., Vicksburg, MS, 1974.)

$$M_x' \, \Delta x' + M_y' \, \Delta y' + M_\alpha' \, \Delta\alpha' + M = 0 \qquad (3.26c)$$

where $\Delta x' = n\Delta x$, $\Delta y' = n\Delta y$, $\Delta\alpha' = n\Delta\alpha$, and n = pile constant, which is generally evaluated from pile load test. In the above equations, X = horizontal load acting on the pile cap, Y = vertical load acting on the pile cap, and M = moment acting on the pile cap (Figure 3.25). X_x', X_y', Y_y', M_x', M_y', and M_α' are called foundation (or pile cap) constants that are evaluated from the equilibrium of forces and moments at the

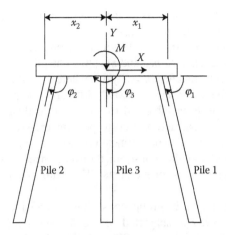

FIGURE 3.25   Two-dimensional pile group: Hrennikoff's Method.

pile head due to constrained displacement (or rotation) applied to the cap [40]. For example, X_x', X_y', and M_x' are obtained from the equilibrium of forces and moment (imparted by the pile to the cap) due to $\Delta x = 1$ unit, $\Delta y = 0$, and $\Delta \alpha = 0$. Likewise, M_x', M_y', and M_α' can be obtained from the equilibrium of forces and moment (imparted by the pile to the cap) at the pile head due to $\Delta \alpha = 1$ unit, $\Delta x = 0$, and $\Delta y = 0$. As shown by Hrennikoff [40], the pile constants can be expressed in terms of pile location (x), pile inclination (φ), and pile constants (n, t_δ, m_δ, and m_α) as follows:

$$X_x' = (X_x/n) = -[\cos^2 \varphi + r_1 \sin^2 \varphi] \tag{3.27a}$$

$$X_y' = (X_y/n) = -(1/2)\,(1 - r_1)\sin 2\varphi \tag{3.27b}$$

$$M_x' = (M_x/n) = -(1/2)\,(1 - r_1)\,x \sin 2\varphi + r_2 \sin \varphi \tag{3.27c}$$

$$Y_y' = (Y_y/n) = -[\sin^2 \varphi + r_1 \cos^2 \varphi] \tag{3.27d}$$

$$M_y' = (M_y/n) = -[(\sin^2 \varphi + r_1 \cos^2 \varphi)\,x + r_2 \cos \varphi] \tag{3.27e}$$

$$M_\alpha' = (M_\alpha/n) = -[(\sin^2 \varphi + r_1 \cos^2 \varphi)\,x^2 + 2r_2 \times \cos \varphi + r_3] \tag{3.27f}$$

where $r_1 = (t_\delta/n)$, $r_2 = (m_\delta/n)$, and $r_3 = (m_\alpha/n)$. Note that the total foundation constants are obtained by adding the contribution of each pile in the group (Figure 3.25).

Pile forces: Knowing pile cap response ($\Delta x'$, $\Delta y'$, and $\Delta \alpha'$), pile forces can be calculated from the following equations:

$$P = \Delta x' \cos \varphi + \Delta y' \sin \varphi + \Delta \alpha'\,x \sin \varphi \tag{3.28a}$$

$$Q = -r_1\,[\Delta x' \sin \varphi - \Delta y' \cos \varphi - \Delta \alpha'\,x \cos \varphi] + r_2\,\Delta \alpha' \tag{3.28b}$$

$$S = r_2\,[\Delta x' \sin \varphi - \Delta y' \cos \varphi - \Delta \alpha'\,x \cos \varphi] - r_3\,\Delta \alpha' \tag{3.28c}$$

where P = axial pile force, Q = transverse pile force, and S = pile head moment. It should be noted that for hinged piles, $m_\delta = 0$.

Pile constants: The solution of pile groups using this method is essentially based on the pile constants, n, t_δ, m_δ, and m_α [40]. The pile constant, n, can be estimated from the axial pile load test, $n = (P_a/\delta_{la})$, where P_a = allowable axial load and δ_{la} = corresponding pile displacement (axial). Similarly, the pile constant t_δ can be estimated from pile load tests subjected to transverse loading, $t_\delta = (Q_a/\delta_{ta})$, where Q_a = allowable transverse load and δ_{ta} = corresponding transverse displacement. Pile constants t_δ, m_δ, and m_α can also be estimated by considering the pile as a beam on Winkler foundation and applying constrained displacement (or rotation) and calculating the corresponding forces (or moments). The details of the pile constants can be found in Ref. [40].

Numerical example: A pile group consisting of two inclined piles and one vertical pile (Figure 3.25) is subjected to the following loads: horizontal load $X = 5$ kip (22.2 kN), vertical load $Y = 200$ kip (890 kN), and moment $M = 3,000$ kip-in (340 kN-m). The pile constants are $n = 100$ kip/in (175 kN/cm), $t_\delta = 100$ kip/in (175 kN/cm), $m_\delta = 1000$ kip-in/in (445 kN-m/cm), and $m_\alpha = 1000$ kip-in/rad (113 kN-m/rad). Determine the pile cap response and the pile forces.

Solution: From the pile constants given, we have $r_1 = t_\delta/n = 100/100 = 1$, $r_2 = m_\delta/n = 1000/100 = 10$, and $r_3 = m_\delta/n = 1000/100 = 10$. For pile 1: $x_1 = 60$ in, $\varphi_1 = 60°$; for pile 2: $x_2 = -60$ in and $\varphi_2 = 120°$; for pile 3: $x_3 = 0$ in and $\varphi_3 = 90°$. The foundation constants can now be calculated for each pile from Equations 3.27a through 3.27f. The total foundation constants $(X_x', X_y', Y_y', M_x', M_y',$ and $M_\alpha')$ can then be obtained by adding the contribution of each pile. The resulting equilibrium equations for this case can be expressed as follows:

$$-3\,\Delta x' + 0\,\Delta y' + 27.3\,\Delta\alpha' + 5 = 0 \qquad (3.29a)$$

$$0\,\Delta x' - 3\,\Delta y' + 0\,\Delta\alpha' + 200 = 0 \qquad (3.29b)$$

$$27.32\,\Delta x' + 0\,\Delta y' - 8430\,\Delta\alpha' + 3000 = 0 \qquad (3.29c)$$

Solving these equations simultaneously, we have $\Delta x' = 5.06$, $\Delta y' = 66.67$, and $\Delta\alpha' = 0.37$. The corresponding $\Delta x = 5.06/100 = 0.005$ in (0.0127 cm), $\Delta y = 66.67/100 = 0.66$ in (1.68 cm), $\Delta\alpha = 0.37/100 = 0.0037$ rad. Knowing the pile

TABLE 3.5
Pile Head Forces and Moments

Pile Number	Pile 1	Pile 2	Pile 3
Axial force, P (Kip)	79.6	35.8	66.6
Transverse force, Q (Kip)	43.8	−22.8	−1.3
Pile head moment, S (kip-in)	−404.9	261.7	46.8

Note: 1 kip = 4448 N; 1 kip-in = 113 N-m.

cap response, the pile head forces (P and Q) and moment (S) can be calculated from Equations 3.28a, 3.28b, and 3.28c. A summary of the pile forces and moments is given in Table 3.5.

3.5.5   Example 3.5: Model Retaining Wall—Active Earth Pressure

Figure 3.26 shows a model retaining wall, which retains a mixture of aluminum rods having two different diameters (1.6 and 3 mm), but an uniform length (5.0 cm) [42,43]. The retaining wall has a height of 48 cm. The average density of the compacted backfill rods was 2.21 gf/cm³ (0.0217 N/cm³). Tests were conducted to measure active pressure on the wall by rotating the wall about a hinge at the bottom (Figure 3.26). The analysis of this problem may not be possible by using conventional Rankine's or Coulomb's theories, and a computer (FE) method was warranted.

The coefficient of horizontal earth pressure at rest, K_o, was found to be about 0.98, which was obtained by assuming hydrostatic pressure distribution. The changes in the vertical and horizontal coordinates with respect to the initial locations of the rods were measured. The final settlement of the backfill surfaces and maximum shear strains in the backfill were computed from these measurements.

Testing and constitutive model: A number of biaxial tests were conducted on these aluminum rods to find the constitutive model parameters. Figure 3.27 shows a cross-sectional view of the test setup containing biaxial specimens, $15 \times 15 \times 5$ cm. A vertical stress of magnitude 527.4 gf/cm² (5.17 N/cm²) was applied at the top. Figure 3.28 shows typical results in terms of stress ratio σ_1/σ_2 and volumetric strain ε_v versus maximum shear strain γ_m. A plot of the maximum or peak σ_1/σ_2 versus σ_2 yielded the angle of friction, φ, of the retained material to be about 29.3°.

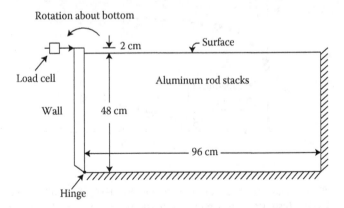

FIGURE 3.26   Model retaining wall with aluminum rod backfill. (From Ugai, K. and Desai, C.S., Application of Hierarchical Plasticity Model for Prediction of Active Earth Pressure Tests, Report, Dept. of Civil Eng. and Eng. Mechanics, Univ. of Arizona, Tucson, AZ, USA, 1994. With Permission.)

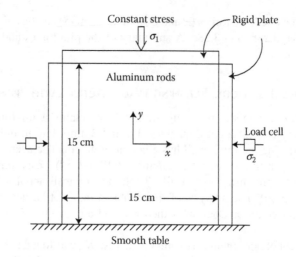

FIGURE 3.27 Bi-axial test for determination of constitutive parameters. (From Ugai, K. and Desai, C.S., Application of Hierarchical Plasticity Model for Prediction of Active Earth Pressure Tests, Report, Dept. of Civil Eng. and Eng. Mechanics, Univ. of Arizona, Tucson, AZ, USA, 1994. With Permission.)

The HISS-δ_1 plasticity model [26,44], which includes the nonassociated flow rule, was used to characterize the behavior of backfill metal rods. The details of the HISS-δ_1 model are given in Appendix 1. The elastic moduli E and v were determined from initial portions of the relations in Figure 3.28. The parameters for the HISS-δ_1 model obtained from these tests are given below:

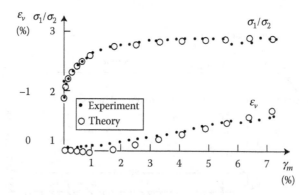

FIGURE 3.28 Comparison between predictions by HISS-δ_1 model and measurements from Biaxial Tests (γ_m = maximum shearstrain). (From Ugai, K. and Desai, C.S., Application of Hierarchical Plasticity Model for Prediction of Active Earth Pressure Tests, Report, Dept. of Civil Eng. and Eng. Mechanics, Univ. of Arizona, Tucson, AZ, USA, 1994. With Permission.)

Elastic:	
E	= 59,400 gf/cm² (583 N/cm²)
v	= 0.28
Plasticity:	
γ	= 0.02835
β	= 0.4923
n	= 6.80
a_1	= 0.0001685
η_1	= 0.7758
κ	= 0.629

The HISS-δ_1 model was validated with respect to the biaxial tests. The back predictions were obtained by integrating the incremental Equation A1.38 in Appendix 1. Typical comparisons between predictions and test data are shown in Figure 3.28, which are considered to be highly satisfactory.

3.5.5.1 Finite Element Analysis

The previous version (SSTIN/SEQ-2DFE) of the FE code (DSC-SST2D) [44] was used to calculate earth pressures and the behavior of the backfill. The FE mesh with boundary conditions is shown in Figure 3.29. The base of the backfill was assumed to be fixed. The smooth side boundary was placed at a distance of 60 cm from the wall. Figure 3.30 shows that the computed and observed displacements compare well with the measurements. In Figure 3.30, δ denotes displacements at the top of the wall (see also Figure 3.32).

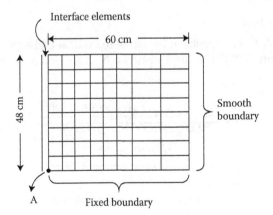

Point A: Free to move in horizontal direction

FIGURE 3.29 Finite element mesh for model retaining wall. (From Ugai, K. and Desai, C.S., Application of Hierarchical Plasticity Model for Prediction of Active Earth Pressure Tests, Report, Dept. of Civil Eng. and Eng. Mechanics, Univ. of Arizona, Tucson, AZ, USA, 1994. With Permission.)

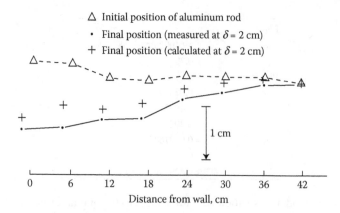

FIGURE 3.30 Settlements of backfill surface. (From Ugai, K. and Desai, C.S., Application of Hierarchical Plasticity Model for Prediction of Active Earth Pressure Tests, Report, Dept. of Civil Eng. and Eng. Mechanics, Univ. of Arizona, Tucson, AZ, USA, 1994. With Permission.)

The FE mesh contained 64, eight-noded isoparametric elements and 225 nodal points. Six-noded interface element was used to model friction between the wall and backfill. It was modeled by using the *thin-layer element* [45] with elastic–plastic Mohr–Coulomb criterion. Special tests, depicted in Figure 3.31, were performed for the behavior of interfaces. A horizontal force, Q, was applied to the interface formed by the specimen of the wall and aluminum rods, under a constant vertical load, P. The average value of interface friction angle was found to be about 15°, and the adhesion was equal to zero. The thickness of the interface elements was adopted as 0.80 mm with Young's modulus, $E = 10,000$ gf/cm² (98.1 N/cm²).

3.5.5.2 Validations

A horizontal load, F, was applied at the top of the wall at a distance of 50 cm from the bottom (Figure 3.26). The displacements, δ, at a distance of 48 cm from the

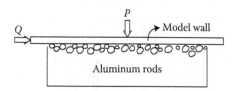

P : Vertical force
Q : Horizontal force at sliding of model wall
$\tan\phi_w = Q/P$
where ϕ_w = friction angle between
aluminum rods and model wall

FIGURE 3.31 Test set-up for friction angle between wall and aluminum rods. (From Ugai, K. and Desai, C.S., Application of Hierarchical Plasticity Model for Prediction of Active Earth Pressure Tests, Report, Dept. of Civil Eng. and Eng. Mechanics, Univ. of Arizona, Tucson, AZ, USA, 1994. With Permission.)

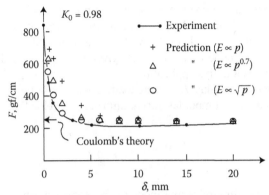

(F: Horizontal active earth force measured at the
point 50 cm high, δ: Displacement of top of wall,
K_0: Coefficient of horizontal earth pressure
at rest, E: Young's modulus, $p = (\sigma_1 + \sigma_2)/2$ at
intial stage)
(1 gf = 0.00981 N)

FIGURE 3.32   Predictions and test data for active force and wall displacements. (From Ugai, K. and Desai, C.S., Application of Hierarchical Plasticity Model for Prediction of Active Earth Pressure Tests, Report, Dept. of Civil Eng. and Eng. Mechanics, Univ. of Arizona, Tucson, AZ, USA, 1994. With Permission.)

bottom were measured. The measured F versus δ relation is shown in Figure 3.32. Also shown in the figure are computed results for three relations assumed between E and p (confining stress):

$$E = k_1 p \tag{3.30a}$$

$$E = k_2 p^{0.7} \tag{3.30b}$$

$$E = k_3 \sqrt{p} \tag{3.30c}$$

The relation (Equation 3.30c) yields the best computations compared to the other two. The force from the Coulomb active earth pressure theory [22] was about 243 gf/cm (2.39 N/cm), which compares very well with the predicted force indicated in Figure 3.32.

We may conclude that the FEM with the HISS plasticity model can provide highly satisfactory predictions for the behavior of retaining walls, such as considered herein.

3.5.6   EXAMPLE 3.6: GRAVITY RETAINING WALL

Clough and Duncan [46] performed a detailed FE analysis for earth pressures on gravity retaining walls. We present here some of the results pertaining to active and passive pressures.

Figure 3.33a shows a section of the concrete retaining wall with sand backfill. The nonlinear elastic hyperbolic model was used to characterize both the backfill and interfaces between the inner side of the wall and backfill, and backfill and foundation. The details of the hyperbolic model are given in Appendix 1. Here, we give a brief description related to this application.

The tangent elastic modulus, E_t, during loading was the same as Equation A1.4b in Appendix 1. The tangent modulus during unloading or reloading is given by

$$E_t = K_{ur} p_a \left(\frac{\sigma_3}{p_a}\right)^n \tag{3.31}$$

where K_{ur} is the unloading/reloading modulus, p_a is the atmospheric pressure constant, and n is a material parameter. The behavior of the backfill was expressed in terms of modified parameters:

Deviatoric modulus, M_D:

$$M_D = \frac{1}{2} \frac{E_t}{(1+v)} \tag{3.32a}$$

and bulk modulus, M_B:

$$M_B = \frac{1}{2} \frac{E}{(1+v)(1-2v)} \tag{3.32b}$$

(a)

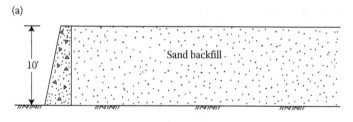

(b)

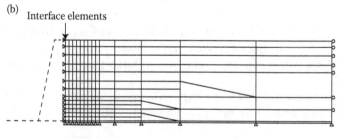

FIGURE 3.33 Gravity wall and FE mesh. (a) Retaining wall and blackfill; (b) finite element mesh. (From Clough, G.W. and Duncan, J.M., *Journal of Soil Mechanics and Foundations Divisions, ASCE*, 97(SM12), Dec. 1971, 1657–1673; Clough, G.W. and Duncan, J.M., Finite Element Analyses of Port Allen and Old River Locks, Report S-69-6, U.S. Army Engineers Waterways Experiment Station, Corps of Engineers, Vicksburg, MS, Sept. 1969. With Permission.)

Here, $E = E_t =$ tangent modulus during loading for a given increment i and $E_{ur} =$ unloading/reloading modulus.

3.5.6.1 Interface Behavior

Descriptions for interface behavior are given in Appendix 1. Here, we present brief descriptions of the interface constitutive relations for the 2-D condition:

$$\begin{Bmatrix} \Delta\tau \\ \Delta\sigma_n \end{Bmatrix} = \begin{bmatrix} k_{st} & 0 \\ 0 & k_n \end{bmatrix} \begin{Bmatrix} \Delta u_r \\ \Delta v_r \end{Bmatrix} \tag{3.33}$$

where τ and σ_n are interface shear and normal stresses, respectively, k_{st} is the tangent shear stiffness, k_n is the normal stiffness, u_r and v_r are relative shear and normal displacements, respectively, and Δ denotes increment. The expression for k_{st} in the hyperbolic form, in the incremental analysis is given by

$$k_{st} = K^j \gamma_w \left(\frac{\sigma_n}{p_a}\right)^n \left(1 - \frac{R_f \tau}{c_a + \sigma_n \tan\delta}\right)^2 \tag{3.34}$$

where K^j is a dimensionless stiffness parameter, n and R_f are found from interface (direct) shear tests [31,46], c_a is the adhesion of the interface, δ is the interface friction angle, and γ_w is the unit weight of water.

In the FE analysis here, the normal stiffness was arbitrarily chosen as a high value (e.g., 10^9 pcf (1.57×10^8 kN/m³)) during compressive state to avoid any overlap at the interface. After shear failure and if the interface was in compression, k_{st} was reduced to a small value but k_n was still kept large. For computed tensile state or a gap (separation) between structural material and soil, both shear and normal stiffnesses were assigned very small values (e.g., 10^{-1} pcf or 17.7 N/m³).

3.5.6.2 Earth Pressure System

Figure 3.33a shows a gravity retaining wall and a backfill system. The FE mesh is shown in Figure 3.33b. Interface elements were provided between the backfill and both wall and rigid base. The initial stress condition, $\{\sigma_o\}$, was established on the basis of the at-rest condition, that is, using K_o, the coefficient of at-rest state, as shown in Table 3.6. Then, the wall was moved away from the backfill or pushed toward the backfill until active or passive earth pressure state was reached, respectively. Table 3.6 shows the parameters for the backfill and interface.

Three analyses were performed by varying the parameters for interfaces between wall and backfill, in particular, the interface roughnesses. They are given below.

Figures 3.34a through 3.34c show results in terms of horizontal (active) wall pressures versus depth for the three different analyses, for which the wall was rotated away from the backfill, given by Δ/H, where H is the height of the wall and Δ is the top movement shown in Figure 3.34. It can be seen that the pressure distributions are nonlinear for the three cases (roughnesses). The active pressure is first reached at the top of the wall. When the outward motion, Δ, becomes equal to 0.0023 H, the active

TABLE 3.6
Parameters for Backfill Materials and Interface in Earth Pressure System

Material	Parameter	Value
Medium-dense backfill	Unit weight	$\gamma = 100$ lb/ft³ (1600 kg/m³)
	Coefficient of at rest earth pressure	$K_o = 0.43$
	Cohesion	$c = 0$ psi (0 kN/m²)
	Angle of friction	$\varphi = 35°$
	Loading modulus	$K = 720$
	Unloading/reloading modulus	$K_{ur} = 900$
	Exponent in modulus	$n = 0.50$
	Failure ratio	$R_f = 0.80$
	Poisson's ratio	$v = 0.30$
Interface: Wall and backfill	Angle of friction	$\delta =$ Variable[a]
	Stiffness number	$K^j =$ Variable
	Exponent	$n =$ Variable
	Failure ratio	$R_f =$ Variable
Interface: Base and backfill	Angle of friction	$\delta = 24°$
	Stiffness number	$K_j = 75,000$
	Exponent	$n = 0.50$
	Failure ratio	$R_f = 0.90$

[a] See description below
1. Smooth interface, that is, $\delta \approx 0$, $K^j = 1.0$, $n = 0.0$, and $R_f = 1.0$.
2. Angle of friction for wall and backfill $= 0.67\ \varphi$ ($\delta = 24°$), $K^j = 40,000$, $n = 1.0$, and $R_f = 0.90$.
3. Angle of friction for wall and backfill $= \varphi = \delta = 35°$, $K^j = 75,000$, $n = 1.0$, and $R_f = 0.9$.

pressure state was reached for almost the entire height of the backfill. Terzaghi [47] reported that the active pressure state for a rough wall was reached at $\Delta/H = 0.0014$ for dense sand, and for $\Delta/H = 0.0084$ for loose sand. Thus, for medium dense sand and rough wall, the value of 0.0014 agrees well with that shown in Figure 3.34c.

Figure 3.35 shows computed relations between horizontal earth pressure force versus movements away (active pressure) and toward (passive pressure) the backfill for smooth wall. It shows the initiation of active and passive pressures and comparisons with classical earth pressure results. Further details are given by Clough and Duncan [48], who show that the FE predictions yield satisfactory results and they compare well with classical solutions. It may be noted that the behavior compared involves specific and limiting conditions regarding earth pressures; however, the classical solution for deformation behavior may not provide such agreement with predictions from the FE analysis with the nonlinear soil model.

3.5.7 EXAMPLE 3.7: U-FRAME, PORT ALLEN LOCK

Duncan and Clough [48,49] presented FE analyses of the U-frame Port Allen lock, located on the right bank of the Mississippi river opposite to Baton Rouge, Louisiana.

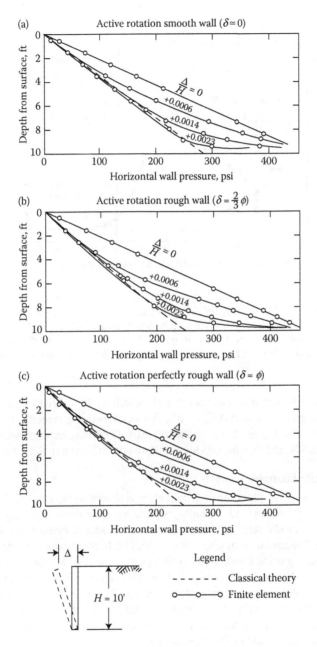

FIGURE 3.34 Distributions of horizontal wall pressures for smooth to rough wall; with displacements, Δ/H. (From Clough, G.W. and Duncan, J.M., *Journal of Soil Mechanics and Foundations Divisions, ASCE*, 97(SM12), Dec. 1971, 1657–1673; Clough, G.W. and Duncan, J.M., Finite Element Analyses of Port Allen and Old River Locks, Report S-69-6, U.S. Army Engineers Waterways Experiment Station, Corps of Engineers, Vicksburg, MS, Sept. 1969. With permission.)

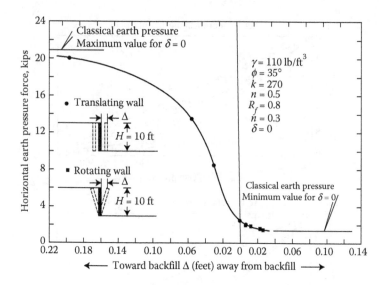

FIGURE 3.35 FE predictions for earth pressure force with wall movements. (From Clough, G.W. and Duncan, J.M., *Journal of Soil Mechanics and Foundations Divisions, ASCE*, 97(SM12), Dec. 1971, 1657–1673; Clough, G.W. and Duncan, J.M., Finite Element Analyses of Port Allen and Old River Locks, Report S-69-6, U.S. Army Engineers Waterways Experiment Station, Corps of Engineers, Vicksburg, MS, Sept. 1969. With permission.)

Figure 3.36 shows the reinforced concrete lock, which is 84 ft (25.6 m) wide and 1200 ft (366 m) long capable of a 50 ft (15.2 m) lift. The lock was instrumented, which provided measurements of wall deflections, strains at various locations, earth pressures, heaves, piezometric heads, and thermal conditions; the details are given in Refs. [48,50].

3.5.7.1 Finite Element Analysis

The FE mesh with the plane strain idealization is shown in Figure 3.37. Incremental FE analyses were performed, including simulation of various states and construction sequences: *in situ* state, dewatering, excavation, and placement of concrete and backfill [49]. The initial or *in situ* soil stresses and fluid pressures at the beginning were computed by using coefficient of earth pressure at rest (K_o); unit weight of soils and piezometric levels shown in Figure 3.38. The values of K_o were estimated from the plasticity index and overconsolidation ratios. The values of overconsolidation ratios obtained were equal to 1.0 for silty soil and 1.50 for the overlying clays, based on various consolidation tests. The initial groundwater table was determined from piezometric data obtained before dewatering.

The excavation profile is shown in Figure 3.39. It was simulated in three steps up to the full depth of 57 ft (17.40 m). The changes in water pressure due to dewatering were included during the second and third steps of excavation simulation. The backfill and re-establishment of regular groundwater states were analyzed during various increments, the final incremental step involved filling of lock with water (Figure 3.40) [48,49].

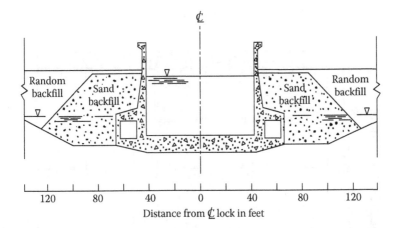

FIGURE 3.36 Cross-section of port allen lock. (Adapted from Duncan, J.M. and Clough, G.W., *Journal of the Soil Mechanics and Foundations Division, ASCE*, 97(SM8), August 1971, 1053–1068; Clough, G.W. and Duncan, J.M., Finite Element Analyses of Port Allen and Old River Locks, Report S-69-6, U.S. Army Engineers Waterways Experiment Station, Corps of Engineers, Vicksburg, MS, Sept. 1969.)

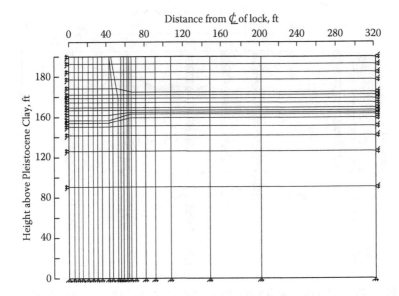

FIGURE 3.37 FE mesh for port allen lock for gravity turn on analysis. (Adapted from Clough, G.W. and Duncan, J.M., Finite Element Analyses of Port Allen and Old River Locks, Report S-69-6, U.S. Army Engineers Waterways Experiment Station, Corps of Engineers, Vicksburg, MS, Sept. 1969.)

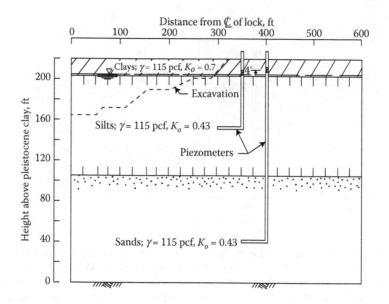

FIGURE 3.38 Initial soil details for port allen lock site. (Adapted from Duncan, J.M. and Clough, G.W., *Journal of the Soil Mechanics and Foundations Division, ASCE*, 97(SM8), August 1971, 1053–1068; Clough, G.W. and Duncan, J.M., Finite Element Analyses of Port Allen and Old River Locks, Report S-69-6, U.S. Army Engineers Waterways Experiment Station, Corps of Engineers, Vicksburg, MS, Sept. 1969.)

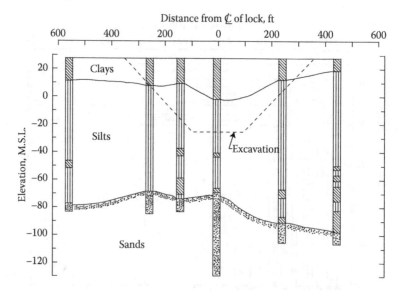

FIGURE 3.39 Soil cross-section and boring logs at port allen lock site. (Adapted from Duncan, J.M. and Clough, G.W., *Journal of the Soil Mechanics and Foundations Division, ASCE*, 97(SM8), August 1971, 1053–1068; Clough, G.W. and Duncan, J.M., Finite Element Analyses of Port Allen and Old River Locks, Report S-69-6, U.S. Army Engineers Waterways Experiment Station, Corps of Engineers, Vicksburg, MS, Sept. 1969.)

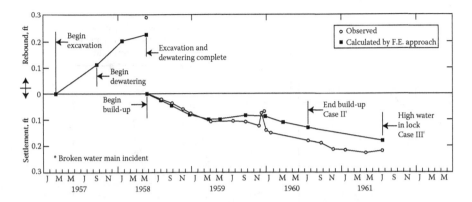

FIGURE 3.40   Predicted and measured settlements and rebound with time. (From Duncan, J.M. and Clough, G.W., *Journal of the Soil Mechanics and Foundations Division, ASCE,* 97(SM8), August 1971, 1053–1068; Clough, G.W. and Duncan, J.M., Finite Element Analyses of Port Allen and Old River Locks, Report S-69-6, U.S. Army Engineers Waterways Experiment Station, Corps of Engineers, Vicksburg, MS, Sept. 1969. With permission.)

3.5.7.2   Material Modeling

The nonlinear elastic simulation of the backfill material (soils) and interfaces between the concrete structure and the backfill was obtained by representing the stress–strain and shear–stress relative displacement curves, by using hyperbolic functions [48,49]. The Poisson's ratio was adopted as 0.30. Appendix 1 gives the details of the hyperbolic models.

The behavior of concrete was assumed to be linear elastic with the elastic modulus of 3×10^6 psi (21×10^6 kN/m^2). The Poisson's ratio was adopted as 0.20.

3.5.7.3   Results

Figure 3.40 shows the variations of computed and observed rebound and settlements with time, at the center line of the lock at the end of construction before the lock was filled (case or stage II′). Figure 3.41 shows comparisons with predictions and observations for deflections of the structure at the end of stage III; the cases or stages are marked in Figure 3.40. Figure 3.42 shows the variation of effective earth pressures with time at the center line of the base slab and the point on the upper lock wall just above the culvert. The observed and computed effective earth pressures for case II′ (Figure 3.40) on the base and lock walls are shown in Figure 3.43. Overall, the FE predictions show good agreement with the field observations.

3.5.8   EXAMPLE 3.8: COLUMBIA LOCK AND PILE FOUNDATIONS

The Columbia Lock, located in a cutoff between miles 131.5 and 134.5 on the Ouachita river near Columbia, Louisiana, was designed as a gravity-type structure; here, the load is transferred mainly through foundation piles [51,52]. The design was performed using the Hrennikoff method [40]. The field data during and after construction were not in agreement with the distribution of loads in the pile groups

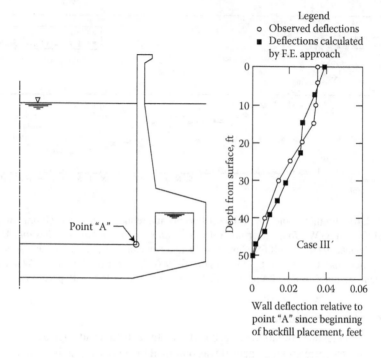

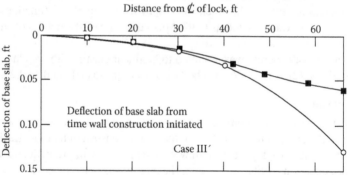

FIGURE 3.41 Deflections of structural components of lock: Case III′: (Figure 3.40). (From Duncan, J.M. and Clough, G.W., *Journal of the Soil Mechanics and Foundations Division, ASCE*, 97(SM8), August 1971, 1053–1068; Clough, G.W. and Duncan, J.M., Finite Element Analyses of Port Allen and Old River Locks, Report S-69-6, U.S. Army Engineers Waterways Experiment Station, Corps of Engineers, Vicksburg, MS, Sept. 1969. With permission.)

computed using the Hrennikoff method. Hence, the analyses were performed using the FEM in which the sequences of construction were simulated as closely as possible to predict the history of settlements and load distribution in the pile groups. First, the *in situ* stresses due to gravity loading were computed. Then, the following sequences were simulated: dewatering, excavation, piles installation, construction for piles and lock, backfilling, filling the lock with water, and the development of uplift pressures.

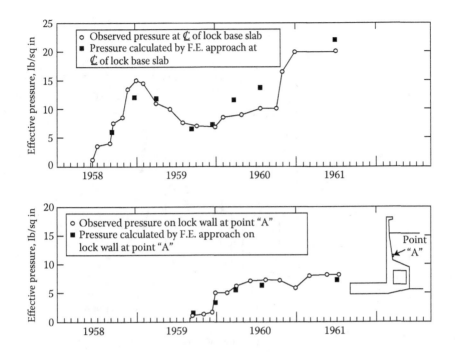

FIGURE 3.42   Predicted and observed effective earth pressures with time for part allen lock. (Adapted from Duncan, J.M. and Clough, G.W., *Journal of the Soil Mechanics and Foundations Division, ASCE*, 97(SM8), August 1971, 1053–1068; Clough, G.W. and Duncan, J.M., Finite Element Analyses of Port Allen and Old River Locks, Report S-69-6, U.S. Army Engineers Waterways Experiment Station, Corps of Engineers, Vicksburg, MS, Sept. 1969.)

Figure 3.44a shows a typical section through the lock chamber with foundation and adjacent soils [51,53]. Monoliths, 10-L and 10-R, which were instrumented and included in these analyses are shown in Figure 3.45; also shown are strain gages placed on the steel H-piles.

The subsoils in the foundation consisted mainly of cohesive back swamp deposits and/or cohesionless substratum deposits beneath the east wall and tertiary deposits interfingered with colluvium and substratum deposits beneath the west wall (Figure 3.44a) [54,55].

3.5.8.1   Constitutive Models

The behavior of soils and interfaces between the lock and soils (backfill) was simulated by using the hyperbolic representation of stress–strain and shear stress–relative displacement curves; see Appendix 1 for the details of the models [31,48,56,57]. The parameters for the hyperbolic model for soils were determined from laboratory (triaxial) tests available in Refs. [51,52]. The parameters for the hyperbolic model for the interfaces were developed based on direct shear tests for similar soils (sands and clays), and with the pile material (concrete and steel) at other sites [30,31]. Table 3.7a shows the hyperbolic parameters for various soils. The parameters for interfaces are given in Table 3.7b.

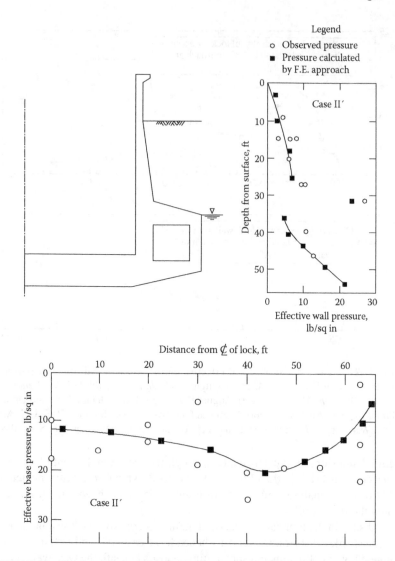

FIGURE 3.43 Predicted and observed effective earth pressures for port allen lock: Case II′ (Figure 3.40). (From Duncan, J.M. and Clough, G.W., *Journal of the Soil Mechanics and Foundations Division, ASCE*, 97(SM8), August 1971, 1053–1068; Clough, G.W. and Duncan, J.M., Finite Element Analyses of Port Allen and Old River Locks, Report S-69-6, U.S. Army Engineers Waterways Experiment Station, Corps of Engineers, Vicksburg, MS, Sept. 1969. With permission.)

The terms in Tables 3.7a and 3.7b are related to the expression for tangent modulus and Poisson's ratio, and shear and normal stiffness for soils and interfaces, respectively. They are given in Appendix 1. The value of K_o^c, the coefficient of earth pressure (Table 3.7a) for clay in compression was computed from values for similar overconsolidated soils [55,56]. Since sufficient volumetric data were not available, the Poisson's ratio, v, for soil was assumed to be constant, around 0.30.

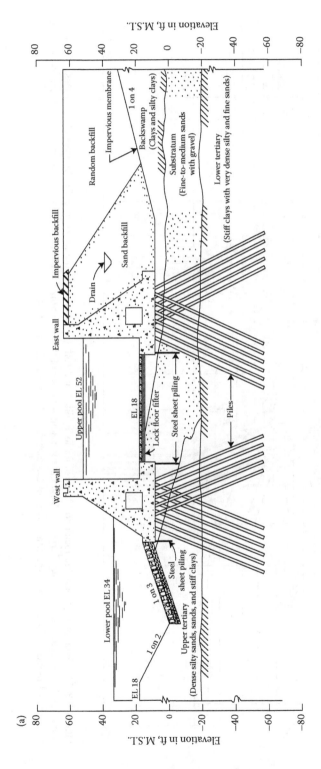

FIGURE 3.44 Columbia lock and FE mesh for lock, piles, and soil. (a) Cross-section and soil profile; (b) finite element mesh. (Adapted from Desai, C.S., Johnson, L.D., and Hargett, C.M., Finite Element Analysis of the Columbia Lock Pile Foundation System, Tech. Report No. S-74-6, U.S. Army Engineers Waterways Expt. Stn., Corps of Engineers, Vicksburg, MS, July. 1974.)

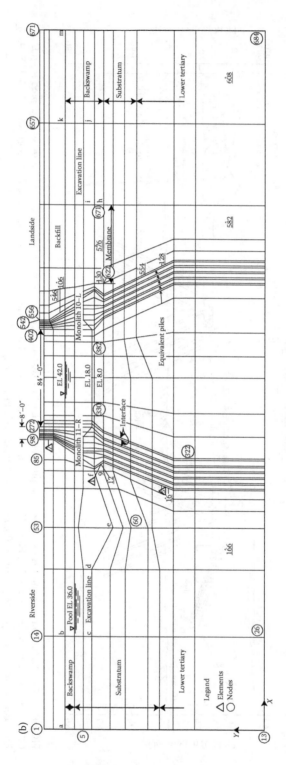

FIGURE 3.44 (continued) Columbia lock and FE mesh for lock, piles, and soil. (a) Cross-section and soil profile; (b) finite element mesh. (Adapted from Desai, C.S., Johnson, L.D., and Hargett, C.M., Finite Element Analysis of the Columbia Lock Pile Foundation System, Tech. Report No. S-74-6, U.S. Army Engineers Waterways Expt. Stn., Corps of Engineers, Vicksburg, MS, July. 1974.)

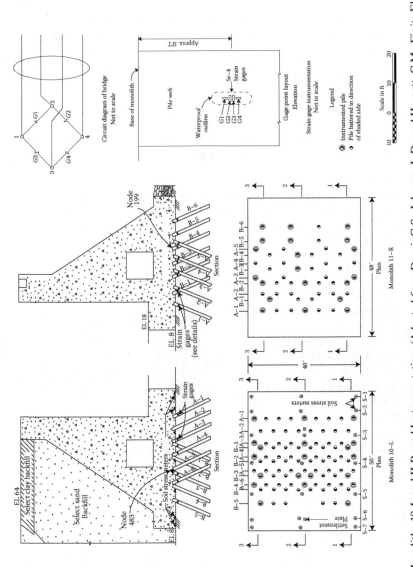

FIGURE 3.45 Moniliths 10-L and II-R, and Instrumentation. (Adapted from Desai, C.S., Johnson, L.D., and Hargett, C.M., Finite Element Analysis of the Columbia Lock Pile Foundation System, Tech. Report No. S-74-6, U.S. Army Engineers Waterways Expt. Stn., Corps of Engineers, Vicksburg, MS, July, 1974.)

TABLE 3.7a
Hyperbolic Parameters and Properties for Different Materials

Material[a]	v	v_f	γ (lb/ft³)	R_f	K_o^c	φ	c	$\bar{K}$	K_u	n
1	0.10	0.10	—	—	—	—	—	—	—	0.0
2	0.30	0.48	120.0	0.90	0.70	26	0	1000	2000	0.0
3	0.30	0.48	58.0	0.90	0.70	26	0	1000	2000	0.0
4	0.30	0.48	53.0	0.85	0.45	40	0	1160	1750	0.5
5	0.30	0.48	58.0	0.85	1.10	30	0	400	800	0.5
6	0.28	0.28	61.0	1.00	1.00	—	—	—	—	—
7	0.28	0.28	66.0	1.00	1.00	—	—	—	—	—
8	0.10	0.10	0.0	—	—	—	—	—	—	0.0
9	0.20	0.20	150.0	—	1.00	—	—	—	—	—
10	0.30	0.48	115.0	0.85	0.45	40	0	580	860	0.5
11	0.48	0.48	62.4	1.00	1.00	—	—	—	—	0.0

1 lb/ft³ = 157.06 N/m³.

Note: v = Poisson's ratio; v_f = Poisson's ratio at failure; γ = unit weight; R_f = failure ratio; K_o^c = coefficient of earth pressure in compression; φ = Mohr–Coulomb (angle of internal friction) parameter; c = Mohr–Coulomb cohesive strength parameter; $\bar{K}$ = hyperbolic loading parameter; K_u = hyperbolic unloading parameter; n = experimentally determined parameter.

[a] Material	Identification
1	Air
2	Backswamp clay above water table
3	Backswamp clay below water table
4	Substratum sand
5	Tertiary clay
6	Equivalent piles—11-R
7	Equivalent piles—10-L
8	Air as replacement
9	Concrete
10	Backfill sand
11	Water

TABLE 3.7b
Parameters for Interfaces

Interface	K_j (lb/ft³)	n	R_f	k_r (lb/ft³)	δ (deg)	c_a
Lock—backfill	7.5×10^4	1.0	0.87	10	33	0.0
Piles—substratum sand	2.5×10^4	1.0	0.87	10	26	0.0
Piles—tertiary clay	2.5×10^4	1.0	0.87	10	20	0.0

1 lb/ft³ = 157.06 N/m³.

When the shear stress exceeded the Mohr–Coulomb strength, the value of the shear stiffness for interface was set equal to the residual stiffness, k_r (Table 3.7b). Before failure, the normal stiffness was set a high value of 10^8 lb/ft^3 (157×10^8 N/m^3); after failure, a low value of 10 lb/ft^3 (1570 N/m^3) was adopted.

The behavior of concrete was assumed to be linear elastic. The Young's modulus E for concrete was adopted as 4×10^8 lb/ft^2 (192×10^8 N/m^2). The value of E for the steel piles was assumed as 4.04×10^9 lb/ft^2 (193×10^9 N/m^2).

The equivalent E for piles in two monoliths, 10-L and 11-R (Figure 3.45), were computed as 13.3×10^7 and 7.5×10^7 lb/ft^2 (636.8×10^7 and 359×10^7 N/m^2) for battered piles inward, and 11.8×10^7 and 6.8×10^7 lb/ft^2 (565×10^7 and 326×10^7 N/m^2) for piles battered outward, respectively. The densities of equivalent piles (6 and 7 in Table 3.7a) were computed by equating the weights of the equivalent piles and the H-piles. The computation of equivalent quantities is given below.

3.5.8.2 Two-Dimensional Approximation

The lock–pile–soil problem is 3-D. However, for this analysis, it was approximated as 2-D. Hence, the FE mesh (Figure 3.44b) was based on plane strain idealization (Figure 3.46) with equivalent material properties to include the effect of piles.

Figure 3.47 shows a schematic representation of the monoliths with two types of piles, battered in and out. We assumed that the main response of piles is in the axial direction. Then the total axial stiffness, S, of the two monoliths can be expressed as

$$ S = \sum_{j=1}^{n} \frac{A_j E_j}{L_j} = (n_i + n_o)\frac{AE}{L} \tag{3.35} $$

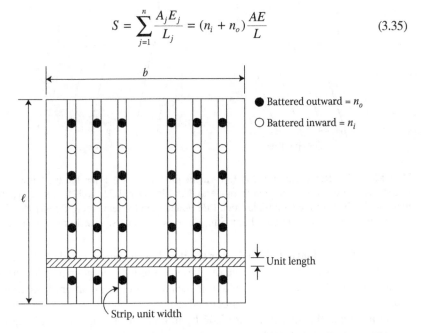

FIGURE 3.46 Schematic of equivalent simulation of monolith. (Adapted from Desai, C.S., Johnson, L.D., and Hargett, C.M., Finite Element Analysis of the Columbia Lock Pile Foundation System, Tech. Report No. S-74-6, U.S. Army Engineers Waterways Expt. Stn., Corps of Engineers, Vicksburg, MS, July. 1974.)

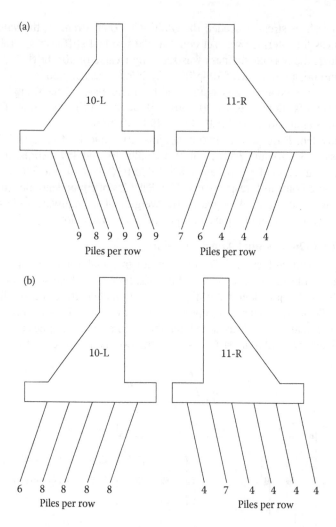

FIGURE 3.47 Representation for Equivalent properties. (a) Piles battered inwards; (b) Piles battered outwards. (Adapted from Desai, C.S., Johnson, L.D., and Hargett, C.M., Finite Element Analysis of the Columbia Lock Pile Foundation System, Tech. Report No. S-74-6, U.S. Army Engineers Waterways Expt. Stn., Corps of Engineers, Vicksburg, MS, July. 1974.)

where E is the modulus of elasticity for H-piles, L is the length of pile (taken as 70 ft or 21.35 m), A is the projected area of pile, n is the total number of piles, n_i is the number of piles battered inward, and n_o is the number of piles battered outward. It is assumed that all piles have equal areas and lengths.

Let us divide the area in Figure 3.46 by m number of strips of unit length. The stiffness per strip of battered in and out piles, s_i and s_o, can be expressed as

$$s_i = \frac{n_i}{m} \cdot \frac{AE}{L} \qquad (3.36a)$$

$$s_o = \frac{n_o}{m} \cdot \frac{AE}{L} \tag{3.36b}$$

The equivalent stiffness of a strip for piles battered in and out, s_{ei} and s_{eo}, can be expressed as

$$s_{ei} = \frac{A_{ei} E_{ei}}{L_{ei}} \tag{3.37a}$$

$$s_{eo} = \frac{A_{eo} \cdot E_{eo}}{L_{eo}} \tag{3.37b}$$

where A_e, E_e, and L_e are the equivalent area, elastic modulus, and length of pile, respectively, and the second subscripts i and o denote pile battered inward and outward, respectively.

Now, two kinds of FE analyses were performed, with equivalent properties, by adopting a strip of unit dimension. Figures 3.47a and 3.47b show the lock walls and piles, battered inward and outward, respectively.

Equivalent properties: There were six rows of piles battered inward in monolith 10-L and five rows in monolith 11-R, (Figure 3.47). The number of piles battered inward are $n_i = 53$ for monolith 10-L, and $n_i = 25$ for monolith 11-R. Also, $A = 0.15$ ft^2 (0.014 m^2), $E = 4.04 \times 10^9$ lb/ft^2 (193 × 10^9 N/m^2), $m = 6$, $A_{ei} = 1 \times 40$ (unit width multiplied by length of the monolith) = 40 ft^2 (3.72 m^2) and $L = L_{ei}$. Therefore, equating Equations 3.36a and 3.37a, we can evaluate the equivalent E_{ei} for monolith 10-L battered inward (Figure 3.47a):

$$E_{ei} = \frac{53}{6} \times \frac{0.149 \times 4.04 \times 10^9}{40} = 13.3 \times 10^7 \text{ lb/ft}^2 \ (637 \times 10^7 \text{ N/m}^2) \tag{3.38a}$$

and for piles battered inward for monolith 11-R:

$$E_{ei} = \frac{25 \times 0.149 \times 4.04 \times 10^9}{5 \times 40} = 7.5 \times 10^8 \text{ lb/ft}^2 \ (359 \cdot 10^8 \text{ N/m}^2) \tag{3.38b}$$

Similarly, for piles battered outward, with number of piles = 38 and 27 (Figure 3.47b):

$$E_{eo} = \frac{38 \times 0.149 \times 4.0 \times 10^9}{5 \times 40} = 11.8 \times 10^7 \text{ lb/ft}^2 \ (565 \times 10^7 \text{ N/m}^2) \tag{3.39a}$$

$$E_{eo} = \frac{27 \times 0.149 \times 4.04 \times 10^9}{6 \times 40} = 6.8 \times 10^7 \text{ lb/ft}^2 \ (326 \times 107 \text{ N/m}^2) \tag{3.39b}$$

Computed results: The computer code developed and used for the analysis of the Port Allen lock by Clough and Duncan [48] was utilized for the analysis of the Columbia lock-pile system; however, a number of modifications and corrections were introduced. For instance, the stresses in interface elements were made consistent with those in adjacent soil elements, and interfaces became operational as soon as the construction sequences were initiated.

The steps simulated in the construction sequences are shown in Table 3.8. The initial or *in situ* stresses induced by the gravity were first computed, by using K_o^c in Table 3.7a. Then, the effect of dewatering due to lowering of the water table was computed. The excavation was performed in three steps. Steel piles were then installed in place of the soil (elements) at appropriate locations. The lock was constructed in three steps. The buildup of water in the lock up to El. 42.0 and development of uplift pressure were simulated in sequence 11 and sequence 12, respectively.

The uplift pressures equal to the hydrostatic pressures due to the head of water in the lock were applied at nodes on the base of monoliths 10-L and 11-R and at the membrane (Figure 3.45). Three iterations for each sequence provided convergent solutions.

Comparisons between the predicted vertical displacements measured at nodes near typical settlement plates in the backfill adjacent to monolith 10-L and the field data are shown in Figure 3.48. The predictions included are related to sequences 4, 7, 9, 10, 11, and 12 (Table 3.8). The agreement between predictions and observations is considered good. According to the observed data, settlements remained essentially the same or decreased in magnitude after water was filled in the lock. Such decrease may be due to the uplift pressures.

TABLE 3.8
Sequences Simulated in Finite Element Analysis

Operation	Details	Sequences
Initial stresses	—	—
Dewatering	From El 34.0[a] to El 5.0	1
Excavation	In three stages:	
	El 58.0 to 42.0	2
	El 42.0 to 26.0	3
	El 26.0 to 8.0	4
Placement of piles	—	5a
Lock construction	In three stages:	
	El 8.0 to 26.0	5b
	El 26.0 to 42.0	6
	El 42.0 to 64.0	7
Backfill	In three stages:	
	El 8.0 to 26.0	8
	El 26.0 to 42.0	9
	El 42.0 to 64.0	10
Filling of the lock	—	11
Development of uplift pressure	—	12

[a] All elevations (El) cited herein are in feet referred to mean sea level. (1 ft = 0.305 m).

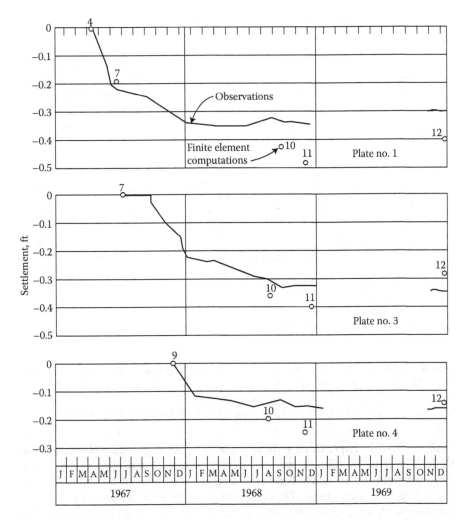

FIGURE 3.48 Comparisons between predicted and measured settlements: Monolith 10-L. (1 ft = 0.305 m). *Note:* Numbers Indicate FE sequences. (Adapted from Desai, C.S., Johnson, L.D., and Hargett, C.M., Finite Element Analysis of the Columbia Lock Pile Foundation System, Tech. Report No. S-74-6, U.S. Army Engineers Waterways Expt. Stn., Corps of Engineers, Vicksburg, MS, July. 1974.)

Figure 3.49 shows computed vertical settlements with construction sequences at Nodes 483 and 199 beneath monoliths 10-L and 11-R, respectively (Figure 3.45); a few observed values are also shown in Figure 3.49. The correlation is considered good. The distributions of loads in piles are compared in Figure 3.50, in two mono-liths for typical steps of construction sequences. Here, the comparisons include FE predictions, field data, and computation by the Hrennikoff method. Overall, the FE predictions compare well with observations, while the results by the Hrennikoff method do not show as good correlation.

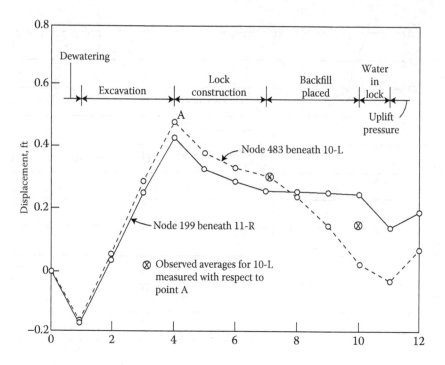

FIGURE 3.49 Settlements versus sequences of construction at typical nodes: 199, 483, Fig. 3.45. (Adapted from Desai, C.S., Johnson, L.D., and Hargett, C.M., Finite Element Analysis of the Columbia Lock Pile Foundation System, Tech. Report No. S-74-6, U.S. Army Engineers Waterways Expt. Stn., Corps of Engineers, Vicksburg, MS, July. 1974.)

Figure 3.51 shows predicted vertical and horizontal stresses in the elements along section Y–Y in the backfill. The vertical stresses indicate a linear variation, $(\sigma_y = \gamma D)$, for a major part of the depth; here, σ_y = vertical stress, γ = unit weight of backfill, and D = depth of overburden. The computed pressures show a significant decrease compared with linear (conventional) distribution in the lower part (Figure 3.51). The occurrence of arching near the base of the wall can be a reason for the decrease. Horizontal stresses, $\sigma_x = K_o^c \gamma D$, are also linear for a major part, but decrease in the lower part of the monolith.

Based on the results from Examples 3.7 and 3.8, it can be concluded that the FEM can provide satisfactory predictions for the analysis and design for similar lock structures.

3.5.9 EXAMPLE 3.9: UNDERGROUND WORKS: POWERHOUSE CAVERN

A powerhouse cavern in the Himalayas was analyzed by using the FE procedure with realistic constitutive models; the latter involved the DSC for rock and rock mass.

The mechanical behavior of the rock mass at the location can be influenced by factors such as origin of rock, faults and folds, discontinuities (joints), and other geoenvironmental issues. Hence, appropriate laboratory and/or field testing are

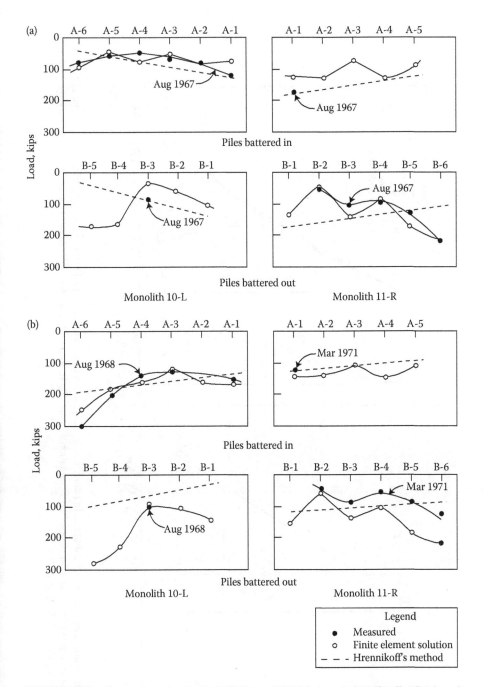

FIGURE 3.50 Comparisons between predictions and field measurement for distribution of loads in piles. (a) End of lock construction-no backfill (case 7); (b) backfill complete (case 10). (Adapted from Desai, C.S., Johnson, L.D., and Hargett, C.M., Finite Element Analysis of the Columbia Lock Pile Foundation System, Tech. Report No. S-74-6, U.S. Army Engineers Waterways Expt. Stn., Corps of Engineers, Vicksburg, MS, July. 1974.)

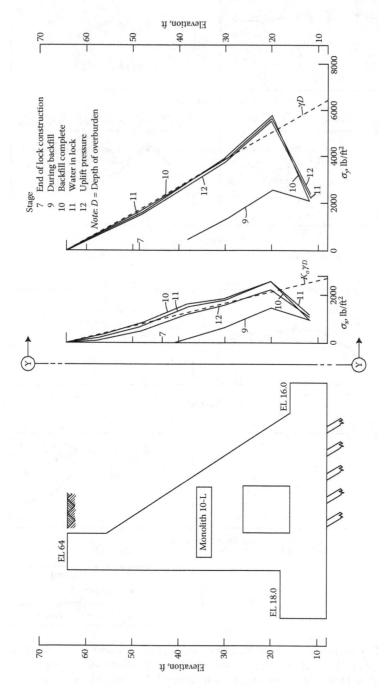

FIGURE 3.51 Computed pressure distributions in backfill with construction sequences. (Adapted from Desai, C.S., Johnson, L.D., and Hargett, C.M., Finite Element Analysis of the Columbia Lock Pile Foundation System, Tech. Report No. S-74-6, U.S. Army Engineers Waterways Expt. Stn., Corps of Engineers, Vicksburg, MS, July, 1974.)

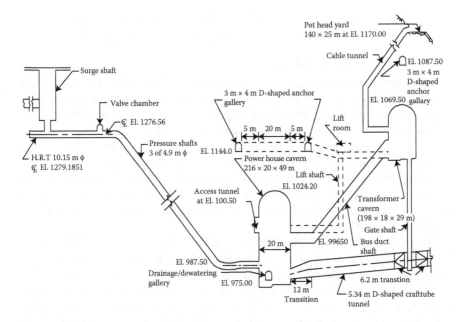

FIGURE 3.52 East-west section of powerhouse cavern. (From Varadarajan, A. et al., *International Journal of Geomechanics, ASCE*, 1(1), 2001, 83–107; Varadarajan, A. et al., *International Journal of Geomechanics, ASCE*, 1(1), 2001, 109–127. With permission.)

required to calibrate the DSC model. The model needs to be implemented in computer (FE) procedures so as to obtain realistic solutions for analysis and design. We present in the following the model for rock mass based on the behavior of intact rock (and joints), validations for laboratory specimen tests, and validation using the FE code, DSC-SST2D [44] for prediction of behavior of the powerhouse cavern located at Nathpa Jhakri, Himachal Pradesh, India. Figure 3.52 shows the east-west section including the surge shaft, pressure shaft, and powerhouse cavern [58–61].

3.5.9.1 Validations

This example is used to validate the constitutive model developed by Desai and coworkers based on the DSC [62]. For level 1 validation, model predictions are compared with the test data from which the (average) parameters were obtained. For level 2 validation, the model predictions are compared with an independent set of test data by using parameters determined from another (previous) set of test data. For level 3 validation, boundary value problems are solved and the results (e.g., FE in which the constitutive model is used) are compared with the measurements (either in the field or simulated in the laboratory) of practical problems.

The DSC model has been validated for the above three levels for a wide range of materials, interfaces, and joints, for example, clays, sands, rocks, concrete, asphalt concrete, metals, alloys, and silicon [62]. A typical validation involving behavior of rocks and analysis of an underground powerhouse cavern is included here.

3.5.9.2 DSC Modeling of Rocks

The descriptions herein are adopted from Refs. [58–60]. The rocks at the site include mainly three types: quartz mica schist, quartz mica schist with quartz veins, and biotite schist [57–60]. The properties of these rocks are given in Table 3.9.

Testing for the rocks was performed using a triaxial device under high confining pressures. Rock specimens of about 5.475 cm diameter and 10.95 cm length were tested under a number of initial confining pressures $\sigma_3 = 0$–45 MPa.

Typical test data for the three rocks under typical confining pressure are shown in Figure 3.53, for quartz mica schist, Figure 3.54 for quartz mica schist with quartz veins, and Figure 3.55 for biotite schist. The material parameters for the rocks are given in Table 3.10.

The predictions were obtained by two methods: (1) integration of DSC incremental, Equation A1.38a in Appendix 1, starting from the initial confining pressure called *single point method* (SPM), and (2) using the FEM with the DSC model. The FE analyses were performed by discretizing the quarter of the specimen (Figure 3.56). Figures 3.53 through 3.55 show a comparison between predictions and laboratory data for the three rock types. It can be seen that the DSC model using these two methods show similar and highly satisfactory correlations between the predictions and laboratory measurements.

3.5.9.3 Hydropower Project

Figure 3.52 depicts the powerhouse caverns, including the surge and pressure shafts. The cavern consists of two major openings, that is, machine (powerhouse) hall 216 m × 20 m × 49 m with the overburden of 262.5 m at the crown, and the transformer hall 198 m × 18 × 29 m (Figure 3.52). The openings are located in the left bank, about 500 m from the Sutlej river. Based on the measurements and analysis, the coefficient of lateral pressure was found to be 0.8035 for the E–W section considered herein [58–61].

The National Institute of Rock Mechanics (NIRM), India instrumented the powerhouse cavern site to measure the movements in the rock mass during various sequences of excavation [63]. The instrumentation included mechanical and remote extensometers. Figure 3.57 shows the instrumentation scheme for a section

TABLE 3.9

Properties of Rocks

Type of Rock	Specific Gravity, G	Dry Density (kN/m³)	Tensile Strength (MPa)
1. Quartz mica schist	2.74	26.0–27.6	8.00
2. Quartz mica schist with quartz veins	2.83	26.0–27.4	10.00
3. Biotite schist	2.82	26.4–27.9	6.00

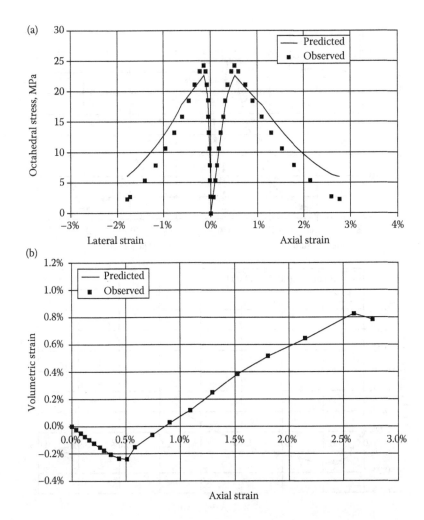

FIGURE 3.53 Comparisons between predicted and observed behavior for Quartz Mica Schist, $\sigma_3 = 30$ MPa. (a) Stress-strain behavior; (b) volume change response. (From Varadarajan, A. et al., *International Journal of Geomechanics, ASCE,* 1(1), 2001, 83–107; Varadarajan, A. et al., *International Journal of Geomechanics, ASCE,* 1(1), 2001, 109–127. With permission.)

in the middle of the cavern; A, B, C, and D denote various instrument sets at various elevations.

The FE procedure using the code DSC-SST2D, which contains the DSC model, was used to analyze the cavern [44]. The rocks in the area of the powerhouse cavern (machine hall) are quartz mica schist and biotite schist [60,61,64]. However, since the former is predominant in the vicinity of the cavern, it is adopted for the FE analysis. Also, the rock mass contains joints and discontinuities, with average rock mass rating (RMR) and tunneling quality index (Q) being 50 and 2.7, respectively. Therefore, the material parameters for the rock mass were obtained by modifying the foregoing

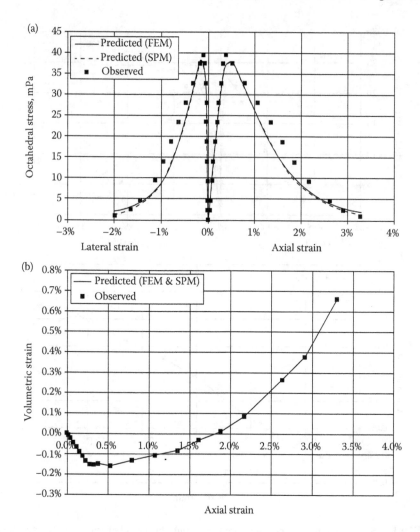

FIGURE 3.54 Comparisons between predicted and observed behavior for Quartz mica schist with Quartz veins, $\sigma_3 = 10$ mPa. (a) Stress-strain behavior; (b) volume change response. (From Varadarajan, A. et al., *International Journal of Geomechanics, ASCE,* 1(1), 2001, 83–107; Varadarajan, A. et al., *International Journal of Geomechanics, ASCE,* 1(1), 2001, 109–127. With permission.)

intact rock parameters. The procedures proposed by Ramamurthy [65] and Bhasin et al. [64] were used for such modifications. The model parameters for both the quartz mica schist (as intact rock) and modified for the jointed rock mass are presented in Table 3.11. As can be seen, the values of E, γ, β, and 3R for the rock mass have decreased compared to the intact rock.

The FE mesh with boundary conditions (Figure 3.58) contains 1167 nodal points and 364 eight-noded isoparametric elements. The initial stresses in the elements were obtained by using $K_o = 0.8035$. The loading is caused primarily by the

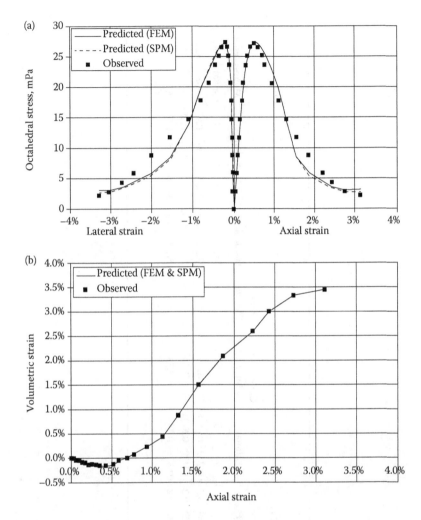

FIGURE 3.55   Comparisons Between predicted and observed behavior for biotite schist, $\sigma_3 = 7.5$ mPa. (a) Stress-strain behavior; (b) volume change response. (Adapted from Varadarajan, A. et al., *International Journal of Geomechanics, ASCE*, 1(1), 2001, 83–107; Varadarajan, A. et al., *International Journal of Geomechanics, ASCE*, 1(1), 2001, 109–127.)

excavation, which is simulated in 12 stages or sequences (Figure 3.59). For each sequence of excavation, the element and nodes to be removed are deactivated from the original mesh. In other words, the stiffness matrices and load vectors of the deactivated elements are not included in the global stiffness matrix and load vectors. The analysis is performed using an incremental iterative procedure [44,66].

Results: The FE results in terms of displacements, strains, and stresses were processed through the commercial code NISA [61,67]. Figure 3.60 shows contours of horizontal displacements around the cavern. The maximum displacement of about 42.6 mm was measured at the face of the cavern. It decreased to about 9.22 mm at

TABLE 3.10
Material Parameters for the Three Rocks

Type of Rock	Elastic, E^a			Ultimate		Cohesive Stress Intercept	Phase Change	Hardening		Disturbance			Mohr–Coulomb	
	$\bar{n}$	K	ν	γ	β	$3R$ (mPa)	n	a_1	η_1	A	Z	D_u	c(mPa)	φ (deg)
Quartz mica schist	0.2645	28269	0.2	0.0202	0.4678	46.99	5.0	0.13E-13	0.6	220.7	1.339	0.97	6.38	22.24
Biotite schist	0.2597	59320	0.2	0.0429	0.6431	57.54	5.0	0.5E-13	0.62	107.4	1.111	0.98	13.11	34.45
Quartz mica schist with quartz veins	0.9806	2641	0.2	0.097	0.74	4.73	5.0	0.15E-11	0.3	147.6	1.1054	0.99	6.37	44.41

a $E = K\sigma_3^{\bar{n}}$, σ_3 = confining stress.

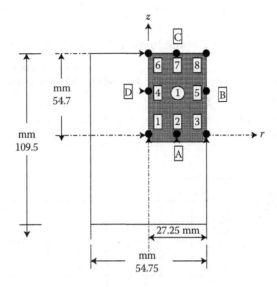

FIGURE 3.56 FE Mesh for simulation of triaxial test.

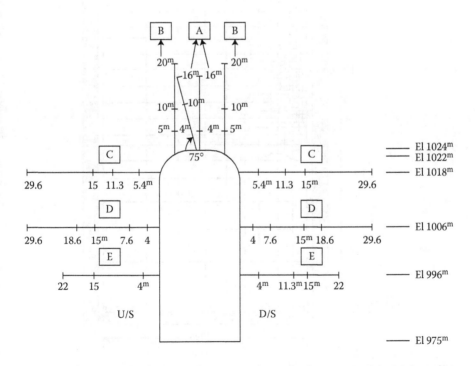

FIGURE 3.57 Instrumentation at mid section near powerhouse cavern. (From Varadarajan, A. et al., *International Journal of Geomechanics, ASCE,* 1(1), 2001, 83–107; Varadarajan, A. et al., *International Journal of Geomechanics, ASCE,* 1(1), 2001, 109–127. With permission.)

TABLE 3.11

Material Parameters for Intact and Jointed Rock Mass (Quartz Mica Schist)

Type	Elasticity		Ultimate		Phase Change	Hardening		Cohesive Stress Intercept (MPa)	Disturbance		
	v	E	γ	β	n	a_1	η_1	$3R$	D_u	A	Z
Intact rock	0.2	8591	0.0202	0.4678	5.0	0.013E–12	0.6	46.99	0.97	220.7	1.339
Rock mass	0.2	6677	0.0135	0.3889	5.0	0.013E–12	0.6	41.9	0.97	220.7	1.339

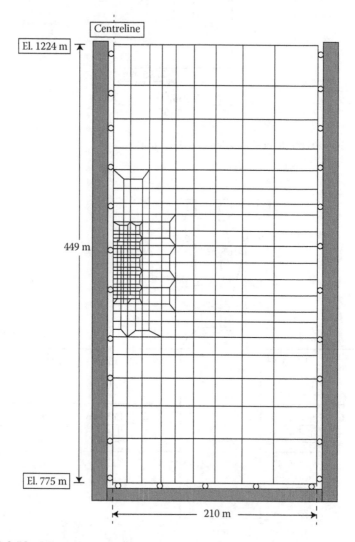

FIGURE 3.58 FE Mesh for Powerhouse cavern and rock mass. (From Varadarajan, A. et al., *International Journal of Geomechanics, ASCE,* 1(1), 2001, 83–107; Varadarajan, A. et al., *International Journal of Geomechanics, ASCE,* 1(1), 2001, 109–127. With permission.)

El. 1024 m
El. 1018 m
El. 1014 m
El. 1010 m
El. 1006 m
El. 1000 m
El. 996 m
El. 991.8 m
El. 987.6 m
El. 983.4 m
El. 979.2 m
El. 975 m

II   I   II
III
IV
V
VI
VII
VIII
IX
X
XI
XII

FIGURE 3.59 Sequences of excavation for powerhouse cavern. (From Varadarajan, A. et al., *International Journal of Geomechanics, ASCE,* 1(1), 2001, 83–107; Varadarajan, A. et al., *International Journal of Geomechanics, ASCE,* 1(1), 2001, 109–127. With permission.)

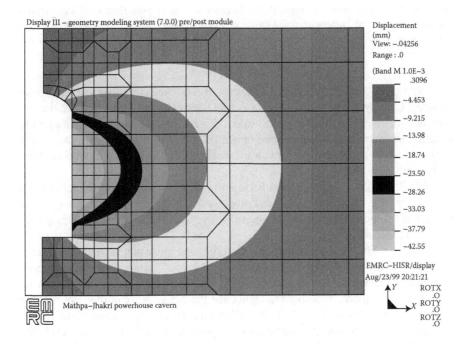

Display III – geometry modeling system (7.0.0) pre/post module

Displacement
(mm)
View: –.04256
Range : .0

(Band M 1.0E–3
        .3096

~–4.453
~–9.215
~–13.98
~–18.74
~–23.50
~–28.26
~–33.03
~–37.79
~–42.55

EMRC–HISR/display
Aug/23/99 20:21:21

Y   ROTX
        .O
X   ROTY
        .O
        ROTZ
        .O

EMRC   Mathpa–Jhakri powerhouse cavern

FIGURE 3.60 Contours of horizontal displacements around powerhouse cavern. (From Varadarajan, A. et al., *International Journal of Geomechanics, ASCE,* 1(1), 2001, 83–107; Varadarajan, A.et al., *International Journal of Geomechanics, ASCE,* 1(1), 2001, 109–127. With permission.)

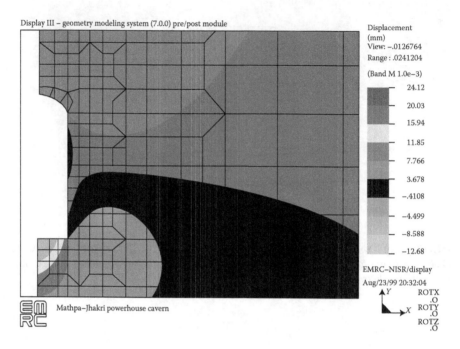

FIGURE 3.61 Contours of vertical displacements around powerhouse cavern. (From Varadarajan, A. et al., *International Journal of Geomechanics, ASCE,* 1(1), 2001, 83–107; Varadarajan, A. et al., *International Journal of Geomechanics, ASCE,* 1(1), 2001, 109–127. With permission.)

the distance of 73 m from the face of the cavern. The predicted displacement by the present FE analysis at the face was about 43.0 mm (Figure 3.60); the predictions compare very well with the measured values. Figure 3.61 shows the contours for the vertical displacements. The predicted upward displacement near the top was about 24.0 mm, while the maximum downward displacement near the bottom was about 12.70 mm.

Table 3.12 shows a comparison between predictions from the FE analysis and observations for displacement at the Powerhouse Cavern boundary, for stages one through six (Figure 3.59). It can be seen that overall, the predictions compare very well with the measurements.

Comments: The behavior of geologic materials and joints play a significant role in the design, construction, and maintenance of underground space. The available models based on elasticity, conventional plasticity, and so on are not capable of allowing the effects of important factors that influence the behavior. A unified constitutive modeling approach called the DSC that allows for the effects of most of the important factors that influence the behavior of underground works has been developed and used (Appendix 1). It is believed that the DSC can provide a general and unique constitutive modeling approach for a wide range of materials and joints, including underground works. Thus, its application potential goes beyond that provided by any other previously available model.

TABLE 3.12

Comparison of Predicted (FEM) and Observed (Instrumentation) Displacements at Boundary of Powerhouse Cavern

| Stage No. | Excavation Stage | | Instrumentation at El. (m) | Deformation (mm) | |
	From El. (m)	To El. (m)		Predicted (FEM)	Observed (Instrumentation)
1	Widening of central drift	1024 (A)	10.4	13–18	
2	Widening of central drift	1022 (B)	12	6–12	
3	1018	1006	1022 (B)	0.6	−1.3 to +2.5
4	1006	1000	1018 (C)	3.5	1–4
5	1000	975	1006 (D)	23.7	10–45
6	983	975	996 (E)	9.4	1–3

3.5.10   EXAMPLE 3.10: ANALYSIS OF CREEPING SLOPES

This example involves the analysis of a creeping slope at Villarbeney Landslide in Switzerland using a nonlinear FE procedure. The FE results are compared with the field measurements [68,69]. Figure 3.62 shows a plan view and cross-sectional dimensions of the slope. Three inclinometers E_0, E_1, and E_2 were installed to measure the movement of the slope. The depths of these inclinometers were 42.75, 54.35, and 37.5 m, respectively. In addition, to monitor the groundwater level and its fluctuations, piezometric cells were installed in borings E_4, E_5, and E_6 (Figure 3.62a). It was seen that the groundwater level was at about 2 m below the ground surface, (Figure 3.62b) and did not vary significantly for the one-year time period considered in the FE analysis. Therefore, the influence of steady-state seepage forces for depths below 2 m was introduced by assuming hydrostatic conditions. The seepage forces from a flow net analysis were superimposed on those due to the weight of the soil. The gravity load (weight of the moving slope) was assumed to increase linearly with depth. In addition to the gravity load and seepage, surface tractions were introduced on the toe to simulate resistance to the movement of the soil (Figure 3.63).

Two types of constitutive models were used to characterize the behavior of associated soils and interfaces. The elastoviscoplastic constitutive model used for characterizing soils accounts for elastic, plastic, and viscous or creep deformations, continuous yielding or hardening, volume change, and stress path effects. It is a version of the HISS family of models, called δ_{vp}, developed by Desai and coworkers [10,44,62,69], and is based on the theory of elastoviscoplasticity by Perzyna [70]. The interface model was derived as a special case from that for the solids, which is a distinct advantage (consistency) in the FE analysis involving both solids and interfaces. Among possible relative motions at interfaces such as sliding, separation or debonding, rebonding, and interpenetration, the translation mode was found to be predominant for the creeping slope [69]. The details of the HISS family of models are given in Appendix 1.

In contrast to a relatively distinct interface that can be assigned in the case of the interface between a structural and a geologic material, in case of a creeping slope, the

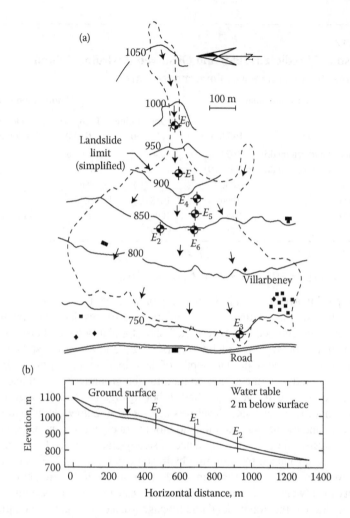

FIGURE 3.62 Villarbeney natural slope. (a) Plan view; (b) slope cross-section. (Adapted from Vulliet, L., Modelisation Des Pentes Naturellcs En Mouvement, These No. 635, Ecole Polytechnique, Federale de Lausanne, Lausanne, Switzerland (in French), 1986; Desai, C.S., Samtani, N.C., and Vulliet, L., Constitutive modeling and analysis of creeping slopes, *Journal of Geotechnical Engineering, ASCE,* 121(1), 1995, 43–56.)

interface representation involves a finite "interface zone." In this zone, near the junction of the moving mass and the underlying stationary parent mass, the variation of translational displacement is much more severe than that in the mass above the interface zone. A schematic of these movements is shown in Figure 3.64. The depth corresponding to OC is the overall depth of the slope that experiences movements, while the depth corresponding to OB involves high levels of relative motions compared to those in the portion CB (Figure 3.64a) [69]. The extent of OB was about 25% of the overall depth of the slope. In view of Figure 3.64a, the thickness of the interface zone in the FE simulation was taken as 6% of the thickness of the moving slope.

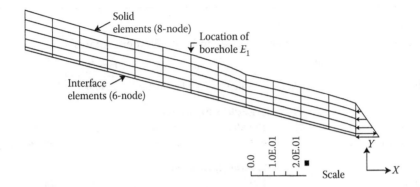

FIGURE 3.63 Typical finite element mesh used: Borehole E_1. (From Desai, C.S., Samtani, N.C., and Vulliet, L., *Journal of Geotechnical Engineering, ASCE*, 121(1), 1995, 43–56. With permission.)

The material parameters for soils and interfaces were obtained from triaxial and simple shear tests, respectively, in the laboratory. A summary of the material parameters is shown in Table 3.13. The soils at locations E_1 and E_2 (Figure 3.62) exhibited similar physical properties with average water content of 20%, plastic limit of 20%, liquid limit of 40%, and dry unit weight of 18.5 kN/m³, and were

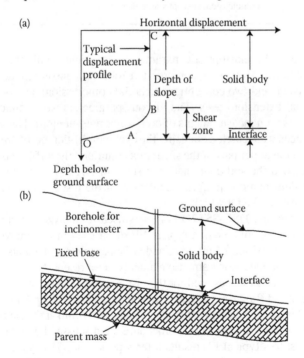

FIGURE 3.64 Idealization of creeping slope. (a) Sliding zones; (b) overall slope. (From Desai, C.S., Samtani, N.C., and Vulliet, L., *Journal of Geotechnical Engineering, ASCE*, 121(1), 1995, 43–56. With permission.)

TABLE 3.13

Parameters for Soil and Interface Used in Finite Element Analysis

Soil			Interface		
Parameter	Symbol	Value	Parameter	Symbol	Value
		(a) Elasticity Parameters			
Modulus	E	10.4 MPa	Normal Stiffness	k_n	8×10^6 kPa/cm
Poisson's ratio	v	0.35	Shear stiffness	k_s	2800 kPa/cm
		(b) Plasticity Parameters			
Ultimate	γ	1.93	Ultimate	γ_i	0.24
Ultimate	β	0.64	—	—	—
Transition	n	2.04	Transition	n_i	2.04
Hardening	a_1	1.47	Hardening	a_i	143
Hardening	n_1	0.06	Hardening	b_i	10
Nonassociative	—	—	Nonassociative	κ	0.57
		(c) Viscous Parameters			
Fluidity	Γ	0.00015/min	Fluidity	Γ_i	0.057/min
Exponent	N	2.58	Exponent	N_i	3.15

Note: All parameters are nondimensional except where indicated.

classified as CL. The elastoplastic parameters associated with the constitutive model were obtained from four compression (one K_o consolidation, one hydrostatic compression, and two conventional triaxial compression), and one extension (reduced triaxial extension) tests. The viscous parameters were found from undrained creep tests with measurements of pore water pressure [68]. The simple shear tests for interfaces were carried out using the cyclic multi-degree-of-freedom shear device in which the upper part of the shear box contained the stiffer base, while the lower part included the soil contained in a stack of circular aluminum rings that allowed shear deformations [62,71]. The details of the test device and test results are given by Desai and Rigby [71].

A 2-D FE program, DSC-SST2D [44], was used to analyze the creeping slope shown in Figures 3.62 and 3.64. A typical FE mesh used in the analysis is shown in Figure 3.63. It consisted of 60 eight-noded isoparametric elements to represent the soil and 12 six-noded isoparametric elements to represent the interface. A variable time-stepping scheme was used for the viscoplastic solution, with $\Omega = 0.02$ and $\theta = 1.2$, where Ω and θ are time-step control constants. The viscoplastic solution was continued until it reached a steady-state condition ($<10^{-9}$/min) at the element integration points. The details of the solution procedure are discussed by Desai et al. [69].

Figure 3.65 shows typical FE results after a period of 354 days. Field observations obtained from the inclinometer E_1 are superimposed on FE predictions for comparison. The velocity was obtained by dividing the difference in displacements by the corresponding time increment when the steady viscoplastic strain rate was

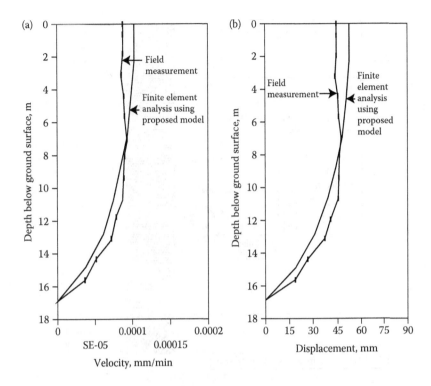

FIGURE 3.65 Comparisons of predictions and observations at Borehole E_1: (a) Velocity; (b) displacement. (From Desai, C.S., Samtani, N.C., and Vulliet, L., *Journal of Geotechnical Engineering, ASCE*, 121(1), 1995, 43–56. With permission.)

reached. It is seen that the FE predictions compare well with the measured velocity and displacement profiles.

3.5.11 EXAMPLE 3.11: TWIN TUNNEL INTERACTION

The construction of tunnels in urban environments often requires tunneling in close proximity to existing tunnels. This example considers the degree of interaction between two tunnels constructed in stiff clay [72]. Two different geometries are considered: one with the tunnels running side by side at the same depth and the other with the tunnels running one above the other along the same vertical axis line in a piggyback fashion (Figure 3.66). Figure 3.67a shows the FE mesh of the twin tunnel for the side-by-side case, whereas Figure 3.67b shows the FE mesh for the piggyback case. Eight-noded plane strain isoparametric elements with reduced integration scheme were used to represent the soil. Three-noded Mindlin beam elements with reduced integration scheme were used to model the tunnel lining [72]. A modified Newton Raphson scheme with a substepping stress point algorithm was used to solve the nonlinear FE equations.

Soil models and parameters: The soil profile used in the FE analysis consists of four different layers: Made Ground, Thames Gravel, London Clay, and Lambeth

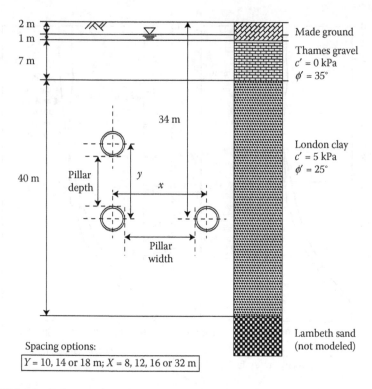

FIGURE 3.66 Twin tunnel configurations considered in FE analysis. (From Addenbrooke, T.I., and Potts, D.M., *International Journal of Geomechanics*, 1(2), 2001, 249–268. With permission.)

Sand (Figure 3.66). The Made Ground layer was modeled as linear elastic, and the Thames Gravel and London Clay were modeled as nonlinear elastic perfectly plastic materials with Mohr–Coulomb yield surfaces and plastic potentials. The preyield behaviors of the Thames Gravel and London Clay were assumed to exhibit stiffness decay with strain level and stiffness variation with mean effective stress p' [72]. In the case of London Clay, high stiffness behavior was reinvoked upon the detection of stress path reversal through monitoring of increment of normalized shear stress and mean stress. Coupled consolidation analysis was employed by considering the construction of the tunnels over a simulated time period of 8 h, with two alternative rest periods of 3 weeks and 7 months. The Made Ground and Gravel were assumed to be nonconsolidating materials. The London Clay was assumed to be a consolidating material, with an isotropic permeability of 1×10^{-10} m/s. The parameters for the decay of stiffness (secant shear modulus, G, and secant bulk modulus, K) with strain level are given in Table 3.14a, where E_d is the deviatoric strain invariant, ε_v is the volumetric strain, and C_1, C_2, C_3, C_4, C_5, C_6, c_1, c_2, c_3, and c_4 are coefficients related to the decay. E_{dmin}, E_{dmax}, ε_{vmin}, and ε_{vmax} provide strain limits and G_{min} and K_{min} provide stiffness limits. The details are given by Addenbrooke and Potts [72].

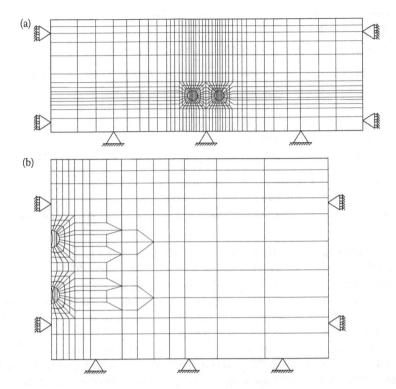

FIGURE 3.67 (a) Finite element mesh for side-by-side configuration (spacing = 12 m; axis depth = 34 m); (b) finite element mesh for piggyback configuration (axis depths 20 m and 34 m). (From Addenbrooke, T.I. and Potts, D.M., *International Journal of Geomechanics*, 1(2), 2001, 249–268. With permission.)

The constitutive model parameters for London Clay are given in Table 3.14b, where a_{LER} and a_{SSR} are parameters defining the size of the linear strain region (LER) and the small strain region (SSR), respectively. The a_{LER} and a_{SSR} were determined from undrained triaxial tests [72]. The tunnel lining was assumed as elastic with Young's modulus and Poisson's ratio values as 28×10^6 kPa and 0.15, respectively. The cross-sectional area and the second moment of area were assumed as 0.168 m²/m and 3.95136×10^{-4} m⁴/m, respectively.

TABLE 3.14a

Material Parameters Defining Decay in Shear Modulus and Bulk Modulus

Strata	C_1	C_2	C_3 (%)	c_1	c_2	$E_{d\,min}$ (%)	$E_{d\,max}$ (%)	G_{min} (kPa)
Thames Gravel	1380	1248	5×10^{-4}	0.974	0.940	8.83346×10^{-4}	0.3464	2000
	C_4	C_5	C_6 (%)	c_3	c_4	$\varepsilon_{v\,min}$ (%)	$\varepsilon_{v\,max}$ (%)	K_{min} (kPa)
	275	225	2×10^{-3}	0.998	1.044	2.1×10^{-3}	0.2	5000

TABLE 3.14b

Properties of London Clay Used in the FE Analysis

Parameter	Value	Parameter	Value
u/p'_{global}	5.851064×10^{-4}	$K_{LER}^{ref}/[p'^{ref}]^{\gamma}$	414.33
B	0.0643	B	1.0
a_{LER}/p'_{global}	2.524468×10^{-3}	γ	1.0
a_{SSR}/p'_{global}	0.1913149	K_{min}	3000.0
$K_{LER}^{ref}/[p'^{ref}]^{\beta}$	214.18	G_{min}	2333.3

Source: From Addenbrooke, T.I. and Potts, D.M., *International Journal of Geomechanics*, 1(2), 2001, 249–268. With permission.

Mohr–Coulomb yield surface parameters, plastic potential parameters, and unit weight values used in the FE analysis are given in Table 3.14c. The details of the constitutive models are given by Addenbrooke and Potts [72].

Boundary conditions and initial stresses: The boundary conditions shown in Figures 3.67a and 3.67b allowed only vertical displacement on vertical boundaries. No vertical and horizontal displacements were allowed at nodes on the horizontal boundary at the bottom of the mesh. The hydraulic boundary conditions assumed no pore pressures in the Made Ground and Thames Gravel at all times. No change in pore pressure was assumed along the remote boundaries and no flow was assumed across a line of symmetry boundary. During the rapid construction of each tunnel (8 h), no flow was permitted across the excavation boundary. During the rest period and construction of the second tunnel, a special boundary was used for the hydraulic behavior of the lining that controlled the lining permeability depending upon whether pore water pressures in the adjacent soils were compressive or tensile.

Tunnel excavation: To simulate tunnel excavation, first, the nodal forces representing the stresses that the soil within the tunnel applied to the tunnel boundary were evaluated. These forces were divided by the number of excavation increments, and the incremental forces were applied in the reverse direction over the prescribed number of increments. The movement of the tunnel boundary was monitored at each increment and used to calculate the volume of soil moving into the tunnel.

TABLE 3.14c

Properties of Various Soils

	Made Ground	Thames Gravel	London Clay
Strength parameters	Linear elastic	$c' = 0$ kPa	$c' = 5$ kPa
		$\varphi' = 35°$	$\varphi' = 25°$
Dilation angle	Linear elastic	$v' = 17.5°$	$v' = 12.5°$
Bulk unit weight (kN/m³)	$\gamma_{dry} = 18$	$\gamma_{sat} = 20$	$\gamma_{sat} = 20$
	$\gamma_{sat} = 20$		

Lining construction was modeled by the activation of the structural beam elements with appropriate mechanical properties. After the activation of the lining construction, the loading boundary condition that models excavation was still applied up to completion of the final increment of excavation. Additional details of the excavation procedure are given by Addenbrooke and Potts [72].

Results: Figure 3.68a shows the surface settlement profile developed above the first tunnel 3 weeks after construction (i.e., just before constructing the second tunnel). Field measurements (between 2 and 6 weeks after the tunnel had passed the instrumented section) were superimposed for comparison. Although the field measurements showed significant scatter, overall, the FE predicted displacements were within the measurements. Figure 3.68b shows the predicted settlements due to the excavation of the second tunnel. The four profiles represent the four different spacing (between tunnels) considered in the analysis. The construction of the second tunnel

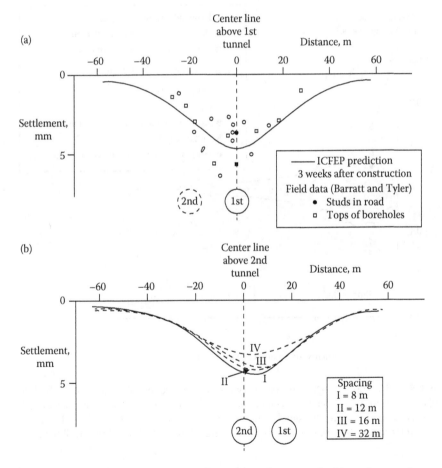

FIGURE 3.68 (a) Surface settlement above 34 m deep tunnel; (b) surface settlement above second tunnel. (From Addenbrooke, T.I. and Potts, D.M., *International Journal of Geomechanics*, 1(2), 2001, 249–268. With permission.)

(a)

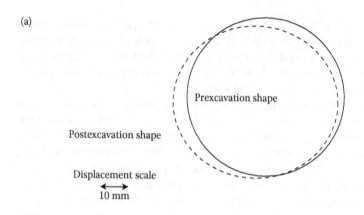

(b)

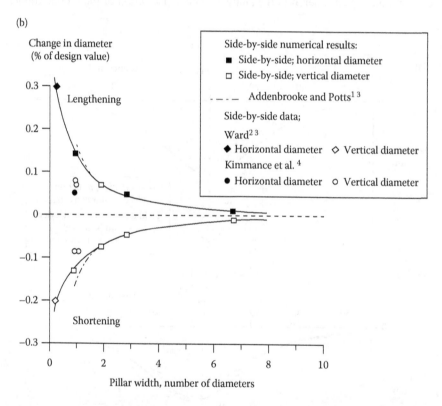

FIGURE 3.69 (a) Distortion to first lining due to construction of second tunnel (both tunnels 34 m deep with a center to center spacing of 8 m); (b) deformation of first tunnel in response to second tunnel for side-by-side configuration. (From Addenbrooke, T.I. and Potts, D.M., *International Journal of Geomechanics*, 1(2), 2001, 249–268. With permission.)

induced asymmetry (or a shift or eccentricity) in the displacement profile. As noted by Addenbrooke and Potts [72], when the ratio between this shift (or eccentricity e) and maximum surface displacement (S_{max}) was 1.0, the settlement profile above the second tunnel was centered over the first tunnel. When this ratio was zero, the settlement profile above the second tunnel was centered over itself. It was noted that for spacing between twin tunnels exceeding 7D, D being the tunnel diameter, the influence of the first tunnel on the second was negligible.

Figures 3.69a and 3.69b show the influence of the excavation of the second tunnel on the lining to the first. The existing lining was pulled to the new excavation and a squatting deformation was induced. Figure 3.69b shows an increase in the horizontal diameter and a decrease in the vertical diameter of the lining due to the excavation of the second tunnel. These distortions reduced with the increase in spacing between the tunnels, as expected.

Figure 3.70 shows the final accumulated surface settlement profiles immediately after the excavation of the second tunnel (upper tunnel constructed first) for different tunnel depths. For each spacing, the excavation of the upper second tunnel resulted in a shallower, wider profile. The difference in profile shape was more noticeable for more closely spaced tunnels as expected.

3.5.12 EXAMPLE 3.12: FIELD BEHAVIOR OF REINFORCED EARTH RETAINING WALL

A geosynthetic-reinforced soil retaining wall using full-height concrete wall facing panel constructed at Tanque Verde Road site for grade-separated interchanges in Tucson, Arizona, USA was analyzed using the finite element method (FEM) with realistic constitutive models for soils and interfaces.

3.5.12.1 Description of Wall

Between November 1984 and 1985, 43 geogrid-reinforced walls were constructed at Tanque Verde Road site for grade-separation interchanges on the Tanque Verde–Wrightstown–Pantano Road project in Tucson, Arizona, USA. This project represents the first use of geogrid reinforcement in mechanically stabilized earth (MSE) retaining walls in a major transportation-related application in North America (Tensar [73]). In this study, the behavior of the instrumented wall panel nos. 26–32 is simulated.

The Tucson wall height was 4.88 m (16.0 ft). The reinforced soil mass was faced with 15.24 cm (6.0 in) thick and 3.05 m (10.0 ft) wide precast reinforced concrete panels. Soil reinforced geogrids were mechanically connected to the concrete facing panels at elevations shown in Figure 3.71, and extended to a length of 3.66 m (12.0 ft). On the top of the wall fill, a pavement structure was constructed that consisted of 10.16 cm (4.0 in) base course covered by 24.13 cm (9.5 in) of Portland cement concrete. Details of the various geometries are reported by Berg et al. [74], Fishman et al. [75,76], and Fishman and Desai [77]; the latter presents a linear finite element analysis of the wall.

The soil reinforcement used was Tensar's SR2 structural geogrid; it is a uniaxial product that is manufactured from high-density polyethylene (HDPE) stabilized

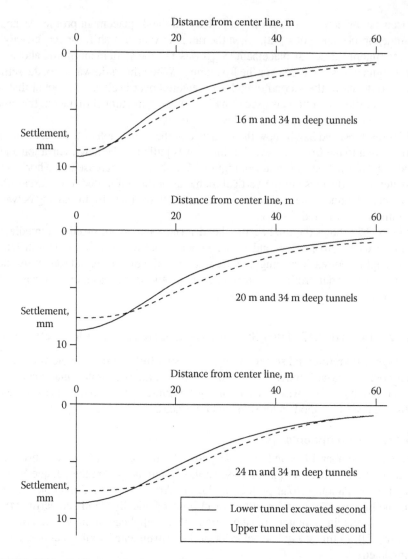

FIGURE 3.70 Surface settlement profiles after completion of both piggyback tunnels. (From Addenbrooke, T.I. and Potts, D.M., *International Journal of Geomechanics*, 1(2), 2001, 249–268. With permission.)

with about 2.5% carbon black to provide resistance to attack by ultraviolet (UV) light [73,78]. The geogrids have maximum tensile strength of 79 kN/m (5400 lb/ft) and a secant modulus in tension at 2% elongation of 1094 kN/m (75,000 lb/ft). The allowable long-term tensile strength based on creep considerations is reported to be 29 kN/m (1986 lb/ft) at 10% strain after 120 years. This value was reduced by an overall factor of safety equal to 1.5 to compute a long-term tensile strength equal to 19 kN/m (1324 lb/ft).

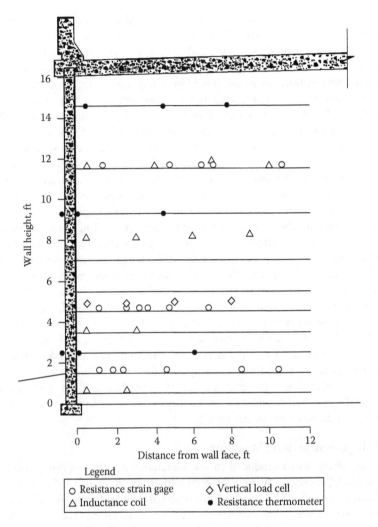

FIGURE 3.71 Locations of instruments for wall panel 26–32. (1 ft = 0.305 m). (Adapted from Tensar Geogrid Reinforced Soil Wall, *Experimental Project 1, Ground Modification Techniques,* FHWA-EP-90-001-005, Department of Transportation, Washington, DC, 1989.)

3.5.12.2 Numerical Modeling

The numerical analysis of the reinforced soil wall was performed using a finite element code called DSC-SST-2D developed by Desai [79]. The program allows for plain strain, plain stress and axisymmetric idealizations including simulation of construction sequences. Various constitutive models, elastic, elasto-plastic (von Mises, Drucker–Prager, Mohr–Coulomb, Hoek–Brown, Critical State, and Cap), hierarchical single surface (HISS), viscoelastic, plastic, and disturbance-DSC (softening) can be chosen for the analysis. The wall was modeled as a plane-strain, two-dimensional problem. Since the Tensar reinforcement is continuous normal to the cross section,

Figure 3.71, the plane strain idealization is considered to be appropriate. Details of the computer analysis with DSC/HISS models are given in Refs. [38,80].

Two finite element meshes, coarse and fine, were used. Figure 3.72a shows the coarse mesh with 184 nodes and 167 elements including 10 wall facing, 18 interface between soil and reinforcement, and 9 bar (for reinforcement) elements. In the coarse mesh, only three layers of reinforcement were considered. The fine mesh contained 1188 nodes, and 1370 elements, including 480 interface, 35 wall facing, and 250 bar elements; it contained 10 layers of reinforcement as in the field. The dimensions for the fine mesh were the same as the coarse mesh; a part of the fine mesh near the reinforcement is shown in Figure 3.72b. It was found that the fine mesh provided satisfactory and improved predictions compared to those from the coarse mesh. Hence, most of the results are presented for the fine mesh; however, typical comparisons from the coarse mesh are included to show the improvement from the fine mesh.

It was assumed that the relative motions between the backfill and reinforcement have significant effect on the behavior. Hence, interface elements were provided between backfill and reinforcement. It was also assumed that the relative motions between wall facing and backfill soil in this problem may not have significant influence; hence, interface elements were not provided. This is discussed later under Displacements.

The meshes involved four-node quadrilateral elements for soil, wall and interfaces, and one-dimensional elements for the reinforcement. As shown in Figure 3.72a, the nodal points at the bottom boundary were fixed, and those on the side boundaries were fixed only in the horizontal direction. The side boundaries were placed at a distance of 2.5 times the length of the reinforcement, and the bottom boundary was placed at a distance of 3.125 times the height of the wall. Such distances and the assumed boundary conditions are considered to approximately simulate the semi-infinite extent of the system.

3.5.12.3 Construction Simulation

The *in situ* stress was introduced in the foundation soil by adopting coefficient, $K_o = 0.4$. Then the backfill was constructed into 11 layers, Figure 3.72b, as was done in the field, Figure 3.71. The soil was compacted in each layer, and the reinforcement was placed on a layer before the next soil layer was installed. The compacted soil in a given layer was assigned the material parameters according to the stress state induced after installing the layer. The completion of the sequences of construction is referred to as "end of construction." Then the surcharge load due to the traffic of 20 kPa was applied uniformly on the top of the mesh, Figure 3.72; this stage is referred to as "after opening to traffic." The concrete pavement was not included in the mesh. However, since it can have an influence on the behavior of the wall, in general, it is desirable to include the pavement.

3.5.12.4 Constitutive Models

The properties of the materials used in the analysis were obtained from experimental results from conventional triaxial compression (CTC) tests for soil backfill, and cyclic multi-degree-of-freedom (CYMDOF) shear tests for interfaces, Desai and Rigby [71]. The test data were used to find the parameters for the DSC/HISS models used for soils and interfaces.

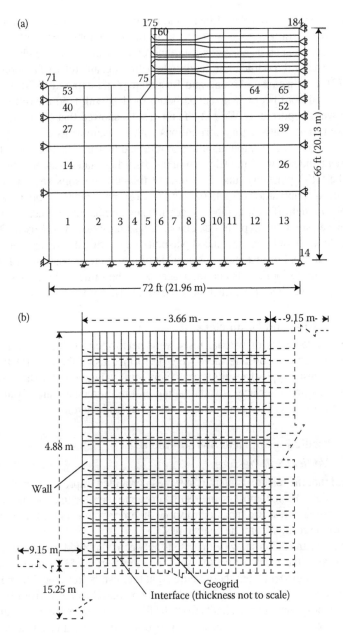

FIGURE 3.72 Coarse and part of fine mesh. (a) Coarse mesh; (b) mesh near geogrid in fine mesh. (Adapted from Desai, C.S. and El-Hoseiny, K.E., *Journal of the Geotechnical and Geoenvironmental Engineering*, 131(6), June 2005, 729–739.)

Soil and Interface Modeling: Linear and nonlinear elastic (e.g., hyperbolic simulation of stress–strain curves) were found to be not suitable for reinforced earth affected by factors such as elastic, plastic and creep strains, volume change, stress path, microstructural modifications leading to softening, and relative motion at interfaces [71,77]. For instance, Ahmad and Basudhar [81] compared FEM predictions using linear and nonlinear elastic (hyperbolic) models with field measurements for reinforced walls reported by Wu [82]. Their results showed that the elastic models used did not provide satisfactory correlations with measurements.

Hence, the soil (backfill, foundation, and retained fill) and the interfaces between reinforcement and soil were modeled using the disturbed state concept (DSC), which included the HISS plasticity model. Details of the DSC for soils and interfaces are given in various publications [13,45,62]; and are also presented in Appendix 1.

The DSC model offers a number of advantages compared to other models such as nonlinear elastic (e.g., hyperbolic), classical plasticity (e.g., von Mises, Drucker–Prager and Mohr–Coulomb), advanced plasticity (e.g., critical and cap) and classical damage (softening). For instance, it is capable of hierarchical accounting for factors presented before, for both soils and interfaces with the same basic framework, with smaller or same number of parameters compared to other available models [62].

3.5.12.5   Testing and Parameters

A comprehensive series of triaxial tests were performed on the soils. The shear tests on reinforcement–soil interfaces were performed using the CYMDOF device. Details of the tests, typical results, parameters, and validations for the DSC/HISS models for soils and interfaces, for the facing and reinforcement are given in Refs. [38,80]; Table 3.15 gives the parameters used in the study herein.

3.5.12.6   Predictions of Field Measurements

3.5.12.6.1   Vertical Soil Pressure

It was found that the results using the finer mesh provided improved correlation with the field test data. Hence, most of the results presented are for the fine mesh; typical results for vertical stress are included to show the improvements from the fine mesh compared to the results from the coarse mesh.

Figures 3.73a and 3.73b show comparisons between computed vertical soil stresses from coarse and fine mesh, respectively, at the elevation = 1.53 m at the end of the construction. It is evident from these figures that the results from the fine mesh show much improved correlation with the field data, compared to those from the coarse mesh. Hence, now onwards the results from the fine mesh are presented and analyzed.

The measured and predicted vertical soil stress near the wall face is less than the overburden value of about 60.30 m (= γh = 18 × 3.35; h = 4.88 − 1.53 m). This can be due to the relative motions between the backfill and reinforcement. It is seen that the vertical stress distribution along the reinforcement layer is nonlinear. The vertical pressure increases in the zone away from the facing panel until reaching a maximum value at a distance of about 1.50 m from the wall face. Thereafter, it shows a decrease. Also, shown in Figure 3.73 are the trapezoidal vertical stress distributions used in the design calculations. The predicted results from finite element analysis

TABLE 3.15

Parameters for Backfill Soil and Interfaces

Material Constant	Symbol	Soil	Interface
Elastic	E or K_n	$f_1\,(\sigma_3)^a$	$f_2\,(\sigma_n)^b$
	ν or K_s	0.3	$f_3\,(\sigma_n)$
Plasticity—ultimate	γ	0.12	2.3
	β	0.45	0.0
Phase change parameter	n	2.56	2.8
Growth parameters	a_1	3.0E–05	0.03
	η_1	0.98	1.0
Nonassociative constant	κ	0.2	0.4
	D_u	0.93	
Disturbance parameters	A	0.37	
	Z	1.60	
Angle of friction and adhesion	$\phi/\delta/c_a$	$\phi = 40°$	$\delta = 34°$
			$c_a = 66$ kPa
Unit weight (field)	γ	18.00 kN/m³	
Coefficient of earth pressure at rest	K_o	0.4	

Source: Desai, C.S. and El-Hoseiny, K.E., *Journal of the Geotechnical and Geoenvironmental Engineering*, 131(6), June 2005, 729–739; El-Hoseiny, K.E, and Desai, C.S., Computer Analysis of Reinforced Earth Retaining Walls with Testing and Constitutive Modeling of Soils and Interfaces, PhD Dissertation and Report, Minufiya University, Minufiya, Egypt; Dept. of Civil Eng. and Eng. Mechanics, Univ. of Arizona, Tucson, AZ, USA, 2004.

[a] $E = 62 \times 10^3 \sigma_3^{0.28}$
[b] K_s (shear stiffness) $= 30 \times 10^3 \sigma_n^{0.28}$; K_n (normal stiffness) $= 18 \times 10^3 \sigma_n^{0.29}$.

agree well with the measured values, but they are not in good agreement with the assumption of a linear distribution of vertical pressure used for the tie-back wedge analysis; such a design assumption neglects interaction in the reinforced wall.

3.5.12.6.2 Lateral Earth Pressure against Facing Panel

The distribution of lateral earth pressure on the wall facing was measured based on the four pressure cells located at or near the wall face, about 0.61, 1.22, 2.44, 3.66, and 4.8 m distance from the bottom of the wall. The earth pressure against the facing panel was obtained in the finite element analysis from the horizontal stress in the soil elements near the facing. This pressure distribution is useful for evaluating the magnitude of the stresses exerted on the facing panels and the tension in the geogrid connection. Figure 3.74 shows the typical predicted and measured lateral soil pressure behind the facing panel after opening to traffic, along with the Rankine distribution. Predicted and measured horizontal soil stresses agree very well. The design procedure assumed that no significant lateral earth pressure should be transferred to the reinforcement. Except at the bottom of the wall, the low value of the horizontal soil stress on the wall panel approximately confirms this assumption.

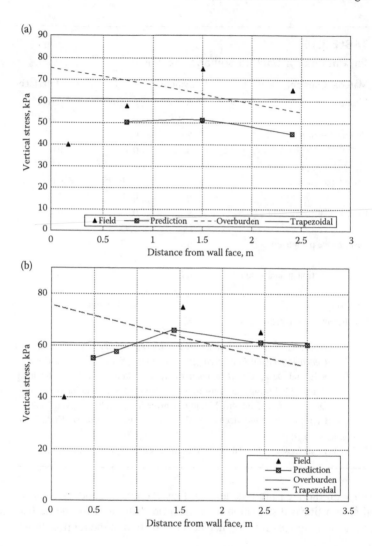

FIGURE 3.73 Comparisons between field measurements and predictions of vertical soil Stresses at elevation 1.53 m at the end of construction. (a) Coarse mesh; (b) fine mesh. (Adapted from Desai, C.S. and El-Hoseiny, K.E., *Journal of the Geotechnical and Geoenvironmental Engineering*, 131(6), June 2005, 729–739.)

3.5.12.6.3 Geogrid Strains

Measured and predicted reinforcement tensile strains at elevations of 1.37 and 4.42 m are shown in Figures 3.75a and 3.75b. Agreement between the measured and predicted values is considered very good. The results demonstrate that tensile strains in the geogrids are less than 0.4% corresponding to 4.4-kN/m load in the geogrid. Comparison of this load to the maximum tensile strength of the geogrid, which is 79 kN/m, indicates that the grids are loaded to about 6.0% of the ultimate strength.

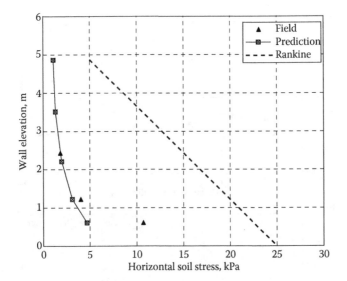

FIGURE 3.74 Comparison of field measurements and predictions for horizontal soil stress after opening to traffic. (Adapted from Desai, C.S. and El-Hoseiny, K.E., *Journal of the Geotechnical and Geoenvironmental Engineering*, 131(6), June 2005, 729–739.)

3.5.12.6.4 Stress Carried by Geogrid

Figure 3.76 shows comparison between measured and predicted results at different elevations for horizontal stress in the geogrid near the wall face. The measurements are obtained from the strain gages installed on the geogrid. The predicted geogrid stresses compare well with the measurements.

3.5.12.6.5 Displacements

Figure 3.77 shows predicted and measured wall movements. The correlation is satisfactory near the lower heights of the wall; however, it is not satisfactory elsewhere. For example, near the top of the wall the predicted value of about 42 mm is not in good agreement with the measured value of about 76 mm. The finite element analysis using linear elastic model reported the maximum displacement of about 30 mm (77); with the present nonlinear soil and interface models, the maximum displacement increased to about 42 mm (Figure 3.77).

A main reason for the discrepancy is considered to be possible errors in the measurements. It is believed that since other measurements compare well with the predictions, the displacements from the finite element predictions can be considered to be reasonable.

The magnitude of the maximum wall displacement, δ_{max}, can be estimated from the following equation, Christopher, et al. [83]:

$$\delta_{max} = \delta_r \times H/75 \tag{3.40}$$

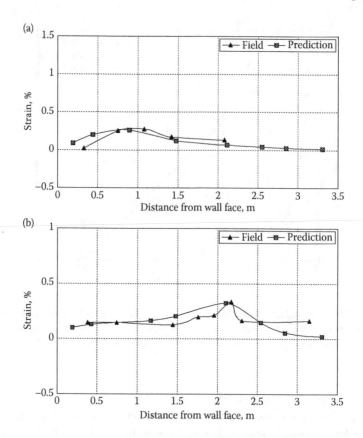

FIGURE 3.75 Comparison between field measurements and predictions for geogrid strains. (a) Elevation = 1.37 m; (b) elevation = 4.42 m. (Adapted from Desai, C.S. and El-Hoseiny, K.E., *Journal of the Geotechnical and Geoenvironmental Engineering*, 131(6), June 2005, 729–739.)

where δ_r = relative displacement found from the chart based on L/H ratio, H = wall height and L = reinforcement length. According to Equation 3.40, the $\delta = \approx 60$ mm, which also does not compare well with the measured value of about 76 mm?

From Figure 3.77, it can be seen that the wall rotates about the toe of the wall. Also, the displacements of the wall and the soil strains are not high [38,80]. The maximum displacement is about 1.5% with respect to the wall height. It appears from this behavior that there is no significant relative motion between the wall and soil for this problem. Hence, it may not be necessary to provide interface elements between the wall and backfill soil for the problem considered herein.

3.5.12.6.6 Comments

From this study, it can be concluded that the use of realistic constitutive models for soils and interfaces is essential for satisfactory predictions of the behavior of geotechnical structures such as reinforced earth retaining walls.

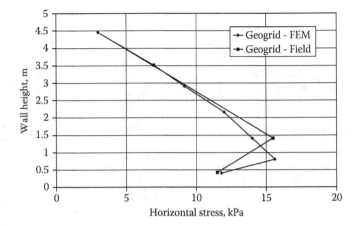

FIGURE 3.76 Comparison of field measurements and predictions for horizontal stress carried by geogrid near wall face. (Adapted from Desai, C.S. and El-Hoseiny, K.E., *Journal of the Geotechnical and Geoenvironmental Engineering,* 131(6), June 2005, 729–739.)

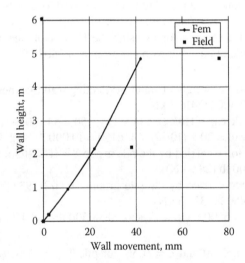

FIGURE 3.77 Comparison between predicted and measured wall face movement after opening to traffic. (Adapted from Desai, C.S. and El-Hoseiny, K.E., *Journal of the Geotechnical and Geoenvironmental Engineering,* 131(6), June 2005, 729–739.)

PROBLEMS

Problem 3.1: Two-Dimensional Pile Group (Inclined Piles)

A 2-D idealization of a water storage tank supported on a pile group is shown in Figure P3.1. The tank has a height of 20 ft (6.09 m) and a wall thickness of 10 in (0.254 m). The thickness of pile cap is 18 in (0.457 m). The water level in the tank is

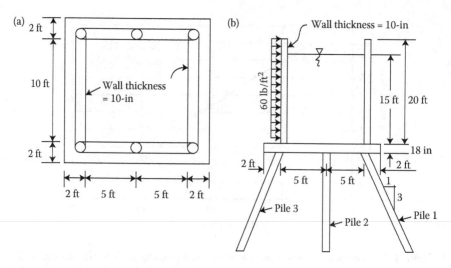

FIGURE P3.1 Two-dimensional pile group (inclined piles).

15 ft (4.57 m). The tank is subjected to a uniform wind load of 60 lb/ft² (2.87 kPa).
Given $r_1 = (t_s/n) = 0.5$, $r_2 = (m_s/n) = 26$ in (10.2 cm), and $r_3 = (m_\alpha/n) = 2700$ in²
(418.5 cm²), determine the pile forces using the Hrennikoff method. The pile cap
and the tank are made of concrete with unit weight of 150 lb/ft³.

> *Partial solution:* Total weight of water (neglecting wall thickness) = 10 × 10 ×
> 15 × 62.4 = 93,600 lb (416.3 kN)
> Total weight of pile cap = (18/12) × 14 × 14 × 150 = 44,100 lb (196.2 kN)
> Weight of tank wall = 20 × (10/12) × 4 × 150 = 10,000 lb (44.5 kN)
> Horizontal wind force carried by the pile group X = (1/2) [60 × 20 × (10 +
> (10/12))] = 6500 lb (28.9 kN)
> Total vertical force carried by the pile group Y = (1/2) (93,600 + 44,100 +
> 10,000) = 73,850 lb (328.5 kN)
> Total moment M = 6500 × (10 × 12 + 9) = 838,500 lb-in (94.7 kN-m)

Using $x = 60$ in (1.52 m) and $\varphi = 71.56°$ for pile 1, $x = 0$ and $\varphi = 90°$ for pile 2,
$x = -60$ in (−1.52 m) and $\varphi = 108.44°$ for pile 3, and the ratios ($r_1 = 0.5$, $r_2 = 26$ in
(10.2 cm), and $r_3 = 2700$ in² (418.5 cm²)), the pile constants (X_x', X_y', M_x', Y_y', M_y',
and M_α) can be evaluated from Equations 3.27a through 3.27f. Knowing the pile
constants and the total forces (X and Y) and moment (M) (at the center of the pile
cap), the pile cap response ($\Delta x'$, $\Delta y'$, and $\Delta \alpha'$) can be evaluated from Equations 3.26a
through 3.26c as follows:

−1.60	0.00	57.31	$\Delta x'$	=	−6500
0.00	−2.90	−0.01	$\Delta y'$	=	−73,850
57.31	−0.01	−16,914.25	α	=	−83,8500

Knowing $\Delta x'$, $\Delta y'$, and $\Delta \alpha'$, the forces (P and Q) and moment (S) can be evaluated from Equations 3.28a through 3.28c, respectively.

Pile Number	Pile 1	Pile 2	Pile 3
Axial force, P (lb)	30,351.2	17,969.6	25,465
Transverse force, Q (lb)	3454.3	−4610.2	−1442.7
Pile head moment, S (lb-in)	−276,712	142,642	−22,066.6

1 kip = 4448 N; 1 kip-in = 113 N-m.

Problem 3.2: Two-Dimensional Pile Group (Vertical Piles)

Determine the pile forces in Problem 3.1 (Figure P3.1) if all three piles are vertical.

Partial solution: Here, all data are the same as in Problem 3.1, except $\varphi = 90°$ for all three piles. For this case, the equilibrium equations become

−1.50	0.00	78.00	$\Delta x'$	=	−6500
0.00	−3.00	0.00	$\Delta y'$	=	−73,850
78.00	0.00	−15,300.00	A'	=	−83,8500

The corresponding pile forces and moments become

Pile Number	Pile 1	Pile 2	Pile 3
Axial force, P (lb)	30,894.7	18,338.6	24,616.7
Transverse force, Q (lb)	−2166.7	−2166.7	−2166.7
Pile head moment, S (lb-in)	−28,379.4	−28,379.4	−28,379.4

1 kip = 4448 N; 1 kip-in = 113 N-m.

REFERENCES

1. Desai, C.S. and Abel, J.F., *Introduction to Finite Element Method*, Van Nostrand Reinhold Company, New York, 1972.
2. Desai, C.S. and Christian, J.T., *Numerical Methods in Geotechnical Engineering*, McGraw-Hill Book Co., New York, NY, USA, 1977.
3. Zienkiewicz, O.C. and Taylor, R.L., *The Finite Element Method*, McGraw-Hill Book Co., New York, NY, USA, 1989.
4. Bathe, K.J., *Finite Element Procedures in Engineering Analysis*, Prentice-Hall, Englewood Cliffs, NJ, USA, 1982.
5. Timoshenko, S., *Advanced Strength of Materials*, Parts I and II, D. Van Nostrand Co., New York, USA, 1952.
6. Abramowitz, M. and Stegun, I.A. (Eds), *Handbook of Mathematical Functions with Formulas, Graphs and Mathematical Tables*, National Bureau of Standards, Applied Math Series 55, Washington, DC, 1964.
7. Desai, C.S. and Siriwardane, H.J., *Constitutive Laws of Engineering Materials*, Prentice-Hall, Englewood Cliffs, NJ, USA, 1984.

8. Chen, W.F. and Han, D.J., *Plasticity for Structural Engineers*, Springer-Verlag, New York, 1988.

9. Roscoe, K.H., Schofield, A. and Wroth, C.P., On the yielding of soils, *Geotechnique*, 8, 1958, 22–53.

10. Desai, C.S., Somasundaram, S. and Frantziskonis, G., A hierarchical approach for constitutive modeling of geologic materials, *International Journal of Numerical and Analytical Methods in Geomechanics*, 10(3), 1986, 225–257.

11. Desai, C.S. and Toth, J., Disturbed state constitutive modeling based on stress-strain and nondestructive behavior, *International Journal of Solids and Structures*, 33(11), 1996, 1619–1650.

12. Mühlhaus, H.B. (Ed.), *Constitutive Models for Materials with Microstructure*, John Wiley, Chichester, UK, 1995.

13. Desai, C.S., and Ma, Y., Modelling of joints and interfaces using the disturbed state concept, *International Journal of Numerical and Analytical Methods in Geomechanics*, 16, 1992, 623–653.

14. Goodman, L.E. and Brown, C.B., Dead load stresses and the instability of slopes, *Journal of the Soil Mechanics and Foundations Division, ASCE*, 89(SM3), May 1963, 103–134.

15. Brown, C.B. and King, I.P., Automatic embankment analysis, *Geotechnique*, XVI(3), Sept. 1966, 209–219.

16. Kulhawy, F.H., Duncan, J.M. and Seed, H.B., Finite Element Analysis of Stresses and Movements in Embankments during Construction, Contract Report, No. TE-69–4, U.S. Army Engineers Waterways Expt. Stn., Nov. 1969.

17. Desai, C.S., Overview, trends and projections: Theory and applications of the finite element method in Geotechnical engineering, *State-of-the-Art Paper, Proceedings of the Symposium on Applications of the Finite Element Method in Geotechnical Engineering*, US Army Waterways Expt. Station, Vicksburg, MS, USA, 1972.

18. Clough, R.W. and Woodward, R.J., Analysis of embankment stresses and deformations, *Journal of the Soil Mechanics and Foundations Division, ASCE*, 93(SM4), July 1967.

19. Chowdhury, R.N., Slope analysis, *Developments in Geotechnical Engineering*, Vol. 22, Elsevier Scientific Publ. Co., Amsterdam, The Netherlands, 1978.

20. Desai, C.S., Solution of Stress-Deformation Problems in Soil and Rock Mechanics Using Finite Element Methods, PhD Dissertation, Dept. of Civil Eng., Univ. of Texas, Austin, TX, 1968.

21. Desai, C.S. and Reese, L.C., Analysis of circular footings on layered soils, *Journal of the Soil Mechanics and Foundations Division, ASCE*, 96(SM4), July 1970, 1289–1310.

22. Terzaghi, K., *Theoretical Soil Mechanics*, John Wiley and Sons, Inc., New York, 1962.

23. Skempton, A.W., The bearing capacity of clays, *Building Research Congress*, 1951, 180.

24. Meyerhof, G.G., Ultimate bearing capacity of foundations, *Geotechnique*, London, 2, 1951, 301.

25. Hashmi, Q.S.E. and Desai, C.S., Nonassociative Plasticity Model for Cohesionless Materials and Its Implementation in Soil-Structure Interaction, Report to National Science Foundation, Dept. of Civil Eng. and Eng. Mechanics, Univ. of Arizona, Tucson, AZ, 1987.

26. Desai, C.S. and Hashmi, Q.S.E., Analysis, evaluation, and implementation of a nonassociative model for geologic materials, *International Journal of Plasticity*, 5(4), 1989, 397–420.

27. Nagaraj, B.K. and Desai, C.S., Modelling of Normal and Shear Behavior in Dynamic Soil-Structure Interaction, Report to National Science Foundation, Dept. of Civil Eng. and Eng. Mechanics, Univ. of Arizona, Tucson, AZ, 1986.

28. Fruco and Associates, Pile Driving and Load Tests: Lock and Dam No. 4, Arkansas River and Tributaries, Arkansas and Oklahoma, United States Army Engineer District, Corps of Engineers, Little Rock, AR, Sept. 1964.

29. Mansur, C.I. and Hunter, A.H., Pile tests—Arkansas river project, *Journal of the Soil Mechanics and Foundations Division, ASCE,* 96(SM5), Sept. 1970, 1545–1582.
30. Desai, C.S., Finite Element Method for Design Analysis of Deep Pile Foundations, Technical Report I, U.S. Army Engineer Waterways Expt. Stn., Vicksburg, MS, 1974.
31. Desai, C.S., Numerical design-analysis for piles in sands, *Journal of the Geotechnical Engineering Division, ASCE,* 100(GT6), June 1974, 613–635.
32. Desai, C.S., Finite Element Method for Analysis and Design of Piles, Misc. Paper S-76-21, U.S. Army Engineer Waterways Expt. Stn., Vicksburg, MS, 1976.
33. Kulhawy, F.H., Duncan, J.M. and Seed, H.B., Finite Element Analyses of Stresses and Movements in Embankments during Construction, Report S-69-8, U.S. Army Engr. Waterways Expt. Stn., Vicksburg, MS, Nov. 1969.
34. Clough, G.W. and Duncan, J.M., Finite Element Analyses of Port Allen and Old River Locks, Report S-69-6, U.S. Army Engr. Waterways Expt. Stn., Vicksburg, MS, Sept. 1969.
35. Ellison, R.D. et al., Load deformation mechanism of bored piles, *Journal of the Soil Mechanics and Foundations Division, ASCE,* 97(SM4), April 1971, 661–678.
36. Robinsky, E.I. and Morrison, C.F., Sand displacement and compaction around model friction piles, *Canadian Geotechnical Journal,* 1(2), March 1964, 81–93.
37. Vesic, A.S., A Study of Bearing Capacity of Deep Foundations, Report, Project B-189, Georgia Inst. of Tech., Atlanta, GA, March 1967.
38. Desai, C.S. and El-Hoseiny, K.E., Prediction of field behavior of geosynthetic reinforced soil wall, *Journal of the Geotechnical and Geoenvironmental Engineering,* 131(6), June 2005, 729–739.
39. Handy, R.L. and Spranger, M.G., *Geotechnical Engineering: Soil and Foundation Principles and Practice,* McGraw-Hill Professional Publishing, New York, 2006.
40. Hrennikoff, A., Analysis of pile foundations with batter piles, *Transactions, American Society of Civil Engineers,* 115, Paper No. 2401, 1950, 351–181.
41. Fang, Y.S. and Ishibashi, I., Static earth pressure with various wall movements, *Journal of Geotechnical Engineering, ASCE,* 112(GT3), 1986, 317–333.
42 Ugai, K. and Desai, C.S., Application of Hierarchical Plasticity Model for Prediction of Active Earth Pressure Tests, Report, Dept. of Civil Eng. and Eng. Mechanics, Univ. of Arizona, Tucson, AZ, USA, 1994.
43. Ugai, K. and Desai, C.S., Application of nonassociative hierarchical model for geologic materials for active earth pressure experiments, *Short Communication, International Journal of Numerical and Analytical Methods in Geomechanics,* 14(8), 1995, 573–580.
44. Desai, C.S., Manual for DSC-SST2D Computer Code for Static and Dynamic Solid, Structure and Soil-Structure Analysis, Reports I, II And III, Tucson, AZ, USA, 1992.
45. Desai, C.S., Zaman, M.M., Lightner, J.G. and Siriwardane, H.J., Thin-layer element for interfaces and joints, *International Journal of Numerical and Analytical Methods in Geomechanics,* 8, 1984, 19–43.
46. Clough, G.W. and Duncan, J.M., Finite element analysis of retaining wall behavior, *Journal of Soil Mechanics and Foundations Divisions, ASCE,* 97(SM12), Dec. 1971, 1657–1673.
47. Terzaghi, K., Large retaining wall tests, I: Pressure on dry sand, *Engineering News Record,* III, Feb. 1934, 136–140.
48. Clough, G.W. and Duncan, J.M., Finite Element Analyses of Port Allen and Old River Locks, Report S-69-6, U.S. Army Engineers Waterways Experiment Station, Corps of Engineers, Vicksburg, MS, Sept. 1969.
49. Duncan, J.M. and Clough, G.W., Finite element analyses of Port Allen Lock, *Journal of the Soil Mechanics and Foundations Division, ASCE,* 97(SM8), August 1971, 1053–1068.
50. Sherman, W.C. and Trahan, C.C., Analysis of Data from Instrumentation Program, Port Allen Lock, Technical Report S-68-7, U.S. Army Engineers Waterways Expt. Stn., Corps of Engineers, Vicksburg, MS, Sept. 1968.

51. U.S. Army Engineer District, Vicksburg, Columbia Lock, Pile Tests-Lock Site, Design Memorandum No. 6, Supplement No. 2, Dec. 1963, Vicksburg, MS.
52. U.S. Army Engineer District, Vicksburg, Columbia Lock, Masonry and Embedded Metals, Design Memorandum No. 6, March 1963, Vicksburg, MS.
53. Sherman, W.C., Jr., The behavior of lock walls supported on batter: Piles, *Proceedings of the 8th International Conference on Soil Mechanics and Foundation Engineering*, Moscow, Oct. 1973.
54. Montgomery, R.L. and Sullivan, A.L., Jr., Interim analysis of data from instrumentation program, Columbia Lock, *A Misc. Paper No. 5-72-30*, Aug. 1972.
55. Worth, N.L. et al., Pile Tests, Columbia Lock and Dam, Ouachita and Black Rivers, Arkansas and Louisiana, Tech. Report No. S-74-6, U.S. Army Engineers Waterways Expt. Stn., Corps of Engineers, Vicksburg, MS, Sept. 1966.
56. Desai, C.S., Johnson, L.D. and Hargett, C.M., Finite Element Analysis of the Columbia Lock Pile Foundation System, Tech. Report No. S-74-6, U.S. Army Engineers Waterways Expt. Stn., Corps of Engineers, Vicksburg, MS, July. 1974.
57. Desai, C.S., Johnson, L.D. and Hargett, C.M., Analysis of pile-supported gravity lock, *Journal of the Geotechnical Engineering Division, ASCE*, 100(GT9), 1974, 1009–1029.
58. Hashemi, M., Constitutive Modeling of a Schistose Rock in the Himalaya, PhD Thesis, Civil Eng., Indian Inst. of Tech., New Delhi, 1999.
59. NSPC, Nathpa-Jhakri Hydroelectric Project, Executive Summary, Nathpa-Jhakri Power Corporation (NJPC), New Delhi, India, 1992.
60. Varadarajan, A., Sharma, K.G., Desai, C.S. and Hashemi, M., Constitutive modeling of a schistose rock in the Himalaya, *International Journal of Geomechanics, ASCE*, 1(1), 2001, 83–107.
61. Varadarajan, A., Sharma, K.G., Desai, C.S. and Hashemi, M., Analysis of a powerhouse cavern in the Himalaya, *International Journal of Geomechanics, ASCE*, 1(1), 2001, 109–127.
62. Desai, C.S., *Mechanics of Materials for Solids and Interfaces: The Disturbed State Concept*, CRC Press, Boca Raton, FL, USA, 2001.
63. NIRM, Rock Mechanics Instrumentation to Evaluate the Long-Term Stability of Powerhouse and Transformer Caverns at Nathpa-Jhakri Power Corporation (NJPC), Final Report, National Institute of Rock Mechanics, Kolar Gold Fields, Karnataka, India, 1997.
64. Bhasin, R., Barton, N., Grimstad, E. and Chryssanthakis, P., Engineering characterization of anisotropic rocks in the Himalayan region for assessment of tunnel support, *Engineering Geology*, 40, 1995, 169–193.
65. Ramamurthy, T., Strength and modulus responses of anisotropic rocks, Chapter 13, *Comprehensive Rock Engineering*, Vol. I, Pergamon Press, Oxford, UK, 1993.
66. Desai, C.S., Sharma, K.G., Wathugala, G.W. and Rigby, D.B., Implementation of hierarchical single surface δ_0 and δ_1 models in finite element procedure, *International Journal of Numerical and Analytical Methods in Geomechanics*, 15(4), 1991, 568–579, 649–680.
67. NISA, *User's Manual*, Engineering Mechanics Research Corporation, Detroit, MI, USA, 1993.
68. Vulliet, L., Modelisation Des Pentes Naturelles En Mouvement, These No. 635, Ecole Polytechnique, Federale de Lausanne, Lausanne, Switzerland (in French), 1986.
69. Desai, C.S., Samtani, N.C. and Vulliet, L., Constitutive modeling and analysis of creeping slopes, *Journal of Geotechnical Engineering, ASCE*, 121(1), 1995, 43–56.
70. Perzyna, P., Fundamental problems is viscoplasticity, *Advances in Applied Mechanics*, 9, 1966, 243–377.
71. Desai, C.S. and Rigby, D.B., Cyclic interface and joint shear device including pore pressure effects, *Journal Geotechnical and Geoenvironmental Engineering, ASCE*, 123(6), 1997, 568–579.

72. Addenbrooke, T.I. and Potts, D.M., Twin tunnel interaction: Surface and subsurface effects, *International Journal of Geomechanics*, 1(2), 2001, 249–268.
73. Tensar Geogrid Reinforced Soil Wall, *Experimental Project 1, Ground Modification Techniques,* FHWA-EP-90-001-005, Department of Transportation, Washington, DC, 1989.
74. Berg, R.R., Bonaparte, R., Anderson, R.P., and Chouery, V.E., Design, construction, and performance of two reinforced soil retaining walls. *Proceedings of the 3rd International Conference on Geotextiles*, Vienna, Austria, 2, 1986, 401–406.
75. Fishman, K.L., Desai, C.S., and Berg, R.R., Geosynthetic-reinforced soil wall: 4-Year history. In *Behavior of Jointed Rock Masses and Reinforced Soil Structures, Transportation Research Record 1330*, TRB, Washington, DC, USA, 1991, 30–39.
76. Fishman, K.L., Desai, C.S., and Sogge, R.L., Field behavior of instrumented reinforced wall, *Journal of Geotechnical Engineering, ASCE*, 119(8), 1993, 1293–1307.
77. Fishman, K.L. and Desai, C.S., Response of a geogrid earth reinforced retaining wall with full height precast concrete facing. *Proceedings, Geosynthetics, -91*, Atlanta, GA, 2, 1991.
78. Tensar Earth-Reinforced Wall Monitoring at Tanque Verde-Wrightstown-Pantano Roads, Tucson, Arizona, Desert Earth Engineering Preliminary Report, Pima County Dept. of Transportation and Flood Control District, Tucson, AZ, USA, 1986.
79. Desai, C.S., DSC-SST-2D: Computer code for static, dynamic, creep and thermal analysis—solid, structure and soil-structure problems, User's Manuals I, II and III, Tucson, AZ, USA, 1998.
80. El-Hoseiny, K.E, and Desai, C.S., Computer Analysis of Reinforced Earth Retaining Walls with Testing and Constitutive Modeling of Soils and Interfaces, PhD Dissertation and Report, Minufiya University, Minufiya, Egypt; Dept. of Civil Eng. and Eng. Mechanics, Univ. of Arizona, Tucson, AZ, USA, 2004.
81. Ahmad, S.M., and Basudhar, P.K., Behaviour of reinforced soil retaining wall under static loading using finite element method, *Proceedings of the 12 IACMAG Conference*, 1–6 October, Goa, India, 2008.
82. Wu, T.J.H., Construction and instrumentation of Denver walls, geosynthetic reinforced soil retaining walls, *Proceedings of the International Symposium on Geosynthetic-Reinforced Soil Walls*, Wu, T. J. H. (Editor), Balkema, Rotterdam, 1992, 21–30.
83. Christopher, B.R., Gill, S.A., Giroud, J.P., Juran, I., Mitchell, J.K., Schlosser, F., and Dunnicliff, J., Reinforced Soil Structures. Volume II, Summary of Research and System Information, Report No. FHWA-RD-89-043, Federal Highway Administration, McLean, VA, November, 1989.

4 Three-Dimensional Applications

4.1 INTRODUCTION

Although we could use 1-D and 2-D idealizations for many problems, some problems require full 3-D analysis; for instance, a cap–pile group–soil system, junction of a dam and a river bank, pavements, and railroad beds. As indicated in Chapter 3, the fully 3-D FE method has been developed and used for solutions of a number of problems. The details of the FE equations for the fully 3-D procedure are presented in Chapter 3. In this chapter, we present some examples of application of the fully 3-D procedure. Also, we provide descriptions of some simplified procedures and applications. The work and publications on the subject are wide; hence, only a limited number of publications are cited, and a limited number of applications are presented.

The cap–pile group–soil problem (Figure 4.1b) provides a simpler example of one of the 3-D problems (Figure 4.1a). A number of procedures based on matrix methods have been developed for the solution of this problem; they often involve spatial solutions based on equilibrium and compatibility conditions with linear or nonlinear soil resistance represented by translational and rotational spring moduli; such procedures are presented in detail in various publications [1–9].

One of the commonly used methods idealizes the pile as a beam-column 1-D element, and simulates the soil resistance using linear or nonlinear modulus of subgrade reaction or soil resistance (p_y–v and p_x–u curves). The pile cap is often assumed to be rigid. The Hrennikoff [1] method is based on this approach for the analysis of pile groups by considering the positions of the pile heads with assumed boundary conditions at the junction of piles and cap (rigid) and linear soil resistance. Reese et al. [5] adopted nonlinear soil resistance for 3-D pile group analysis. O'Neill et al. [8] modified the latter procedure by including the pile–soil-interaction. Chow [9] presented a procedure which includes directly the interaction effects, assumes piles as beam-columns, and considers factors such as pile sizes, nonuniform pile sections, and nonlinear soil behavior.

An alternative and approximate approach called *multicomponent FE procedure*, similar to the matrix methods and its applications, is described below. It could yield satisfactory results for certain problems and provide economical and simplified solutions. Hence, a number of applications are presented using the multicomponent procedure.

A number of researchers have used the boundary element method for analysis of pile group problem [10,11]. Fully 3-D FE procedures have been developed, and used for a number of geotechnical problems [12–30]. The literature and application of the 3-D FE method are very wide and only typical applications are included in this chapter with an emphasis on pile groups.

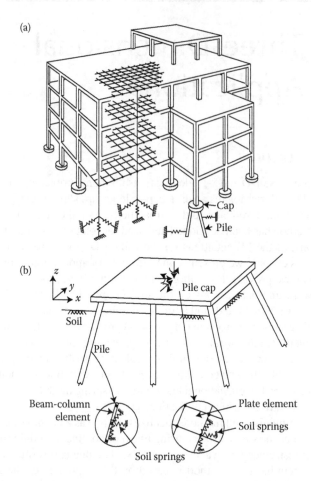

FIGURE 4.1 Multi component system. (a) Frame-slab-column-cap soil springs; (b) pile-cap-pile-foundation.

4.2 MULTICOMPONENT PROCEDURE

In the multicomponent approach [31,32], the cap is represented by plate (slab) elements, including in-plane and bending behaviors; beam-column element for piles with axial and lateral loads and moments, and linear or nonlinear springs for translational and rotational resistance for the soil. Chow [9] derives stiffness properties for beam-column (pile) members by using the matrix method, and assumes the pile cap to be rigid. In the multicomponent approach, the FE approach is used in which the approximation functions for displacements are adopted for the beam-column and the pile cap (treated as a plate or slab). Then, the stiffness properties are derived by using an energy principle. The problem is solved by using only 2-D plate elements, 1-D pile elements, and spring support for soil. Hence, the problem is simplified in comparison to the fully 3-D analysis. Since the assumption of small thickness is involved for plate behavior, the analysis is considered to be approximate; however, the rigid pile cap can be simulated by assigning a high value to the stiffness of the pile cap.

Details of the models for pile, soil resistance, and pile cap in the multicomponent system are presented in the following sections.

4.2.1 PILE AS BEAM-COLUMN

We have considered laterally and axially loaded column in Chapter 2. Figure 4.2 shows the loadings and degrees of freedom (DOFs) for a pile subjected to both lateral and axial loads, causing bending and axial deformations, respectively. To follow usual convention, we have chosen the z-axis as vertical instead of the x-axis as vertical in Chapter 2. The soil can provide resistance, which can be represented by three translational springs, k_x, k_y, and k_z and three rotational springs, $k_{\theta x}$, $k_{\theta y}$, and $k_{\theta z}$ (also see Chapter 2). The displacements in x-, y-, and z-directions are denoted by u, v, and w, respectively. The components u and v are related to the lateral bending behavior, and the component w is related to the axial load.

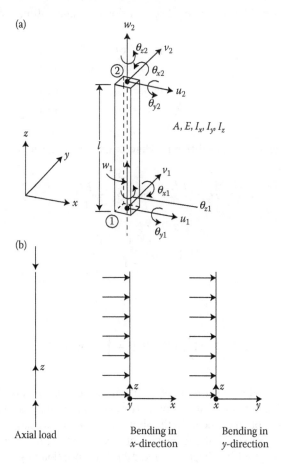

FIGURE 4.2 General beam-column element.

In Chapter 2, we derived bending stiffness equations using cubic displacement approximation. Also, we derived axial stiffness equations using linear displacement approximation. Assuming superposition is valid, we can write the combined stiffness equations as

$$[k]\{q\} = \{Q\} = \begin{Bmatrix} \{Q_b\} \\ \{Q_a\} \end{Bmatrix} \tag{4.1}$$

where the stiffness matrix $[k]$ is given by

$$[k] = \begin{bmatrix} \alpha_x[k_x] & [o] & [o] \\ [o] & \alpha_y[k_y] & [o] \\ o & o & [k_z] \end{bmatrix} \tag{4.2a}$$

which can be derived for the same pile dimension in x- and y-directions as

$$[k_x] = [k_y] = \begin{bmatrix} 12 & 6\ell & -12 & 6\ell \\ & 4\ell^2 & -6\ell & 2\ell^2 \\ \text{symm.} & & 12 & -6\ell \\ & & & 4\ell^2 \end{bmatrix} \tag{4.2b}$$

However, α_x and α_y can be different. Here α_x and α_y are given by

$$\alpha_x = \frac{EI_y}{\ell^3} \quad \text{and} \quad \alpha_y = \frac{EI_x}{\ell^3} \tag{4.2c}$$

$$[k_z] = \frac{AE}{\ell} \begin{bmatrix} 1 & -1 \\ -1 & 1 \end{bmatrix} \tag{4.2d}$$

and

$$\{Q\}^T = A \int_o^\ell [N]^T \{\bar{X}\} \, dz + \int_o^\ell [N]^T \{\bar{T}\} \, dz \tag{4.2e}$$

where E is the modulus of elasticity, A is the cross-sectional area of the element of beam-column, I_x and I_y are moments of inertia about the x- and y-axes, respectively, and ℓ is the length of an element. In the above formulation, we have not considered torsion (about the z-axis) in the formation of beam-column; however, a derivation for torsion is presented in Chapter 2, and a brief description is given later in this chapter.

4.2.2   PILE CAP AS PLATE BENDING

4.2.2.1   In-Plane Response

For the in-plane loading, the element of the cap can be treated as 2-D with plane stress idealization for which an element (quadrilateral) can be adopted (Figure 4.3a). The plate is assumed to be thin. The element for plane stress possesses eight nodal DOF in the x- and y-directions. The approximation function for a four-noded quadrilateral element (Figure 4.3a) can be expressed as [33]

$$u = N_1 u_1 + N_2 u_2 + N_3 u_3 + N_4 u_4$$
$$v = N_1 v_1 + N_2 v_2 + N_3 v_3 + N_4 v_4 \qquad (4.3)$$

$$\{u\} = \begin{Bmatrix} u \\ v \end{Bmatrix} = [N_p]\{q_p\} \qquad (4.4)$$

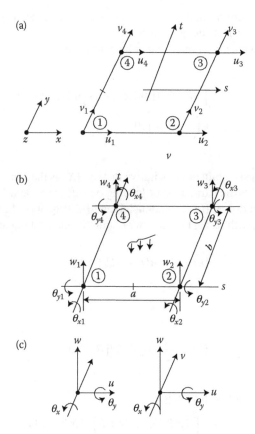

FIGURE 4.3   Plate element: (a) in-plane or membrane behavior; (b) bending behavior; and (c) compatibility between plate and beam-column.

where $\{q_p\}$ is the vector of nodal displacements, and $[N_p]$ is the matrix of interpolation functions for in-plane deformations given by

$$N_i = \frac{1}{4}(1 + ss_i)(1 + tt_i) \tag{4.5}$$

where $i = 1, 2, 3, 4$ and s and t are local coordinates (Figure 4.3a); note that the origin for the local coordinates s, t is assumed as the center of the element.

For the plane stress condition and loading in the plane, the strain vector, $\{\varepsilon\}$, is given by

$$\{\varepsilon\} = \begin{Bmatrix} \varepsilon_x \\ \varepsilon_y \\ \gamma_{xy} \end{Bmatrix} = [B_p]\{q_p\} \tag{4.6}$$

where ε_x, ε_y, and γ_{xy} are normal and shear strains in the plane $(x-y)$ and $[B_p]$ is the strain-displacement transformation matrix.

Let the potential energy of the element be expressed as

$$\pi_p = \frac{h}{2}\int_A \{\varepsilon\}^T[C]\{\varepsilon\}\mathrm{d}x\mathrm{d}y - h\int_A \{u\}^T[\bar{X}]\mathrm{d}x\mathrm{d}y$$
$$- h\int_{S_1} \{u\}^T\{\bar{T}\}\,\mathrm{d}S \tag{4.7}$$

where $[C]$ is the stress–strain or constitutive matrix, $\{\bar{X}\}$ is the vector of body forces, $\{\bar{T}\}$ is the vector of surface tractions or loading on surface S_1, and h is the thickness of the plate. Substitution of $\{u\}$ (Equation 4.4) and $\{\varepsilon\}$ (Equation 4.6) in Equation 4.7 and minimizing π_p with respect to $\{q_p\}$, we obtain the element equation

$$[k]\{q_p\} = \{Q\} = \{Q_1\} + \{Q_2\} \tag{4.8a}$$

where

$$[k] = h\int_A [B_p]^T[C][B_p]\mathrm{d}x\mathrm{d}y \tag{4.8b}$$

$$\{Q\} = \{Q_1\} + \{Q_2\}$$
$$= h\int_A [N]^T\{\bar{X}\}\mathrm{d}x\mathrm{d}y + h\int_{S_1} [N]^T\{\bar{T}\}\mathrm{d}S \tag{4.8c}$$

and [C] is the constitutive or stress–strain relation matrix for the plane stress condition:

$$[C] = \frac{E}{1-v^2} \begin{bmatrix} 1 & v & 0 \\ v & 1 & 0 \\ 0 & 0 & \dfrac{1-v}{2} \end{bmatrix}$$ (4.8d)

where E and v are the elastic modulus and the Poisson's ratio, respectively. The size of stiffness matrix $[k]$ will be 8×8, and of the load vector $\{q_p\}$ 8×1. Equation 4.8d is valid under the assumption that the material in the cap is linearly elastic and isotropic.

4.2.2.2 Lateral (Downward) Loading on Cap-Bending Response

The lateral (vertical) load on the cap will cause bending effects, which can be considered as plate or slab bending. Each point in the plate element will possess three DOFs, vertical (translational) and two rotational about the x- and y-axis (Figure 4.3b) if torsional rotation is not considered. The latter is considered in Chapter 2.

The differential equation for the bending of the isotropic plate can be expressed as [33]

$$D\left(\frac{\partial^4 w^*}{\partial x^4} + \frac{2\partial^4 w^*}{\partial x^2 \partial y^2} + \frac{\partial^4 w^*}{\partial y^4}\right) = p(x,y)$$ (4.9)

where w^* is the "exact" transverse (vertical) displacement for the differential equation, and $D = Eh^3/12(1-v^2)$, h is the thickness of the plate, and $p(x,y)$ is the applied load on the plate. For soil spring support, the right-hand side can be replaced by $p(x,y) - k_o u$, where k_o is subgrade modulus representing the soil support.

The subject of FE analysis for plate bending is very wide [34–37]. Here, we used the formulation proposed by Bogner et al. [37], in which the bicubic Hermitian interpolation function has been used:

$$\begin{aligned} w(x,y) &= N_{x1}N_{y1}w_1 + N_{x2}N_{y1}\theta_{x1} + N_{x1}N_{y2}\theta_{y1} \\ &\quad + N_{x3}N_{y1}w_2 + N_{x4}N_{y1}\theta_{x2} + N_{x3}N_{y2}\theta_{y2} \\ &\quad + N_{x3}N_{y3}w_3 + N_{x4}N_{y3}\theta_{x3} + N_{x3}N_{y4}\theta_{y3} \\ &\quad + N_{x1}N_{y3}w_4 + N_{x2}N_{y3}\theta_{x4} + N_{x1}N_{y4}\theta_{y4} \\ &= [N_1 N_2 N_3 N_4 \ldots N_{12}]\{q_b\} \\ &= [N_b]\{q_b\} \end{aligned}$$ (4.10)

where $[N_b]$ is the matrix of interpolation functions for bending deformations and $\{q_b\}^T$ is the vector of nodal displacement and rotations. The interpolation functions are given by

$$
\begin{aligned}
N_{x1} &= 1 - 3s^2 + 2s^3, & N_{y1} &= 1 - 3t^2 + 2t^3, \\
N_{x2} &= as(s-1)^2, & N_{y2} &= bt\,(t-1)^2, \\
N_{x3} &= s^2(3-2s), & N_{y3} &= t^2(3-2t), \\
N_{x4} &= as^2(s-1), & N_{y4} &= bt^2(t-1).
\end{aligned}
\tag{4.11}
$$

where s and t are local coordinates, and (a, b) are the dimensions of the plate element (Figure 4.3b) $s = x/a$, $0 \le s \le 1$, and $t = y/b$, $0 \le t \le 1$; note that the origin of the local coordinates is at the bottom left corner of the element. In Equation 4.10, an interpolation function corresponding to a DOF is obtained by multiplying two interpolation functions in x and y directions, respectively. For example, for the DOF, w_1 (No. 1):

$$
N_1 = N_{x1}N_{y1} = (1 - 3s^2 + 2s^3)(1 - 3t^2 + 2t^3)
\tag{4.12}
$$

which satisfies the definition of an interpolation function (e.g., N_1) that it bears a value of unity for the DOF it pertains to, and zero for all other DOFs. It is evident from Equation 4.12 that $N_1 = 1$ corresponding to w_1 and zero for all other DOFs (Equation 4.10). The rotational DOF is obtained by taking derivative of w with respect to x (and y). Note that in Equation 4.10, we have not included the rotation around the z-axis. It is considered in Section 4.2.4.

Now the strain (gradient)–unknown (displacement) relation for plate bending is given by [32–34]

$$
\{\varepsilon\} = -z
\begin{Bmatrix}
\dfrac{\partial^2 w}{\partial x^2} \\[2mm]
\dfrac{\partial^2 w}{\partial y^2} \\[2mm]
\dfrac{\partial^2 w}{\partial x \partial y}
\end{Bmatrix}
=
\begin{Bmatrix}
w_{xx} \\ w_{yy} \\ 2w_{xy}
\end{Bmatrix}
= -z[B_b]\{q_b\}
\tag{4.13}
$$

where the strain-displacement transformation matrix $[B_b]$ is obtained by finding appropriate derivatives of w (Equation 4.10). The constitutive or stress–strain relation for linear elastic and isotropic material can be written as

$$
\{\sigma\} =
\begin{Bmatrix}
M_{xx} \\ M_{yy} \\ M_{xy}
\end{Bmatrix}
= \frac{Eh^3}{12(1-v^2)}
\begin{bmatrix}
1 & v & 0 \\
v & 1 & 0 \\
0 & 0 & \dfrac{1-v}{2}
\end{bmatrix}
\begin{Bmatrix}
w_{xx} \\ w_{yy} \\ w_{xy}
\end{Bmatrix}
$$

$$
= [C]\{\varepsilon\}
\tag{4.14}
$$

where w_{xx} and w_{yy} are the second derivative of w with respect to x and y, respectively, and w_{xy} is the second derivative with respect to x and y.

The potential energy for the bending deformation is given by

$$\pi_p = \frac{h}{2}\int_A \{\varepsilon\}^T [C]\{\varepsilon\} \mathrm{d}x\mathrm{d}y - h\int_A \{\bar{X}\}^T \{u\}\mathrm{d}x\mathrm{d}y$$
$$- \int_{S_1} \{\bar{T}\}^T \{u\}\mathrm{d}S \tag{4.15}$$

Minimization of π_p with respect to $\{q_b\}$, leads to

$$[k_b]\{q_b\} = \{Q_b\} \tag{4.16a}$$

where

$$[k_b] = h\int_A [B_b]^T [C][B_b]\mathrm{d}x\mathrm{d}y \tag{4.16b}$$

$$\cdot\{Q_b\} = h\int_A [N]^T \{\bar{X}\}\mathrm{d}x\mathrm{d}y + \int_{S_1} [N]^T \{\bar{T}\}\mathrm{d}S \tag{4.16c}$$

The details of the deviation of $[k_b]$ and $\{Q_b\}$ are given in Ref. [37].

4.2.3 ASSEMBLAGE OR GLOBAL EQUATIONS

The element equations for the beam-column (Equation 4.1), for the in-plane behavior of the plate (Equation 4.16a), and for the bending behavior of the plate (Equation 4.16) can be assembled by using the *direct stiffness method*. In this method, the interelement compatibility for displacements and rotations at common nodes are satisfied. The resulting equations can be written as

$$[K]\{r\} = \{R\} \tag{4.17}$$

where $[K]$ is the assemblage stiffness matrix, $\{r\}$ is the vector of nodal DOF for the entire structure, and $\{R\}$ is the nodal load vector for the entire structure. Now, the boundary conditions in terms of displacements and rotations can be introduced in Equation 4.17. Such modified equations can be solved by using an appropriate procedure, for example, Gauss elimination.

4.2.4 TORSION

The torsion about the z-axis can be considered due to the rotation about the z-axis θ_z, which can be assumed as linear along the element

$$\theta_z = (1 - r)\theta_{z1} + r\theta_{z2} \tag{4.18a}$$

where r is the local coordinate along the z-axis. The energy function, π_p (Equation 4.15), can be modified by adding the following term:

$$\ell \int_o^1 GI_z (w'')^2 \, dS \tag{4.18b}$$

where ℓ is the length of the element, G is the shear modulus, I_z is the polar moment of inertia, and w'' is the second derivative of w about the z-axis. The stiffness matrix for torsion, $[k_T]$, can be derived as

$$[k_T] = \frac{GI_z}{\ell} \begin{bmatrix} 1 & -1 \\ -1 & 1 \end{bmatrix} \tag{4.18c}$$

Then, the stiffness matrix for torsion can be added to Equation 4.17. The torsion aspect is not included in the computer code STFN-FE [38].

4.2.5 Representation of Soil

As in the case of pile (Chapter 2), the soil can be replaced by three equivalent springs, k_x, k_y, and k_z for translation and $k_{\theta x}$, $k_{\theta y}$, and $k_{\theta z}$, for rotation (Figure 4.2). For example, for the translational springs, the p_x–u and p_y–v curves for lateral behavior can be developed by using procedures described in Chapter 2. The p–w curves for the axial behavior can also be developed by using procedures in Chapter 2. The spring constants at given nodal points can be added to the diagonal elements (Equation 4.17) for specific DOFs. The behavior of soil can be assumed linear or nonlinear; for the latter, we can use the Ramberg–Osgood model.

4.2.6 Stress Transfer

Sometimes, an uplift or separation between the structure and the soil can occur, depending on the loading and geometry of plates and beam-columns. The soil–structure system usually starts with compressive stress at the junction between the structure and soil. During loading, the zone under uplift may experience tensile stresses. If the induced tensile stress is greater than the adhesive or tensile strength of the interface, then the separation between the structural elements and adjoining soil element could take place. The procedure for simulating the loss of contact and stress transfer [32,38,39] is given below, in which the excess tensile stress is distributed to the adjoining zones (elements). Figure 4.4a shows a typical beam on foundation problem. The procedure is summarized below:

1. Obtain the solution, for example, using FEM, and identify displacements at all interface nodes, and compute resistance p by multiplying the displacement by relevant stiffness (k).
2. If p is negative, that is, it has changed its sign, the tensile stress condition has occurred at certain nodes (Figure 4.4b).

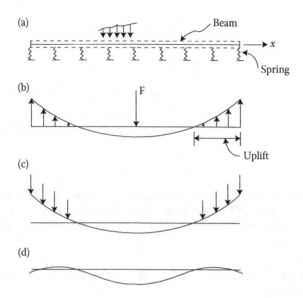

FIGURE 4.4 Representation of stress transfer procedure: (a) idealization of beam and foundation; (b) deflected shapes and zones of uplift; (c) applied equilibrating forces; and (d) final deflected shape after equilibrium.

3. At every node at the soil–structure interface where an upward lift or separation has occurred due to the tensile condition, apply an equal and opposite compressive force (Figure 4.4c), which is the product of the displacement at that node and the spring constant corresponding to that displacement.
4. Apply the new force vector as additional load to the system equation and obtain the new displacements. The total system displacement vector is the sum of the displacement vectors from Steps 1 and 4.
5. Repeat Steps 2 through 4 until convergence. Convergence is satisfied if the nodal p is equal to or less than the tensile strength of the interface (Figure 4.4d).

4.3 EXAMPLES

We now present a number of examples, including analytical and/or laboratory or field validations, for both the approximate multicomponent method and the fully 3-D method. As indicated before, various publications [1–9] can be consulted for the matrix methods and their applications.

4.3.1 EXAMPLE 4.1: DEEP BEAM

Figure 4.5 shows a reinforced concrete deep beam resting on a very deep brick wall [39,40]. Three loads, $P_1 = 40$ t (tons), $P_2 = 100$ t, and $P_3 = 20$ t, act at the locations shown (Figure 4.5). For the linear elastic analysis, the following moduli of

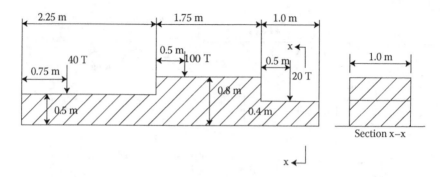

FIGURE 4.5 Deep beam on elastic foundation. (From Cheung, Y.K. and Nag, D.K., *Geotechnique*, 18, 1968, 250–260; Cheung, Y.K., *Numerical Methods in Geotechnical Engineering*, Desai, C.S. and Christian, J.T. (Eds), McGraw-Hill Book Company, New York, 1977. With permission.)

elasticity are used: for the beam, $E_b = 2.1 \times 10^6$ T/m^2, and for the brick wall foundation, $E_f = 3 \times 10^5$ T/m^2.

The deep beam is treated approximately as a beam-column resting on the Winkler foundation. For the FE analysis, the beam is divided into 20 elements of equal length, that is, the length of each element is 0.25 m. The equivalent spring modulus or stiffness, k, for the foundation can be obtained by using equations in Refs. [39–41].

Accordingly, the moments of inertia were calculated for 0.50, 0.80, and 0.40 m depths, and they were found to be 0.01040, 0.04267, and 0.0054 m^4, respectively. The equivalent soil moduli were found to be 344, 498, and 688 kN/m, respectively [38–40]. The average value of $k = 510$ kN/m. A parametric study was also performed in which k was varied. Results for each k were compared with the displacement provided by Cheung and Nag [39,40]. The value of k that yielded displacement results close to those presented in Refs. [39,40] was found to be in the range of 490–540 kN/m (Figure 4.6a); the average value is about 510 kN/m. In Refs. [39,40], the results were obtained by using the FE method with the plane stress idealization. The contact pressure distribution under the deep beam computed by using $k = 510$ kN/m is shown in Figure 4.6b in comparison to that presented in Ref. [39]. The results in Figure 4.6 indicate that the approximate multicomponent analysis using the code STFN-FE [3] can provide satisfactory results.

4.3.2 EXAMPLE 4.2: SLAB ON ELASTIC FOUNDATION

Figure 4.7 shows a $10 \times 10 \times 1$ ft ($3.05 \times 3.05 \times 0.305$ m) slab, partially fixed on the edges, and subjected to a point load of 80,000 lbs (356 kN) at the center. The subgrade modulus for the foundation of the slab is assumed as 40 pci (11.0 N/cm^3). The modulus of elasticity and the Poisson's ratio of the slab are assumed as 3×10^6 psi (2×10^6 N/cm^2) and 0.30, respectively. Figure 4.8 shows the FE mesh used in the analysis. The following results are obtained by using the STFN-FE code [38]:

1. Displacements and moments at various sections, without soil support (Figures 4.9a and 4.9b, respectively);

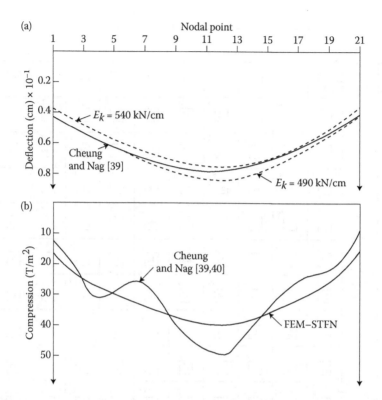

FIGURE 4.6 Predictions and comparisons for deep beam. (a) Displacements under deep beam; (b) contact pressure under deep beam. (From Cheung, Y.K. and Nag, D.K., *Geotechnique*, 18, 1968, 250–260; Cheung, Y.K., *Numerical Methods in Geotechnical Engineering*, Desai, C.S. and Christian, J.T. (Eds), McGraw-Hill Book Company, New York, 1977. With permission.)

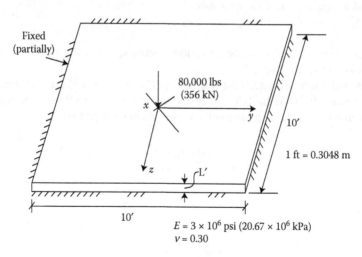

FIGURE 4.7 Centrally loaded slab.

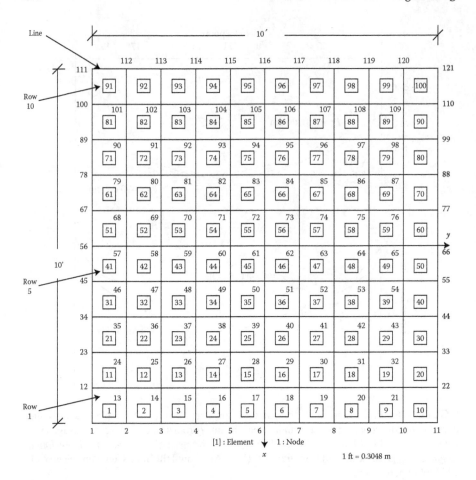

FIGURE 4.8   Finite element mesh for slab.

2. Displacements and moments at various sections, with soil support (Figures 4.10a and 4.10b, respectively);

3. The maximum displacement from the STFN-FE code without soil support is about 0.0131607 in (0.03340 cm). It compares very well with the maximum obtained from the following closed-form solution [34]:

$$u_{max} = \frac{0.0056 \times P \times a^2}{D} \tag{4.19}$$

where

$$D = \frac{Eh^3}{12\,(1-v^2)}$$

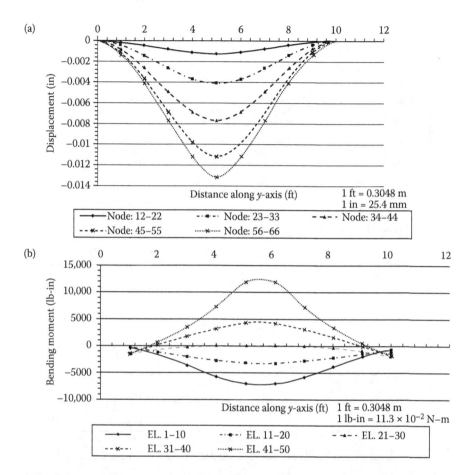

FIGURE 4.9 Results for slab without soil support: (a) displacements; (b) bending moments.

a is the width = 120 in (305 cm), h is the thickness = 12 in (30.5 cm), and P is the load = 80,000 lbs (456 kN). The maximum displacement (0.01359 in = 0.0345 cm) obtained from Equation 4.19 compares very well with the above FE prediction. The maximum bending moment from the FE predictions for no soil support is 12,000 lb-in (1356 N-m). The maximum displacement from the FE predictions for the case with soil support is 0.0130166 (0.03306 cm), and the maximum bending moment is about 12,000 lb-in (1356 N-m). Thus, there is no significant difference between the predictions for with and without soil support. This may be due to the low value of the soil modulus.

4.3.3 EXAMPLE 4.3: RAFT FOUNDATION

A raft foundation of 20×20 m with thickness equal to 90 cm is shown in Figure 4.11. The following material properties are used in the analysis: elastic modulus of

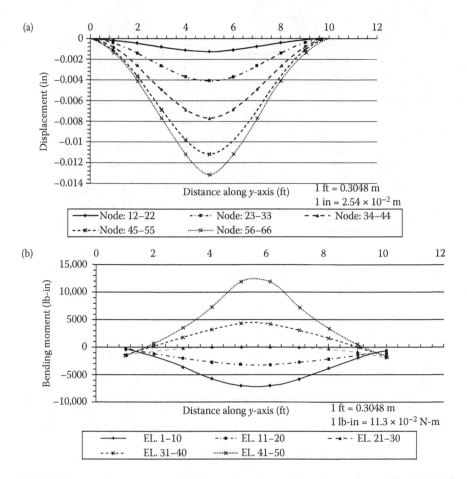

FIGURE 4.10 Results for slab with soil support: (a) displacements; (b) bending moments.

raft, $E = 2.07 \times 10^7$ kN/m², Poisson's ratio of raft, $v = 0.15$, modulus of subgrade reaction, $k_o = 29.4$ kN/cm³.

The raft supports 12 exterior columns, 4 interior columns, and an elevator shaft. The loads carried by columns and elevator shafts are 200, 175, and 750 tons (1780, 1558, and 6675 kN), respectively.

The code STFN-FE was used to solve the problem. The FE mesh contains 64 plate elements and 81 nodes (Figure 4.11). The results from the FE analysis in terms of displacements along the line connecting nodes 64 through 72 are compared (Figure 4.12) with those from the finite difference method [42]. The two results correlate very well.

4.3.4 EXAMPLE 4.4: MAT FOUNDATION AND FRAME SYSTEM

Figure 4.13 shows a mat foundation that supports two frames subjected to two vertical loads of 136.1 T (1211 kN) and horizontal loads of 45.4 T (404 kN), as shown in Figure 4.14 [43]. The geometrical and material properties are as follows:

Frame	
Beam length	= 6.1 m
Column length	= 4.575 m
$I_{xx} = I_{yy}$	= 1.86 × 10⁶ cm⁴
E	= 2.07 × 10⁷ kN/m²
Mat	
Length	= 8.53 m
Width	= 8.53 m
Thickness	= 45.7 cm
E	= 2.07 × 10⁷ kN/m²
v	= 0.15
Central opening	= 3.66 m × 1.22 m

The foundation involves two types of vertically separated and uniformly distributed soils (Figure 4.13) whose properties are

$$k_{o1} = E_{s1} = 27.2 \text{ kN/cm}^3$$
$$k_{o2} = E_{s2} = 13.6 \text{ kN/cm}^3$$

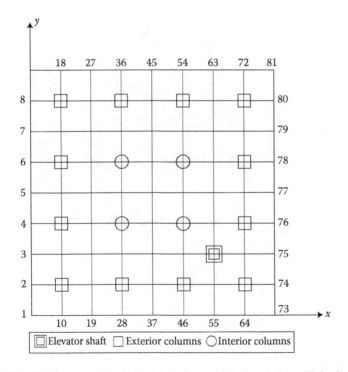

FIGURE 4.11 Finite element mesh for slab (raft) on elastic foundation with loaded columns in superstructure. (From Anandakrishnan, U. et al., *Design Manual for Raft Foundations*, Dept. of Civil Eng., Indian Institute of Technology, Kanpur, India, 1971. With permission.)

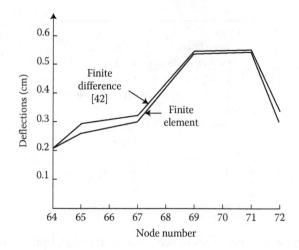

FIGURE 4.12 Deflections along line 64–72 in Figure 4.11.

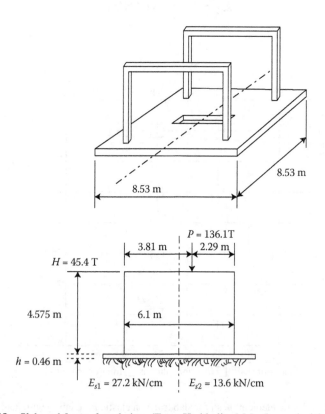

FIGURE 4.13 Slab and frame foundation. (From Haddadin, M.J., *Journal of the American Concrete Institute*, 68(2), 1971, 9945–9949. With permission.)

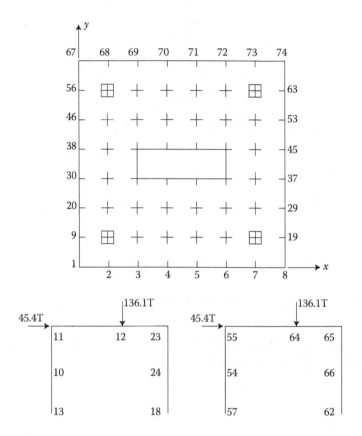

FIGURE 4.14 Finite element mesh for slab-frame problem (Figure 4.13).

The FE mesh in the finite element analysis using the for STFN-FE software is shown in Figure 4.14.

The FE results in terms of contact pressure are shown in Figure 4.15; they include results from, with, and without stress transfer approaches. Haddadin [43] presented and solved the problem by using similar FE method (results are also shown in Figure 4.15). It can be seen that the results obtained from the approximate (STFN-FE) analysis are comparable with the results from the method in Ref. [43].

4.3.5 EXAMPLE 4.5: THREE-DIMENSIONAL ANALYSIS OF PILE GROUPS: EXTENDED HRENNIKOFF METHOD

In this example, we analyze the displacement and rotational behavior of a pile group using a simplified approach, which is based on the equilibrium and compatibility conditions between the cap and the piles in the group [1]. The cap is assumed to be rigid and the piles are assumed to be hinged to the cap. Also, it is assumed that allowable loads, both axial (P_a) and transverse (Q_a), are known from pile load tests or from

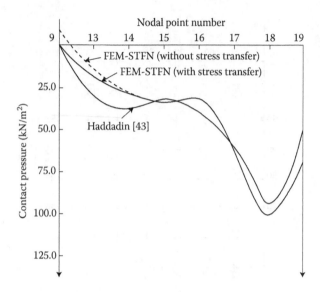

FIGURE 4.15   Contact pressure along line 9–19 in Figure 4.14.

numerical simulations. The axial load, P, and the transverse load, Q, carried by a pile are directly proportional to the pile head displacements:

$$P = nd_n \tag{4.20}$$

$$Q = td_t \tag{4.21}$$

where d_n = axial displacement, d_t = transverse displacement of the pile head, and n and t are pile constants. These constants are defined as forces with which the pile acts on the cap due to a unit displacement given to the cap along the respective direction. Assuming that the pile cap's displacements are Δx, Δy, and Δz along the x, y, and z axes, respectively, and its rotations are α_x, α_y, and α_z about the x, y, and z axes, respectively, the equilibrium equations for the pile cap can be expressed in the following form [3]:

$$a_{11}\Delta x' + a_{12}\Delta y' + a_{13}\Delta z' + a_{14}\alpha'_x + a_{15}\alpha'_y + a_{16}\alpha'_z + P_x = 0 \tag{4.22a}$$

$$a_{21}\Delta x' + a_{22}\Delta y' + a_{23}\Delta z' + a_{24}\alpha'_x + a_{25}\alpha'_y + a_{26}\alpha'_z + P_y = 0 \tag{4.22b}$$

$$a_{31}\Delta x' + a_{32}\Delta y' + a_{33}\Delta z' + a_{34}\alpha'_x + a_{35}\alpha'_y + a_{36}\alpha'_z + P_z = 0 \tag{4.22c}$$

$$a_{41}\Delta x' + a_{42}\Delta y' + a_{43}\Delta z' + a_{44}\alpha'_x + a_{45}\alpha'_y + a_{46}\alpha'_z + M_x = 0 \tag{4.22d}$$

$$a_{51}\Delta x' + a_{52}\Delta y' + a_{53}\Delta z' + a_{54}\alpha'_x + a_{55}\alpha'_y + a_{56}\alpha'_z + M_y = 0 \tag{4.22e}$$

$$a_{61}\Delta x' + a_{62}\Delta y' + a_{63}\Delta z' + a_{64}\alpha'_x + a_{65}\alpha'_y + a_{66}\alpha'_z + M_z = 0 \quad (4.22f)$$

where P_x, P_y, and P_z are equivalent point loads acting at the centroid of the cap, and M_x, M_y, and M_z are external moments on the cap, and a_{ij} are pile cap constants. In Equation 4.22, $\Delta x' = n\,\Delta x$, $\Delta y' = n\,\Delta y$, $\alpha'_x = n\,\alpha_x$, $\alpha'_y = n\,\alpha_y$, and so on, called reduced cap movements, are used for convenience (to express the cap constants a_{ij} as ratios). It can be shown by Betti's law that the set of simultaneous equations above is symmetrical (i.e., $a_{ij} = a_{ji}$).

As shown by Aschenbrenner [3], the pile cap constants, a_{ij}, can be obtained by summing the contribution of forces exerted by each pile to the cap due to a unit displacement (constrained) or rotation (constrained) of the cap. For example, the cap constants a_{11}, a_{12}, a_{13}, and so on in Equation 4.22a can be obtained by considering the equilibrium of forces and moments exerted by each pile to the cap due to $\Delta x = 1$. Likewise, the cap constants a_{41}, a_{42}, a_{43}, and so on in Equation 4.22d can be obtained by considering the equilibrium of forces and moments exerted by each pile to the cap due to $\alpha_x = 1$ [3]. For an arbitrary pile, k, whose head is located at x_{A_k} and y_{A_k} (or α_{A_k} and ρ_{A_k}) and whose inclinations are α_k, β_k, and γ_k (Figure 4.16), the pile cap constants are given in Table 4.1. For piles with two axes of symmetry, the pile constants are given in Table 4.2. For properly established (optimized) loads, $r = (Q_a/P_a) = (t/n)$.

The solution of Equation 4.22 for a given set of loadings (P_x, P_y, and P_z) and moments (M_x, M_y, and M_z) will yield reduced cap displacements $\Delta x' = n\,\Delta x$, $\Delta y' = n\,\Delta y$, $\Delta z' = n\,\Delta z$, and rotations $\alpha'_x = n\,\alpha_x$, $\alpha'_y = n\,\alpha_y$, $\alpha'_z = n\,\alpha_z$; here, positive values of displacements correspond to positive directions of the axes, and the right-hand rule is used for rotations about the axes. The axial pile force (P_k) for pile k can be computed from the following equation:

$$P_k = (n\,d_n)_k = -\Delta x' \cos\alpha_k - \Delta y' \cos\beta_k - \Delta z' \cos\gamma_k$$

$$- \alpha'_x\, y_{A_k} \cos\gamma_k + \alpha'_y\, x_{A_k} \cos\gamma_k - \alpha'_z\, \rho_{A_k} \cos\varepsilon_k \quad (4.23a)$$

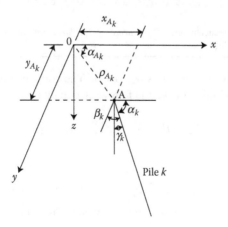

FIGURE 4.16 Pile geometry and nomenclature.

TABLE 4.1
Matrix Coefficients for Nonsymmetrical Pile Arrangement

Matrix Coefficient	Batter Piles	Vertical Piles
a_{11}	$-\cos^2 \alpha_k + r \sin^2 \alpha_k$	$-r$
a_{12}	$(r-1)\cos \alpha_k \sin \beta_k$	0
a_{13}	$(r-1)\cos \alpha_k \cos \gamma_k$	0
a_{14}	$(r-1)\cos \alpha_k \cos \gamma_k y_{A_k}$	0
a_{15}	$(1-r)\cos \alpha_k \cos \gamma_k x_{A_k}$	0
a_{16}	$\{-\cos \alpha_k \cos \varepsilon_k + r(\sin \alpha_{A_k} + \cos \alpha_k \cos \varepsilon_k)\}\rho_{A_k}$	$r y_{A_k}$
a_{22}	$-\cos^2 \beta_k + r \sin^2 \beta_k$	$-r$
a_{23}	$(r-1)\cos \beta_k \sin \gamma_k$	0
a_{24}	$(r-1)\cos \beta_k \cos \gamma_k y_{A_k}$	0
a_{25}	$(1-r)\cos \beta_k \cos \gamma_k x_{A_k}$	0
a_{26}	$-\{\cos \beta_k \cos \varepsilon_k + r(\cos \alpha_{A_k} - \cos \beta_k \cos \varepsilon_k)\rho_{A_k}$	$-r x_{A_k}$
a_{33}	$-\cos^2 \gamma_k + r \sin^2 \gamma_k$	-1
a_{34}	$-(\cos^2 \gamma_k + r \sin^2 \gamma_k)y_{A_k}$	$-y_{A_k}$
a_{35}	$(\cos^2 \gamma_k + r \sin^2 \gamma_k)x_{A_k}$	x_{A_k}
a_{36}	$(r-1)\cos \gamma_k \cos \varepsilon_k \rho_{A_k}$	0
a_{44}	$-(\cos^2 \gamma_k + r \sin^2 \gamma_k)y_{A_k}^2$	$-y_{A_k}^2$
a_{45}	$(\cos^2 \gamma_k + r \sin^2 \gamma_k)x_{A_k} y_{A_k}$	$x_{A_k} y_{A_k}$
a_{46}	$(r-1)\cos \gamma_k \cos \varepsilon_k \rho_{A_k} y_{A_k}$	0
a_{55}	$-(\cos^2 \gamma_k + r \sin^2 \gamma_k)x_{A_k}^2$	$-x_{A_k}^2$
a_{56}	$(1-r)\cos \gamma_k \cos \varepsilon_k \rho_{A_k} x_{A_k}$	0
a_{66}	$\{\cos \alpha_k \cos \varepsilon_k - r(\sin \alpha_{A_k} + \cos \alpha_k \cos \varepsilon_k)\rho_{A_k} y_{A_k}\}$ $- \{\cos \beta_k \cos \varepsilon_k + r(\cos \alpha_{A_k} - \cos \beta_k \cos \varepsilon_k)\rho_{A_k} x_{A_k}\}$	$-r \rho_{A_k}^2$

Source: Adapted from Aschenbrenner, R., *Journal of the Structural Engineering Division, ASCE*, 93(ST1), 1967, 201–219.

where $\cos \varepsilon_k = -\sin \alpha_{Ak} \cos \alpha_k + \cos \alpha_{Ak} \cos \beta_k$. The corresponding transverse force (Q_k) in pile k can be obtained from the following equation:

$$Q_k = r[(d'_{tx})^2 + (d'_{ty})^2 + (d'_{tz})^2]^{\frac{1}{2}} \qquad (4.23b)$$

where

$$d'_{tx} = -\Delta x' \sin^2 \alpha_k + \Delta y' \cos \alpha_k \cos \beta_k + \Delta z' \cos \alpha_k \cos \gamma_k + \alpha'_x \cos \alpha_k \cos \gamma_k y_{A_k}$$
$$- \alpha'_y \cos \alpha_k \cos \gamma_k x_{A_k} + \alpha'_z (\sin \alpha_{ak} + \cos \alpha_k \cos \varepsilon_k)\rho_{A_k}$$

$$(4.23c)$$

TABLE 4.2
Matrix Coefficients for Piles with Two Planes of Symmetry
(x–z and y–z Planes)

Matrix Coefficient	Batter Piles	Vertical Piles
a_{11}	$-4(\cos^2 \alpha_k + r \sin^2 \alpha_k)$	$-4r$
a_{15}	$4(1 - r)\cos \alpha_k \cos \gamma_k x_{A_k}$	0
a_{22}	$-4(\cos^2 \beta_k + r \sin^2 \beta_k)$	$-4r$
a_{24}	$4(r - 1)\cos \beta_k \cos \gamma_k y_{A_k}$	0
a_{33}	$-4(\cos^2 \gamma_k + r \sin^2 \gamma_k)$	-4
a_{44}	$-4(\cos^2 \gamma_k + r \sin^2 \gamma_k) y_{A_k}^2$	$-4y_{A_k}^2$
a_{55}	$-4(\cos^2 \gamma_k + r \sin^2 \gamma_k) x_{A_k}^2$	$-4x_{A_k}^2$
a_{66}	$4[\{\cos \alpha_k \cos \varepsilon_k - r(\sin \alpha_{A_k} + \cos \alpha_k \cos \varepsilon_k)\rho_{A_k} y_{A_k}\}$ $- \{\cos \beta_k \cos \varepsilon_k + r(\cos \alpha_{A_k} - \cos \beta_k \cos \varepsilon_k)\rho_{A_k} x_{A_k}\}]$	$-4r\rho_{A_k}^2$

$$a_{12} = a_{13} = a_{14} = a_{16} = a_{23} = a_{25} = a_{26} = a_{34} = a_{35} = a_{36} = a_{45} = a_{46} = a_{56} = 0$$

Source: Adapted from Aschenbrenner, R., *Journal of the Structural Engineering Division, ASCE,* 93(ST1), 1967, 201–219.

$$d'_{ty} = \Delta x' \cos \alpha_k \cos \beta_k - \Delta y' \sin^2 \beta_k + \Delta z' \cos \beta_k \cos \gamma_k + \alpha'_x \cos \beta_k \cos \gamma_k y_{A_k}$$
$$- \alpha'_y \cos \beta_k \cos \gamma_k x_{A_k} + \alpha'_z(-\cos \alpha_{ak} + \cos \beta_k \cos \varepsilon_k)\rho_{A_k}$$

(4.23d)

$$d'_{tz} = \Delta x' \cos \alpha_k \cos \gamma_k + \Delta y' \cos \beta_k \cos \gamma_k - \Delta z' \sin^2 \gamma_k - \alpha'_x \sin^2 \gamma_k y_{A_k}$$
$$+ \alpha'_y \sin^2 \gamma_k x_{A_k} + \alpha'_z \cos \gamma_{ak} \cos \varepsilon_k \rho_{A_k}$$

(4.23e)

For a vertical pile ($\alpha_k = \beta_k = \varepsilon_k = 90°$), the expressions for the pile forces can be simplified as follows:

$$P_k = (n\, d_n)_k = -\Delta z' - \alpha'_x y_{A_k} + \alpha'_y x_{A_k}$$

(4.23f)

$$d'_{tx} = -\Delta x' + \alpha'_z y_{Ak}$$

(4.23g)

$$d'_{ty} = -\Delta y' - \alpha'_z x_{Ak}$$

(4.23h)

$$d'_{tz} = 0$$

(4.23i)

Pile constants: As noted earlier, the pile constants n and t can be obtained from pile load tests in which the allowable axial load (P_a), lateral load (Q_a), and the corresponding axial displacement $(d_n)_a$ and lateral displacement $(d_t)_a$ are measured

$(n = (d_n)_a/(P_a)$, and $t = (d_t)_a/(Q_a))$. An approximate value of t can be determined by considering the pile as a beam on elastic foundation, $t = 0.5\,K_s D\lambda^{-1}$, where K_s = coefficient of subgrade reaction, D = pile diameter, and $\lambda = [(K_s D)/(4EI)]^{\frac{1}{4}}$, where E = modulus of elasticity of pile's material, and I = moment of inertia of the pile's cross section.

Numerical example: A pile group consisting of four inclined piles is shown in Figure 4.17. The pile cap is subjected to a vertical load $P_z = 300$ kips (1334 kN) at A, and a lateral load $P_x = 20$ kips (89 kN) at B (Figure 4.17). The other data are given below:

Length of each pile = 60 ft (18.3 m)
Diameter of each pile = 10 in (25.4 cm)
Cap thickness = 20 in (50.8 cm)
Allowable axial load for each pile P_a = 100 kips (445 kN)
Ratio $r = Q_a/P_a = 0.108$
Pile constant n = 300 kip/in (3390 kn/cm)

Determine the deflections and rotations of the cap. Also, determine the pile forces.

Solution:

Weight of pile cap = 12 ft × 12 ft × (20/12) ft × 130 lbf/ft³ = 36 kips (160 kN)
Total vertical load $P_z = 300 + 36 = 336$ kips (1495 kN)
Moment about the x-axis $M_x = 300$ kips × 12 in = 3600 kip-in (113 N-m)
Moment about the y-axis $M_y = -300$ kips × 12 in = -3600 kip-in (-113 N-m)

Pertinent geometric properties of the pile group used in the computation of the pile cap constants, a_{ij}, are summarized in Table 4.3.

Using the aforementioned properties, cap constants $(a_{ij};$ see Table 4.2) and loads, the equilibrium equations for the pile group can be expressed as follows:

$$-1.87\,\Delta x' + 84.78\,\alpha_y' + 20 = 0 \qquad (4.24a)$$

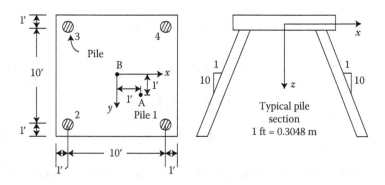

FIGURE 4.17   Pile cap geometry, location, and inclination of piles.

TABLE 4.3

Geometric Properties Used in the Computation of Pile Constants

Pile Number	α_{ak} (o)	ρ_{Ak} (in)	x_{Ak} (in)	y_{Ak} (in)	α_k (o)	β_k (o)	γ_k (o)
1	45	84.85	60	60	84.29	90	5.71
2	135	84.85	-60	60	95.71	90	5.71
3	225	84.85	-60	-60	95.71	90	5.71
4	315	84.85	60	-60	84.29	90	5.71

$$-1.73\,\Delta y' = 0 \tag{4.24b}$$

$$-15.86\,\Delta z' + 336 = 0 \tag{4.24c}$$

$$-57{,}091.4\,\alpha'_x + 3600 = 0 \tag{4.24d}$$

$$84.78\,\Delta x' - 57091.4\,\alpha'_y - 3600 = 0 \tag{4.24e}$$

$$0.\,\alpha'_z = 0 \tag{4.24f}$$

Solving these equations, the following displacements and rotations of the pile cap are obtained:

$\Delta x' = 33.62;\ \Delta x = \Delta x'/n = 33.62/300 = 0.112$ in (0.28 cm)
$\Delta y' = 0;\ \Delta y = \Delta y'/n = 0$
$\Delta z' = 84.75;\ \Delta z = \Delta z'/n = 84.75/300 = 0.2825$ in (0.72 cm)
$\alpha'_x = 0.25;\ \alpha_x = \alpha'_x/n = 0.25/300 = 0.000833$ rad
$\alpha'_y = -0.2;\ \alpha_y = \alpha'_y/n = -0.2/300 = -0.000666$ rad
$\alpha'_z = 0;\ \alpha_z = 0$

The corresponding pile forces, P and Q, can be obtained from Equations 4.23a and 4.23b, respectively (see Table 4.4).

TABLE 4.4

Pile Forces

Pile Number	$P = n.\,d_n$ (Equation 4.23a)	d'_{tx} (Equation 4.23c)	d'_{ty} (Equation 4.23d)	d'_{tz} (Equation 4.23e)	Q (kip) (Equation 4.23b)
1	-114.8	-22.2	0	2.22	2.41
2	-83.96	-41.98	0	-4.2	4.56
3	-53.85	-38.98	0	-3.9	4.23
4	-84.69	-25.2	0	2.52	2.73

4.3.6 EXAMPLE 4.6: MODEL CAP–PILE GROUP–SOIL PROBLEM: APPROXIMATE 3-D ANALYSIS

A model cap–pile group–soil system, consisting of vertical and batter piles (3:1), subjected to vertical and horizontal loads, was tested in the laboratory [45]. Figures 4.18a and 4.18b show the plan and cross-sectional dimensions, respectively. The pile group was installed in a sand bin 4.0 ft (1.22 m) in diameter with a depth of 4.0 ft (1.22 m).

The pile cap, made of Hydrocal, was 15×9 in (38.10×22.86 cm) in plan dimensions and 2.5 in (6.35 cm) thick. The modulus of elasticity of Hydrocal was assumed to be 2.0×10^6 psi (1.38×10^3 kN/cm^2) and the Poisson's ratio was assumed to be 0.30. The properties of the piles made of hollow aluminum tubing are given below:

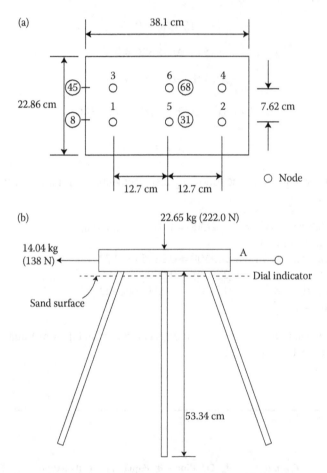

FIGURE 4.18 Pile cap–pile group. *Note*: Same as in Figure 4.24a with different units. (a) Plan; (b) side view. (From Alameddine, A.R. and Desai, C.S., Finite Element Analysis of Some Soil–Structure Interaction Problems, Report, VA Tech., 1979; Fruco and Associates, Pile Driving and Loading Tests: Lock and Dam No. 4, Arkansas River and Tributaries, Arkansas and Oklahoma, Report, U.S. Army Corps of Engineers District, Little Rock, Sept. 1964. With permission.)

$E = 9.75 \times 10^6$ psi $(6.73 \times 10^3$ kN/m²$)$, $I_x = I_y = 1.54 \times 10^{-3}$ in⁴ $(6.41 \times 10^{-2}$ cm⁴$)$, diameter $= 0.50$ in $(1.27$ cm$)$, length $= 21$ in $(53.34$ cm$)$. The inner diameter of the pile was 1.067 cm and the cross-sectional area $A = 0.373$ cm².

The pile cap was 1.27 cm above the sand surface. The internal angle of friction of the sand was 36.4° and density $= 104.0$ pcf $(16.28$ kN/cm³$)$. One vertical load (P) and one horizontal load (H) were applied (Figure 4.18b); their values are given below:

$$P = 50 \text{ lbs } (222.40 \text{ N}), \quad H = 31 \text{ lbs } (137.8 \text{ N})$$

The problem was analyzed by using the approximate multicomponent system discussed earlier [31,32,38]. Here, the cap was treated as a plate, piles as 1-D beam-column (Chapter 2), and the soil was represented using nonlinear springs. Figure 4.19 shows the p_y–v curves for the sand [31,32]; they were developed by using the empirical equations described in Chapter 2 [45]. The parameters for the R–O model are listed in Table 4.5 and they $(E_{ti}$ and $P_u)$ vary linearly with depth. Parameter E_{tf} is assumed to be zero and the exponent as unity, and they were assumed to be constant with depth. Since the piles are divided into 10 elements with 11 nodes, 11 values are given in Table 4.5.

The FE mesh for the cap and the six beam-columns contained a total number of 88 nodes and 78 elements (18 for the cap and 60 for six piles). Since the loads were applied at the center of the cap and the middle of the left side of the cap, and no nodal points exist at those locations, we applied half of the vertical load at nodes 31 and 68 each and half of the horizontal load at nodes 8 and 45 each.

In the laboratory, the system was subjected to the vertical and horizontal loads consecutively. First, the loads of 111 N were applied vertically at nodes 31 and 68;

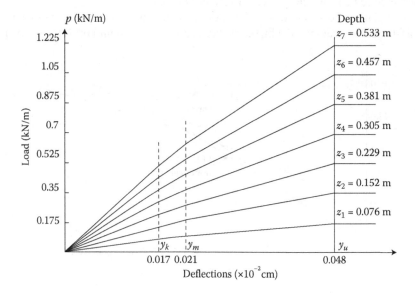

FIGURE 4.19 Resistance–displacement $(p_y$–$v)$ curves for sand. (From Alameddine, A.R. and Desai, C.S., Finite Element Analysis of Some Soil–Structure Interaction Problems, Report, VA Tech., 1979. With permission.)

TABLE 4.5
Parameters for Sand: R–O Model

Depth (cm)	E_{ti} (kN/cm²)	P_u (kN/cm)	E_{tf} (kN/cm²)	M
0	0	0	0	1.0
6.35	21.59	0.123	0	1.0
12.70	43.18	1.245	0	1.0
19.05	64.77	0.368	0	1.0
25.40	86.36	0.490	0	1.0
31.75	107.95	0.613	0	1.0
38.10	129.59	0.735	0	1.0
44.45	151.13	0.858	0	1.0
50.80	172.72	0.980	0	1.0
57.15	194.31	1.103	0	1.0
63.50	215.90	1.226	0	1.0

this load caused horizontal (measured) displacement at point A (Figure 4.18b) of the cap of about 8.0×10^{-4} in (2.03×10^{-3} cm). The code STFN-FE gave a horizontal displacement at point A of 4.3×10^{-5} in (1.09×10^{-4} cm), which was much lower than the above-measured value. Since the other computed values compared well with the measurements, the discrepancy may be due to experimental error.

Now, the horizontal loads of 68.9 N were applied at nodes 8 and 45 in 10 increments. Figure 4.20 shows the computed displacements versus applied load in comparison with the observed values for linear and nonlinear models; the correlation is considered to be satisfactory except in the initial zone.

The axial load distributions in the piles computed using STFN-FE are shown in Table 4.6 and are compared with those obtained from the Hrennikoff [1] method and

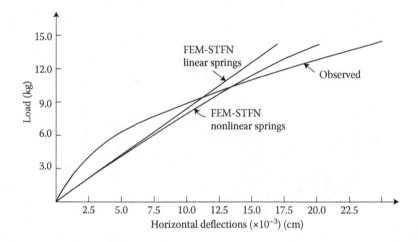

FIGURE 4.20 Lateral load versus deflections 1 kg = 9.81 N. (From Alameddine, A.R. and Desai, C.S., Finite Element Analysis of Some Soil–Structure Interaction Problems, Report, VA Tech., 1979. With permission.)

TABLE 4.6
Comparison of Axial Load Distribution in the Pile Group (kg)

Pile	Experimentally Observed	Predicted by Hrennikoff [1]	Predicted by Reese et al. [5,7]	Predicted by STFN-FE 5 Step	1 Step
1	8.15	7.29	6.75	7.26	6.92
3	5.80	7.29	6.75	6.56	6.40
5	2.95	4.30	3.81	4.42	4.58
6	2.45	4.30	3.81	4.40	4.55
4	1.26	0.18	0.82	0.07	0.74
2	1.38	0.18	0.82	0.69	1.15

Reese et al. [5]; these results are adopted from Bowles [7], who used the method by Reese et al. [5]. The results in Table 4.6 show that the three methods yield satisfactory results; however, the current method (STFN-FE) yields somewhat improved correlation with the test data compared to the other two methods.

Figure 4.21 shows the distribution of axial loads in piles versus depth obtained from the STFN-FE analysis. The distribution of moments along the depth is shown in Figure 4.22a, while the variation of moments along a typical section in the pile cap is shown in Figure 4.22b.

The deflected shape of the system by STFN-FE, at the end of loads, is shown in Figure 4.23. The displacement in final range (Figure 4.20) is close to the measured ones, although, in the initial range, they are not close. This may be due to the

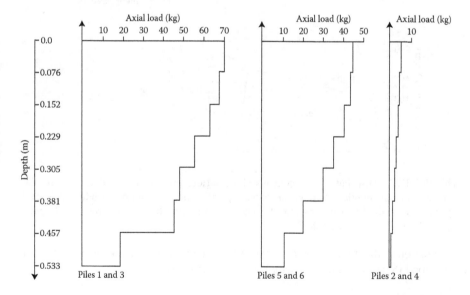

FIGURE 4.21 Distribution of axial load in various piles. (From Alameddine, A.R. and Desai, C.S., Finite Element Analysis of Some Soil–Structure Interaction Problems, Report, VA Tech., 1979. With permission.)

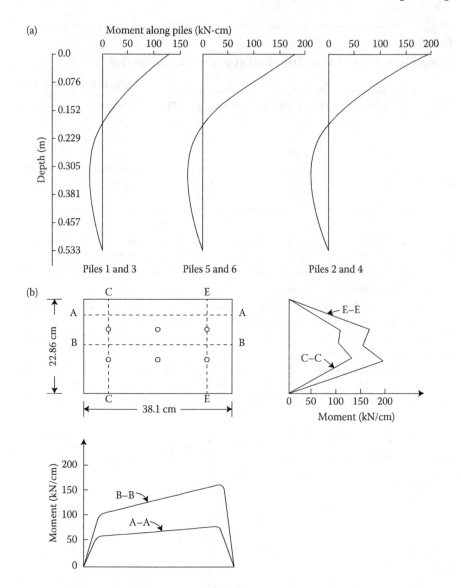

FIGURE 4.22 Distributions of moments along various piles, and typical sections in pile cap. (a) Moments in piles; (b) moments along typical sections. (From Alameddine, A.R. and Desai, C.S., Finite Element Analysis of Some Soil–Structure Interaction Problems, Report, VA Tech., 1979. With permission.)

assumption of the initial slope E_{ti}, Table 4.5, which was obtained by using empirical equation [46] rather than from actual stress–strain curves.

4.3.6.1 Comments

The computed horizontal displacement of the cap by using the STFN-FE code is about 0.020 cm (Figure 4.20); while by the Hrennikoff method, it is about 0.0193 cm.

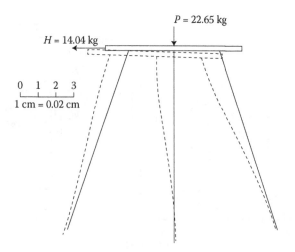

FIGURE 4.23 Deflected shape of cap and pile group. (From Alameddine, A.R. and Desai, C.S., Finite Element Analysis of Some Soil–Structure Interaction Problems, Report, VA Tech., 1979. With permission.)

The measured displacement is 0.0224 cm, which compares well with both. However, the displacement obtained from the current method is somewhat closer to the measured value.

When linear springs are used, the STFN-FE gives a horizontal displacement of about 0.0175 cm, which is much lower than the measured one than that by the nonlinear method. The Hrennikoff method assumes that the soil behavior is constant with depth. In the current method, soil behavior varied linearly with depth (Table 4.5), which yields more realistic predictions.

4.3.7 EXAMPLE 4.7: MODEL CAP–PILE GROUP–SOIL PROBLEM—FULL 3-D ANALYSIS

The multicomponent method, which is an approximate method, was used to analyze the model problem in Example 4.6. The same problem is solved in this example by using a full 3-D procedure [17,18].

Figures 4.24a and 4.24b show the details of the pile group, and the mesh configuration containing fully 3-D elements for piles, cap, and soil (Chapter 3). Interfaces between piles and soil are simulated by using the thin-layer element [47].

Sufficient (laboratory) tests were not available to define the behavior of the soil and the interface. Hence, the parameters for soil and interfaces were estimated from laboratory tests for similar sand with similar grain size distribution [48–50].

4.3.7.1 Properties of Materials

Pile: Cross-sectional area $(0.57 \times 0.36 \text{ in}) = 0.21 \text{ in}^2$ (1.32 cm^2); $E = 2.77 \times 10^6$ psi $(19 \times 10^6 \text{ kPa})$; $v = 0.20$.

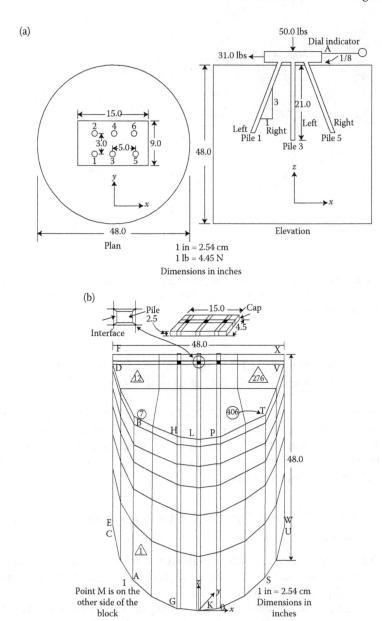

FIGURE 4.24 Cap–pile group–soil and FE mesh. (a) Details of pile group. (From Muqtadir, A. and Desai, C.S., Three-Dimensional Analysis of Cap-Pile-Soil Foundations, Report, Dept. of Civil Eng. and Eng. Mechanics, The Univ. of Arizona, Tucson, AZ, 1984; Fruco and Associates, Pile Driving and Loading Tests: Lock and Dam No. 4, Arkansas River and Tributaries, Arkansas and Oklahoma, Report, U.S. Army Corps of Engineers District, Little Rock, Sept. 1964. With permission.) (b) Schematic of FE mesh: cap-pile-soil. (From Muqtadir, A. and Desai, C.S., Three-Dimensional Analysis of Cap-Pile-Soil Foundations, Report, Dept. of Civil Eng. and Eng. Mechanics, The Univ. of Arizona, Tucson, AZ, 1984. With permission.)

cap: $E = 2.0 \times 10^6$ psi (13.8×10^6 kPa); $v = 0.30$.

Sand: The hyperbolic nonlinear elastic mode (see Appendix 1 for details) was adopted for the sand and its properties and model parameters are given below:

Unit weight of sand, $\gamma_s = 104$ pcf (1670 kg/m^3); effective size = 0.89 mm; coefficient of uniformity = 2.36; relative density = 72%; angle of internal friction for sand, $\varphi = 36.4°$; angle of interface friction between steel pile and sand, $\delta = 23.0°$; coefficient of earth pressure at rest, $K_o = 0.33$.

Hyperbolic parameters (sand): $K' = 570.0$; $n = 0.625$; $R_f = 0.85$.

4.3.7.2 Interface Element

Available test results from direct shear tests (see Appendix 1) on sand–steel interface were used to estimate the behavior of the sand–aluminum interface. The relation between shear modulus, G_i, of the interface and the normal stress, σ_n, was adopted as $G_i = \beta \sigma_n$, where β is a constant; its value was estimated to be 1.25. The thin-layer interface element [47] was used with thickness (t) of about 0.10 in (0.254 cm). Then, the shear stiffness, k_s, of the interface can be obtained approximately by dividing G_i by thickness, t. If we assume that the elastic modulus of the interface is approximately equal to that of the sand, we can obtain the normal stiffness, k_n, by dividing it by t. However, usually the normal stiffness is assumed to be high (of the order of 10^8) before separation due to tensile stress; then, it is reduced to a small value of the order of 10% of the initial value. After sliding, the shear stiffness is reduced to about 10% of the initial value.

Loading: In the laboratory test, a vertical load equal to 50 lb (222.4 N) was first applied at the middle of the cap. A lateral load equal to 31 lb (138 N) was then applied in a number of increments. For the FE linear analysis, the vertical and horizontal loads are applied in one increment. For the nonlinear analysis, the vertical load is applied first in two increments, and then the horizontal load is applied in six increments.

Results: The predicted load–displacement curves for the pile cap at point A (Figure 4.24a) by linear and nonlinear analyses are shown in Figure 4.25a. The observed behavior is also plotted for comparison. The computed displacements corresponding to total horizontal load (31 lbs = 139 N) by the linear and nonlinear methods are 0.0082 in (0.0208 cm) and 0.0090 in (0.0229 cm), respectively. The Hrennikoff's method, which assumes linear elastic soil, gives a displacement of 0.0074 in (0.019 cm) for the total horizontal load. Thus, the current 3-D procedure yields the displacement closer to the measured value of about 0.0096 in (0.0224 cm). The displacement by the Hrennikoff's method is closer to the 3-D linear elastic analysis because the Hrennikoff's method is also based on the linear elastic behavior.

However, the predicted and measured behavior in Figure 4.25a, in early ranges, do not show good agreement; this occurrence is the same as that in Example 4.6. This can be due to experimental errors and the discrepancy in stress–strain parameters, which were determined from tests on the sand from the Jonesville Lock site (48). Although the grain size distributions were similar, this sand was not the same as that used in the model test. Figure 4.25b shows the deformed shapes for the cap–pile system. The trend here is similar to that in Figure 4.23; however, the magnitudes from the 3-D analysis are somewhat different.

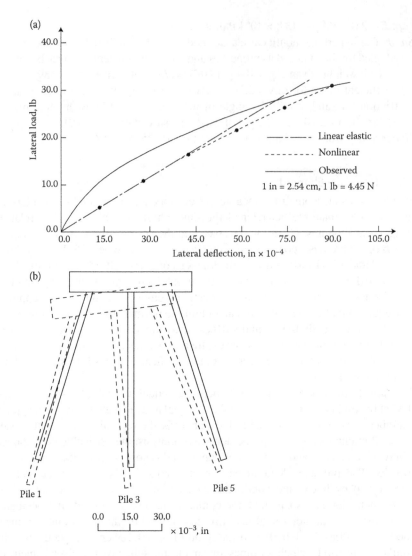

FIGURE 4.25 Comparison between computed and observed movements, and deflected shape (1) lb = 4.448 N, 1 inch = 2.54 cm. (a) Lateral movement of pile cap at the point A, Figure 4.24a; (b) deflected shape at CUVD, Figure 4.24b after application of total loads. (From Muqtadir, A. and Desai, C.S., Three-Dimensional Analysis of Cap-Pile-Soil Foundations, Report, Dept. of Civil Eng. and Eng. Mechanics, The Univ. of Arizona, Tucson, AZ, 1984. With permission.)

4.3.8 EXAMPLE 4.8: LATERALLY LOADED PILES—3-D ANALYSIS

3-D FE analysis was performed for a laterally loaded pile, which was tested in the laboratory [12,45,49]. Figure 4.26 shows a plan view of the piles tested. A single solid steel pile (A20) with cross section 0.50 × 0.50 in (1.3 × 1.3 cm), embedded to a depth of 24 in (61 cm) in sand, was adopted here. The adopted pile was subjected to

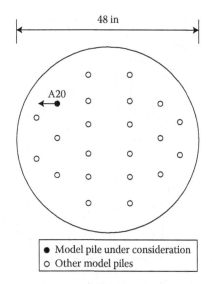

FIGURE 4.26 Test tank and model piles. (Pile A20 is analyzed here; arrow indicates direction of loading.) 1 inch = 2.54 cm. (From Desai, C.S. and Appel, G.C., Three-dimensional FE analysis of laterally loaded structures, *Proceedings of the 2nd International Conference on Numerical Methods in Geomechanics*, Vol. 1, ASCE, Sept. 1976. With permission.)

a total lateral load of 6 lbs (27 N), monotonically increasing in one pound (4.45 N) increments.

The foundation sand had the following properties:

Angle of friction = 36.4°
Density, γ = 104 pcf (1667 kg/m^3)
Effective size, D_{10} = 0.89 mm
Coefficient of uniformity = 2.36
Relative density = 72%

Loading: In the laboratory tests, the lateral displacement of the base of the pile, after each load increment of 1.0 lb (4.448 N), was measured at the ground level. The load applied was 6.0 lbs (15.24 N).

4.3.8.1 Finite Element Analysis

The FE analysis was performed using (1) interface elements and (2) no interface elements. Figure 4.27 shows the FE mesh for half domain, with details of interface elements given in Figure 4.28. The mesh with interface elements contained 148 nodes and 70 elements, including 10 interface elements (Figures 4.27 and 4.28). The mesh without interface elements consisted of 120 nodes and 60 brick elements.

No adequate laboratory tests were available to define the behavior of the soil and the interface. Hence, a parametric study was performed in which the elastic modulus, E, was varied, based on the experience with similar sands [12,49–51]. The variation of E in the parametric study is given below:

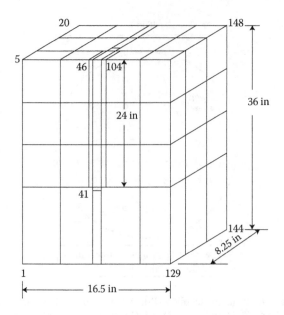

FIGURE 4.27   Finite element mesh with interface elements. (From Desai, C.S. and Appel, G.C., Three-dimensional FE analysis of laterally loaded structures, *Proceedings of the 2nd International Conference on Numerical Methods in Geomechanics*, Vol. 1, ASCE, Sept. 1976. With permission.)

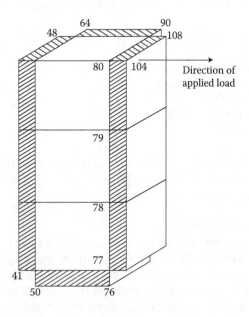

FIGURE 4.28   Details of interface (joint) elements (shaded). (From Desai, C.S. and Appel, G.C., Three-dimensional FE analysis of laterally loaded structures, *Proceedings of the 2nd International Conference on Numerical Methods in Geomechanics*, Vol. 1, ASCE, Sept. 1976. With permission.)

E_{sand} (psf)	31743	35975	46556	63486
E_{sand} (Pa)	(1.5×10^6)	(1.7×10^6)	(2.2×10^6)	(3.0×10^6)

The approximate value of $E = 36,000$ psf $(1.72 \times 10^6$ Pa) was adopted because it yielded the displacements of about 2.5×10^{-3} in $(6.35 \times 10^{-3}$ cm), 7×10^{-3} in $(18 \times 10^{-3}$ cm), and 14×10^{-3} in $(36 \times 10^{-3}$ cm) for applied loads equal to 1.016 $(4.448$ N), 2 lbs $(8.90$ N), and 4.0 lbs $(17.80$ N), respectively, which compared closely with the observed values at those loads (see Figure 4.29).

The modulus, E, for steel, was adopted equal to 4.3×10^9 psf $(2 \times 10^{11}$ Pa). The Poisson's ratios for the sand and steel were adopted as 0.30 and 0.20, respectively.

For the interface between the sand and the steel pile, the following properties were adopted based on the results in Refs. [12,49,50].

Shear stiffness $k_{sx} = 3.2 \times 10^5$ kg/m³ $(31.4 \times 10^5$ N/m³)
Shear stiffness $k_{sy} = 3.2 \times 10^5$ kg/m³ $(31.4 \times 10^5$ N/m³)
Normal stiffness $k_{nz} = 1.6 \times 10^{10}$ kg/m³ $(15.7 \times 10^{10}$ N/m³)

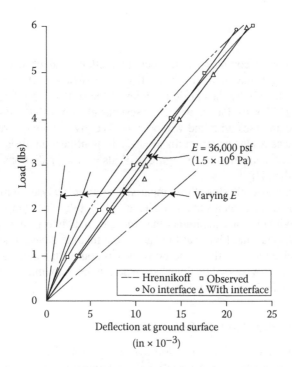

FIGURE 4.29 Parametric study and comparisons of buff displacement (pile A-20) (1 inch = 2.54 cm, 1 lb = 4.448 N). (From Desai, C.S. and Appel, G.C., *Three-dimensional FE analysis of laterally loaded structures, Proceedings of the 2nd International Conference on Numerical Methods in Geomechanics*, Vol. 1, ASCE, Sept. 1976. With permission.)

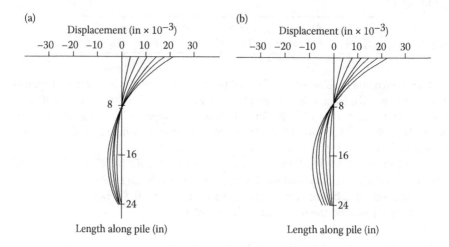

FIGURE 4.30 Distribution of displacements along pile (A-20) under load increments (1 inch = 2.54 cm). (a) Without interface; (b) with interface elements. (From Desai, C.S. and Appel, G.C., Three-dimensional FE analysis of laterally loaded structures, *Proceedings of the 2nd International Conference on Numerical Methods in Geomechanics*, Vol. 1, ASCE, Sept. 1976. With permission.)

4.3.8.2 Results

Figure 4.29 shows a comparison between predictions and observations for the load–displacement behavior at the base. It can be seen that the predictions are influenced significantly by the elastic modulus of the soil. The prediction for about $E = 36,000$ psf (1.72×10^6 Pa) shows the closest correlation with the test data. It can be seen that the analyses with and without interface give very close results. It may be noted that these results are from linear analysis; nonlinear analysis may give different results. The comparisons in Figure 4.29 also show predictions by using the Hrennikoff method [1].

Figures 4.30a and 4.30b show the comparison between analyses, with and without interface, for the displacements along the length of the pile. The computed values of displacements with interface are higher than those without interface in the lower part of the pile, and somewhat higher at the base. Note that the analysis here involves a linear elastic behavior. For other conditions such as nonlinear soil behavior, loading, and geometry, the provision of the interface generally leads to improved and realistic results; for example, Example 4.9 below.

4.3.9 EXAMPLE 4.9: ANCHOR–SOIL SYSTEM

Full-scale field tests for an anchor–soil system were performed by Scheele [52] near Munich, Germany. A cross-sectional elevation of the test pit and details regarding the anchor are shown in Figures 4.31a and 4.31b, respectively. The length of the grouted anchor was 2.0 m, and the diameter of grouted body was about 90 mm. The diameter of the steel bar was 32.0 mm. The total length was

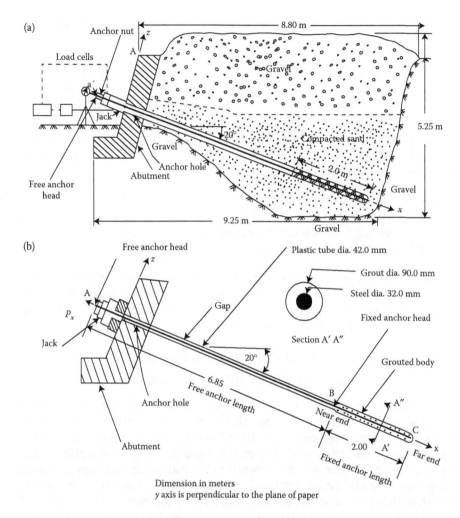

FIGURE 4.31 (a) Cross section of anchor-soil system; (b) anchor components (Adapted from Muqtadir, A. and Desai, C.S., Three-Dimensional Analysis of Cap-Pile-Soil Foundations, Report, Dept. of Civil Eng. and Eng. Mechanics, The Univ. of Arizona, Tucson, AZ, 1984).

about 8.85 m; hence, the free length of the steel bar beyond the grouted anchor was about 6.85 m.

The sand in the test pit was compacted to desired density; the surcharge over the compacted sand consisted of gravel of about 2.0 m depth. The density of the compacted sand was 2.04 g/cm³ and the uniformity coefficient was about 9.0 [17–19,52,53].

4.3.9.1 Constitutive Models for Sand and Interfaces

Both linear elastic and elastic–plastic models were adopted for the sand [17,19,53]. The constitutive parameters for the sand were determined from a comprehensive series of multiaxial tests using the cubical device [19]. The parameters for the linear elastic model are given below:

Material →	Sand	Gravel	Concrete	Steel
E (kN/m²)	1.12×10^5	1.12×10^6	20.7×10^6	20.7×10^7
v	0.36	0.30	0.17	0.30

An earlier version of the single surface yield function was used for the sand; a description of this model is provided in Appendix 1. The properties and parameters for the plasticity model are given below [17,19,53]:

Mass density of sand	2.04 g/cc
Angle of friction of sand, φ	40.0°
Angle of sliding friction between concrete and sand	37.0°
Coefficient of earth pressure at rest	0.56

Single surface plasticity model: The symbols for the following parameters refer to the earlier yield function [17,18], which is somewhat different from the HISS yield function in Appendix 1, Equation A1.28. Hence, the parameters below are different from those for the HISS model, Equation A1.28:

$$\alpha = 0.212; \quad \gamma = 25.28 \text{ kN/m}^2; \quad k = 0.0 \text{ kN/m}^2; \quad \beta_a = 0.000278;$$

$$\eta_1 = 1.161; \quad \beta_b = 0.805; \quad \eta_2 = 0.704; \quad \beta_c = 0.0055; \quad \eta = 0.74$$

Interface element: The interface between the anchor and the sand was modeled using the thin-layer element [47]. The value of the shear modulus, G_i, for the interface between anchor and sand was obtained from direct shear tests [17,19]; however, its magnitudes were found to be too small because the interface was affected significantly by the grout around the anchor. Hence, a parametric study was performed using the FE method in which the value of G_i was varied from 1000 to 75,000 kN/m². Then, from comparisons of the computed forces in the anchor with that observed in the field, it was found that the value of $G_i = 41,200$ kN/m² yielded close correlation. This value is similar in magnitude to the shear modulus of the sand. The Poisson's ratio for the interface was adopted as 0.36.

Figure 4.32 shows the FE mesh; owing to the symmetry, only one-half of the anchor–soil system was discretized. It contained 580 nodes and 352 eight-noded isoparametric elements. The x-axis coincided with the center line of the anchor and the y-axis was perpendicular to the plane of the paper. The DOFs on the front face marked by abcdefghijklm and the back face by ABCDEFGHIJKLM were constrained in the y-direction. The DOFs at nodes on the bottom boundary, marked abcdeEDCBA, were fixed in the z-direction. The DOFs on the side efgGFE were constrained in the x-direction. The DOFs on the bottom face of the abutment were constrained in the x-direction. Note the capital symbols in parentheses relate to the nodes on the opposite (back) face.

Loading: A pullout force of 250 kN was applied in the field test, in five equal increments by using an annular jack located between the nut at the anchor head and the

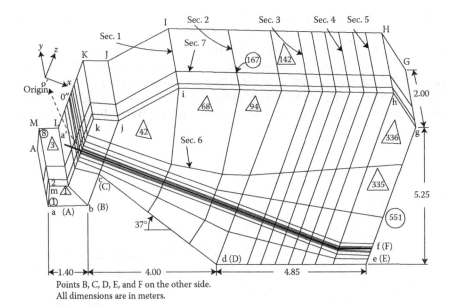

FIGURE 4.32   Finite element mesh for anchor-soil problem. (From Muqtadir, A. and Desai, C.S., Three-Dimensional Analysis of Cap-Pile-Soil Foundations, Report, Dept. of Civil Eng. and Eng. Mechanics, The Univ. of Arizona, Tucson, AZ, 1984. With permission.)

abutment. The same total load was applied in one increment along the negative x-direction at the free anchor head, and an equal magnitude of reaction force was applied along the positive x-direction to the abutment in the zone where the jack was located.

Figure 4.33a shows the load versus displacement curves predicted from the linear and nonlinear FE analyses in comparison with those observed in the field, at the anchor head, that is, point A in Figure 4.31b. Similar comparisons for point B on the anchor (Figure 4.31b) are shown in Figure 4.33b. Figure 4.33c presents the comparisons for computed axial force distributions along the anchor and observed in the field for typical load levels equal to 50, 150 and 250 kN. It can be seen from Figures 4.33a through 4.33c that the nonlinear analyses with the single surface plasticity model gives improved correlations with the field data. The three results (Figures 4.33a through 4.33c) show that the predictions with the use of the interface lead to, in general, improved results.

4.3.10   EXAMPLE 4.10: THREE-DIMENSIONAL ANALYSIS OF PAVEMENTS: CRACKING AND FAILURE

The four-layered flexible pavement structure considered here is shown in Figure 4.34. The DSC model with HISS plasticity was used for the first layer, that is, asphalt concrete. The base, subbase, and subgrade materials are modeled using the HISS plasticity model. Details of the DSC and HISS plasticity models are given in Appendix 1. The model parameters are shown in Table 4.7 [22,25,54] for materials in four layers. The FE mesh is shown in Figure 4.35. The analysis was performed by using code DSC-SST3D [55].

Loading: The corner area close to the center (Figure 4.34) was subjected to two types of loading: (a) monotonic load up to 200 psi (1.4 MPa)—it was applied in 50 increments, and (b) repetitive loading (Figure 4.36) with amplitudes of load denoted by P. The repetitive load–unload–reload cycles were applied sequentially; however, the time dependence was not included. The accelerated procedure (AP) was used in which full FE computations were performed for each load amplitude up to reference cycles, for example, N_r (= 20); details of the AP are given in Ref. [56]. After N_r

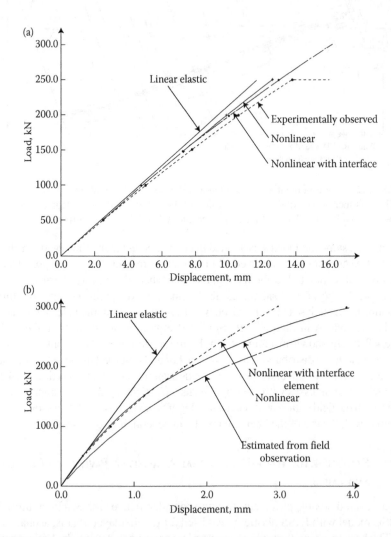

FIGURE 4.33 Comparisons of load displacement curves and axial load distributions along anchor. (a) At free anchor head, point A, Figure 4.31b; (b) at fixed anchor head, point B, Figure 4.31b; and (c) axial force distribution along the fixed anchor. (From Muqtadir, A. and Desai, C.S., Three-Dimensional Analysis of Cap-Pile-Soil Foundations, Report, Dept. of Civil Eng. and Eng. Mechanics, The Univ. of Arizona, Tucson, AZ, 1984. With permission.)

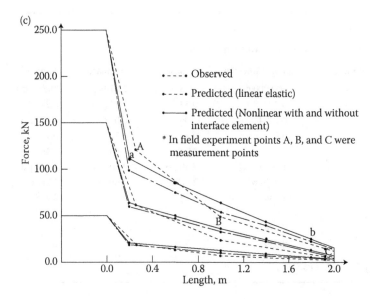

FIGURE 4.33   (continued) Comparisons of load displacement curves and axial load distributions along anchor. (a) At free anchor head, point A, Figure 4.31b; (b) at fixed anchor head, point B, Figure 4.31b; and (c) axial force distribution along the fixed anchor. (From Muqtadir, A. and Desai, C.S., Three-Dimensional Analysis of Cap-Pile-Soil Foundations, Report, Dept. of Civil Eng. and Eng. Mechanics, The Univ. of Arizona, Tucson, AZ, 1984. With permission.).

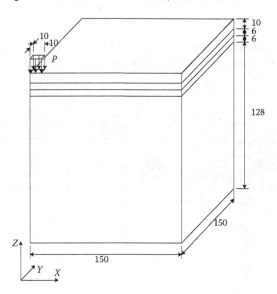

FIGURE 4.34   Four-layered pavement system (dimension in inches; 1 inch = 2.54 cm). (Adapted from Desai, C.S., Mechanistic pavement analysis and design using unified material and computer model, Keynote Paper, *Proceedings of the 3rd International Symposium on 3-D Finite Element for Pavement Analysis, Design and Research*, Amsterdam, The Netherlands, 2002.)

TABLE 4.7

Parameters for Pavement Materials

Parameter	Asphalt Concrete	Base	Subbase	Subgrade
E	500,000 psi	56,533 psi	24,798 psi	10,013 psi
v	0.3	0.33	0.24	0.24
γ	0.1294	0.0633	0.0383	0.0296
β	0.0	0.7	0.7	0.7
n	2.4	5.24	4.63	5.26
3R	121 psi	7.40 psi	21.05 psi	29.00 psi
a_i	1.23E–6	2.0E–8	3.6E–6	1.2E–6
η_1	1.944	1.231	0.532	0.778
D_u	1			
A	5.176			
Z	0.9397			

1 psi = 6.895 kPa.

cycles, the deviatoric trajectory or accumulated plastic strains (ξ_D) for a given cycle, N, was computed using the following equation:

$$\xi_D(N) = \xi_D(N_r) \left(\frac{N}{N_r} \right)^b \tag{4.25}$$

where b is the parameter depicted in Figure 4.37. The disturbance, D, was computed at a given cycle by using the following equation:

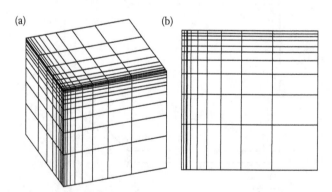

(a)   (b)

FIGURE 4.35 Three-dimensional mesh for pavement system. (a) 3-D mesh and (b) 2-D cross-section. (Adapted from Desai, C.S., Mechanistic pavement analysis and design using unified material and computer model, Keynote Paper, *Proceedings of the 3rd International Symposium on 3-D Finite Element for Pavement Analysis, Design and Research*, Amsterdam, The Netherlands, 2002.)

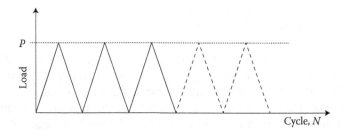

FIGURE 4.36 Schematic of repetitive loading. (Adapted from Desai, C.S., Mechanistic pavement analysis and design using unified material and computer model, Keynote Paper, *Proceedings of the 3rd International Symposium on 3-D Finite Element for Pavement Analysis, Design and Research*, Amsterdam, The Netherlands, 2002.)

$$D = D_u [1 - \exp(-A \{\xi_D(N)^Z\}]\qquad(4.26)$$

where D_u, A, and Z are disturbance parameters. With the above information, we can compute the cycle at failure N_f depending upon the criterion chosen for critical disturbance, D_c:

$$N = N_r \left[\frac{1}{\xi_D(N_r)} \left\{ \frac{1}{A} \ell n \left(\frac{D_u}{D_u - D} \right) \right\}^{\frac{1}{z}} \right]^{\frac{1}{b}} \qquad(4.27)$$

in which for $D = D_c$, $N = N_f$. The value of D_c can be found from test data; for the material (asphalt concrete), $D_c = 0.80$ was adopted [22,25], which implies that after

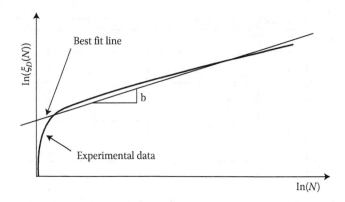

FIGURE 4.37 Plastic strain trajectory versus number of cycles for accelerated analysis. (Adapted from Desai, C.S., Mechanistic pavement analysis and design using unified material and computer model, Keynote Paper, *Proceedings of the 3rd International Symposium on 3-D Finite Element for Pavement Analysis, Design and Research*, Amsterdam, The Netherlands, 2002.)

the growth of microcracking, fractures can initiate around D_c, and for $D_c \geq 0.80$, fractures in the material (FEs) grow further.

Results: Figures 4.38a and 4.38b show permanent (plastic) displacements, which may lead to rutting, at the surface for the load = 100 (0.70 MPa) and 200 psi (1.40 MPa), respectively. Figure 4.39 shows contours of disturbance at load = 200 psi (1.4 MPa); for this monotonic load, the maximum disturbance reached is about 0.024. This low value means that no microcracking and fractures occur under such monotonic load. However, for repetitive load with similar or lower amplitudes, microcracking and fracture would occur; this is illustrated below.

Figures 4.40a through 4.40c show contours of disturbance after 10, 1000, and 20,000 cycles (Figure 4.36) under a load amplitude of 70 psi (480 kPa) and $b = 1.0$ (Equation 4.27). At and after about 20,000 cycles, the critical disturbance, D_c, and its greater values, that is, $D_c \geq 0.80$ occur, and a portion of the pavement is considered to experience fracture and failure and consequent rutting (Figure 4.40c).

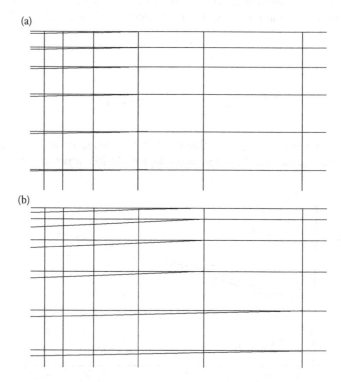

(a)

(b)

FIGURE 4.38 Permanent deformations at loading steps (1 psi = 6.89 kPa, 1 inch = 2.54 cm). (a) Step = 25 (load = 100 psi) displacements are scaled by 10; (b) step = 50 (load = 200 psi) displacements are scaled by 10. (Adapted from Desai, C.S., Mechanistic pavement analysis and design using unified material and computer model, Keynote Paper, *Proceedings of the 3rd International Symposium on 3-D Finite Element for Pavement Analysis, Design and Research*, Amsterdam, The Netherlands, 2002.)

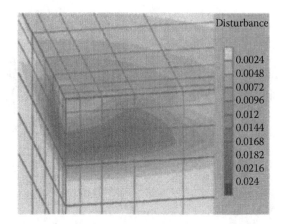

FIGURE 4.39 Contours of disturbance after 50 steps (load = 200 psi; 1 psi = 6.89 kPa). (Adapted from Desai, C.S., Mechanistic pavement analysis and design using unified material and computer model, Keynote Paper, *Proceedings of the 3rd International Symposium on 3-D Finite Element for Pavement Analysis, Design and Research*, Amsterdam, The Netherlands, 2002.)

4.3.11 Example 4.11: Analysis for Railroad Track Support Structures

Instrumented sections of railroad tracks were tested in the field at the Transportation Test Center (TTC), Pueblo, Colorado; Figure 4.41 shows a typical UMTA (Urban Mass Transportation Administration) test section with various instruments [57].

Linear analysis: 3-D linear elastic FE analyses were performed first by assuming linear elastic properties for all components; the parameters are given in Table 4.8 [15,57–60].

4.3.11.1 Nonlinear Analyses

Nonlinear 3-D FE analyses were also performed by treating rail (metal) and concrete as linear elastic, subballast using the cap model, ballast using the variable moduli model, and subgrade sand using the modified cam-clay model [15,59,60]; see Appendix 1 for details on various models. The parameters for the models were obtained from comprehensive truly triaxial tests on specimens of subgrade silty sand, subballast, and wood tie obtained from the field test section; details are given in Refs. [57–60]. A list of parameters is given in Table 4.9.

Interface behavior: Behavior of interfaces, that is, junctions between tie and ballast, and between other components can influence the behavior of the entire track support system. In this study, the thin-layer element [47] was used for the interfaces shown in Figure 4.42 with the FE mesh of the system. The parameters for nonlinear elastic behavior of the interfaces were obtained from tests conducted using the CYMDOF device [58]; they are given below:

E = 30,000 psi (208.5 kN/m^2)
v = 0.30
G = 225.0 psi (1553 kN/m^2)

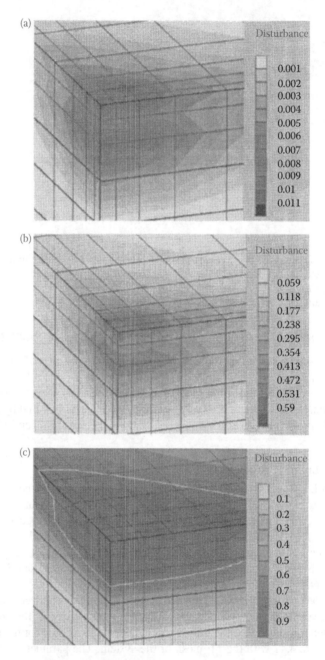

FIGURE 4.40 Contours of disturbance at various cycles: amplitude of load (P) = 70 psi; $b = 1.0$ (1 psi = 6.89 kPa). (a) $N = 10$ cycles; (b) $N = 1000$ cycles; and (c) $N = 20,000$ cycles ($D \geq 0.8$ inside white curve). (Adapted from Desai, C.S., Mechanistic pavement analysis and design using unified material and computer model, Keynote Paper, *Proceedings of the 3rd International Symposium on 3-D Finite Element for Pavement Analysis, Design and Research*, Amsterdam, The Netherlands, 2002.)

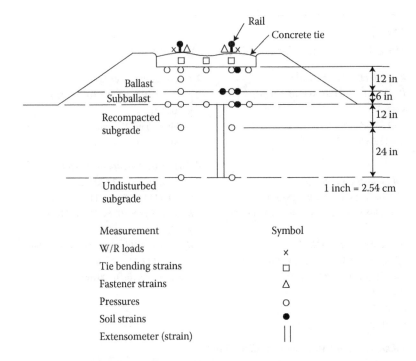

FIGURE 4.41 UMTA test section and location of instruments. (Adapted from Stagliano, T.R. et al., Pilot Study for Definition of Track Component Load Environments, Final Report, Kaman Avidlyne, Burlington, MA, July 1980; Siriwardane, H.J. and Desai, C.S., Nonlinear Soil-Structure Interaction Analysis for One-, Two- and Three-Dimensional Problems Using Finite Element Method, Report to DOT-Univ. Research, Va Tech, Blacksburg, VA, USA, 1980.)

The value of the shear modulus, G, was obtained from tests using the CYMDOF device with the following equation:

$$G = \frac{\tau \times t}{u_r} \tag{4.28}$$

where τ = shear stress, u_r = relative shear displacement, and t = finite (small) thickness of the interface; here, the thickness is assumed to be 0.30 in (0.51 cm).

TABLE 4.8

Material Properties Used in Linear Analysis

	Rail	Tie	Ballast	Subballast	Subgrade (Sand)
E psi	3×10^6	5×10^6	30,000	20,000	5000
(kN/m²)	(207×10^6)	(34.5×10^6)	(207×10^3)	(138×10^3)	(34.5×10^3)
ν	0.35	0.2	0.40	0.35	0.45

TABLE 4.9

Material Properties for Nonlinear UMTA Section

Rail (Steel): Linear Elastic

$E = 30 \times 10^6$ psi (207.0×10^6 kN/m²)

$v = 0.3$

$I = 71.4$ in⁴ (2971.9 cm⁴)

$A = 11.65$ in² (75.16 cm²)

Length = 108 in (274.32 cm)

Subballast: Cap Model

$E = 20,000$ psi (138×10^3 kN/m²)

$D = 0.308 \times 10^{-4}$ psi⁻¹ (0.446×10^{-4} kPa⁻¹)

$v = 0.40$

$\alpha = 26.0$ psi (179.4 kN/m²)

$\gamma_m = 0.0833$ lb/in³ (0.00230 kg/cm³)

$\gamma = 21.5$ psi (148.35 kN/M²)

$R = 1.24$

$\beta = 0.02$

$W = 0.035$

Tie Concrete: Linear Elastic

$E = 4.2 \times 10^6$ psi (29×10^6 kN/m²)

$v = 0.20$

Width = 10 in (25.4 cm)

Depth = 6 in (15.24 cm)

Ballast: Variable Moduli Model

$K_0 = 4 \times 10^3$ psi (27.6 kN/m²)

$G_0 = 1.7414 \times 10^3$ psi (11.83 kN/m²)

$K_1 = -5.4917 \times 10^3$ psi (11.83 kN/m²)

$\gamma_1 = 2.5593$

$K_2 = 2.536 \times 10^7$ psi (17.50×10^7 kN/m²)

$\gamma_2 = -60.172$

$\gamma_m = 0.0613$ lb/in³ (0.00169 kg/cm³)

Subgrade (Sand): Modified Cam-Clay Model

$M = 1.24$

$E_0 = 12,000$ psi (82.8×10^3 kN/m²)

$\lambda_c = 0.014$

$v = 0.28$

$\kappa = 0.0024$

$\gamma_m = 0.081$ lb/in³ (0.00224 kg/cm³)

$e_0 = 0.340$

The FE mesh for quarter domain of test section (Figure 4.41) is shown in Figure 4.42. The boundary conditions were adopted as follows:

From a parametric study, it was found that the location of the end boundary at a distance of about 150 in (381 cm) from the bottom of ballast for the bottom boundary, and from the centerline for the side boundary, provided satisfactory solutions. Therefore, the bottom boundary was provided at a distance of 150 in (381 cm) and side boundary at a distance of 250 in (635 cm). The nodes on the bottom boundary were assumed to be constrained in the x-, y-, and z-directions. On the side boundary, the nodes were assumed to be constrained in the x-direction, and smooth or free in the y- and z-directions. The boundary in the longitudinal direction was assumed free in x-, y-, and z-directions.

Loading: A load of 16,000 lbs (71.2 kN), equivalent of a static wheel load of 13,100 lbs (58.3 kN), was applied on the rail above the central tie (Figure 4.42).

Results: The prediction from numerical computations for vertical displacement at the subgrade was found to be 0.0202 in (0.0513 cm). This compares very well with the observed displacement of 0.021 in (0.0533 cm) at the subgrade [57]. The seating

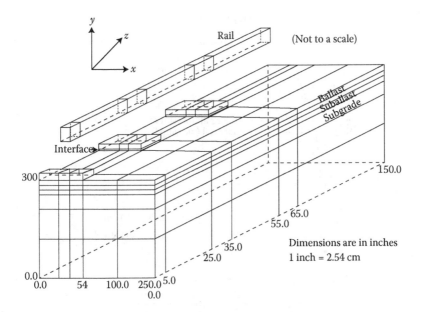

FIGURE 4.42 Finite element mesh for UMTA section. (Adapted from Siriwardane, H.J. and Desai, C.S., Nonlinear Soil-Structure Interaction Analysis for One-, Two- and Three-Dimensional Problems Using Finite Element Method, Report to DOT-Univ. Research, Va Tech, Blacksburg, VA, USA, 1980.)

load computed by integrating the vertical stresses in the tie elements (Figure 4.42) at integration points just below the rail was found to be about 50% of the applied wheel load. This value compares well with that reported from the field data [57].

Figure 4.43 shows comparisons between the FE predictions and observations for vertical stress, below the inner and outer rails of the test section, including the average values for the inner and outer rails, and predictions by 3-D nonlinear analysis. The correlation between predictions and test data is not satisfactory, particularly in the zone below the surface. This could be due to reasons such as errors and inconsistencies in measurements, and constitutive models used. However, the trends in Figure 4.43 are considered to be satisfactory, in light of the facts that the predicted displacements and seating load above provided excellent correlation.

4.3.12 EXAMPLE 4.12: ANALYSIS OF BURIED PIPELINE WITH ELBOWS

In this example, 3-D FE analysis is used to study the interaction between a buried pipe with elbows and the surrounding soil due to potential ground displacement [29]. Specifically, the effect of opening and closing modes of the elbow section (Figure 4.44) for different initial pipe bending angles (α) is investigated. For a pipe with 90° elbow, if the bending angle after deformation, α_{after}, is larger than 90°, it is considered "opening mode." If α_{after} is less than 90°, it is considered "closing mode." A commercial FE software, ABAQUS, is used in analyzing the pipe, which is modeled using

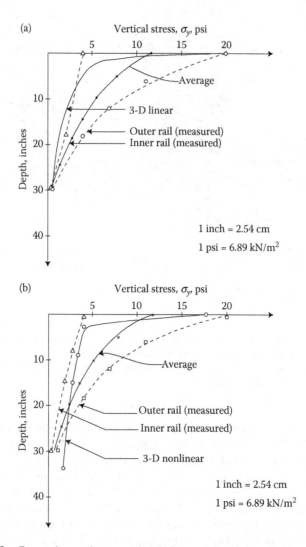

FIGURE 4.43 Comparisons of computed and observed data for vertical stresses; (a) linear analysis; (b) nonlinear analysis. (From Siriwardane, H.J. and Desai, C.S., Nonlinear Soil-Structure Interaction Analysis for One-, Two- and Three-Dimensional Problems Using Finite Element Method, Report to DOT-Univ. Research, Va Tech, Blacksburg, VA, USA, 1980. With permission.)

3-D linear shell elements, whereas the soil is modeled using 3-D solid continuum elements. The FE mesh and model geometry for $H/D = 4$ and $\alpha = 90°$ is shown in Figure 4.45, where H = embedment depth and D = pipe diameter. Since the analysis considers only the case in which the elbow is subjected to lateral loading in the direction perpendicular to the maximum curvature of the elbow, only half of the pipe is modeled [29]. The FE mesh consists of 6300 solid continuum elements to represent soil, and 32 shell elements are used in the circumferential direction to represent the

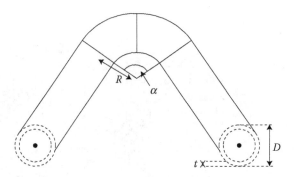

FIGURE 4.44 Schematic diagram of pipeline with elbow. (Adapted from Cheong, T.P., Soga, K., and Robert, D. J., *Journal of Geotechnical and Geoenvironmental Engineering*, ASCE, 137(10), 2011, 939–948.)

pipe. The left, right, front, and back sides of the wall boundaries are constrained in the horizontal direction perpendicular to the corresponding plane; the bottom boundary is constrained in all directions; and the top boundary is set as a free surface [29].

The model pipe is displaced laterally by imposing a prescribed horizontal displacement on all nodes of the pipe, permitting free vertical movement. The resulting contact between the pipe and the surrounding soil is modeled using contact elements that allow slip and separation, with friction coefficient $\mu = 0.32$ for medium dense sand and $\mu = 0.4$ for dense sand. Two different constitutive models are used to represent the soil: an elastic–perfectly plastic Mohr–Coulomb model with nonassociated flow rule, and a more advanced model, called Nor–Sand model, that accounts for strain hardening and softening behavior and stress-dilatancy [44]. The input parameters for the pipe geometry are shown in Table 4.10, whereas the soil properties are used in the analysis, are summarized in Tables 4.11 and 4.12 for the Mohr–Coulomb model and the Nor–Sand model, respectively.

The distribution of normal force, N (kN/m), along the pipe length for different relative pipe displacements, is shown in Figure 4.46 for $\alpha = 90°$ and $H/D = 4$. The angle θ corresponds to an individual strip of the pipe element (Figure 4.45b), and the dotted line separates the elbow and the straight portion of the pipe. It is seen that the peak force occurs at $\theta = 11°$ in the case of medium dense sand. For the case of dense sand, the peak force occurs at $\theta = 0°$, which indicates that in case of medium dense sand, the soil tends to slip near the maximum curvature when the pipe is pulled laterally. This mechanism is not observed in the case of dense sand because the high dilative nature restricted any sliding movement along the elbow [29]. The maximum dimensionless peak forces, $N_{max}/(\gamma HDL)$, or N_q are shown in Figure 4.47, as a function of H/D. It is seen that N_q at the elbow section of the pipe is greater (40–140%) than that along the straight portion, for a given H/D. Also, an elbow with a larger α yields a lager N_q due to greater localized soil deformation. Larger peak forces are seen for a deeper pipe with larger α than a shallower pipe with smaller α [29]. Additional results and details are given by Cheong et al. [29]. The results from such analysis are useful for designing a buried pipe with elbow.

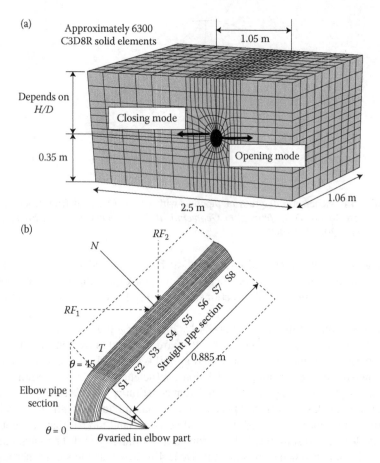

FIGURE 4.45 Finite element mesh used: (a) idealization of soil; (b) idealization of elbow. (From Cheong, T.P., Soga, K., and Robert, D. J., *Journal of Geotechnical and Geoenvironmental Engineering, ASCE*, 137(10), 2011, 939–948. With permission.)

TABLE 4.10

Pipe Properties Used in the Analysis

Pipe diameter, D	102 mm
Wall thickness, t	6.4 mm
D/t ratio	16
Radius of curvature, R	1.5 D
Elbow angle, α	90°, 45°, 22.5°

Source: From Cheong, T.P., Soga, K., and Robert, D.J., *Journal of Geotechnical and Geoenvironmental Engineering*, ASCE, 137(10), 2011, 939–948. With permission.

TABLE 4.11
Mohr–Coulomb Model Input Parameters

Parameter	Medium Dense Sand	Dense Sand
Dry unit weight, γ_{dry}	16.4 kN/m³	17.7 kN/m³
Effective unit weight, γ'	10.4 kN/m³	11.2 kN/m³
Void ratio, e	0.669	0.548
Peak friction angle, φ_{peak}	35°	44°
Dilation angle, ψ	5°	16.3°
Poisson's ration, υ	0.3	0.3
Young's modulus, E (varies with H)	1143–2950 kPa	1414–3650 kPa
Cohesion, c' (varies with H)	0.1–0.5 kPa	0.1–0.5 kPa
Earth pressure coefficient, K_o	1.0	1.0

Source: From Cheong, T.P., Soga, K., and Robert, D.J., *Journal of Geotechnical and Geoenvironmental Engineering*, ASCE, 137(10), 2011, 939–948. With permission.

4.3.13 EXAMPLE 4.13: LATERALLY LOADED TOOL (PILE) IN SOIL WITH MATERIAL AND GEOMETRIC NONLINEARITIES

For certain geotechnical problems, it may be necessary to perform 3-D analysis by including both material and geometric nonlinearities. Such problems can include certain mining and tunneling problems such as multiple underground excavations, tillage tools moving in soil, structures on soft soil foundations, caving of floor and crown in underground construction, and sea floor movements for offshore structures.

TABLE 4.12
Nor–Sand Model Input Parameters

Parameter	Input Value
Shear modulus constant (A)	300
Pressure exponent (n)	0.5
Poisson's ration (υ)	0.3
Critical state ratio (M)	1.25
Maximum void ratio (e_{max})	0.852
Minimum void ratio (e_{min})	0.497
N value in flow rule	0.2
Hardening parameter (H)	1000
Maximum dilatancy coefficient (χ)	3.5

Source: From Cheong, T.P., Soga, K., and Robert, D.J., *Journal of Geotechnical and Geoenvironmental Engineering*, ASCE, 137(10), 2011, 939–948. With permission.

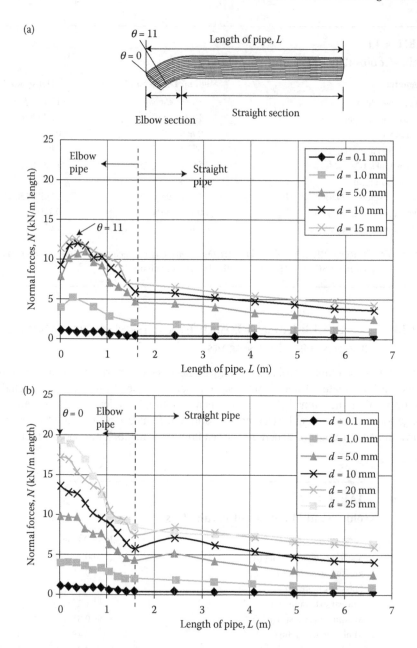

FIGURE 4.46 Variation of normal force on pipe: (a) medium dense sand; (b) dense sand. (From Cheong, T.P., Soga, K., and Robert, D. J., *Journal of Geotechnical and Geoenvironmental Engineering, ASCE*, 137(10), 2011, 939–948. With permission.)

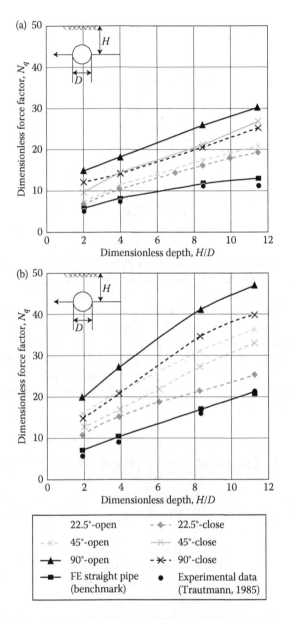

FIGURE 4.47 Dimensionless peak force for different H/D: (a) medium dense sand; (b) dense sand. (From Cheong, T.P., Soga, K., and Robert, D. J., *Journal of Geotechnical and Geoenvironmental Engineering, ASCE*, 137(10), 2011, 939–948. With permission.)

A 3-D FE formulation with both nonlinearities was developed for the analysis of tillage tool moving in (artificial) soils. The numerical predictions were compared with the results from a laboratory test device that simulated the tillage tool movements in soils. We present below a brief description of the FE procedure.

The updated Lagrangian formulation was based on the virtual work principle [14,16,61,62]:

$$\int_V \sigma_{ij}^{n+1} \delta\varepsilon_{ij}^{n+1} dV^{n+1} = \delta W^{n+1} \tag{4.29}$$

where

$$\delta W^{n+1} = F_i^{n+1} \delta u_i^{n+1} + \int_{V(n+1)} \overline{F}_i^{n+1} \delta u_i^{n+1} dV^{n+1}$$

$$+ \int_{S_1} T_i^{n+1} \delta u_i^{n+1} dS^{n+1} \tag{4.30}$$

and σ_{ij}^{n+1} is the Cauchy (real) stress tensor at load step $n+1$ (Figure 4.48) $\delta\varepsilon_{ij}^{n+1}$ is the first variation of the Almansi (real) strain at load step $n+1$, F_i^{n+1} is total concentrated or point load at step $n+1$, $\overline{F}_i^{n+1}$ is the body force vector at step $n+1$, δu_i^{n+1} is the first variation of displacements at step $n+1$, S_1 is the part of the boundary where traction or surface load, T_i^{n+1} is applied, and V^{n+1} and S^{n+1} are current volume of the body (element) and surface, respectively, at $n+1$. Appropriate 2-D or 3-D approximation function was defined for displacement ($\{u\}$) for an element; the strains ($\{\varepsilon\}$) are developed from the displacement function. Then, the substitution of $\{u\}$ and $\{\varepsilon\}$ in Equation 4.29 leads to the following incremental element equations [14,16,61,62]:

$$([k_L^n] + [k_{NL}])\{\Delta q\} = \{Q^{n+1}\} - \left\{ \int_{V(u)} [B_L^n]\{\sigma_n^n\} dV^n = \{Q^{n+1}\} - \{Q_0\} \right\} \tag{4.31}$$

where

$$[k_L^n] = \int_{V(n)} [B_L^n]^T [C^n][B_L^n] dV^n \tag{4.32a}$$

$$[k_{NL}^n] = \int_{V(n)} [B_{NL}^n]^T [\sigma_n^n][B_{NL}^n] \tag{4.32b}$$

$[B_L^n]$ = Linear strain displacement transformation matrix at step n (4.32c)

that is computed at configuration V^n

$[B_{NL}^n]$ = Nonlinear strain displacement matrix at step n (4.32d)

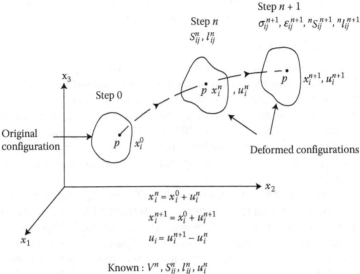

FIGURE 4.48 Generic configurations of body. (Adapted from Phan, H.V., Desai, C.S., and Perumpal, J.V., Geometric and Material Nonlinear Analysis of Three-Dimensional Soil-Structure Interaction, Report to National Science Foundation, Dept. of Civil Eng., VA Tech, Blacksburg, VA 1979.)

$$[\sigma_n^n] = \text{Updated total stress matrix at step } n \qquad (4.32e)$$

$$[C^n] = \text{Tangent constitutive matrix at step } n \qquad (4.32f)$$

$$\{F^{n+1}\} = \text{Total nodal force applied at step } n+1 \qquad (4.32g)$$

$$[k_L^n] = \text{Conventional tangent stiffness matrix at step } n \qquad (4.32h)$$

$$[k_{NL}^n] = \text{Geometric stiffness matrix at step } n \qquad (4.32i)$$

$$\{\Delta q\} = \text{Incremental nodal displacement vector from step } n \text{ to } n+1 \qquad (4.32j)$$

$$\{Q_0^n\} = \text{Internal nodal force vector at step } n \qquad (4.32k)$$

We solve Equation 4.31 to compute the incremental displacement, $\{\Delta q\}$, from step n to $n+1$.

Soil: Artificial soil used in the laboratory tests, described subsequently, exhibit nonlinear behavior, including irreversible or plastic deformations. Hence, three plasticity models, conventional Drucker–Prager, critical state, and cap, were considered; details of these models are given in Appendix 1.

4.3.13.1 Constitutive Laws

The artificial soil used consisted of sand, clay, and spindle soil [14,16,63]. This soil was used to reduce moisture changes in the soil before and during testing. The soil was placed in a large soil-bin for testing of movements of tillage tools (Figure 4.49).

The soil specimens were tested using cylindrical, triaxial, and multiaxial devices [14,16,63]—the latter with $4 \times 4 \times 4$ in ($10 \times 10 \times 10$ cm) cubical samples. The tests were performed at various initial confining pressures and stress paths (Appendix 1). On the basis of the laboratory tests, the cap model [64] was proposed for the behavior of the artificial soil.

Typical stress–strain tests in terms of octahedral stress (τ_{oct}) and octahedral strain (γ_{oct}) under various stress paths, for example, conventional triaxial extension (CTE), simple shear (SS), and triaxial compression (TC) are shown in Figure 4.50a. The plastic potential and yield surfaces are shown in Figures 4.50b and 4.50c, respectively. It can be seen from Figures 4.50b and 4.50c, respectively, that the plastic potential and yield surface are approximately the same. Hence, the associated flow rule was adopted with the yield surface according to the cap model (Appendix 1). The parameters for the cap model were determined on the basis of various stress–strain data, and are given below:

$E = 4000$ psi (27,600 kPa); $v = 0.35$
$C_t = 0.17$ psi (1.17 kPa); $\phi = 35.0°$
$A = 5.60$ psi (38.64 kPa); $\beta = 0.11$
$C = 5.60$ psi (38.64 kPa); $B = 0.062$ psi^{-1} (0.009 kPa^{-1})
$R = 2.00$; $W = 0.18$; $D = 0.05$ psi^{-1} (0.0072 kPa^{-1})

FIGURE 4.49 Laboratory soil bin-tillage tool test equipment. (Adapted from Phan, H.V., Desai, C.S., and Perumpal, J.V., Geometric and Material Nonlinear Analysis of Three-Dimensional Soil-Structure Interaction, Report to National Science Foundation, Dept. of Civil Eng., VA Tech, Blacksburg, VA 1979.)

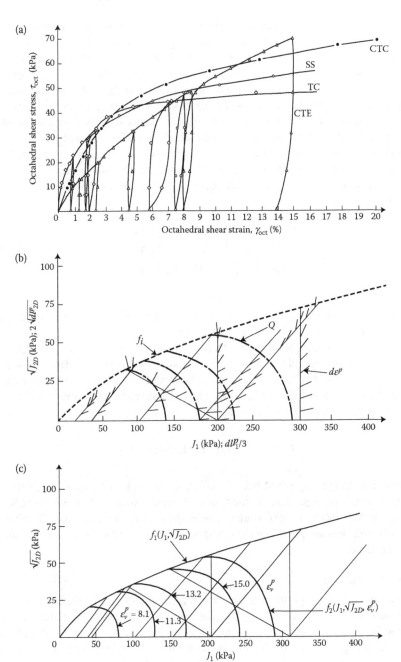

FIGURE 4.50 Behavior of artificial soil under various stress paths. (a) Octahedral shear versus strain; (b) plastic strain increment vectors and potential surfaces; and (c) yield surfaces-contours of volumetric plastic strains. (Adapted from Phan, H.V., Desai, C.S., and Perumpal, J.V., Geometric and Material Nonlinear Analysis of Three-Dimensional Soil-Structure Interaction, Report to National Science Foundation, Dept. of Civil Eng., VA Tech, Blacksburg, VA 1979.)

The above symbols of the cap moved may be different from those in Appendix 1; details of the expressions for the cap model are given in Ref. [16].

Tensile stress redistribution: The movement of a tool causes a considerable amount of cracks and fractures in soil and separation from the soil. In this FEM study, the excess tensile stress induced in the soil elements around the tool was computed at every stage of incremental analysis. The excess stresses are equal to the computed stresses minus the tensile strength of the soil. The excess stresses were converted to nodal forces and redistributed in the soil during the iterative steps [14,16].

Interface between soil and tool: Relative motions occur at interfaces, and the mechanism of interface deformation is different from the surrounding solid (soil) material. The incremental constitutive equation for the interface is given by

$$
\begin{Bmatrix} \Delta \tau_x \\ \Delta \tau_y \\ \Delta \sigma_n \end{Bmatrix} = \begin{bmatrix} k_{sx} & 0 & 0 \\ 0 & k_{sy} & 0 \\ 0 & 0 & k_{nz} \end{bmatrix} \begin{Bmatrix} \Delta u_{rx} \\ \Delta u_{ry} \\ \Delta u_{rz} \end{Bmatrix}
\tag{4.33}
$$

where k_{sx}, k_{sy} = shear stiffnesses, k_{nz} = normal stiffness, τ_x, τ_y = shear stresses, σ_n = normal stress, and Δ denotes an increment.

A series of laboratory tests were performed using a direct shear device with 2×2 in (5.08×5.08 cm) samples, under various (four) normal stresses from $\sigma_n = 2.5$–10.00 psi (17.25–69.00 kN/m²). A linear elastic model with Mohr–Coulomb strength for initiation of slip at the interface was adopted; the material parameters obtained from the shear tests are given below [14,16]:

$$
k_{sx} = k_{sy} = 200 \text{ pci } (54{,}330 \text{ kN/m}^3)
$$

Adhesive strength, $c_a = 0.42$ pci (114 kN/m³), and angle of friction, $\delta = 24.0°$

The normal response was characterized by high stiffness, $k_{nz} = 5 \times 10^{10}$ kN/m³, during the slip mode, $\sigma_n \geq 0$. Under tensile stress state, $\sigma_n < 0$, a small value of $k_{nz} = 50$ kN/m³ was adopted. The shear response in the x- and y-directions is assumed to be the same.

4.3.13.2 Validation

A series of tests were performed by changing the width and inclinations and soil properties. The tests were conducted in the artificial soils placed in a large soil-bin test facility (Figure 4.49) [63]. The tool was partly driven into the soil and then moved by using a special device mounted on the tool carriage. The displacements during the tool movement were measured at various times.

Finite element mesh: The FE mesh for half of the soil mass in the test bin is shown in Figure 4.51, by assuming symmetry along the longitudinal center line. The displacements (at nodes) orthogonal to the soil-bin were assumed to be zero at the side boundaries and the center line. The vertical displacements were assumed to be zero at the bottom boundary. Thus, finite boundaries were defined at the sides and bottom of the bin. However, the soil was assumed to be "infinite" in the longitudinal direction; the longitudinal boundaries were placed at distance equal to about 2 times the thickness of the soil mass.

The incremental load was applied at the nodes on the top of the tool. Since the tool was considered to be rigid in relation to the soil, the force was assumed to be applied at the surface level.

Figures 4.52a and 4.52b show FE predictions and measurements for typical cases, for example, (1) vertical tool pushed against soil mass and (2) tool at the inclination of 45° to the ground surface. It can be seen that the computed results compare very

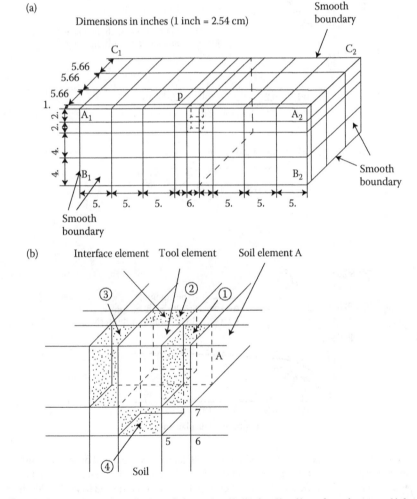

FIGURE 4.51   (a) Finite element mesh for tool-soil; (b) details of interface elements. (Adapted from Phan, H.V., Desai, C.S., and Perumpal, J.V., Geometric and Material Nonlinear Analysis of Three-Dimensional Soil-Structure Interaction, Report to National Science Foundation, Dept. of Civil Eng., VA Tech, Blacksburg, VA 1979.)

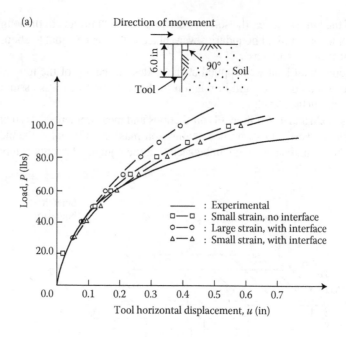

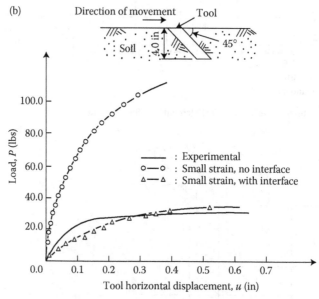

FIGURE 4.52 Comparisons of predictions and test data: (a) vertical tool pushed against soil mass (6 in tool); (b) tool at 45° to horizontal (4 in tool) (1 inch = 2.54 cm, 1 lb = 4.448 N). (Adapted from Phan, H.V., Desai, C.S., and Perumpal, J.V., Geometric and Material Nonlinear Analysis of Three-Dimensional Soil-Structure Interaction, Report to National Science Foundation, Dept. of Civil Eng., VA Tech, Blacksburg, VA 1979.)

well with the measurements when interface elements were provided. However, the predictions deviate from measurements if no interface elements are used.

The FE formulation provided for geometric nonlinearity including small and large strains. However, magnitudes of the computed strains (ε_{xx}, ε_{yy}, ε_{zz}, τ_{xy}, τ_{yz}, τ_{zx}) did not exhibit large strains. Hence, although the tool experiences larger displacements, the strains in the soil, in the vicinity of the tool were small, of the order of 0.50% [14,16]. Thus, in this specific problem, it may not be necessary to allow for large strains.

4.3.14 EXAMPLE 4.14: THREE-DIMENSIONAL SLOPE

In this example, the 3-D FE analysis was used to evaluate slope stability. 2-D plane idealizations are commonly used for slope stability analysis for simplicity. All slope failures are, however, 3-D in nature, particularly when slopes with complex geometries and loading conditions are involved [65]. According to previous studies, the results (factor of safety (FOS)) from 2-D analyses are generally conservative than those from the 3-D analyses [66]. The actual stability and geometry cannot be appropriately considered without the third dimension.

3-D limit equilibrium method with a strength-reduction technique has been used for analyzing the stability of 3-D slopes [66]. A major limitation of the 3-D limit equilibrium method is the lack of a suitable way to locate the critical 3-D slip surface [67]. 3-D FE analysis with a strength-reduction technique can simultaneously provide the FOS and the critical slip surface (location and shape). Also, it provides other important information such as stresses, deformations, and progressive failure. Moreover, complex geometries and loading conditions are accurately represented by the 3-D FE analysis.

In a traditional slope stability analysis, the FOS is defined as the ratio of the average shear strength of the soil to the average shear stress developed along the critical sliding surface or slip surface. In the 3-D FE analysis employed in this example, a shear strength-reduction technique is used to calculate the FOS in terms of reduced shear strength parameters C'_R and φ'_R as follows:

$$\tan \varphi'_R = (\tan \varphi')/\text{SRF} \tag{4.34a}$$

$$C'_R = C'/\text{SRF} \tag{4.34b}$$

where SRF is a strength-reduction factor and C' and φ' are the original shear strength parameters. The FOS is obtained by increasing the SRF gradually until the reduced strength parameters (C'_R and φ'_R) bring the slope to a limit equilibrium state. In this process, the SRF is assumed to apply equally to both C' and $\tan \varphi'$. When the slope reaches the limit equilibrium state, the FOS and the SRF become equal [65–67]. With respect to FE analysis, the limit equilibrium is accompanied by a dramatic increase in nodal displacements and nonconvergence of solution.

FE analysis and material properties: A commercial FE software ABAQUS was used for the analysis of the slope [66]. Figure 4.53 shows the FE meshes (considering

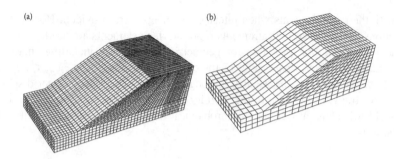

FIGURE 4.53 Three-dimensional finite element mesh used in the slop stability analysis: (a) fine mesh; (b) coarse mesh. (From Nian, T.-K., Huang, S.S., and Chen, G.-Q., *Canadian Geotechnical Journal*, 49, 2012, 574–588. With permission.)

symmetry) used in the analysis. Both meshes (coarse and fine) used 20-noded hexahedral elements for the discretization of the slope. An elastic–perfectly plastic constitutive model with Mohr–Coulomb failure criterion and no tensile strength and zero dilation was used to represent the soil behavior [65]. A cross section of the slope with free surface and weak layer, along with the dimensions of the slope, is shown in Figure 4.54 [65]. The specifics of the geometric and material properties are given in Table 4.13. The slope was first analyzed using an extended Spencer's method. Four cases were considered in the analysis: (1) homogeneous slope; (2) nonhomogeneous slope with a thin weak layer; (3) homogeneous slope with piezometric line; and (4) nonhomogeneous slope with a piezometric line.

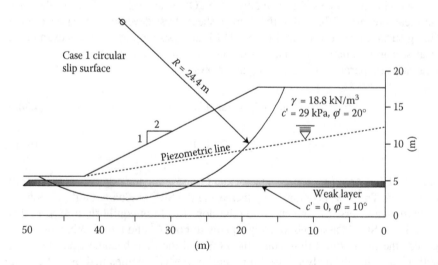

FIGURE 4.54 Cross-sectional view of 3-D slope with critical failure surface. (From Nian, T.-K., Huang, S.S., and Chen, G.-Q., *Canadian Geotechnical Journal*, 49, 2012, 574–588. With permission.)

TABLE 4.13
Material Properties Used

Material Properties	Upper Soil Layer	Weak Layer
Friction angle, φ' (°)	20	10
Cohesion, c' (kPa)	29	0
Dilation angle, ψ (°)	0	0
Young's modulus, E' (kPa)	1×10^4	1×10^4
Poisson's ratio, υ'	0.25	0.25
Unit weight, γ (kN/m³)	18.8	19.5

4.3.14.1 Results

For Case 1, it is assumed that the slope is homogeneous, and the presence of the weak layer (Figure 4.54) is neglected. The FOS obtained from the 3-D FE analysis was 2.15, which was in good agreement with the solutions obtained from other solutions. For example, for the same slope, Zhang [68] obtained a FOS of 2.122 using a 3-D limit equilibrium analysis and Griffiths and Marquez [66] reported 2.17 using a 3-D FEM. The corresponding deformed FE mesh at failure (nonconvergence) is shown in Figure 4.55a. It can be seen that both the failure mechanism and the location and shape of the failure surface are similar to those obtained by the 3-D limit equilibrium method (Figure 4.54).

For Case 2, with a weak layer in the slope, the FOS obtained from the 3-D FE analysis with the SRF was 1.59, which compared well with the value (1.553) reported by Zhang [68] using the 3-D limit equilibrium method and with the value (1.58) reported

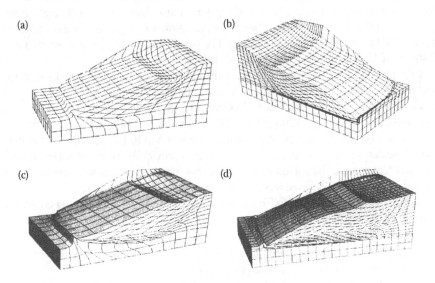

(a)

(b)

(c)

(d)

FIGURE 4.55 Deformed mesh for different cases: (a) case (1); (b) case (2); (c) case (3); and (d) case (4). (From Nian, T.-K., Huang, S.S., and Chen, G.-Q., *Canadian Geotechnical Journal*, 49, 2012, 574–588. With permission.)

by Griffiths and Marquez [66] using 3-D FEM. The corresponding deformed FE mesh is shown in Figure 4.55b, which shows a "dog-leg" path in the X–Y plane similar to that reported in Ref. [66]. The failure mechanism for this case was very different from that seen for case 1, which indicates the importance of considering the third dimension.

For Case 3, a homogeneous slope with a piezometric line, as shown in Figure 4.54, was considered. The FOS obtained from the strength reduction-based on the 3-D FE analysis was 1.86, which was close to the 3-D limit equilibrium solution reported by Zhang [68]. The average of 2-D methods, including Bishop's method, Janbu's simplified method, Janbu's rigorous method, and Morgenstern–Price method, was 1.799 [69]. Compared with case 1, the FOS for this case decreased significantly (from 2.15 to 1.86) due to the presence of groundwater. Also, the critical slip surface became deeper (Figures 4.54 and 4.55c). The deformed FE mesh in Figure 4.55c shows that the shape and location of the critical slip surface from the 3-D FE analysis are similar to those obtained from the 3-D limit equilibrium method.

The FOS for Case 4 obtained from the shear strength reduction-based FE approach was 1.298, which was much lower than the values obtained for other cases. This FOS was lower than that (1.441) reported by Zhang [68] using the 3-D limit equilibrium analysis, but more closer to the average (1.258) of the 2-D methods. From the deformed FE mesh in Figure 4.55d, it is evident that the shape and location of the critical slip surface for this case were similar to those for Case 2.

PROBLEMS

Problem 4.1: Cap–Pile Group–Soil

Figure P4.1 shows the section of a pile cap–pile group subjected to vertical and lateral loads and moments as shown. The width of the cap is 30 ft (9.15 m) and the length perpendicular to the plane of the paper is also 30 ft (9.15 m). There are six piles, each 60.0 ft (18.3 m) long; they are situated on a straight line at a distance of 2.5 ft (0.763 m) from the edge of the cap. The two sets of three piles are situated at a distance of 12.5 ft (3.8 m) from the center line in the x–y plane.

Piles are made of reinforced concrete with a cross section 15×15 in $(0.38 \times 0.38$ m) and length of 60 ft (18.3 m). The modulus of elasticity of concrete, $E = 3 \times 10^6$ psi $(2.1 \times 10^4$ MPa). The foundation soil is uniform in depth, and the soil stiffnesses were found to be $k_x = k_y = 1200$ psi (8.3 MN/m^2) and $k_z = 2400$ psi (16.6 MN/m^2).

It can be assumed that there is a distance of about 6 in (0.153 m) between the bottom of the cap and soil; hence, no soil resistance is provided on the bottom of the cap. The problem can also be solved by assuming that the cap interacts with the soil; then we can assume that soil resistance, k_z, can be applied to the cap bottom.

Divide the cap and piles into FEs, and solve by using STFN-FE or other computer code. Find the results in terms of

1. Nodal displacements, here, draw the deformed shape of the cap and piles, and identify the maximum displacement
2. Axial forces, moments, and transverse force in the piles
3. Moments and shear forces along typical cap sections
4. Plot soil resistance (p) versus the depth for the three piles

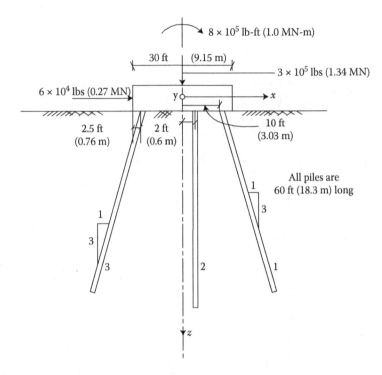

FIGURE P4.1   Cap-pile group-soil system.

By using appropriate design codes, perform design analysis for the piles and soil, using factors of safety of 1.50 and 1.80 for concrete and soil, respectively.

Problem 4.2: Cap–Pile Group–Soil

Figure P4.2 shows a pile cap consisting of nine piles with length of 20 m each. The properties of piles, pile cap, and soil resistance are given below: approximate load–displacement and moment–rotation curves are given in Figure P4.3. The translational and rotational soil spring moduli are computed below based on the plots in Figure P4.3. The properties of the pile (HP 360-152) are obtained from steel manufacturer's handbooks, for example, Bethlehem Steel Corp (Figure P4.4).

Pile Details: HP 360-152, Figure P4.4:

Area	$= 19.4 \times 10^{-3}$ m^2
Moments of inertia:	
$J(I_z)$	$= 2 \times 10^{-6}$ m^4
I_x	$= 437 \times 10^{-6}$ m^4
I_y	$= 158 \times 10^{-6}$ m^4
Depth, d	$= 356$ mm (0.356 m)
Width, w	$= 376$ mm (0.376 m)
Elastic modulus, E	$= 200 \times 10^6$ kN/m^2 (200,000 MPa)
Poisson's ratio, v	$= 0.15$

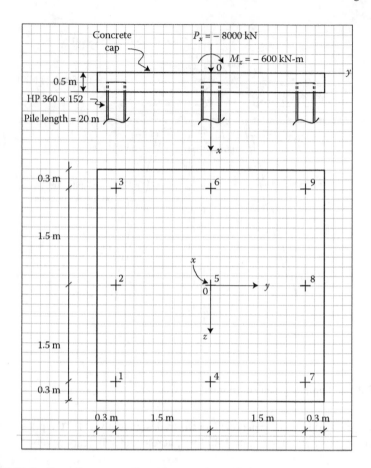

FIGURE P4.2 Cap-pile group-soil problem: section and plan.

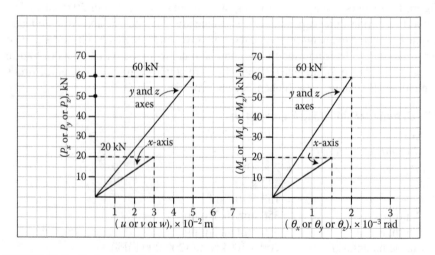

FIGURE P4.3 Spring moduli for soil resistance.

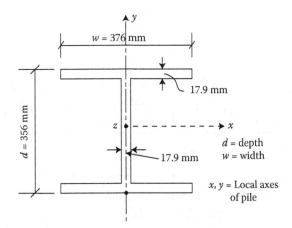

FIGURE P4.4 Details of HP-360-152 piles.

The pile is divided into 20 elements of length 1.0 m each, and the cap is divided into two sets of elements: (1) within the boundaries of the piles, 0.50×0.50 m size for the 3×3 m region, and (2) outside the boundaries: (i) four elements at corners 0.30×0.30 size, and (ii) 24 elements at 0.50×0.30 and 0.30×0.50 sizes (Figure P4.5). The soil is represented by three translational and two rotational springs, which are described below. The torsional spring is not considered.

Soil Properties

Soil resistance is represented by six spring moduli, three translational, and three rotational; the values in x-, y-, and z-directions are evaluated from approximate responses assumed in those directions (Figure P4.3).

The modulus k (kN/m³) is derived by dividing by the appropriate dimension (e.g., $d = 0.376$ m) and 1 m:

$$\bar{k} = \frac{60}{0.05} = 1200 \text{ kN/m}$$

$$k_y = \frac{1200}{0.376 \times 1} = 3191 \text{ kN/m}^3$$

and

$$k_z = \frac{1200}{0.356 \times 1} = 3371 \text{ kN/m}^3$$

$$\bar{k}_\theta = \frac{60}{0.02} = 3000 \text{ kN-m/rad}$$

$$k_{\theta z} = \frac{3000}{0.356 \times 1} = 8427 \text{ kN-m/rad/m}^2$$

$$k_{\theta y} = \frac{3000}{0.376 \times 1} = 7979 \text{ kN-m/rad/m}^2$$

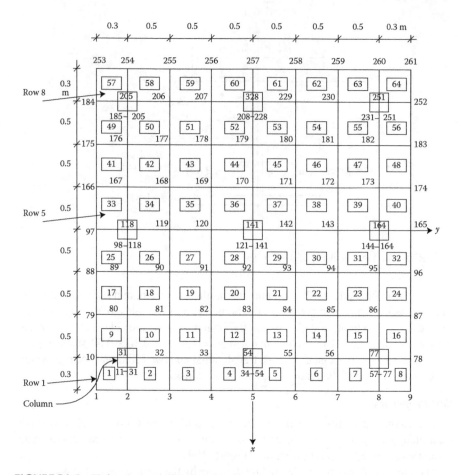

FIGURE P4.5 Finite element mesh (proposed).

Now, the moduli for axial (x) spring and rotational (torsional) spring can be calculated as

$$\bar{k}_x = \frac{20}{0.03} = 667 \text{ kN/m}$$

$$k_x = \frac{667}{C \times 1}$$

where C is the circumference of pile given by (assuming rectangular effective surface)

$$C = 2\,(0.356 + 0.376) = 1.464 \text{ m}$$

Thus

$$k_x = \frac{667}{1.461 \times 1} = 457 \text{ kN/m}^3$$

The torsional spring modulus

$$\bar{k}_{\theta x} = \frac{20}{0.0015} = 13{,}333 \text{ kN-m/rad}$$

$$k_{\theta x} = \frac{13{,}333}{1.461 \times 1} = 9126 \text{ kN-m/rad/m}^2$$

Note: The STFN-FE code does not include torsional behavior. Hence, $k_{\theta x}$ is not be used.

Pile Cap Properties

Concrete pile cap:

$E = 25 \times 10^6 \text{ kPa}$
$v = 0.3$
Thickness = 0.5 m

Using the code STFN-FE [37] or other suitable code, obtain the following:

a. Pile displacement and forces
b. Pile moments and shear forces
c. Pile cap displacement, moments, and shear forces

Note: The STFN code allows use of the Ramberg–Osgood model for soil resistances (translation and rotation). Since we consider a constant moduli, they can be obtained, approximately, by setting the final modulus E_{tf} SKP(I) = 0.0, ultimate load p_u (SPP(I) = kx 10^3 (high value), and exponent m ROE (I) = 1.00.

Problem 4.3: Cap–Pile Group–Soil

A pile cap of dimensions 10 ft × 20 ft (3.05 m × 6.1 m) and thickness 2 ft (0.61 m) is supported on four batter piles, as shown in Figure P4.6. The analysis of superstructure shows that a vertical load of 400 kips (1780 kN) and a horizontal load of 100 kips (445 kN) act at the center of the cap. The sectional view of piles 1 and 2 on the X–Z plane is shown in the same figure. Piles 3 and 4 are identical to piles 1 and 2, with respect to inclination. For simplicity, the hinge condition is assumed between the cap and the pile. Assuming that the pile can be considered rigid, determine the displacements and rotations of the cap using the extended Hremmikoff's method, described in this chapter. Also, determine the pile forces. The other data are given below:

Unit weight of concrete = 150 lb/ft^3 (24 kN/m^3)
Pile constants t_δ = 50 kip/in (87.5 kN/cm), and n = 100 kip/in (175 kN/cm)
Ratio, $r = t_\delta/n$ = (50 kip/in)/(100 kip/in) = 0.5

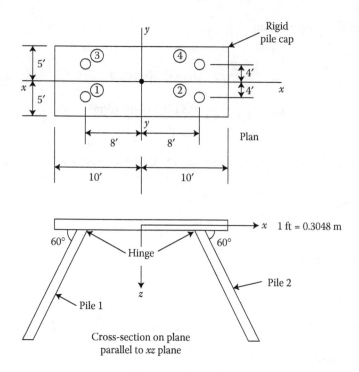

FIGURE P4.6   Pile geometry and location.

Partial Solution

The parameters defining pile geometry (inclination) and pile head location for this problem are tabulated below. These values along with the value of r can be used to determine the pile cap constants a_{ij} from Table 4.2 (because of symmetry with respect to the x- and y-axes). Here, $P_x = 100$ kips (445 kN) and $P_z = 400$ kips (1780 kN), $P_y = 0$, $M_x = M_y = M_z = 0$. Knowing the pile head constants and loading, the pile cap movements can be determined using Equations 4.22a through 4.22f. The pile forces can then be computed using Equations 4.23a through 4.23e.

Pile number	α_{ak} (o)	ρ_{A_k} (in)	x_{A_k} (in)	y_{A_k} (in)	α_k (o)	β_k (o)	γ_k (o)
1	26.56	107.33	96	48	60	90	30
2	153.44	107.33	−96	48	120	90	30
3	206.56	107.33	−96	−48	120	90	30
4	333.44	107.33	96	−48	60	90	30

1 in = 2.54 cm.

Problem 4.4. Cap–Pile Group–Soil Problem No. 3: Load at the Center Only

Solve Example 4.3 assuming that the vertical load is applied at the center of the cap.

Partial solution: Here, the pile cap constants, a_{ij}, would remain unchanged. Also, both moments would be zero (i.e., $M_x = 0$ and $M_y = 0$). For this case, the resulting equilibrium equations can be expressed as follows:

$$
\begin{bmatrix}
-1.87 & 0.00 & 0.00 & 0.00 & 84.78 & 0.00 \\
0.00 & -1.73 & 0.00 & 0.00 & 0.00 & 0.00 \\
0.00 & 0.00 & -15.86 & 0.00 & 0.00 & 0.00 \\
0.00 & 0.00 & 0.00 & -57091.40 & 0.00 & 0.00 \\
84.78 & 0.00 & 0.00 & 0.00 & -57091.40 & 0.00 \\
0.00 & 0.00 & 0.00 & 0.00 & 0.00 & 0.00
\end{bmatrix}
\begin{Bmatrix}
\Delta x' \\
\Delta y' \\
\Delta z' \\
\alpha'_x \\
\alpha'_y \\
\alpha'_z
\end{Bmatrix}
=
\begin{Bmatrix}
-20.00 \\
0.00 \\
-336.00 \\
0.00 \\
0.00 \\
0.00
\end{Bmatrix}
$$

The corresponding pile cap displacements and rotations would be as follows:

$\Delta x'$	45.89
$\Delta y'$	0.00
$\Delta z'$	84.75
$\alpha x'$	0.00
$\alpha y'$	0.07
$\alpha z'$	0.00

Pile 1 (kip)		Pile 2 (kip)		Pile 3 (kip)		Pile 4 (kip)	
P_k	−84.83	P_k	−83.83	P_k	−83.83	P_k	−84.83
Q_k	4.06	Q_k	5.89	Q_k	5.89	Q_k	4.06

1 kip = 4.448 kN.

REFERENCES

1. Hrennikoff, A., Analysis of pile foundations with batter piles, *Transactions of ASCE*, 115, 1950, 351–389.
2. Prakash, S., Behavior of Pile Groups Subjected to Lateral Loads, PhD Thesis, Dept. of Civil Eng., Univ. of Illinois, Urbana, IL, 1962.
3. Aschenbrenner, R., Three-dimensional analysis of pile foundations, *Journal of the Structural Engineering Division, ASCE*, 93(ST1), 1967, 201–219.
4. Saul, W.E., Static and dynamic analysis of pile foundations, *Journal of the Structural Engineering Division, ASCE*, 94(ST5), 1968, 1077–1100.
5. Reese, L.E., O'Neill, M.W., and Smith, R.E., Generalized analysis of pile foundations, *Journal of the Soil Mechanics and Foundations Division, ASCE*, 96(SM1), 1970, 235–246.
6. Butterfield, R. and Banerjee, P.K., The problem of pile group—Pile cap interaction, *Geotechnique (London)*, 21(2), 1971, 135–142.

7. Bowles, J.E., *Analytical and Computer Methods in Foundation Engineering*, McGraw-Hill Book Company, New York, 1974.

8. O'Neill, M.W., Ghazzaly, O.I., and Ha, H.B., Analysis of three-dimensional pile groups with nonlinear soil response and pile-soil-pile interaction, *Proceedings of the 9th Offshore Technical Conference*, 2, 1977, 245–256.

9. Chow, Y.K., Three-dimensional analysis of pile groups, *Journal of Geotechnical Engineering, ASCE*, 113(6), 1987, 637–651.

10. Banerjee, P.K., and Driscoll, R.M.C., Three-dimensional analysis of raked pile groups, *Proceedings of the Institution of Civil Engineers*, U.K., Part 2, 61, 1976, 653–671.

11. Polous H.G., An approach for the analysis of offshore pile groups, *Proceedings of the International Conference on Numerical Methods in Offshore Piling*, London, UK, 1980, 119–126.

12. Desai, C.S. and Appel, G.C., Three-dimensional FE analysis of laterally loaded structures, *Proceedings of the 2nd International Conference on Numerical Methods in Geomechanics*, Blacksburg, VA, Vol. 1, ASCE, Sept. 1976.

13. Wittke, W., Static analysis of underground openings in jointed rock, Chapter 18 in *Numerical Methods in Geotechnical Engineering*, Desai, C.S and Christian, J.T. (Editors), McGraw Hill Book Co., New York, 1977.

14. Phan, H.V., Desai, C.S., and Perumpal, J.V., Geometric and Material Nonlinear Analysis of Three-Dimensional Soil-Structure Interaction, Report to National Science Foundation, Dept. of Civil Eng., VA Tech, Blacksburg, VA 1979.

15. Desai, C.S. and Siriwardane, H.J., Numerical models for track support structures, *Journal of the Geotechnical Engineering Division, ASCE*, 108(GT3), March 1982, 461–480.

16. Desai, C.S., Phan, H.V., and Perumpal, J.V., Mechanics of three-dimensional soil-structure interaction, *Journal of the Engineering Mechanics Division, ASCE*, 108(EM5), 1982, 731–747.

17. Muqtadir, A. and Desai, C.S., Three-Dimensional Analysis of Cap-Pile-Soil Foundations, Report, Dept. of Civil Eng. and Eng. Mechanics, The Univ. of Arizona, Tucson, AZ, 1984.

18. Muqtadir, A. and Desai, C.S., Three-dimensional analysis of a pile-group foundation, *International Journal of Numerical and Analytical Methods in Geomechanics*, 10, 1986, 41–58.

19. Scheele, F., Desai, C.S., and Muqtadir, A., Testing and modeling of 'muniuch' sand, soils and foundations, *Journal of the Japanese Society of Soil Mechanics and Foundation Engineering*, 26(3), 1986, 1–11.

20. Elgamal, A.W. and Adel-Ghaffar, A.M., Elasto-plastic seismic response of 3-D dams: Application, *Journal of Geotechnical Engineering, ASCE*, 113(11), 1987, 1309–1325.

21. Komiya, K., Soga, K., Akagiu, H., Hagiwara, T., and Bolton, M.D., Finite element modeling of excavation and advancement processes of a shield tunneling machine, *Soils and Foundations*, Japanese Geotech. Society, 39(3), 1999, 37–52.

22. Desai, C.S., Mechanistic pavement analysis and design using unified material and computer model, Keynote Paper, *Proceedings of the 3rd International Symposium on 3-D Finite Element for Pavement Analysis, Design and Research*, Amsterdam, The Netherlands, 2002.

23. Tanaka, T., Harada, D., Masukawa, D., and Mori, H., Dynamic and pseudo-static failure analysis of Embankment Dams, *Proceedings of the 4th International Conference on Dam Engineering*, Nanjing, China, 2004, 75–88.

24. Kasper, T., and Meschke, G., A 3D finite element simulation for TBM tunneling in soft ground, *International Journal of Numerical and Analytical Methods in Geomechanics*, 28(4), 2004, 1441–1460.

25. Desai, C.S., Unified DSC constitutive model for pavement materials with numerical implementation, *International Journal of Geomechanics, ASCE*, 7(2), March/April 2007, 83–101.

26. Liu, H.Y., Small, J.C., and Carter, J.P., Full 3D modelling for effects of tunnelling on existing support systems in the Sydney Region, *Tunnelling and Underground Space Technology*, 23(4), 2008, 399–420.
27. Wang, D., Hu, Y., and Randolph, M., Three-dimensional large deformation finite-element analysis of plate anchors in uniform clay. *Journal of Geotechnical And Geoenvironmental Engineering*, 136(2), 2010, 355–365.
28. Ling. H.I., Yang, S., Leshchinsky, D., Liu, H., and Burke, C., Finite element simulations of full scale modular block reinforced soil retaining wall under earthquake loading, *Journal of Engineering Mechanics, ASCE*, 136(5), 2010, 653–661.
29. Cheong, T.P., Soga, K., and Robert, D. J., 3D FE analysis of buried pipeline with elbows subjected to lateral loading, *Journal of Geotechnical and Geoenvironmental Engineering, ASCE*, 137(10), 2011, 939–948.
30. Xue, F., Ma, J., and Yan, L., Three-dimensional FEM analysis of bridge pile group in soft soils, *Proceedings, GeoHunan, 2011*, Hunan, China, 2011.
31. Alameddine, A.R. and Desai, C.S., Finite Element Analysis of Some Soil–Structure Interaction Problems, Report, VA Tech., 1979.
32. Desai, C.S., Kuppusamy, T., and Alameddine, R., Pile cap-pile group-soil interaction, *Journal of the Structural Engineering Division, ASCE*, 107(ST5), 1981, 877–834.
33. Desai, C.S. and Abel, J.F., *Introduction to Finite Element Method*, Van Nostrand Reinhold Company, New York, 1972.
34. Timoshenko, S. and Woinowsky-Kriger, S., *Theory of Plates and Shells*, McGraw-Hill Book Company, New York, 1959.
35. Zienkiewicz, O.C. and Taylor, R.L., *The Finite Element Method*, McGraw-Hill Book Company, New York, 1989.
36. Bathe, K.J., *Finite Element Procedures in Engineering Analysis*, Prentice-Hall, Englewood Cliffs, NJ, 1982.
37. Bogner, F.K., Fox. R.L., and Schmit, L.A. Jr., The generation of inter-element-compatible stiffness and mass matrices by the use of interpolation formulas, *Proceedings of the 1st Conference on Matrix Methods in Structural Mechanics*, Wright-Patterson Air Force Base, Ohio, Dec. 1969.
38. Desai, C.S., *Structure-Foundation Analysis: User's Manual for Code STFN-FE*, Tucson, AZ, 1983.
39. Cheung, Y.K. and Nag, D.K., Plates and beams on elastic foundations: Linear and non-linear behavior, *Geotechnique*, 18, 1968, 250–260.
40. Cheung, Y.K., Beams, slabs and pavements, Chapter 5 in *Numerical Methods in Geotechnical Engineering*, Desai, C.S. and Christian, J.T. (Eds), McGraw-Hill Book Company, New York, 1977.
41. Durelli, A.J., Parks, V.J., Mok, C.C., and Lee, H.C., Photoelastic study of beams on elastic foundations, *Journal of the Structural Engineering Division, ASCE*, 95(ST8), Aug. 1969, 1713–1725.
42. Anandakrishnan, U., Kuppusamy, T., and Krishnaswamy, N.R., *Design Manual for Raft Foundations,* Dept. of Civil Eng., Indian Institute of Technology, Kanpur, India, 1971.
43. Haddadin, M.J., Mats and combined footings: Analysis by the finite element method, *Journal of the American Concrete Institute,* 68(2), 1971, 9945–9949.
44. Jefferies, M.G, Nor-sand: A simple critical state model for sand, *Geotechnique*, 43(1), 1993, 91–103.
45. Fruco and Associates, Pile Driving and Loading Tests: Lock and Dam No. 4, Arkansas River and Tributaries, Arkansas and Oklahoma, Report, U.S. Army Corps of Engineers District, Little Rock, Sept. 1964.
46. Reese, L.C., Cox, W.R., and Koop, F.D., Analysis of laterally loaded piles in sand, *Proceedings of the 6th Offshore Technology Conference*, Houston, TX, 1974.

47. Desai, C.S., Zaman, M.M., Lightner, J.G., and Siriwardane, H.J., Thin-layer elements for interfaces and joints, *International Journal of Numerical and Analytical Methods in Geomechics*, 8(1), 1984, 19–43.

48. Desai, C.S., Finite Element Method for Analysis and Design of Piles, Report, U.S. Army Corps of Engineering Division, Vicksburg, MS, 1976.

49. Desai, C.S., Finite Element Procedure and Code for Three-Dimensional Soil-Structure Interaction, Report, VPI-E-75.27, Dept. of Civil Eng., VA Tech, Blacksburg, VA, 1975.

50. Desai, C.S., Numerical design analysis for piles in sands, *Journal of the Geotechnical Division, ASCE*, 100(GT6), June 1974, 1008–1029.

51. Desai, C.S., Johnson, L.D., and Hargett, C.M., Analysis of pile-supported gravity lock, *Journal of the Geotechnical Engineering Division, ASCE*, 100(9), Sept. 1974, 1009–1029.

52. Scheele, F., Tragfähigheit von Verpressankern in Nichtbindigem Bodex, Neue Erkenntrises durch Dehnungsmessungen im Verankerungs-bereich, Dissertation, Lehrstuhl für Grundbau und Bodenmechanik, TO München, Germany, 1981.

53. Desai, C. S., Muqtadir, A., and Scheele, F., Interaction analysis of anchor-soil system, *Journal of Getechnical Engineering*, 112(5), 1986, 537–553.

54. Desai, C.S., *Mechanics of Materials and Interfaces: The Disturbed State Concept*, CRC Press, Boca Raton, FL, 2001.

55. Desai, C.S., *DSC-SST3D Code for Three-Dimensional Coupled Static, Repetitive and Dynamic Analysis: User's Manual I to III*, Tucson, AZ, 2000.

56. Desai, C.S. and Whitenack, R., Review of models and the disturbed state concept for thermomechanical analysis in electronic packaging, *Journal of Electronic Packaging, ASME*, 123, 2001, 1–15.

57. Stagliano, T.R., Mente, L.J., Gadden Jr. E.C., Baxter, B.W., and Hale, W.K., Pilot Study for Definition of Track Component Load Environments, Final Report, Kaman Avidlyne, Burlington, MA, July 1981.

58. Desai, C.S., Siriwardane, H.J., and Janardhaman, R., Load Transfer and Interaction in Track-Guideway Systems, Report to Dept. of Transport., Univ. Research, Washington, DC, Dept. of Civil Eng., VA Tech, Blacksburg, VA, June 1980.

59. Siriwardane, H.J. and Desai, C.S., Nonlinear Soil-Structure Interaction Analysis for One-, Two- and Three-Dimensional Problems Using Finite Element Method, Report to DOT-Univ. Research, Va Tech, Blacksburg, VA, USA, 1980.

60. Janardhaman R. and Desai, C.S., Three-dimensional testing and modeling of ballast, *Journal of Geotechnical Engineering, ASCE*, 109(6), June 1983, 783–796.

61. Bathe, K.J., Ramm, E., and Wilson, E.L., Finite element formulation for large deformation dynamic analysis, *International Journal for Numerical Methods in Engineering*, 9, 1975, 353–386.

62. Oden, J.T., Finite element formulation for problems of finite deformation and irreversible thermodynamics of nonlinear continua—A survey and extension of recent development, *Proceedings of the Conference on Recent Advances in Matrix Methods of Structural Analysis and Design*, Univ. of Alabama Press, AL, 1971, 693–724.

63. Perumpral, J.V. and Desai, C. S., A generalized model for a soil-tillage tool interaction, *Proceedings of the Conference American Society of Agricultural Engineers*, St. Joseph, Michigan, USA, 1979.

64. DiMaggio, F.L. and Sandler, I.S., Material model for granular soil, *Journal of Engineering Mechanics Division, ASCE*, 97 (EM3), June 1971, 935–950.

65. Nian, T.-K., Huang, S.S., and Chen, G.-Q., Three-dimensional strength reduction finite element analysis of slopes: Geometric effects, *Canadian Geotechnical Journal*, 49, 2012, 574–588.

66. Griffiths, D.V. and Marquez, R.M., Three-dimensional slope stability analysis by elasto-plastic finite elements, *Geotechnique*, 57(6), 2007, 537–546.

67. Wei, W.B., Cheng, Y.M., and Li, L., Three-dimensional slope failure analysis by the strength reduction and limit equilibrium methods, *Computers and Geotechnics*, 36(1–2), 2009, 70–80.
68. Zhang, X., Three-dimensional stability analysis of concave slopes in plan view, *Journal of Geotechnical Engineering*, *ASCE*, 114(6), 1988, 658–671.
69. Freedlund, D.G. and Krahn, J., Comparison of slope stability methods of analysis, *Canadian Geotechnical Journal*, 14(3), 1977, 429–439.

5 Flow through Porous Media

Seepage

5.1 INTRODUCTION

Fluid (water) flowing through porous soils and rocks under and in the vicinity of loaded engineering structures causes coupled effects exhibited by interacting deformation and fluid (pore) water pressures. The resulting fluid pressures cause changes in the effects of mechanical loading, and thereby influence the stability of soil–structure interaction systems. For some problems, we could assume that the skeleton of the geologic medium experiences no deformations. The water flowing through the pores of such rigid medium, which is often called *seepage*, causes forces or pressures that are to be evaluated for the analysis and design of geotechnical structures. In this chapter, we consider seepage and its relation to the stability of geotechnical structures.

5.2 GOVERNING DIFFERENTIAL EQUATION

The general governing differential equation (GDE) for 3-D seepage through a porous medium is given by [1–8]

$$\frac{\partial}{\partial x}\left(k_x \frac{\partial \phi}{\partial x}\right) + \frac{\partial}{\partial y}\left(k_y \frac{\partial \phi}{\partial y}\right) + \frac{\partial}{\partial z}\left(k_z \frac{\partial \phi}{\partial z}\right) + \bar{Q} = n\frac{\partial \phi}{\partial t} \tag{5.1a}$$

where k_x, k_y, and k_z are the coefficients of permeability in the x-, y-, and z-directions, respectively, $\phi = p/\gamma + y$ is the fluid head or potential, p is the pressure, γ is the unit weight of water, y is the elevation head, n is the effective porosity or specific storage (sometimes it could be replaced by specific storage S), t is the time, and $\bar{Q}$ is the applied fluid flux. Equation 5.1 is based on various assumptions such as that the flow is continuous and irrotational, the fluid is incompressible and homogeneous, the material is "rigid," capillary and inertia effects are negligible, the magnitudes of velocities are small, the Darcy's law holds good, and x, y, and z are the principal directions for permeability.

If the time dependence does not occur, the right-hand side in Equation 5.1 vanishes. The resulting equation relates to the steady-state seepage, which for isotropic media, results in the well-known Laplace equation

$$\nabla^2 \phi = 0 \tag{5.1b}$$

where ∇^2 is the Laplacian operator.

Some of the problems in geotechnical engineering involving steady and time-dependent (transient) seepage are shown in Figures 5.1a through 5.1d for 2-D idealization. The impervious sheet pile walls or a dam constructed into or resting on a porous foundation in which the upstream and downstream heads do not change represents *steady confined* seepage (Figure 5.1a) because it does not involve phreatic or free water surface. Figure 5.1b shows an example of *transient confined* flow, which involves time-dependent flow of water by the pump action, but which may not involve a free surface (FS). The case of *steady unconfined* or FS flow is shown in Figure 5.1c, in which the upstream and downstream heads do not change, but there exists a FS in the porous structure (dam); here, the foundation of the dam is often considered to be impervious. The general case of *transient unconfined* seepage in Figure 5.1d involves time dependence because of the change in heads with time (e.g., upstream) and also the FS in the riverbank, dam, or embankment. The change in heads may involve rise, steady head, and drawdown in the reservoir or river levels.

5.2.1 BOUNDARY CONDITIONS

Various boundary conditions exist for different categories of seepage (Figure 5.1). They are stated below:

1. Head or potential boundary condition:

$$\phi = \bar{\phi}(t) \tag{5.2}$$

on part of the boundary, B_1; here, the over bar denotes the known quantity
2. Flow boundary condition:

$$k_x \frac{\partial \phi}{\partial x} \ell_x + k_y \frac{\partial \phi}{\partial y} \ell_y + k_z \frac{\partial \phi}{\partial z} \ell_z + \bar{q}_n(t) = 0 \tag{5.3}$$

on the part of the boundary, B_2, where the intensity of flow or flux is specified; ℓ_x, ℓ_y, and ℓ_z are direction cosines of the outward normal to the boundary, and $\bar{q}_n(t)$ is the intensity of specified fluid flux.
3. Steady unconfined flow (Figure 5.1c):

$$\phi = \bar{\phi}_u (= D) \quad \text{on the upstream boundary } [1\text{--}2] \tag{5.4a}$$

$$\phi = \bar{\phi}_d (= d) \quad \text{on the downstream boundary} [4\text{--}5] \tag{5.4b}$$

$$\phi = y \quad \text{on the free or phreatic surface } [2\text{--}3],$$
$$\text{and surface of seepage } [3\text{--}4] \tag{5.4c}$$

$$\frac{\partial \phi}{\partial n} = 0 \text{ on free surface FS, where } n \text{ is the normal to the surface} \tag{5.4d}$$

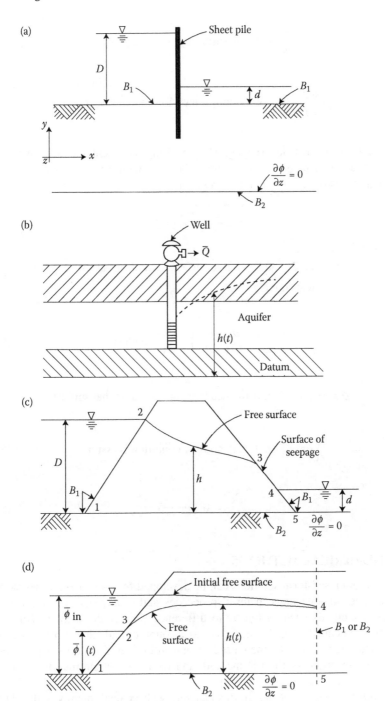

FIGURE 5.1 Categories of seepage. (a) Steady confined flow; (b) transient confined flow; (c) steady unconfined flow; and (d) transient unconfined flow.

$$\frac{\partial \phi}{\partial n} \leq 0 \quad \text{on surface of seepage [3–4]} \tag{5.4e}$$

$$\frac{\partial \phi}{\partial y} = 0 \quad \text{on bottom impervious boundary, } B_2 \tag{5.4f}$$

4. Transient or unsteady seepage (Figure 5.1d): Transient seepage involves continuous changes or movements of the FS with time, and the relevant boundary conditions can be expressed as

$$\phi = \overline{\phi}_u(t) \quad \text{on 1–2} \tag{5.5a}$$

$$\phi = y(t) \quad \text{on 2–3 and 3–4, free surface FS and surface of seepage, respectively} \tag{5.5b}$$

$$\frac{k}{n}\left(\frac{\partial \phi}{\partial y} - \frac{\partial \phi}{\partial x} \cdot \frac{\partial h}{\partial t}\right) = \frac{\partial h}{\partial t} \quad \text{on 3–4} \tag{5.5c}$$

$$\phi = \overline{\phi}_d \quad \text{on 4–5, if the heads are specified on that surface} \tag{5.5d}$$

$$\frac{\partial \phi}{\partial x} = 0 \quad \text{on 4–5, if no flow condition is assumed} \tag{5.5e}$$

$$\frac{\partial \phi}{\partial y} = 0 \quad \text{on 1–5} \tag{5.5f}$$

5.3   NUMERICAL METHODS

Analytical or closed-form solutions can be developed for specialized cases such as isotropic and homogeneous material properties and steady-state condition [1–4]. Numerical methods such as the finite difference (FD), finite element (FE), and boundary element (BE) can be used so as to account for realistic factors that cannot be accounted for usually by the closed-form solutions. In this book, we will present brief descriptions of the FD method, and then concentrate mainly on the FE method.

One-dimensional problems: In this chapter, we have dealt mainly with 2-D and 3-D seepage problems. However, the 1-D idealization may provide satisfactory results for certain problems such as a rectangular flow domain. Some details of 1-D (unconfined) seepage are given in Appendix A in this chapter.

5.3.1 FINITE DIFFERENCE METHOD

We consider first the case of 2-D steady-state confined flow problems. For this case, Equation 5.1a will reduce to the following form, assuming the coefficients of permeability to be the same:

$$k_x \frac{\partial^2 \phi}{\partial x^2} + k_y \frac{\partial^2 \phi}{\partial y^2} + \bar{Q} = 0 \tag{5.6}$$

Equation 5.6 can be expressed in the (central) FD form as (Figure 5.2)

$$k_{x(i,j)} \frac{\phi_{i+1,j} - 2\phi_{i,j} + \phi_{i-1,j}}{\Delta x^2} + k_{y(i,j)} \frac{\phi_{i,j+1} - 2\phi_{i,j} + \phi_{i,j-1}}{\Delta y^2} + \bar{Q}_{i,j} = 0 \tag{5.7a}$$

$$\beta_x \phi_{i+1,j} - 2(\beta_x + \beta_y)\phi_{i,j} + \beta_x \phi_{i-1,j} + \beta_y \phi_{i,j+1} + \beta_y \phi_{i,j-1} + \bar{Q}_{i,j} = 0 \tag{5.7b}$$

where $\beta_x = k_{x(i,j)}/\Delta x^2$ and $\beta_y = k_{y(i,j)}/\Delta y^2$. Equation 5.7 can be written at all nodal points (Figure 5.2) in the domain of the problem, which results in a set of simultaneous equations with ϕ as unknown at all node points.

5.3.1.1 Steady-State Confined Seepage

Here, the boundary conditions can be introduced in the simultaneous equations expressed for all nodes (i, j) (Figure 5.2). The given potential conditions, for example,

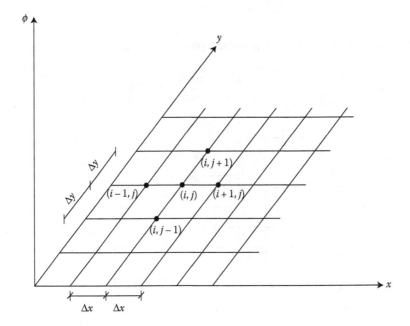

FIGURE 5.2 Finite difference mesh.

$\phi = \bar{\phi}$ can be applied to all nodes on the upstream and downstream boundaries (A′–A, A–B, E–F, and F–F′ (Figure 5.3). The boundaries A–A′ and F–F′ represent truncated lines at sufficient distance from the structure to represent approximately the infinite medium. The flow boundary conditions are applied to the bottom boundary (A′–F′) and the driven sides of the structure B–B′and E–E′. Since such boundaries are assumed to be impervious, the following FD equations can be used.

5.3.1.1.1 x-Direction (Vertical Boundary)

For this case, the boundary condition (gradient in the x-direction) can be expressed as

$$\frac{\phi^*_{i+1,j} - \phi_{i-1,j}}{2\Delta x} = 0 \tag{5.8a}$$

where the superscript * denotes a hypothetical point, and point (i, j) denotes a typical point P on the boundary. Therefore, $\phi_{i+1,j} = \phi_{i-1,j}$ is substituted for the equations for nodes on the vertical end boundary (Figure 5.3). Similarly, for a boundary at the horizontal direction (Figure 5.3), the boundary condition can be incorporated as described below.

5.3.1.1.2 y-Direction (Horizontal Boundary)

The gradient in the y-direction is given by

$$\left(\frac{\partial \phi}{\partial y}\right)_P \approx \frac{\phi^*_{i,j-1} - \phi_{i,j+1}}{2\Delta y} = 0 \tag{5.8b}$$

Therefore, the head on the hypothetical point can be expressed as

$$\phi^*_{i,j-1} = \phi_{i,j+1}$$

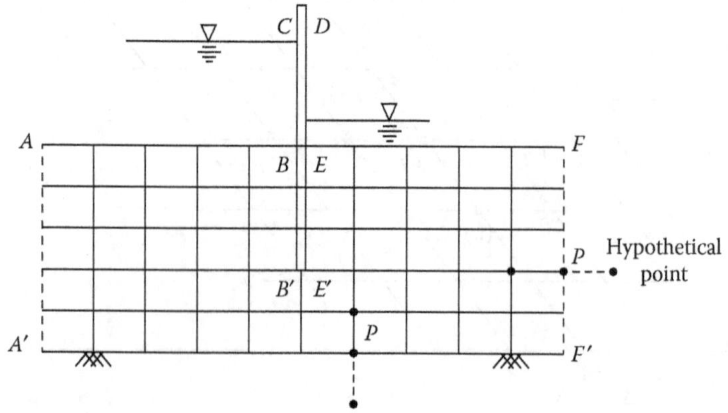

FIGURE 5.3 Schematic of finite difference mesh for steady confined seepage.

5.3.1.2 Time-Dependent Free Surface Flow Problem

Desai and Sherman [8] presented a 2-D FD procedure by using specialized 2-D, nonlinear equations [1,3,8–11]:

$$\frac{k_x}{2}\frac{\partial^2 h^2}{\partial x^2} + \frac{k_y}{2}\frac{\partial^2 h^2}{\partial y^2} = n\frac{\partial h}{\partial t} \tag{5.9a}$$

where h is the fluid potential on the FS. It is based on a number of assumptions such as laminar flow, incompressible fluid, rigid soil skeleton, and invariant k_x, k_y, and n with time. The Dupuit's assumption is used in deriving Equation 5.9a, and the term h^2 satisfies approximately the basic governing equation [3]. If the mean head $\bar{h}$ is assumed, a linearized equation is obtained [9–11].

$$\bar{h}\left(k_x\frac{\partial^2 h}{\partial x^2} + k_y\frac{\partial^2 h}{\partial y^2}\right) = n\frac{\partial h}{\partial t} \tag{5.9b}$$

Equation 5.9a or 5.9b can be used to solve the time-dependent problem. Here, we can use the FD method with implicit or explicit formulation. The implicit formulation is usually stable and provides satisfactory results. The explicit method often suffers from instability and may provide less accurate results. Here, we present the alternating direction explicit procedure (ADEP) [12–14], which is found to provide a computationally stable and efficient scheme for arbitrary values of timewise subdivision. We first give a brief description of the implicit procedure.

The boundary conditions associated with Equation 5.9 can be expressed as (Figure 5.4)

1. $h(x,y,o) = 0$ for initially dry soil   (5.10a)
2. $h(x,y,t) = f(t)$ on upstream face   (5.10b)
3. $\dfrac{\partial h}{\partial y} = 0$ at impervious base   (5.10c)
4. $h(x,y,t) =$ elevation head at FS   (5.10d)

Boundary conditions similar to Equation 5.10b can also occur on the downstream face of a dam. Also, a surface of seepage can occur for unconfined seepage along

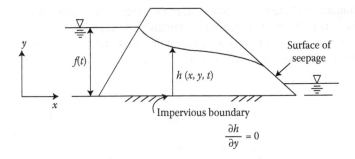

FIGURE 5.4 Schematic of steady free surface seepage.

the downstream face; here, the head (h) is equal to the elevation head. The surface of seepage during drawdown is discussed later.

5.3.1.3 Implicit Procedure

In the implicit procedure, the differential equations are expressed in the finite difference form at time $t+1$. Hence, the resulting simultaneous equations, expressed in terms unknown at $t+1$, need to be solved at $t+1$. Implicit methods are usually more time consuming, but yield relatively accurate and stable results compared to the explicit procedures.

We can express Equation 5.1a, 5.9a, or 5.9b for 2-D idealization by writing the FD equations at time $t+1$ (Figure 5.2). For example, Equation 5.9b will give, assuming equal Δx and Δy

$$\frac{\bar{h}}{n}\left[k_x \frac{h_{i-1,j,t+1} - 2h_{i,j,t+1} + h_{i+1,j,t+1}}{\Delta x^2} + k_y \frac{h_{i,j-1,t+1} - 2h_{i,j,t+1} + h_{i,j+1,t+1}}{\Delta y^2} \right]$$
$$= \frac{h_{i,j,t+1} - h_{i,j,t}}{\Delta t} \tag{5.11}$$

Here, k_x, k_y, and n are related to the node ($i, j, t+1$) and $h_{i,j,t}$ is known because the solution for the end of the previous time step has already been computed. As can be seen, Equation 5.11 will result in a set of simultaneous equations in which the unknown would be h at the nodes at time $t+\Delta t$. This is the implicit formulation and is often found to be stable and accurate. However, it can be time consuming since the solution of the simultaneous equations will be required at each time step.

5.3.1.4 Alternating Direction Explicit Procedure (ADEP)

In an explicit procedure, the differential equations are expressed in a difference form such that the unknowns at $t+1$ are functions of only the known values of head at time t. Hence, it results in an economic procedure involving the solution of the unknown at $t+1$ for nodes, one by one. At the same time, such procedures can suffer from stability problems.

The ADEP [12–14] is a specially devised explicit procedure in which the solution at $t+1$ is expressed in terms of known values at t and $t+1$; the latter are available from initial conditions and at some specific nodes at $t+1$. Hence, the ADEP results in the formulation where the solutions can be obtained at each node, one by one; that is, there is no need for a solution of the simultaneous equations.

Figure 5.5 shows the FD nets at two time levels, t and $t+1$. By using the nets, the ADEP procedure for the nonlinear equation, Equation 5.9a can be expressed as

$$h_{i,j,t+1} = h_{i,j,t} + \beta_x \left(\frac{h_{i-1,j,t+1}^2 - h_{i,j,t+1}^2}{\Delta x_1} - \frac{h_{i,j,t}^2 - h_{i+1,j,t}^2}{\Delta x_2} \right)$$
$$+ \beta_y \left(\frac{h_{i,j+1,t+1}^2 - h_{i,j,t+1}^2}{\Delta y_1} - \frac{h_{i,j,t}^2 - h_{i,j-1,t}^2}{\Delta y_2} \right) \tag{5.12a}$$

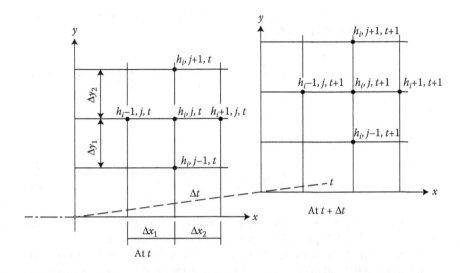

FIGURE 5.5 Finite difference approximation for ADEP.

where $\beta_x = k_x \, \Delta t/[n(\Delta x_1 + \Delta x_2)]$ and $\beta_y = k_y \, \Delta t/[n(\Delta y_1 + \Delta y_2)]$. Equation 5.12a can be expressed in a quadratic form as

$$ah^2_{i,j,t+1} + bh_{i,j,t+1} + c = 0 \tag{5.12b}$$

where Δx and Δy are spatial intervals, Δt is the time interval, a, b, and c are known constants and functions of β_x and β_y.

The ADEP FD model for the linearized equation (Equation 5.9b) is expressed below. Let us express the FD equations for the three derivatives in Equation 5.9b (Figure 5.5) as

$$\frac{\partial^2 h}{\partial x^2} = \frac{(h_{i-1,j,t+1} - h_{i,j,t+1})/\Delta x_1 - (h_{i,j,t} - h_{i+1,j,t})/\beta_x \Delta x_1}{(1/2)\,(\Delta x_1 + \beta_x \Delta x_1)}$$

$$= \frac{2}{\Delta x_1^2\,(1 + \beta_x)}\left(h_{i-1,j,t+1} - h_{i,j,t+1} - \frac{h_{i,j,t}}{\beta_x} + \frac{h_{i+1,j,t}}{\beta_x}\right) \tag{5.13a}$$

Similarly

$$\frac{\partial^2 h}{\partial y^2} = \frac{2}{\Delta y_1^2\,(1 + \beta_y)}\left(h_{i,j-1,t} - h_{i,j,t} - \frac{h_{i,j,t+1}}{\beta_y} + \frac{h_{i,j+1,t+1}}{\beta_y}\right) \tag{5.13b}$$

and

$$\frac{\partial h}{\partial t} = \frac{h_{i,j,t+1} - h_{i,j,t}}{\Delta t} \tag{5.13c}$$

where $\Delta x_2 = \beta_x \Delta x_1$ and $\Delta y_2 = \beta_y \Delta y_1$. Now, the substitution of Equations 5.13a, 5.13b, and 5.12c in Equation 5.9b leads to

$$h_{i,j,t+1} = \frac{C}{D} h_{i,j,t} + \frac{A}{D}\left(h_{i-1,j,t+1} + \frac{1}{\beta_x} h_{i+1,j,t} \right) + \frac{B}{D}\left(h_{i,j-1,t} + \frac{1}{\beta_y} h_{i,j+1,t+1} \right) \qquad (5.13d)$$

where

$$A = \frac{2 \cdot k_x \bar{h} \Delta t}{n \Delta x_1^2 (1 + \beta_x)}, \quad B = \frac{2 k_y \bar{h} \Delta t}{n \cdot \Delta y_1^2 (1 + \beta_y)}, \quad C = 1 - \frac{A}{\beta_x} - B, \quad D = 1 + A + \frac{B}{\beta_y}$$

If the increments in the x- and y-directions are equal, that is, $\Delta x_1 = \Delta x_2$ and $\Delta y_1 = \Delta y_2$ or $\beta_x = \beta_y = 1$, Equation 5.13 will be significantly simplified.

In Equation 5.13d, two time levels are used, and at each of the two time levels, only one (unknown) head (h) from each direction is included. In the ADEP, a proper choice of the starting point needs to be made, for example, at the upstream face, where $h = f(t)$ is prescribed for all times. Then, $h_{i,j,t+1}$ is the only unknown; therefore, it can be computed explicitly. Equations 5.12b and 5.13d are sequentially applied point by point, in either the x- or y-direction. It has been found that the ADEP is more suitable and computationally stable compared to some other FD schemes, and can also be extended to 3-D seepage.

5.3.1.4.1 Rise of External Head

When the fluid rises on the upstream side of the structure such as a dam, slope, riverbank, canal, and so on, the highest point of the (reservoir) water level at the intersection of the slope coincides with the intersection of the FS and the slope. On the other hand, when drawdown occurs, that is, when the fluid level decreases (e.g., on the upstream side), the point of intersection of the fluid level in the reservoir and the slope is lower than the exit point of the FS (Figure 5.6a). The distance between the intersections is called the *surface of seepage*. A procedure for computing the surface of seepage during drawdown, that is, distance A–B in Figure 5.6a is described below [8,10,15].

5.3.1.4.2 Computation of Exit Point and Surface of Seepage

The method of fragments by Pavlovskii [3,15,16] is described here to find the approximate location of the exit point (Figures 5.6a and 5.6b). Figure 5.6a shows the locations of the FS at time levels t and $t + 1$ or $t + \Delta t$ during drawdown on the upstream side. The time interval Δt from t to $t + 1$ is divided into a number of small time intervals, $\Delta \tau$, for example, for $\Delta t = 100$ s, $\Delta \tau$ can be 0.1 or smaller.

Now, the quantity of fluid flowing out of the upstream face is equated to the amount of fluid contained between the FSs at two time levels $\Delta \tau$ apart. The flow out per unit length, ΔQ, assuming it to be essentially horizontal, is given by

$$\Delta Q = -k_x \left[h_e (t + \Delta \tau) - h_d (t + 1) \right] \tan \alpha (1 + \log \lambda) \Delta \tau \qquad (5.14a)$$

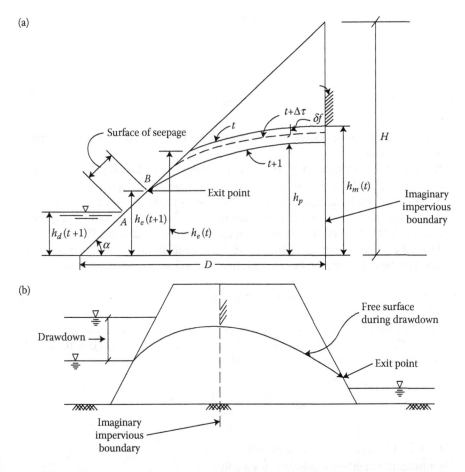

FIGURE 5.6 Surface of seepage. (a) Surface of seepage and exit point; (b) drawdown and imaginary boundary.

where $\lambda = h_e (t + \Delta\tau)/[h_e (t + \Delta\tau) - h_d (t + 1)]$, and α is the angle of the slope. The corresponding volume change, ΔV, is expressed as

$$\Delta V = n [h_m(t) - h_e (t + \Delta\tau)] \cot \alpha \delta f + n \cot \alpha \, (\delta f)^2 \qquad (5.14b)$$

where δf denotes the fall of the FS during time $\Delta\tau$. Now, equating ΔQ and ΔV, we can derive δf as

$$\delta f = \frac{-a + \sqrt{a^2 + 4b}}{2} \qquad (5.14c)$$

where $a = h_m(t) - h_e(t + \Delta\tau)$ and $b = \Delta Q \tan \alpha/n$.

For the special case of $\alpha = 90°$ ΔQ and ΔV are given by

$$\Delta Q = k_x \frac{h_m^2(t) - h_d^2 \delta (t + 1)}{2D} \qquad (5.15a)$$

$$\Delta V = nD\delta f \tag{5.15b}$$

and

$$\delta f = \frac{\Delta Q}{nD} \tag{5.15c}$$

where D is the distance between the entrance toe and the location of the maximum head (Figure 5.6a).

To apply the above method, an impervious (vertical) boundary is required (Figure 5.6a). An imaginary location of such a boundary is usually adopted as the vertical through maximum head $h_m(t)$. The locations of such a boundary for situations like a (river) bank, and drawdown in a dam are shown in Figures 5.6a and 5.6b, respectively, which are the vertical boundaries at the maximum head in the FS.

Once the value of the fall of the FS, δf, corresponding to $\Delta \tau$ are computed, the location of the exit point can be obtained from the following recursive equation:

$$h_e\,(t_i + \Delta \tau) = h_e\,(t_i) - \delta f_i \tag{5.16}$$

where t_i lies between t and $t + 1$, which includes iterations, for example, for $\Delta t = 100$ s. The last value, $h_e(t + 1)$, gives the exit head for the current time $t + 1$. Then, the length A–B of the surface of seepage can be found as

$$AB = \frac{h_e(t + 1) - h_d(t + 1)}{\sin \alpha} \tag{5.17}$$

5.3.1.4.3 Upstream Boundary Heads

The upstream boundary heads can be specified as

$$h\,(x, x\tan\alpha, t + 1) = h_d\,(t + 1) \text{ for points below } h_d\,(t + 1) \tag{5.18a}$$

$$h\,(x, x\tan\alpha, t + 1) = \text{elevation head along the surface of seepage} \tag{5.18b}$$

$$h\,(x, x\tan\alpha, t + 1) = \frac{h_e(t + 1) + h_m(t)}{2} \text{ for points above } h_e\,(t + 1) \tag{5.18c}$$

which are arbitrarily chosen.

5.3.1.4.4 Boundary Conditions at Downstream Face

The foregoing procedure for the surface of seepage can also be used for the downstream face. Then, the heads on the downstream face can be specified similar to Equations 5.18. Sometimes for long river banks (see Figure 5.9), an approximate method can be used. Here, the zero head is assumed at a distance of one Δx outside the (downstream) exit face, and a linear head variation is assumed from the outside point to the point one Δx inside the exit face.

5.3.1.4.5 Location of Free Surface

The entire domain (see Figure 5.9) is divided into a FD mesh (Figure 5.5). Then, Equation 5.12 is used to find values of $h(t + 1)$ at all nodes subject to the upstream and downstream boundary conditions. The approximate location of the FS at a given time is obtained on the basis of the computed heads, by finding points at which the computed head equals the elevation head. We can compare computed total heads, at various nodes along (vertically) inclined lines such as line $a–a$ in Figure 5.7, with their elevation heads. During such a comparison, when the total head is smaller than the elevation head, it implies that the FS will lie between the previous (f_1) and current (f_2) points (Figure 5.7). Then, the point on the FS along $a–a$ can be computed by a linear interpolation between points 1 and 2. The procedure is repeated for all lines in the FD (or FE) mesh. The FS is obtained by joining the FS points along various lines. This idea has been used in the residual flow procedure (RFP) with the FE method [17–21]; it is described later.

An alternative procedure [15,22] can be developed to define the FS at a given time $(t + 1)$ (Figure 5.6a). Here, it is required to evaluate the value of $h_e(t)$, which is used with computed exit head $h_e(t + 1)$, to draw the FS using a parabola. The expression for finding $h_m(t)$ was presented by Newlin and Rossier [22]:

$$\frac{kt}{Hn}\tan^2\alpha = C\left(\frac{h_d}{H - h_d}\right)\ell n\left[\left(1 + \frac{h_m - h_d}{2H - h_d}\right)\frac{h_d}{h_m}\right] \qquad (5.19)$$

where C is a factor found experimentally; its value varying between 0.30 and 0.70 was obtained from model tests with various slopes [22], $h_d(t)$ is the fluid head in the reservoir, $h_e(t)$ is the exit head of the FS, H is the vertical dimension, and h_m is the highest head on the FS. The values of h_m at any time t can be found from Equation 5.19, and often by plotting $(kt/Hn)\tan^2\alpha$ versus $(h_m - h_d)/(2H - h_d)$. Once the values of exit point $h_e(t)$ (Equation 5.16) and the corresponding $h_m(t)$ (Equation 5.19) are found, the FS can be defined by a parabola given by

$$h_p = h_m - (h_m - h_e)\left(\frac{x}{A}\right)^2 \qquad (5.20)$$

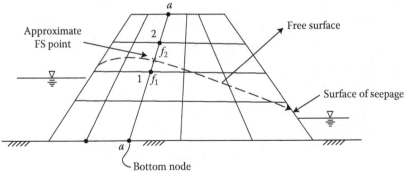

FIGURE 5.7 Location of points on free surface.

where h_p is the head at any point on the FS, h_e is the exit point at time t, and $A = (H - h_e) \, ctn \, \alpha$.

5.3.2 EXAMPLE 5.1: TRANSIENT FREE SURFACE IN RIVER BANKS

The fluctuations of levels in a river or reservoir of a dam, particularly, drawdown in which the water level decreases at a rapid rate, can influence the stability of the riverbank or dam. For instance, high gradients in fluid head occur around the exit point of the FS (Figure 5.6) and may cause liquefaction leading to failure.

A series of experiments were performed using the Hele–Shaw viscous flow model [23], with different slope angles, histories of rise, steady state, and drawdown in the upstream side [8,10,11]. A schematic of the model is shown in Figure 5.8. The viscous flow model was large, about 300 cm long and 50 cm high, in which the silicon fluid, which is stable under the effect of temperature, was used. The level in the reservoir in the model was changed by pumping fluid at selected rates, with the use of a special device to permit the fall or drawdown of the fluid level. The FS in the model was monitored during the rise, steady state, and drawdown in the water level. A typical variation of the water level for the model with a slope of $\alpha = 45°$ is shown in Figure 5.9; the variation in head is shown in the upper right corner. Typical measured FSs during rise, steady state, and drawdown are shown at different time levels (Figures 5.9a through 5.9c). The computed FSs using the FD ADEP procedure are also shown

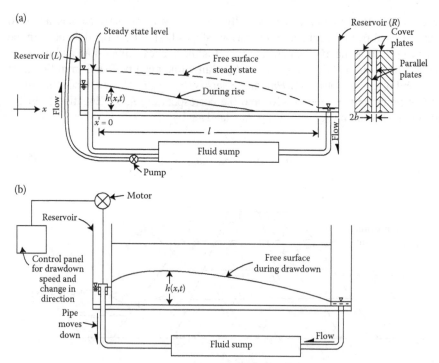

FIGURE 5.8 Schematic of rise and fall (drawdown) in viscous flow model. (a) Rise in external fluid level; (b) drawdown in external fluid level.

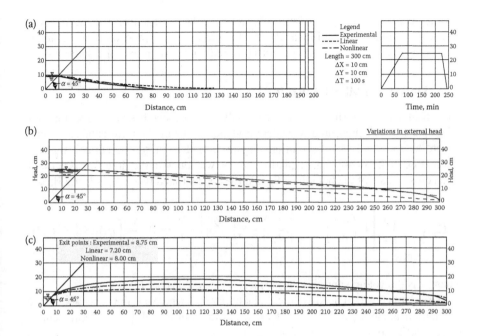

FIGURE 5.9 Comparisons between linear and nonlinear predictions with viscous flow model measurements, $\alpha = 45°$. (a) After 30 min; (b) after 225 min; and (c) after 240 min.

in these figures; they are labeled as linear and nonlinear. It was found that the ADEP with the nonlinear Equation 5.9a gives improved predictions compared to those with the linearized Equation 5.9b.

We can construct flow nets during the variation of upstream head, at different time levels, toward the analysis and design of the structure. Figures 5.10a and 5.10b show flow nets at typical time levels after 30 min during rise, and after 240 min during

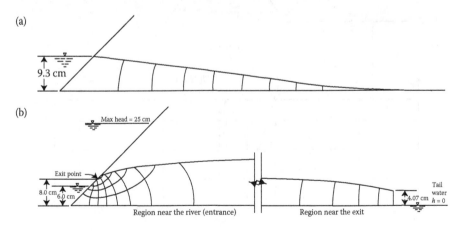

FIGURE 5.10 Typical (approximate) flow nets for viscous flow model, $\alpha = 45°$, during rise and drawdown. (a) After 30 min rise; (b) after 240 min (drawdown).

drawdown. Such flow nets can be used to compute the seepage forces induced on the structure, leading to its stability analysis. The computer (FD) procedure can be modified for factors such as nonhomogeneous (layered) and anisotropic soil conditions [8].

5.4 FINITE ELEMENT METHOD

We consider the FE formulation for the steady-state seepage first; then, the transient FS seepage will be presented. We adopt Equation 5.1a with the right-hand side equal to zero for the steady-state condition (Figure 5.11). The energy functional, Ω_p, corresponding to the steady-state equation, can be written with the fluid flux $\bar{q}$ per unit area across a part of the boundary S_2, and the concentrated flux, $\bar{Q}$ [5–7,24,25], as

$$\Omega_p(\phi) = \int_V \frac{1}{2}\left[k_x \left(\frac{\partial \phi}{\partial x}\right)^2 + k_y \left(\frac{\partial \phi}{\partial y}\right) + k_x \left(\frac{\partial \phi}{\partial z}\right) \right] - \int_V \bar{Q}\phi\,dV - \int_{S_2} \bar{q}\,\phi\,dS \qquad (5.21)$$

where V is the volume and S is the surface. We now consider the 2-D case.

We express the total head or potential $\phi\,(= p/\gamma + y)$, where p is the pressure, y is the elevation head, and γ is the water density. For the four-noded quadrilateral element (Figure 5.12), ϕ can be expressed as [5,24]

$$\phi(x, y) = \sum_{i=1}^{4} N_i \phi_i = [N]\{q\} \qquad (5.22)$$

where N_i are interpolation functions $= 1/4(1 + ss_i)(1 + tt_i)$, s and t are the local coordinates, ϕ_i are the nodal fluid heads, and $\{q\}$ is the vector of nodal fluid heads.

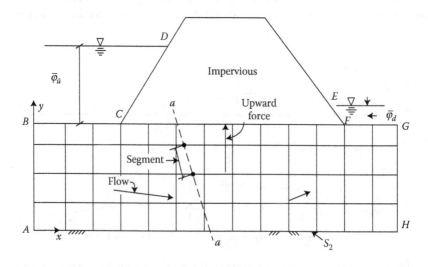

FIGURE 5.11 Confined steady seepage through foundation of impervious dam.

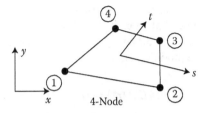

FIGURE 5.12   Four node quadrilateral element.

By substituting ϕ and the gradients of ϕ with respect to x and y in Equation 5.21 and equating the variation of Ω_p to zero, we obtain [5,24,25] the element equations as

$$[k_\phi]\,\{q\} = \{Q\} \tag{5.23}$$

where

$$[k_\phi] = \int_A [B]^T[R][B]\,\mathrm{d}A$$

$$\{Q\} = \int_A [N]^T \bar{Q}\,\mathrm{d}A + \int_{S_2} [N]^T \bar{q}\,\mathrm{d}S$$

$[R] = \begin{bmatrix} k_x & 0 \\ 0 & k_y \end{bmatrix}$ is the principal permeability matrix, A is the area of the element, and S_2 is the part of the surface on which $\bar{q}$ is applied. Equation 5.23 can be used for steady-state conditions, including confined and unconfined (FS) seepage; the latter requires additional considerations, as described later.

5.4.1   CONFINED STEADY-STATE SEEPAGE

Consider the steady confined seepage in Figure 5.11. Here, the flow occurs in the confined space through the foundation subjected to applied heads $\bar{\phi}_u$ and $\bar{\phi}_d$ on the upstream and downstream sides, respectively. The FE mesh involves only the confined zone and the upstream heads are applied on the boundaries AB, BC, and CD, and the downstream heads are applied on boundaries EF, FG, and GH. The bottom boundary and the structure are assumed to be impervious. In the above FE procedure, the nodes on the boundary such as A–H are assumed to be "free," that is, there is no applied head on that impervious boundary; this implies the impervious surface under the FEM formulation above.

The equations for all elements are now assembled to lead to the global or assemblage equations, given by

$$[K_\phi]\{r\} = \{R\} \tag{5.24a}$$

where $[K_\phi]$ is the global permeability ("stiffness") matrix, $\{r\}$ is the global nodal head vector, and $\{R\}$ is the global applied "load" vector.

The global equations are modified by introducing the heads at the boundary nodes, which results in the modified global equations

$$[K_\phi^*]\{r^*\} = \{R^*\} \tag{5.24b}$$

Here, the superscript denotes modified matrix and vectors. The solution of Equation 5.24b leads to the computation of heads at all the nodal points in the mesh, except the boundary nodes where the heads are specified and known. Once the heads are determined, we can draw curves in the mesh of equal heads or potentials, and then lines normal to the equipotential lines, to obtain the flow net. The latter can be used to compute force on the base of the dam (Figure 5.11). In addition, we can also compute velocities and quantity of flow, as described below.

5.4.1.1 Velocities and Quantity of Flow

The velocities, say, at the nodes or the centroid of the elements, are computed as

$$\begin{Bmatrix} v_x \\ v_y \end{Bmatrix}^e = -\begin{bmatrix} k_x & 0 \\ 0 & k_y \end{bmatrix}^e [B\,(s,t)]\{q\}^e \tag{5.25a}$$

$$\{v\} = -[R][B]\{q\} \tag{5.25b}$$

where $B(s,t)$ denotes the evaluation of the components of matrix $[B]$ at the desired (integration) points with local coordinates, s and t, and superscript e denotes the element. The quantity of flow across the normal section a–a (Figure 5.11) in the mesh can be computed as

$$\{Q_f\} = -\{v\}\,\{A\}^T \tag{5.26}$$

where $\{Q_f\}$ is the vector of the components of flow in the x- and y-directions, and $\{A\}$ is the vector containing component areas normal to the velocity components v_x and v_y.

For given sections, for example, $(a$–$a)$ in Figure 5.11, the total quantity of flow in the chosen direction can be found as

$$Q_f = \sum_{i-1}^{N} \Delta Q_{fi} \tag{5.27}$$

where ΔQ_{fi} is the flow across the segment, which is a part of section a–a, and N are the total number of segments along a–a (Figure 5.11).

5.4.2   Example 5.2: Steady Confined Seepage in Foundation of Dam

This and the subsequent Example 5.3 are adopted from Gong and Desai [26]. Figure 5.13a shows an impervious dam resting on a nonhomogeneous foundation. The lower and upper layers are isotropic with the coefficient of permeability $k_x = k_y = k$ and $k_x = k_y = 2\,k$, respectively, with $k = 10$ m/day. The applied heads on the upstream and downstream dam surfaces are 10.0 and 2.0 m, respectively. The FE mesh consists of quadrilateral elements with nodes and elements as 105 and 80, respectively. The computer code SEEP-2DFE [27] was used to solve the problem; any other suitable code can be used.

Results. The contours of equal heads or potentials are plotted in Figure 5.13b; then, the flow net is obtained by drawing lines (curves) such that they are orthogonal to the equipotential lines.

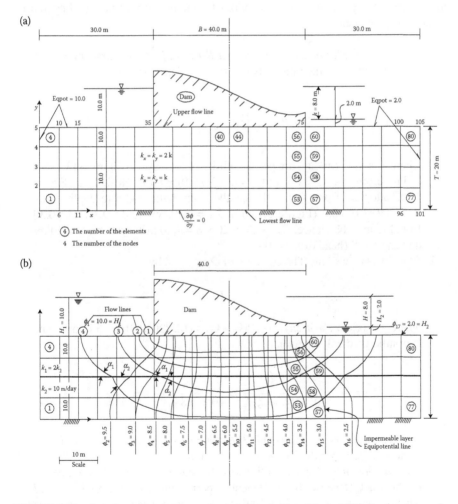

FIGURE 5.13   Steady confined seepage through foundation of dam. (a) Finite element mesh; (b) flow net in foundation.

TABLE 5.1
Comparisons for Quantity of Flow

	FEM (m³/day)	Closed-Form Methods (m³/day)		
		Polubarinova-Kochina	Directly by Darcy's Law	Flow Net
Q_f (m³/day)	46.55	48.0	41.67	41.00

The hydraulic gradients, useful for the analysis and design, to the x, y, and normal directions are denoted by i_x, i_y, i_n, where $i_n = \sqrt{i_x^2 + i_y^2}$. Table 5.1 shows comparisons between the predictions for the quantity of flow from the current FE analysis and the closed-form solutions [1,3]. The results correlate well. A brief description of the closed form, Darcy's, and flow net solutions for the quantity of flow and hydraulic gradients is given below.

1. *Closed-form solution by Polubarinova-Kochina [1,3]:* We first compute ε, which is related to the ratio of two permeabilities, from

$$\tan \pi\varepsilon = \sqrt{\frac{k_2}{k_1}} = \sqrt{\frac{10}{20}} = \frac{\sqrt{2}}{2}$$

where k_1 and k_2 (m/day) are permeabilities of two layers (Figure 5.13).
Therefore, $\pi\varepsilon = 35.26°$ and $\varepsilon = (35.26/180) = (1/5)$. Now, from Figure 5.14, adopted from Ref. [3], for the dam with width-to-depth ratio, $B/T = (40/20) = 2.0$ (Figure 5.13a) and $\varepsilon = 0.20$, we find $(Q_f/k_1 \, \phi) = 0.30$ from Figure 5.14. Hence, $Q_f = 0.30 \times 20 \times 8 = 48.0$ m³/day; here, $k_1 = 20$ m/day and $h = 8$ (head drop $= 10 - 2$).
2. *Flow by Darcy's law:* The equation of Q_f is given by

$$Q_f = k \, i \, T$$

where $k = (k_1 d + k_2 d)/(2d) = (1/2)(k_1 + k_2) = (1/2)(20 + 10) = 15$ m/day. The hydraulic gradient i is given by

$$i = \frac{h}{L} = \frac{h}{B + \xi_1 T + \xi_2 T} = \frac{8}{40 + 0.88 \times 20} = \frac{8}{57.6}$$

where parameters $\xi_1 = \xi_2 = 0.44$ [3] and $h =$ loss of head $= 10 - 2 = 8.0$ ft.
Hence, $Q_f = 15 \times (8/57.6) \times 20 = 41.67$ m³/day.
3. *Flow by flow nets:* Figure 5.13b shows the equipotential and flow lines in the flow net. Because the coefficient of permeability k_1 in layer 1 is 20 m/day and that in layer 2 is 10 m/day, the flow nets in layer 1 can be drawn as square, while those in layer 2 as rectangular with width-to-length ratio of 2.

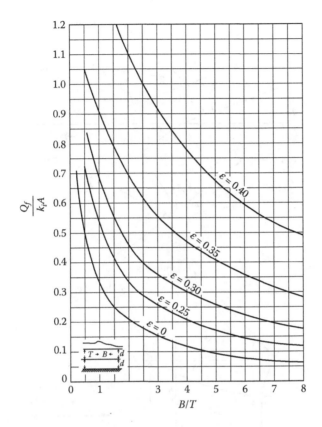

FIGURE 5.14 Determination of quantity of flow. (Adapted from Polubarinova-Kochina P. Ya., *Theory of Ground Water Movement*, translated by De Wiest, R.J.M., Princeton University Press, Princeton, NJ, 1962; Harr, M.E., *Groundwater and Seepage*, McGraw-Hill Book Co., New York, 1962.)

The flow net at the junction of the two layers can be developed by using the following ratio (Figure 5.15a):

$$\frac{k_1}{k_2} = \frac{\tan \alpha_2}{\tan \alpha_1}$$

From Figure 5.15a, $\alpha_1 = 20°$ and $\alpha_2 = 36°$ for the flow line No. 3 (Figure 5.13b). Similarly, $\alpha_1 = 36°$ and $\alpha_2 = 55°$ can be obtained for flow line No. 4. The quantity of flow through the foundation layer is given by [1,3]:

$$Q_f = k_1 \frac{N_f}{N_d} h$$

where $N_f = 4.10$ is the approximate number of flow paths, $N_d = 16$ is the number of potential drops (Figure 5.13b), and h is the loss of head $= 10.0 - 2.0 = 8.0$ m.

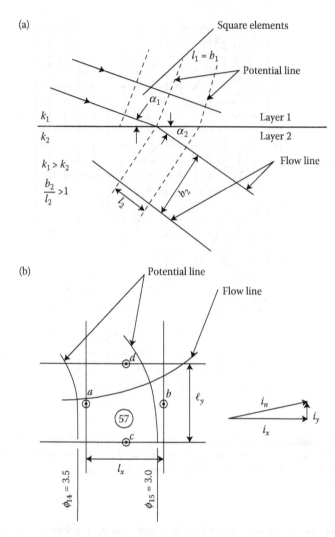

FIGURE 5.15 Flow channels and gradient calculation. (a) Flow channels at boundary between two layers with different coefficients of permeability; (b) Calculation of gradient for element.

Hence

$$Q_f = \frac{20 \times 4.10 \times 8}{16} = 41.0 \text{ m}^3/\text{day}$$

Thus, the quantity of flow by various methods (Table 5.1) compares well to those from the FE method, from which the flow nets were developed.

5.4.2.1 Hydraulic Gradients

Hydraulic gradients are useful for analysis, design, and stability of the structure. Since the critical gradient may usually occur in the soil near and below the front toe of the

TABLE 5.2

Gradients for Typical Elements

Number of the Element	$\Delta\phi_x$ (m)	ℓ_x (m)	$i_x = \dfrac{\Delta\phi_x}{\ell_x}$	$\Delta\phi_y$ (m)	ℓ_y (m)	$i_y = \dfrac{\Delta\phi_y}{\ell_y}$	$i_n = \sqrt{i_x^2 + i_y^2}$
57	$3.48 - 3.00 = 0.48$	5.0	0.096	$3.20 - 3.10 = 0.10$	5.0	0.02	0.098
59	$2.93 - 2.50 = 0.43$	5.0	0.086	$2.84 - 2.45 = 0.39$	5.0	0.078	0.116
60	$2.45 - 2.25 = 0.20$	5.0	0.040	$2.45 - 2.0 = 0.45$	5.0	0.09	0.098

dam, we calculate gradients near that zone, that is, in elements 57, 59, and 60 (Figure 5.13b). As an example, the gradients in element 57 are computed as (Figure 5.15b):

$$i_x = \frac{\Delta\phi_x}{\ell_x} = \frac{\phi_a - \phi_b}{\ell_x} = \frac{3.48 - 3.00}{5.0} = 0.096$$

$$i_y = \frac{\Delta\phi_y}{\ell_y} = \frac{\phi_c - \phi_d}{\ell_y} = \frac{3.20 - 3.10}{5} = 0.02$$

Hence

$$i_n = \sqrt{i_x^2 + i_y^2} = \sqrt{0.096^2 + 0.02^2} = 0.098$$

Here, i_x, i_y, and i_n are the gradients in the x-, y-, and normal directions, respectively, $\Delta\phi_x$ and $\Delta\phi_y$ are drops in potential along the x- and y-directions, respectively, ϕ_a, ϕ_b, ϕ_c, and ϕ_d are potentials as shown in Figure 5.15b at the midpoints of each side of the element obtained by interpolation. Table 5.2 shows the computation for the hydraulic gradients for elements 57, 59, and 60.

It can be seen that the gradients are much smaller than 1.0; hence, no instability can be expected.

5.4.3 STEADY UNCONFINED OR FREE SURFACE SEEPAGE

Figure 5.16 shows a schematic of time-independent (steady) FS seepage in a dam. This problem is nonlinear and involves additional boundary conditions on the FS, b–c, as follows:

$$\phi = y \; (p = 0) \tag{5.28a}$$

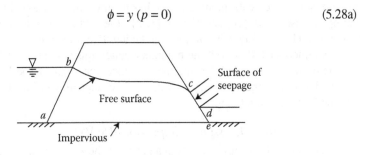

FIGURE 5.16 Steady unconfined or free surface seepage.

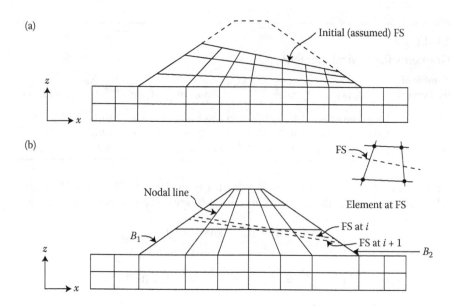

FIGURE 5.17 Finite element meshes for variable and invariant mesh procedures. (a) Variable mesh; (b) invariant mesh.

$$\frac{\partial \phi}{\partial n} = 0 \tag{5.28b}$$

Similarly, on the surface of seepage, c–d, we can write

$$\phi = y \ (p = 0) \tag{5.28c}$$

As a result, the solution and the determination of FS seepage requires an iterative solution. There are two main methods for the solution: (1) variable mesh (VM) and (2) invariant or fixed mesh (Figure 5.17).

5.4.3.1 Variable Mesh Method

The VM method was proposed by Taylor and Brown [28] and Finn [29]. This method, sometimes with modifications, has been used by various investigators, for example, Desai et al. [30] and France [31]. In the VM method, a location of the FS is assumed (Figure 5.17a). Then, through an iterative procedure, the solution for the FS is obtained such that both the boundary conditions (Equations 5.28a and 5.28b) are approximately satisfied. During the iterations, the assumed mesh is successively modified to satisfy the boundary conditions. The steps in Schemes 5.1 and 5.2 in the VM method are given below.

5.4.3.1.1 Scheme 5.1: Assumed Impervious FS: VM

1. Assume an FS profile and adopt a mesh for the zone below the assumed FS (Figure 5.17a).

2. Assume there is no flow across the FS. Then, obtain the heads at the nodal points by solving Equation 5.24b with the upstream and downstream boundary conditions.

3. Compare the computed heads for the nodes on the FS with their elevations, y. If the assumed FS is correct, the computed heads for the FS nodes will be approximately equal to the elevation heads because the pressure at those nodes is zero.

4. If the condition in Ref. [3] is not satisfied, we need to modify the mesh for the next iteration. First, evaluate the sum of the differences, Δr, between the elevations and computed nodal heads on the FS. Then, multiply the sum by a tolerance factor in the range of 0.100–0.0001; here, we have assumed the tolerance factor $= 0.01$, which is denoted by ε. Compare Δr with ε. If $\Delta r - \varepsilon$ is equal to or less than zero, stop the iterations; otherwise, go to step 5 below.

5. Modify the assumed (previous) FS by changing the coordinates of the nodes on the FS. The coordinates of each node on the FS can be changed in the vertical direction by the amount $\Delta r_i^* = \lambda \Delta r$, where λ is a factor, which can be adopted as 0.50.

6. Now, modify the coordinates in the mesh based on the change, Δr_i^*, where i is the vertical or inclined nodal lines in the mesh; here, the angle with the horizontal or the nodal lines can be used. Often, only selected nodes in the vicinity of the FS are modified so as to minimize any distortions in the mesh. The modification may not be required for the mesh in the foundation, for example, Figure 5.17a.

7. Obtain the FE solution for nodal heads using the revised mesh.

8. Repeat Steps 3 and 4. If $(\Delta r - \varepsilon)$ is approximately zero (a small value), stop the procedure with the last solution.

9. If $(\Delta r - \varepsilon)$ near to zero is not satisfied, repeat the foregoing steps.

5.4.3.1.2 Scheme 5.2: Assumed Heads on Free Surface (VM)

In the foregoing VM method, we assumed no flow across the FS that satisfied the boundary condition in Equation 5.28b. Then, the nodal heads were computed in the entire domain, including the FS. In such a procedure, it is required to verify that the other boundary condition (Equation 5.28a) is satisfied during the iteration in the VM method.

On the other hand, in Scheme 2, we assume that the nodal heads on the FS are equal to their elevation (y) heads; thus, $\phi_i = y_i$ on the FS are specified together with the other boundary conditions. In this case, it is required to satisfy, during the iterations, that there was no flow across the FS; let us express that the normal velocity across the FS vanishes, that is

$$v_n = v_{nx} + v_{ny} = 0 \qquad (5.29)$$

where v_n is the velocity (at a point) normal to the FS, and v_{nx} and v_{ny} are normal components of the actual velocities given by

$$v_x = -k_x \frac{\partial \phi}{\partial x}, \quad v_y = -k_y \frac{\partial \phi}{\partial y} \qquad (5.30)$$

If the condition in Equation 5.29 is not satisfied (approximately), we perform the following procedure for the modification of the mesh:

Compute average velocities:

$$\bar{v}_x = \frac{v_x^m + v_x^{m+1}}{2} \tag{5.31a}$$

$$\bar{v}_y = \frac{v_y^m + v_y^{m+1}}{2} \tag{5.31b}$$

where m denotes a node on the FS (Figure 5.18). The velocities are computed for all FS nodes except at the entrance and exit faces as

$$\bar{v}_x = v_x^m \quad \text{and} \quad \bar{v}_y = v_y^m$$

Now, the normal velocity is found as

$$v_n = v_x \sin \theta + v_y \cos \theta$$

and the corresponding displacement, u_n, is obtained as

$$u_n = v_n \times \Delta t$$

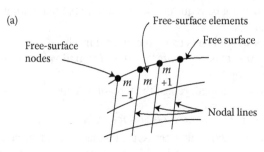

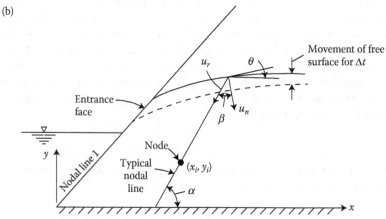

FIGURE 5.18 Movement of free surface. (a) FD mesh; (b) free surface.

Now, compute the displacement, u_r (Figure 5.18) as

$$u_r = \frac{u_n}{\cos \beta}, \quad \beta = \frac{\pi}{2} - \alpha + \theta$$

and

$$u_x = u_r \cos \alpha$$

$$u_y = u_r \sin \alpha$$

Finally, the coordinates of the mesh are revised as

$$x_i^j = x_i^{j-1} + u_x$$
$$y_i^j = y_i^{j-1} + u_y$$

where i denotes a nodal point in the mesh and j denotes an iteration.

The procedure is repeated until Equation 5.29 is approximately satisfied. If Δt is small, about one to three iterations are sufficient for an acceptable solution.

The above scheme involving the assumed heads equal to the elevation head for the FS nodes is relatively more stable and involves lower levels of mesh distortions compared to the scheme that assumes impervious FS; see subsequent examples, for example, Example 5.3.

5.4.4 Unsteady or Transient Free Surface Seepage

The FE equations, based on Equation 5.1 with the time-dependent terms on the right-hand side, can be derived as [10,11,25,32]

$$[c]\{\dot{q}\} + [k_\phi]\{q\} = \{Q(t)\} \tag{5.32}$$

where $[c]$ is the element property (porosity) matrix, $[k_\phi]$ is the element property (coefficient of permeability) matrix, $\{q\}$ is the nodal head vector, $\{Q(t)\}$ is the time-dependent applied forcing function vector, and the over dot denotes the derivative with respect to time.

By writing the first time derivatives in vector $\{q\}$ in the FD form [24] (Figure 5.19), we obtain

$$\dot{\phi}(\tau) = \frac{\partial \phi}{\partial t} \approx \frac{\phi_{t+\Delta t} - \phi_t}{\Delta t} \tag{5.33}$$

where τ is the time level between t and $t + \Delta t$ and Δt is the time step. The substitution of Equation 5.33 in Equation 5.32 leads to

$$\frac{1}{\Delta t}([c] + [k_\phi])_t \{q\}_{t+\Delta t} = \{Q(t + \Delta t)\} + \frac{1}{\Delta t}[c]_t\{q\}_t \tag{5.34}$$

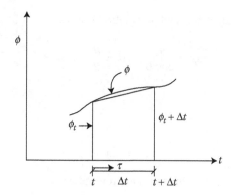

FIGURE 5.19 Finite difference approximation for first time derivative.

Equation 5.34 can be solved for nodal heads at $t + \Delta t$ by introducing the specified boundary conditions at time $t + \Delta t$, say, on upstream and downstream boundaries, and the initial conditions, given by nodal heads at time $t = 0$:

$$\phi(x,y,0) = \bar{\phi}(x,y) \tag{5.35}$$

where the overbar denotes the known or specified heads at time $t = 0$. For problems involving FSs, we can use the foregoing VM method for each time level, $t + \Delta t$, by using scheme 1 or 2. Alternatively, we can use the residual flow procedure (RFP), described subsequently.

5.4.5   EXAMPLE 5.3: STEADY FREE SURFACE SEEPAGE IN HOMOGENEOUS DAM BY VM METHOD

Figure 5.20 shows the FE mesh for an earth dam, resting on an impermeable foundation [26], which is solved by using SEEP2D-FE [27]. The applied heads on the upstream and downstream are 32 and 4.5 m, respectively. The soil in the dam is considered to be sandy loam with its isotropic coefficient of permeability k $(k_x = k_y) = 0.10$ m/day. Figure 5.20 also shows the initially assumed FS as a straight line. The mesh below the assumed FS consists of 56 elements and 75 nodes.

The VM method was used with applied nodal heads on the initial FS equal to elevation heads. Figures 5.21a, 5.21b, and 5.22c show the modified mesh, nodal heads, and FS after three iterations, respectively. Figure 5.22 shows the equipotential and flow lines for the computed nodal potential according to (final) FS (Figure 5.21c); Figure 5.22 also shows a comparison between results of current and Schaffernak's methods [3].

Table 5.3 shows the quantities of flow normal to the section passing through elements 45, 46, 47, and 48, and through the section passing through elements 21, 22, 23, and 24 (Figure 5.20). Table 5.3 also shows the flow computed by the conventional Schaffernak method [3,33] and by the flow net method. The values of flow by the three methods compare very well. Table 5.4 shows the values of hydraulic gradients

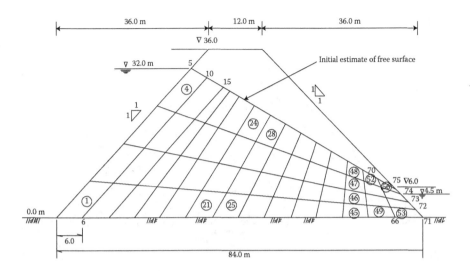

FIGURE 5.20 Initial mesh for the steady free surface flow in a dam.

in various elements. It can be seen that the hydraulic gradient is nearer to or greater than unity in elements 50, 51, and 52; hence, the possibility of instability or piping may exist in the zone near the exit point on the downstream face.

5.4.6 EXAMPLE 5.4: STEADY FREE SURFACE SEEPAGE IN ZONED DAM BY VM METHOD

Figure 5.23 shows a zoned earth dam with a central core [34], which was solved by using SEEP2D-FE [27]. The coefficients of permeabilities k ($k_x = k_y$) of the core and shell materials are assumed to be equal to 0.002 m/day and 0.100 m/day, respectively.

The steady heads on the upstream and downstream are 30 and 4 m, respectively. The FE mesh is shown in Figure 5.23. The FS is evaluated by using the VM method, in which the initial FS is assumed as a straight line (Figure 5.23). The final and converged FS are shown in Figure 5.24. The quantity of flow normal to Section 1 (Figure 5.23) near the downstream face was found to be 0.028 m³/day.

5.4.7 EXAMPLE 5.5: STEADY FREE SURFACE SEEPAGE IN DAM WITH CORE AND SHELL BY VM METHOD

Figure 5.25 shows a dam with a core and porous zone as the toe down. This problem was solved by the VM method by Taylor and Brown [28]. The dam is about 440 ft (134.20 m) wide and 100 ft (30.5 m) high. The ratio of the coefficients of permeability of the core to the shell material was 1:4.

The steady FS solution with the initially assumed FS as composed of straight lines (Figure 5.25) was obtained by using the present code SEEP-2DFE [27] for the upstream head equal to 95 ft (29 m) and the downstream head as zero. The final computed FS is shown in Figure 5.25. The solution [with upstream head equal to

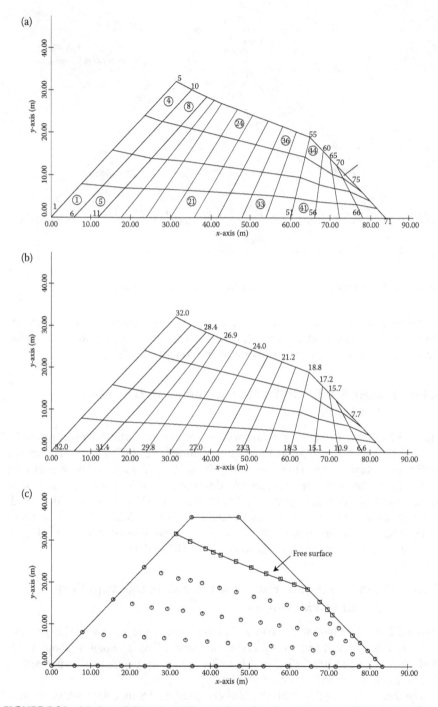

FIGURE 5.21 Mesh, nodal heads and free surface after three iterations. (a) Mesh after three iterations; (b) nodal heads after three iterations; and (c) free surface after three iterations.

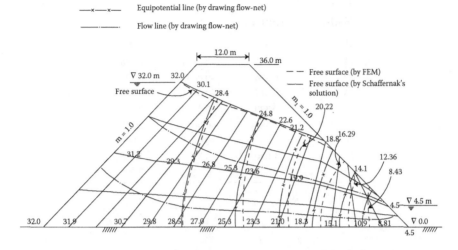

FIGURE 5.22 Comparison of free surface obtained by FE method and conventional method, and flow net.

100 ft (30.5 m)] by Taylor and Brown [28] is also shown in this figure. The two solutions compare very well.

5.4.8 EXAMPLE 5.6: STEADY CONFINED/UNCONFINED SEEPAGE THROUGH COFFERDAM AND BERM

Figure 5.26a shows a cofferdam that is assumed to be impervious, resting on a three-layered foundation with a berm on the downstream side; the middle layer has higher

TABLE 5.3
Quantity of Flow

Section	FEM Number of the Element	Q_f (m³/day)	Schaffernak Solution Q_f (m³/day)	Flow Net Q_f (m³/day)
1-1	45	0.2621		
	46	0.2602		
	47	0.2310		
	48	0.1789		
	Total	0.9322	0.882	0.898
2-2	21	0.1722		
	22	0.2065		
	23	0.2342		
	24	0.2496		
	Total	0.8626		

TABLE 5.4
Hydraulic Gradients

Element	I_x	I_y	I_n
21	0.265	−0.0196	0.266
22	0.294	−0.0607	0.300
23	0.317	−0.0934	0.331
24	0.330	−0.0111	0.349
49	0.787	0.0699	0.790
50	0.989	0.0346	0.990
51	1.050	−0.325	1.100
52	0.901	−0.429	0.998

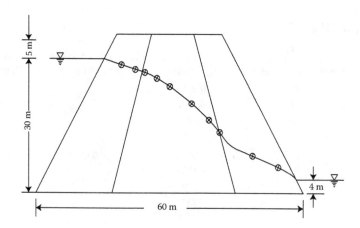

FIGURE 5.23 Zoned dam and initial mesh.

FIGURE 5.24 Final free surface in zoned dam.

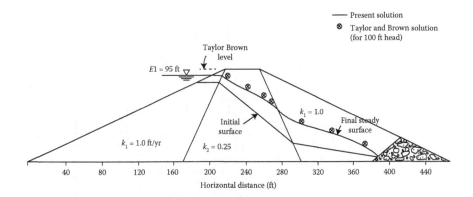

FIGURE 5.25 Zoned dam and comparisons with Taylor and Brown [28] solution (1 ft = 0.3048 m). (From Taylor, R.L. and Brown, C.B., *Journal of Hydraulics Divisions, ASCE*, 93(HY2), 1967, 25–33. With permission.)

permeability [35]. Details of this problem were provided by the St. Louis District, U.S. Corps of Engineers for the structure related to the Lock Dam No. 26, Mississippi River, near Alton, Illinois.

Figure 5.26b shows the FE mesh used for the foundation and berm in which the vertical left- and right-hand boundaries were fixed at distances of 150 ft (46.0 m) and 160 ft (53.30 m) from the upstream and downstream faces of the cofferdam, respectively.

The nodes on the vertical and horizontal surfaces on the upstream side were subjected to the head of 133 ft (40 m). A head of 54 ft (16.3 m) was applied on the downstream horizontal and vertical surfaces.

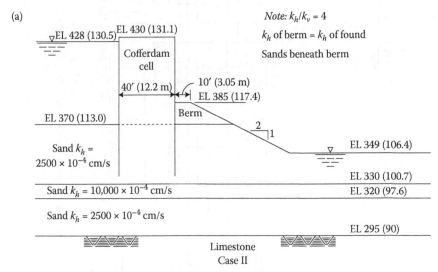

Note: Alternative metric units are given in the parenthesis.

FIGURE 5.26 Cofferdam and finite element results. (a) Cofferdam and soil properties; (b) mesh, computed equipotentials and free surface.

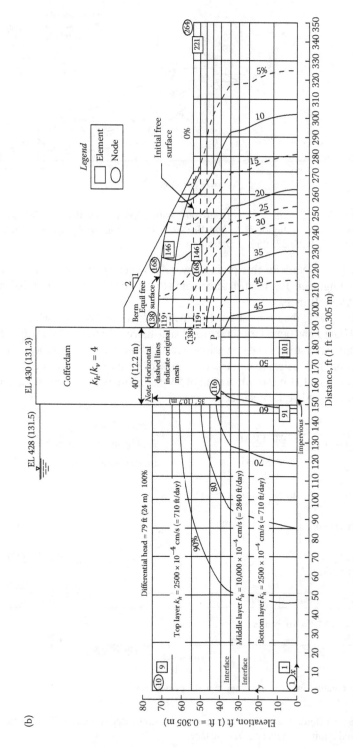

FIGURE 5.26 (continued) Cofferdam and finite element results. (a) Cofferdam and soil properties; (b) mesh, computed equipotentials and free surface.

5.4.8.1   Initial Free Surface

The flow in the berm represents a FS seepage condition. In this analysis, the initial FS was assumed to be at the level of the base of the berm, which is shown as the top dashed lines in Figure 5.26b at the base of the berm. Then, to reach equilibrium, the water rises as a mound in the berm until it reaches the equilibrium FS (Figure 5.26b) when $\phi = y$, the elevation head.

The FE procedure involved the assumption of the FS to be impervious. Then, the computer solutions yielded potential heads at the nodes on the FS. As described before, the iterative procedure involved satisfaction of the condition that $\phi_i = y_i$. The initially horizontal FS at the base of the berm experienced movements (upward) during iterations; some nodes on the FS approached the inclined face of the berm. When the latter happened, the potentials (ϕ) at those points on the *surface of seepage* were set equal to the elevation head.

Figure 5.26b also shows computed equipotential lines; it can be seen that the equipotential lines involve discontinuities at the layer interfaces. The dissipation of head from the upstream to point P in the right bottom corner of the cofferdam is about 50% of the applied head difference of 79 ft (= 133 − 54) (24 m). The final location of the computed FS was found to satisfy the condition of equal potential drops [33]. Thus, the above results are considered to be satisfactory.

The knowledge of the remaining head at point P, flow net, and the associated gradients can be used for the stability and design of the cofferdam.

5.5   INVARIANT MESH OR FIXED DOMAIN METHODS

The VM method (Figure 5.17a), described before has been used for solutions of some free or phreatic surface problems. However, it can suffer from certain limitations such as irregular mesh, resulting in uneven FSs, instability in computations for such meshes, and nonhomogeneous or layered soil masses when the FS crosses the junctions of the layers. The invariant mesh (IM) or fixed domain method avoids such difficulties and is based on the mesh for the entire zone of the problem (Figure 5.17b). Hence, in the IM method, the need for modifying the mesh is avoided. A description of the IM method and typical example problems are presented in the following.

There are mainly two formulations available for the IM method: (1) variational inequality (VI) method and (2) RFP. The VI method has been presented by Baiocchi [36,37], Alt [38,39], Duvant and Lions [40], and Lions and Stampucchia [41]; further developments and the use of this method have been presented by various investigators [42–44].

The VI method has been used for the solution of various problems in mechanics, for example, solid mechanics and fluid flow including FSs. Its mathematical formulation can be relatively complex, and its use for available practical problems in FS seepage is rather limited. RFP developed by Desai [17] and associates [18–21] is relatively simple. Hence, we concentrate on the RFP and its applications for FS seepage.

The IM method is based on the RFP concept and involves progressive correction of the FS by using the *residual or correction* vector, which acts like the "residual or initial load" in nonlinear (FE) analysis of structures [5,25]. Procedures proposed for the saturated–unsaturated flow by Bouwer [45], Freeze [46], Neuman [47], and Cathie and Dungar [48] involve similar considerations.

5.5.1 RESIDUAL FLOW PROCEDURE

The basic idea of the IM method with the RFP was used around 1971 by Desai and Sherman [8] while using the FD method for FS seepage. The development of the FE-RFP with the IM method was presented by Desai [17]. It has been applied for both steady and transient FS seepage and been presented in various publications [18–21]. Westbrook [49] has presented similarities between the VI method and the RFP, and Bruch [43] has presented a detailed review of both formulations. A brief description of the RFP is presented below.

Equation 5.1 can be considered as the GDE for flow through the domain Ω_1 (Figure 5.27). In the RFP, the domain Ω_1 is extended into Ω_2 so that Equation 5.1 is assumed to hold, approximately, in the entire Ω, by introducing the definition of the coefficient of permeability k as follows:

$$k(p) = \begin{cases} k_s & \text{in } \Omega_1 \\ k_{us} = k_s - f_1(p) & \text{in } \Omega_2 \end{cases} \tag{5.36a}$$

where k_s is the saturated permeability, k_{us} is the unsaturated permeability, p is the pressure head, and $f(p)$ is a smooth function of the pressure head p.

The relation between the coefficient of permeability k and the pressure head can be presented as in Figure 5.28a [18–21,50]. As a simplification, Equation 5.36a can be expressed as

$$k(p) = \begin{cases} k_s & \text{in } \Omega_1 \ (p \geq 0) \\ k_s/\lambda & \text{in } \Omega_2 \ (p < 0) \end{cases} \tag{5.36b}$$

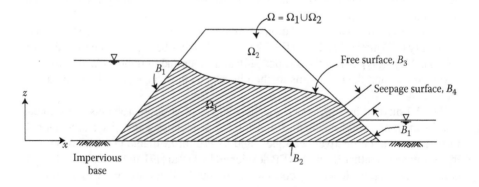

FIGURE 5.27 Schematic of seepage through an earthdam.

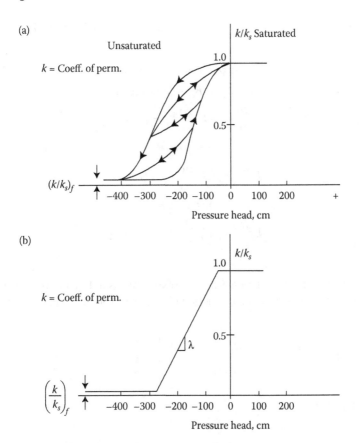

FIGURE 5.28 Relations for saturated and unsaturated permeabilities (a) general; (b) simplified.

where λ is a number, for example, 1000. The effective porosity (specific storage) $n(S)$ can also be expressed as the function of p as for $k(p)$ in Equation 5.36a:

$$n(p) = \begin{cases} n_s & \text{in } \Omega_1 \\ n_{us} = n_s - f_2(p) & \text{in } \Omega_2 \end{cases} \qquad (5.37)$$

The relations between k and p and n and p will be nonlinear. For example, Figure 5.28a shows the relation between k (or k/k_s) and p. However, in the RFP, the relation can be simplified approximately as shown in Figure 5.28b. It has been found that the use of such a linearized relation with a small value of $(k/k_s)_f$ in comparison to the fully saturated values, of the order of $1/1000 \, k_s$, can lead to satisfactory and convergent solutions [18]; such a simplified form has been used in Ref. [51].

The boundary conditions (for 2-D section) are given by (Figure 5.27):

$$\phi = \bar{\phi}_u \text{ on } B_1 \text{—upstream}$$

$$\phi = \bar{\phi}_d \text{ on } B_1 \text{—downstream}$$

$$k_x \frac{\partial \phi}{\partial x} \ell_x + k_y \frac{\partial \phi}{\partial y} \ell_y = \bar{q}_n \text{ on } B_2$$

$$\phi = y \text{ on } B_3 \qquad\qquad (5.38)$$

$$k_x \frac{\partial \phi}{\partial x} \ell_x + k_y \frac{\partial \phi}{\partial y} \ell_y = 0 \text{ on } B_3$$

$$\phi = y \text{ on } B_4$$

$$k_x \frac{\partial \phi}{\partial x} \ell_x + k_y \frac{\partial \phi}{\partial y} \ell_y \le 0 \text{ on } B_4$$

The terms k and n in Equation 5.1 can be expressed as in Equations 5.36 and 5.37, and the corresponding variational function U for the 3-D domain can be expressed as

$$U = \frac{1}{2} \int \left[(k_s - f_1) \left\{ \left(\frac{\partial \phi}{\partial x}\right)^2 + \left(\frac{\partial \phi}{\partial y}\right)^2 + \left(\frac{\partial \phi}{\partial z}\right)^2 \right\} - 2\left[\bar{Q} - (n_s - f_2)\frac{\partial \phi}{\partial t} \right] \phi \right\} dV$$

$$- \int_{B_2} \bar{q}_n \, \phi \, dB \qquad\qquad (5.39)$$

where $\bar{q}_n$ is the specified intensity of flow and V denotes the volume of the element.

5.5.1.1 Finite Element Method

In the FE method, the nodal head, ϕ, over an element, quadrilateral or "brick," can be expressed as

$$\phi = [N] \{q\} \qquad\qquad (5.40a)$$

where $[N]$ is the matrix of interpolation functions for a 2-D quadrilateral with bilinear variation of ϕ; $[N]$ will be a 1×4 row vector consisting of interpolation or shape or basis functions as

$$N_i = \frac{1}{4}(1 + ss_i)(1 + tt_i), \quad i = 1,2,3,4 \qquad\qquad (5.40b)$$

where s and t are local coordinates (Figure 5.12). For the 3-D element with eight nodes (Figure 5.29) $[N]$ will be a 1×8 row vector.

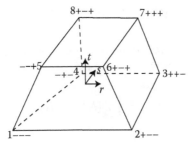

FIGURE 5.29 3-D eight-noded element.

The gradient, $\{g\}$, relation (for 2-D element) can be expressed as

$$\{g\} = \begin{Bmatrix} \dfrac{\partial \phi}{\partial x} \\[2mm] \dfrac{\partial \phi}{\partial y} \end{Bmatrix} = [B]\{q\} \tag{5.41}$$

where $[B]$ is the gradient-nodal head transformation matrix.

Substitution of $\{\phi\}$, Equations 5.40, and 5.41, and taking a variation of U with respect to $\{q\}$ and equating it to zero leads to the following element equations [20,21]:

$$\int_V ([B]^T [R][B] - [B]^T [f_1][B]) \{q\}\, dV + [p_s]\{\dot{q}\} = \int_V [N]^T \{\bar{Q}\} dV + \int_{B_2} [N]^T \bar{q}_n dB$$

$$- \int_V [p_{us}]\{\dot{q}\}\, dV \tag{5.42}$$

Equation 5.42 can be written in the matrix form as

$$[k_s]\{q\} + [p_s]\{\dot{q}\} = \{Q\} + [k_{us}]\{q\} + [p_{us}]\{\dot{q}\} = \{Q\} + \{Q_r\} \tag{5.43}$$

where

$$[k_s] = \int_V [B^T][R][B]\, dV \tag{5.44a}$$

$$[k_{us}] = \int_V [B]^T [f_1][B] dV \tag{5.44b}$$

$$\{Q\} = \int_V [N]^T \{\bar{Q}\} dV + \int_{B_2} [N]^T \{\bar{q}_n\} dB \tag{5.44c}$$

$$\{Q_r\} = [k_{us}]\{q\} + [p_{us}]\{\dot{q}\} \tag{5.44d}$$

Here, $\{Q_r\}$ is called the *residual or correction* flow vector. The terms related to the variation of n (Equation 5.37) are given by

$$[p_s] = \int_V n_s [N]^T [N] dV \tag{5.44e}$$

as the porosity matrix at saturation, and

$$[p_{us}] = \int_V f_2 [N]^T [N] dV \tag{5.44f}$$

as the unsaturated or residual porosity matrix, and $\{\dot{q}\}$ is the vector of the time-dependent nodal fluid heads. The permeability matrix and function f_1 in Equation 5.44 are given by

$$[R] = \begin{bmatrix} k_x & 0 & 0 \\ 0 & k_y & 0 \\ 0 & 0 & k_z \end{bmatrix} \tag{5.45a}$$

$$[f_1] = \begin{bmatrix} f_{1x} & 0 & 0 \\ 0 & f_{1y} & 0 \\ 0 & 0 & f_{1z} \end{bmatrix} \tag{5.45b}$$

5.5.1.2 Time Integration

Using the simple Euler (backward) scheme, we can derive the FE equations at time $t + \Delta t$ (Figure 5.19):

$$[\bar{k}]\{q\}_{t+\Delta t}^i = \{Q\}_{t+\Delta t} + \{Q_r\}^{i-1} \tag{5.46}$$

where

$$[\bar{k}] = [k_s] + \frac{1}{\Delta t}[p_s] \tag{5.47a}$$

and

$$\{Q_r\}^{i-1} = [k_{us}]\{q\}_t^{i-1} + \frac{1}{\Delta t}[p_s]\{q\}_t \tag{5.47b}$$

where $i = 1, 2, 3, \ldots$ denote the iterations for a given $t + \Delta t$, and Δt is the time step. The term $[p_{us}]\{\dot{q}\}$ is dropped assuming that it is relatively small.

5.5.1.3 Assemblage Global Equations

The element Equation 5.46 can be used to generate equations for all elements. They are assembled by ensuring that the nodal heads are compatible at common nodes of the neighboring elements. The assemblage or global equations are expressed as

$$[K_s]\{r\} = \{R\} + \{R_r\} \tag{5.48}$$

where $[K_s]$ is the global material property matrix, $\{r\}$ is the global nodal head vector, $\{R\}$ is the global vector of applied forcing functions at nodes, and $\{R_r\}$ is the global residual flow or correction vector.

Equation 5.48 is solved for nodal heads at $(t + \Delta t)$ by introducing the initial and boundary conditions. In the case of transient problem, the initial condition is introduced at time $t = 0$ as follows:

$$\phi(x,y,z,0) = \bar{\phi}(x,y,z) \tag{5.49}$$

where $\bar{\phi}$ is the applied nodal heads at $t = 0$.

For the first iteration $(i = 1)$, the vector $\{Q_r\}$ is assumed to be zero, and the solution of Equation 5.48, after the introduction of the initial and boundary conditions, provides nodal head values for the entire domain. Note that the properties at saturation are used for all iterations. In the second iteration, the vector $\{Q_r\}$ is computed by using Equation 5.47b. The following convergence criterion can be used to terminate iterations at time step $t + \Delta t$:

$$\frac{\left|\left(\sum_{j=1}^{M}\phi_m\right)^i - \left(\sum_{j=1}^{M}\phi_m\right)^{i-1}\right|}{\left|\left(\sum_{j=1}^{M}\phi_m\right)^{i-1}\right|} \leq \varepsilon \tag{5.50}$$

where M is the number of nodes on the FS, and ε is a small nonnegative number; a value of $\varepsilon = 0.005$ can be used.

Equation 5.48 can be easily specialized for steady FS seepage. The following procedure for the location of the FS is applicable for both steady and transient FS seepage.

5.5.1.4 Residual Flow Procedure

In the RFP, we compute nodal heads by solving Equation 5.48, which is considered to apply for the whole domain, Ω, containing both saturated and unsaturated zones. Hence, we need to seek a correction for the nodal heads proportional to the difference in saturated and unsaturated properties, which is usually done by using saturated and unsaturated permeabilities. Such corrected nodal heads would contain the FS, which satisfies the boundary conditions (Equations 5.4c and 5.4d).

The iterative numerical algorithm for the solution for steady seepage using equations at the element level is described below:

Iteration	Solution	Comments
$i = 0$	$[k_s]\{q\}^0 = \{Q\}$	At $i = 0, \{Q_r\} = 0$
$i = 1$	$[k_s]\{q\}^1 = \{Q\} + \{Q_r\}^0$	
$i = 2$	$[k_s]\{q\}^2 = \{Q\} + \{Q_r\}^1$	

$$(5.51)$$

$$\cdot \qquad \cdot$$
$$\cdot \qquad \cdot$$
$$\cdot \qquad \cdot$$

$$i = n - 1 \quad [k_s]\{q\}^{n-1} = \{Q\} + \{Q_r\}^{n-2}$$
$$i = n \qquad [k_s]\{q\}^n = \{Q\} + \{Q_r\}^{n-1}$$

where $\{Q_r\}$ for steady FS seepage is $[k_{us}] \{q\}$, and i denotes a step in the iterative procedure. In the above procedure, we solve the equations by assuming that the entire domain is saturated, that is, using only $[k_s]$. Since a part of the domain, above the FS, is unsaturated or partially saturated, this assumption indeed introduces an error (residual), which should be corrected. By using the computed heads at $i = 0$, we evaluate $\{Q_r\}^0$. Then, we can obtain the solution for nodal heads at $i = 1$. The procedure is continued till the following criterion is satisfied:

$$\frac{\left|\{q\}^i - \{q\}^{i-1}\right|}{\left|q^{i-1}\right|} \leq \varepsilon \tag{5.52}$$

where ε is a nonnegative small number. In the above procedure, the computation of $\{Q_r\} = [k_{us}]\{q\}$ requires the evaluation of $[k_{us}]$, that is, the values of unsaturated permeability, $k_s - f_1(p)$.

By using the computed heads, we find the approximate location of FS point on a mesh line such as $a-a$ (Figure 5.7) by the examination of successive nodes, starting from the bottom. Then, the location of the FS point is achieved by finding two consecutive nodes such that the first node (1) has the head greater than its elevation head, and the second node (2) has the head smaller than its elevation head (Figure 5.7). If the computed head is greater than the elevation head, the nodes have a positive pressure, that is, $\phi - y$, and their location is below the FS. When one of the nodes has a positive pressure and the subsequent node has a negative pressure (i.e., it is in the unsaturated zone), we find the approximate location of the point along that nodal line for the 2-D seepage (Figure 5.7). Often, linear interpolation is used to find the location of the FS node on a nodal line (Figure 5.30). The FS is obtained by joining such approximate points along the nodal lines.

Also, the elements with nodes that have negative pressures are used to compute $\{Q_r\}$, the residual or correction load vector (Equation 5.44d). The unsaturated permeability in an element (e) is obtained from relations in Equation 5.36a, corresponding to the pressure p^e

$$p^e = \phi_i^e - y_i^e \tag{5.53}$$

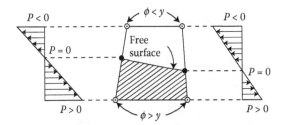

FIGURE 5.30   Interpolation for points on free surface.

where e is any element in the unsaturated zone. As indicated in the iterative solution, the residual vector $\{Q_r\}$ is modified progressively until convergence, which denotes a converged or equilibrated FS.

According to Equation 5.46, for the transient seepage problem

$$[\bar{k}]\{q\}^i_{t+\Delta t} = \{Q\}_{t+\Delta t} + \{Q_r\}^{i-1} \tag{5.46}$$

which has the form similar to Equation 5.51, for the steady seepage problem. Hence, the procedure for steady FS can also be used for transient FS problems, by assuming $\{Q_r\} = 0$ for the first iteration. In other words, the same procedure, as for the steady FS, can be used to locate the FS at every time step $t + \Delta t$.

The procedure for the 3-D seepage for the location of FS is depicted in Figure 5.31, in which Figure 5.31a shows a typical element. Figure 5.31b shows the main possibilities for the FS to intersect an element. The 12 nodal lines in an element along which we seek the points of zero pressure are shown in Figure 5.31a. The latter are obtained by identifying two adjacent nodes between which the pressure head $(p/\gamma = \phi - y)$ changes from positive to negative. This is achieved by a linear interpolation shown in Figure 5.31c.

5.5.1.5   Surface of Seepage

The surface of seepage occurs often along the downstream side between the exit point of the FS and the point showing the downstream head ϕ_d. It can also occur on the upstream face during the drawdown. It has been reported that appropriate prediction of the location of the surface of seepage can be important for a reliable and convergent solution [46,52]. Among a number of ways, the artificial flow-deflecting zone [53] is used where the seepage surface occurs as in Figure 5.32. The deflecting zone consists of a layer or strip of finite thickness of high permeability (Figure 5.32) in which the head, ϕ, along A–B is prescribed as zero. Then, the flow lines are deflected along the surface.

5.5.1.6   Comments

The invariant (IM) method with the FE-RFP provides improved solutions compared to those by the VM method. Also, the IM method can handle factors such as

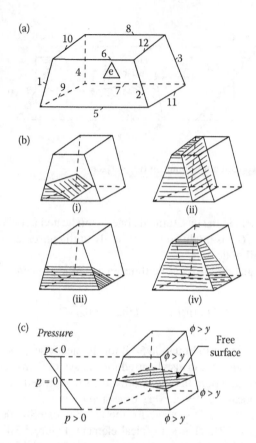

FIGURE 5.31 Free surface points for 3-D seepage. (a) Typical element and nodal lines; (b) possible intersection of free surface in partially saturated elements; and (c) location of the free surface in partially saturated element.

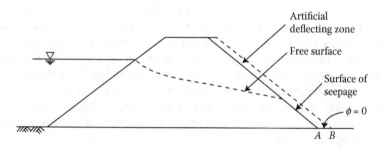

FIGURE 5.32 Model for surface of seepage.

nonhomogeneities (layered system), anisotropy, and arbitrary shapes that influence the FE solution. Some of these factors are identified in the later applications.

5.6 APPLICATIONS: INVARIANT MESH USING RFP

5.6.1 Example 5.7: Steady Free Surface in Zoned Dam

Figure 5.33 shows a zoned dam composed of materials with different coefficients of permeability [54]. The problem was solved by using STRESEEP-2DFE [27]. A drainage blanket with very high permeability is provided near the toe of the dam. The dam is subjected to the upstream head of 50 ft (15.25 m) and no downstream head. In using the RFP, the FE mesh was provided over the entire zone. The computed steady FS is shown in the figure labeled as "Numerical." The FS was also computed using the graphical solution procedure [3]. The numerical prediction compares very well with the graphical procedure.

5.6.2 Example 5.8: Transient Seepage in River Banks

The stability of river banks, dams, and slopes is affected significantly by transient seepage caused by rise, steady state, and drawdown in the water levels. The transient FS flow was simulated in the laboratory using the parallel plate Hele–Shaw model and numerical predictions were obtained by using the RFP.

Figure 5.34a shows the Hele–Shaw parallel plate or viscous flow model (VFM) [23,55], designed and constructed for simulating the seepage in the banks of the Mississippi river [8,54]. The Hele–Shaw model was about 335 cm long with an average gap between the parallel plates of about 0.20 cm. The coefficient of permeability, k_s, for the model in which silicon fluid was used was found to be 0.32 cm/s, and the value of specific storage, $n(S)$, was found to be 0.10. The ratio of $(k/k_s)_f$ was assumed to be 0.1 with the slope λ to be 0.09 (Figure 5.28b).

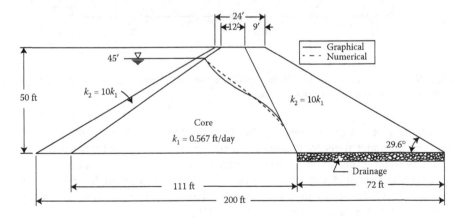

FIGURE 5.33   Steady free surface in zoned dam (1 ft = 0.3048 m).

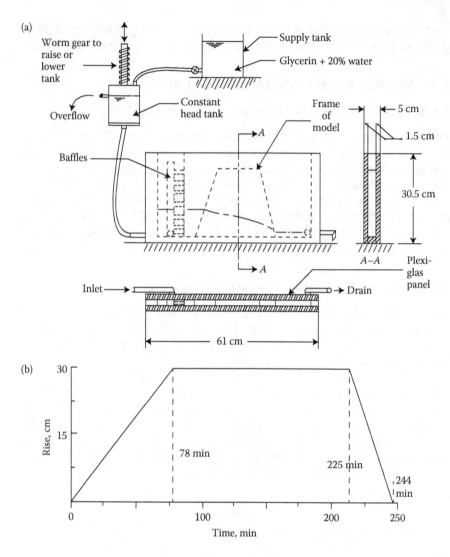

FIGURE 5.34 Viscous flow model and head variation. (a) Schematic of viscous flow (Hele–Shaw) model; (b) typical variation of head.

Parallel plates with different upstream slope angles were used for monitoring seepage under transient conditions involving rise, steady, and downstream in upstream water levels. In the example considered here, the angle of inclination was 45° to the horizontal. A typical history of the variation of upstream heads is shown in Figure 5.34b. The locations of the phreatic or FSs were recorded photographically during the transient head variations (Figure 5.34) [8,11,56].

Figure 5.35 shows the FE mesh for the entire domain of the model. The predictions from the FE-RFP procedure were compared with measurements of FSs at different time levels, 30, 60, 80, 225, 240, and 244 minutes during rise, steady state,

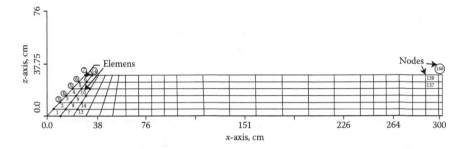

FIGURE 5.35 Finite element mesh for viscous flow model (VFM).

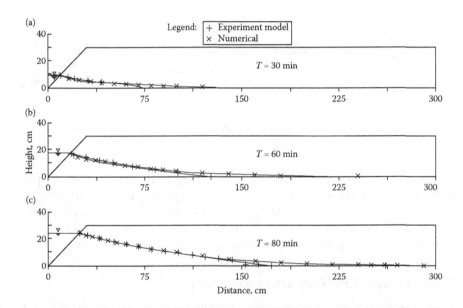

FIGURE 5.36 Comparisons of predictions with measurements during rise in VFM. (a) $T =$ 30 min; (b) $T = 60$ min; and (c) $T = 80$ min.

and drawdown (Figures 5.36 and 5.37). The correlation between predictions and measurements is considered very good. The RFP used here is found to provide as good as or improved predictions as compared to those from the FD ADEP [8].

5.6.3 EXAMPLE 5.9: COMPARISONS BETWEEN RFP AND VI METHODS

The predictions between the RFP [20,21,57] and VI [43,44] methods have been compared for a number of problems. One typical example for seepage through a homogeneous 3-D model dam (Figure 5.38a) is presented here [44]. The lower face of the dam shows the inlet size, which has a constant height of 10 m. The drainage side has a constant height of 2 m. The FE mesh (Figure 5.38b) consists of 160 elements and 280 nodes.

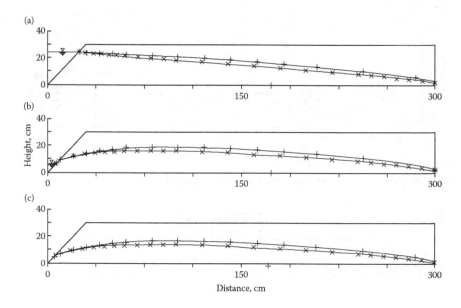

FIGURE 5.37 Comparisons of predictions with measurements during steady state and drawdown in VFM. (a) $T = 225$ min; (b) $T = 240$ min; and (c) $T = 244$ min.

The steady FS analyses were performed using the RFP [20,57] and the VI [44] methods. Figure 5.38b also shows the steady FSs obtained by using the RFP. Figure 5.39 shows the FS results obtained by using the VI method [44]. The two predictions are not shown on a single plot because of the difficulty in plotting from the published paper and the potential loss of clarity. However, both results correlate very well.

5.6.4 EXAMPLE 5.10: THREE-DIMENSIONAL SEEPAGE

To validate the RFP, a 3-D laboratory model was designed and constructed [20,21,57]. The details of the (schematic) model are shown in Figure 5.40. The model contains two components: (1) outside box made of Plexiglas panels, and (2) wire meshes that were placed inside the panels. The wire meshes were used to simulate the sloping sides of the model dam sections and also provided barriers between different sizes of glass beads that simulated nonhomogeneous zones. Glass beads of sizes 1.0 and 3.0 mm were used to simulate the granular soil. The glass beads were coated with silicon by using a commercial spray to reduce the capillary effect. The glass beads were packed in the model dam at a given density (see below).

The permeability of the glass beads in the model were determined by using a special rectangular (2-D) plexiglas model. The specific storage S_s was calculated using the available equations and the compressibility of glass beads. The details for both are given in Refs. [20,21,57]. The values of the permeability coefficients for 1.0 and 3.0 mm glass beads were found to be about 0.04 and 0.125 cm/s at densities of 1.92 g/cm² and 1.46 g/cm², respectively, and the specific storages of about 0.0008 and 0.0005 cm⁻¹, respectively.

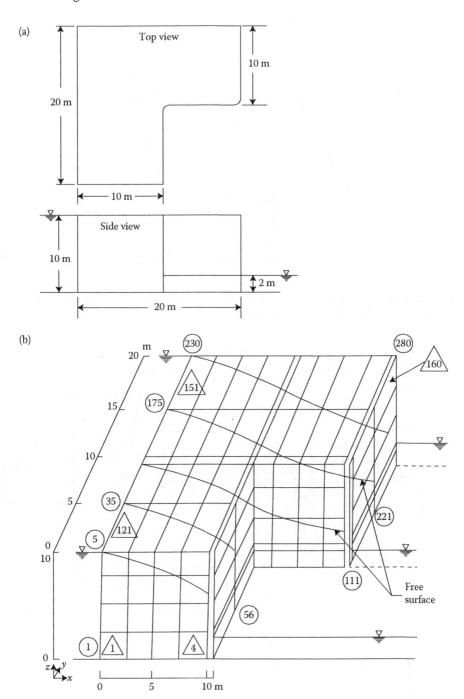

FIGURE 5.38 Model and computed free surface by RFP. (a) 3-D model. (From Caffrey, J. and Bruch, J.C., *Advances in Water Resources*, 12, 1979, 167–176. With permission.) (b) Mesh and free surfaces.

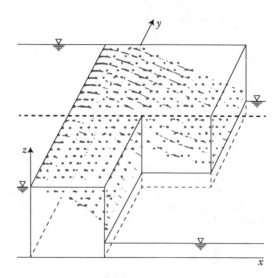

FIGURE 5.39   Free surfaces by VI. (From Caffrey, J. and Bruch, J.C., *Advances in Water Resources*, 12, 1979, 167–176. With permission.)

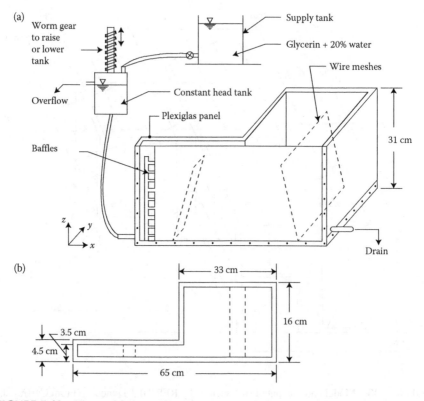

FIGURE 5.40   Views of 3-D nonhomogeneous material model. (a) 3-D schematic; (b) top view.

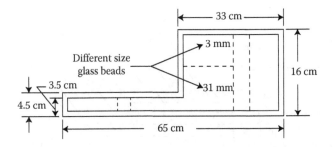

FIGURE 5.41   Plan of model with nonhomogeneity.

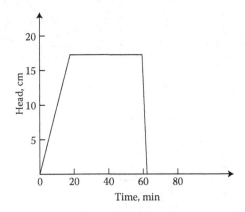

FIGURE 5.42   Variation of head with time.

Tests were performed with various compositions of glass beads, for example, homogeneous and nonhomogeneous. Results for one typical model dam with nonhomogeneous composition are presented here. Figure 5.41 shows the plan view of the nonhomogeneous dam, and Figure 5.42 shows the variation of fluid head with time. Figure 5.43 shows the FE mesh for various zones of the 3-D model.

Figures 5.43a through 5.43d show the comparisons between FE-RFP predictions and test data for the FS on various sections at different times during rise. Figures 5.44a through 5.44d show the comparisons during the drawdown phase. The computations were obtained using the Euler scheme with $\Delta t = 0.50$ min. The comparisons between predictions and test data show very good correlations.

5.6.5   EXAMPLE 5.11: COMBINED STRESS, SEEPAGE, AND STABILITY ANALYSIS

Geotechnical structures are often subjected simultaneously to both loading and fluid flow through the skeleton of the geologic media. Such a problem requires coupled analysis in which the effects of deformation and seepage are considered together; this coupled analysis is presented in Chapter 7. Here, we consider an approximate way to include coupling by superimposing the effects of stress and seepage. Figure 5.45 shows a schematic of the effects of (a) deformation and stress and (b) seepage.

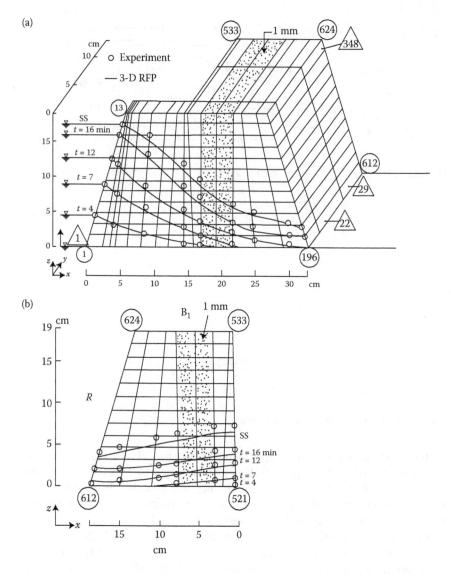

FIGURE 5.43 Comparisons of predictions and measurements during rise and steady state for nonhomogeneous dam (F = Front, B = Back, S = Side). (a) At front section, F; (b) at back section B_1; (c) at back section B_2; and (d) at side section, S. Here SS denotes steady state.

In the FE method for stress analysis, we divide the entire domain into a FE mesh. To add the effect of seepage, it would be useful and desirable to use the same mesh; the RFP is ideally suited for such analysis (Figures 5.45a and 5.45b).

For analysis and design, we use the results of the above combination for stability of slopes, dams, riverbanks, and so on. Such analyses for certain practical (field) problems were presented in Refs. [18,19]. In such analyses, the procedures described before have been used. For stress analysis, the FE equations are expressed as

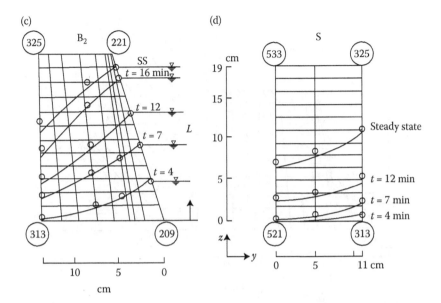

FIGURE 5.43 (continued) Comparisons of predictions and measurements during rise and steady state for nonhomogeneous Dam (F = Front, B = Back, S = Side). (a) At front section, F; (b) at back section B_1; (c) at back section B_2; and (d) at side section, S. Here SS denotes steady state.

$$[k]\{q\} = \{Q\} + \{F_s\} = \{F\} \tag{5.54}$$

where $[k]$ is the element "stiffness" matrix, $\{q\}$ is the vector of nodal displacement, $\{Q\}$ is the vector of external forces, $\{F_s\}$ is the vector of seepage forces, and $\{F\}$ is the vector of total forces. The procedure for deriving the seepage forces is given below.

First, the following equation is used to compute the effect of the body (or weight) forces:

$$[k_o]\{q_o\} = \{Q_o\} \tag{5.55}$$

where o denotes the initial condition. The initial stress vector $\{\sigma\}_o$ is computed from displacements, $\{q_o\}$, by using the value of the coefficient of earth pressure to define horizontal stresses. The permeability coefficient equal to 0.0674 in/s (0.171 cm/s) and the porosity equal to 0.886 were adopted [19,58].

Now, the FE-RFP seepage analysis (steady or transient) is performed to obtain values of fluid heads at the nodes. The force vector $\{F_s\}$ due to the fluid head is computed on the basis of computed nodal heads. At each time step, we use the following equation to compute the changes in displacements $\{\Delta q\}^i$ due to the changes in seepage forces, $\{\Delta Q_s\}^i$:

$$[k]^i \{\Delta q\}^i = \{F\}^i - \{F\}^{i-1} = \{\Delta Q_s\}^i \tag{5.56}$$

where i denotes iteration and $\{F\}^i$ is the load vector due to both body and seepage forces.

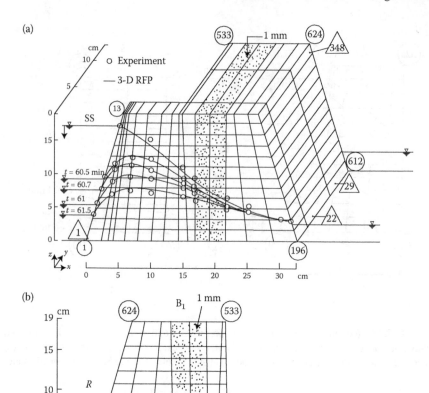

FIGURE 5.44 Comparisons of predictions and measurements during drawdown. (a) At front section, F; (b) at back section B_1; (c) at back section B_2; and (d) at side section, S.

The vector $\{\Delta Q_s\}$ in Equation 5.56 is evaluated as follows:

Consider a soil element at two time levels t_1 and t_2 (Figure 5.46). Then, the change in seepage force vector between two time levels is given by

$$\{\Delta Q_s\} = \{Q_s\}_2 - \{Q_s\}_1 \tag{5.57a}$$

where

$$\{Q_s\}_1 = \left\{ \begin{array}{c} \gamma_d V_{d1} + (\gamma_s - \gamma_w) V_{s1} + Q_{y1} \\ F_{x1} \end{array} \right\} \tag{5.57b}$$

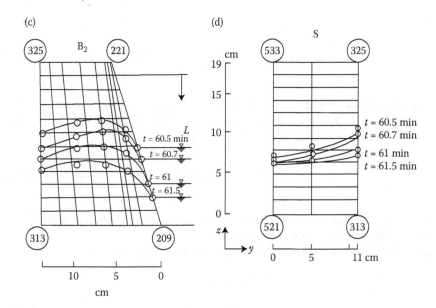

FIGURE 5.44 (continued) Comparisons of predictions and measurements during drawdown. (a) At front section, F; (b) at back section B_1; (c) at back section B_2; and (d) at side section, S.

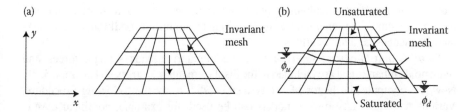

FIGURE 5.45 Superposition of effects of external loads and seepage. (a) Stress analysis: external loads; (b) seepage analysis: invariant mesh.

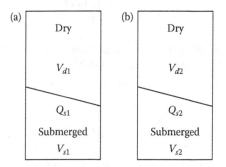

FIGURE 5.46 Soil element at two time levels, t_1; and t_2. (a) At t_1; (b) at t_2.

and

$$\{Q_s\}_2 = \left\{ \begin{matrix} \gamma_d V_{d2} + (\gamma_s - \gamma_w) V_{s2} + Q_{y2} \\ F_{x2} \end{matrix} \right\} \tag{5.57c}$$

where γ_d is the unit weight of dry soil, γ_s is the unit weight of saturated soil, V_d is the volume of dry soil, V_s is the volume of saturated soil, and Q_x and Q_y are components of the seepage force vector $\{Q_s\}$, which is computed as

$$\{F_s\} = \int_V [B]^T \{p\} \, dV \tag{5.58}$$

where $[B]$ is the transformation matrix and $\{p\}$ is the vector of fluid pressure heads computed from the FE analysis as $p = \varphi - y$.

The current values of the displacement and stresses are evaluated as

$$\{q\}^i = \{q\}^{i-1} + \{\Delta q\}^i \tag{5.59a}$$

$$\{\sigma\}^i = \{\sigma\}^{i-1} + \{\Delta\sigma\}^i \tag{5.59b}$$

where i denotes an iteration.

The construction of geotechnical structures often involves installing various layers or zones in a sequence. The STRESEEP-2DFE code [27] with the RFP allows the inclusion of the effect of embankment sequences on the stress-deformation analysis. Figure 5.47 shows the schematic of such sequences: (a) initial condition, (b) embankment sequences, and (c) locations of FS.

Stability and Factor of Safety: The effects of external loads, seepage forces, and sequential embankment are included in the displacements and stresses (Equation 5.59). Now, we compute the factor of safety against sliding by the following procedure. Various material (constitutive) models can be used, for example, nonlinear elastic, hyperbolic [5,59], and elastoplastic Drucker–Prager model [60]. The details of these models are given in Appendix 1.

For the hyperbolic model, the factor of safety over an element (e) is

$$(FS)_e = \frac{\bar{c} + \bar{\sigma}_n \tan \bar{\phi}}{\tau} \tag{5.60}$$

where $\bar{c}$ is the cohesive strength, $\bar{\sigma}_n$ is the normal stress, and $\bar{\phi}$ is the angle of friction; the overbar denotes an effective term, and τ is the induced shear stress. For the elastic–plastic Drucker–Prager model, the factor of safety for an element is given by

$$(FS)_e = \frac{k - \alpha J_1}{\sqrt{J_{2D}}} \tag{5.61}$$

where α and k parameters are related to cohesion c and the angle of friction ϕ, J_1 is the first invariant of the stress tensor, and J_{2D} is the second invariant of the deviatoric

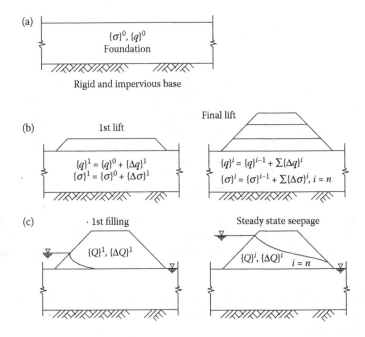

FIGURE 5.47 Schematic of sequential embankment contruction and seepage. (a) Initial condition; (b) embankment sequences; and (c) typical locations of free surface.

stress tensor. The overall factor of safety is expressed as the ratio of resisting shear strength to the total shear stress along a given slip surface as follows:

$$FS = \frac{\sum_{i=1}^{n}(FS)_e \, a_e}{A} \qquad (5.62)$$

where a_e is the part of slip surface intersecting element e, A is the total area of the slip surface, and n is the number of elements on the slip surface.

The stress-seepage code STRESEEP-2DFE [27] was used to solve a number of problems [19]. Here, we present one example related to the Oroville dam [61]. The dam consists of a shell, transition zone, and core (Figure 5.48); the FE mesh used is shown in Figure 5.49. A simplified hydrograph showing the variation of the height of the water in the reservoir is shown in Figure 5.50 in the upper right.

The rockfill dam has a height of 770 ft (235 m) and a base dimension of 3600 ft (1098 m). The dam was constructed in October 1967, and the filling of the reservoir started in November 1967, reaching a height of 746 ft (227.5 m) in June 1969.

The material parameters for the hyperbolic model used in the analysis are shown in Table 5.5 and are adopted from Refs. [19,61]; the values of coefficient of permeability and specific storage are added for the seepage analysis. Brief details of the hyperbolic model are given below (it is also described in Appendix 1), and the parameters are shown in Table 5.5. In the RFP, the FE mesh is constructed for the entire dam (Figure 5.49). Figure 5.50 shows the computed FS at different levels

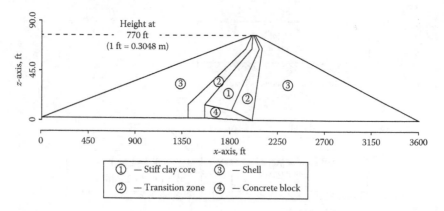

FIGURE 5.48   A section of Oroville dam (1 ft = 0.3048 m). (Adapted from Nobari, E.S. and Duncan, J.M., Effects of Reservoir Filling on Stresses and Movements in Earth and Rockfill Dams, Report No. S-72-2, U.S. Army Engineers Waterways Experiment Station, Vicksburg, MS, 1972.)

during the rise. By using the RFP, Nobari and Duncan [61] assumed that the FS surface was horizontal and existed only in the shell and the transition zone, which may not be realistic.

Equations for the hyperbolic model are given below [5,61]. Parameters are given in Table 5.5. Details are given in Appendix 1.

Tangent Young's modulus:

$$E_t = \left[1 - \frac{R_f(1 - \sin\phi)\,(\sigma_1 - \sigma_3)}{2c\cos\phi + 2\sigma_3\sin\phi}\right]^2 K'p_a\left(\frac{\sigma_3}{p_a}\right)^n$$

Tangent Poisson's ratio:

$$\nu_t = \frac{G - F\log(\sigma_3/p_a)}{(1 - A)^2}$$

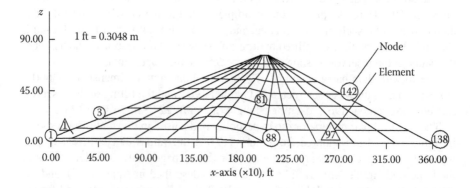

FIGURE 5.49   Finite element mesh of Oroville dam (1 ft = 0.3048 m).

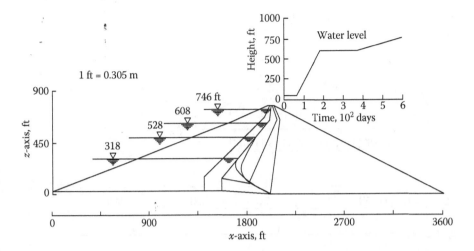

FIGURE 5.50 Computed locations of free surface during reservoir filling and variation of head (1 ft = 0.3048 m).

$$A = \frac{(\sigma_1 - \sigma_3)d}{K' p_a (\sigma_3/p_a)^n [1 - ((R_f (\sigma_1 - \sigma_3)(1 - \sin\phi))/(2c\cos\phi + 2\sigma_3 \sin\phi))]}$$

The stresses were first computed by using FE Equation 5.55 with only the gravity load. Then, the seepage effect was introduced by solving Equation 5.54.

TABLE 5.5
Soil Parameters for Analysis of Oroville Dam[a]

Parameter (1)	Shell Dry (2)	Wet (3)	Transition Dry (4)	Wet (5)	Core (6)	Concrete (7)
Cohesion, in psf, c	0	0	0	0	2620	432,000
Friction angle, in degrees, ϕ.	46.3	44.8	46.3	44.8	25.1	0
Modulus number, K'	2030	1690	1800	1500	345	145,600
Modulus exponent, n	0.34	0.30	0.34	0.30	0.76	0
Failure ratio, R_f	0.86	0.85	0.86	0.85	0.88	1
Poisson's ratio parameters:						
G	0.35	0.31	0.35	0.31	0.30	0.15
F	0.14	0.12	0.14	0.12	−0.05	0
d	10.10	9.40	10.1	9.4	3.83	0
Coefficient of permeability, in feet per day	1000		800.0		0.01	10^{-9}
Specific storage	0.0		0.0		0.0	0.0

[a] The parameters except the last two are adapted from Ref. [61] and occur in the above equations.
Note: D_r = relative density = 90%. 1 psf = 47.88 N/m², 1 ft/day = 0.3048 m/day.

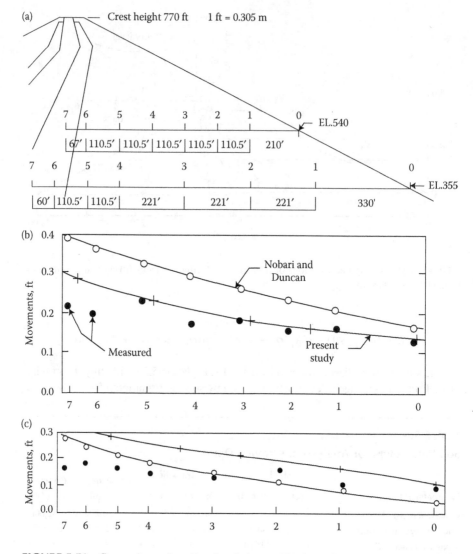

FIGURE 5.51 Comparison of predicted and observed horizontal movements at two elevations (1 ft = 0.348 m). (a) Location of measurement devices; (b) horizontal movements at EL.540; and (c) horizontal movements at EL.355.

Figure 5.51 shows comparisons of horizontal displacements at two different locations, at El. 355 and 540, using the stress-seepage procedure [19], those by Nobari and Duncan [61] and field measurements. The comparison of the displacements for points through a section in the core is shown in Figure 5.52. The present predictions show very good correlation with the field measurements [61]. Also, they are closer to the field data compared to those by Nobari and Duncan [61]; this may be due to the inclusion of the FS (Figure 5.50). The computed displacements in the core (Figure 5.52) appear satisfactory; however, the present predictions are different from those from Ref. [61].

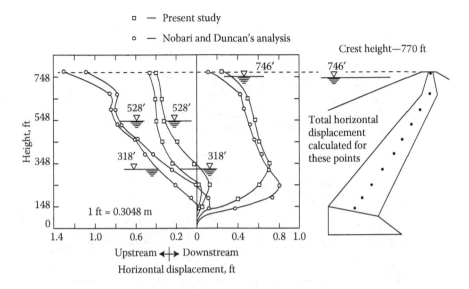

FIGURE 5.52   Comparison of predicted movements at section in core (1 ft = 0.3048 m).

5.6.6   EXAMPLE 5.12: FIELD ANALYSIS OF SEEPAGE IN RIVER BANKS

A comprehensive computer analysis was performed to predict experimental results using the viscous flow (Hele–Shaw) model, and field behavior of seepage at various locations on the Mississippi river [10,11]. Such seepage analyses are warranted because the stability of slopes is affected by the rise and drawdown due to the fluctuations in the river levels. A typical example involving field results at the section called Walnut Bend 6 is included below.

Figure 5.53a shows a cross section at Walnut Bend and boring log [11]. Figure 5.53b shows a history of measured water levels at the Walnut Bend during a part of the year 1965 [62]; it includes variation of heads in piezometers A and B, which were installed in the wells in the bank (Figure 5.53a) at El. 174 and 154, respectively. The river bank at the location consists of mostly silty fine sand (ML). The coefficient of permeability and porosity of the soil at Walnut Bend 6 were estimated as 10×10^{-4} cm/s to 20×10^{-4} cm/s (2.84–5.68 ft/day) and 0.40, respectively. These values were adopted from various investigations performed to obtain values of permeability and porosity [63,64].

Since drawdown can cause severe hydraulic gradients resulting in instability and failure in soil, we consider the drawdown in the river level from April 30 to May 30, 1965 (Figure 5.53b). The river level fell from about El. 187.5 to El. 167.5, that is, 20.0 ft (6.1 m) at an average rate of about 0.67 ft/day. Assuming that the section was porous across the river and that the flood stayed long enough for the FS to develop, the steady FS was adopted as the initial FS (Figure 5.54). The analysis was performed using the VM FE method. Thus, the section in Figure 5.53a was idealized as in Figure 5.54, and the FE mesh contained 48 elements and 63 nodes. On the basis of the observations of the model tests [11], the infinite river bank was represented by a finite region, which was extended to 400 ft (122 m) from the toe of the bank.

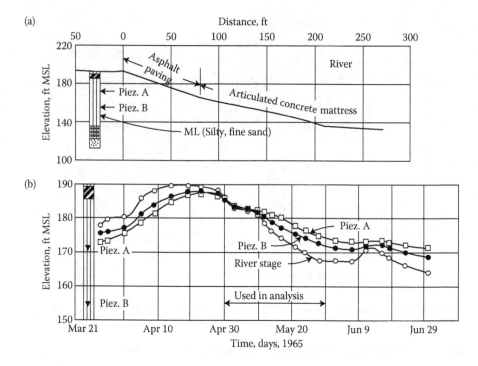

FIGURE 5.53 Cross-section and river stages at Walnut Bend 6. (a) Cross section at Walnut Bend 6 and boring log; (b) river stages and piezometer heads.

Hence, the distance, from the end of the drawdown, was about 13 times the fall during the drawdown of about H = 20 ft (6.10 m). The bottom boundary was placed at a distance of about 3.4 H measured from the final drawdown point. The end boundary was assumed to be impervious, on which the heads were allowed to change. A time interval of $\Delta t = 0.25$ day was used for the analysis.

Figure 5.55 shows locations of FSs at typical time levels of 20 and 30 days during the drawdown for values of $k = 10 \times 10^{-4}$ and 20×10^{-4} cm/s. The locations of the piezometers A and B at El. 174 and 154 are marked in Figure 5.54a.

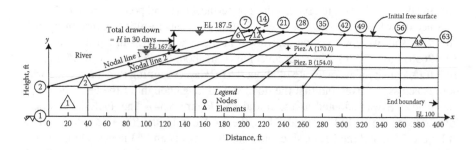

FIGURE 5.54 Finite element mesh for idealized section (1 ft = 0.3048 m).

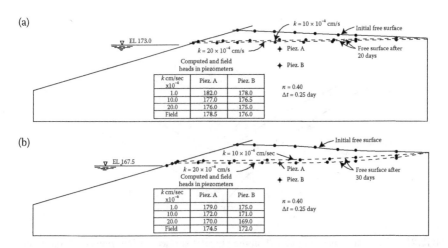

FIGURE 5.55 Comparison of computed and field data, Walnut Bend 6. (a) After 20 days; (b) after 30 days.

Figure 5.55 also shows the computed values of heads in the piezometers in comparison with the field observations; the results for $k = 1 \times 10^{-4}$ cm/s are also shown in tables. The computed values were averages of the heads at the nodes in the neighborhood of locations of the piezometers. The computed values of the heads show good agreement with the field data. In view of the precision that can be obtained in estimating k and n values and measured data for the heads in the field, the agreement is considered highly satisfactory. Similar analyses and agreements were obtained for other sections, for example, King's Point along the Mississippi river [11]. Hence, it can be concluded that the FE method can be used for reliable analyses for the stability of soil slopes such as river banks, dams, and embankments.

5.6.7 EXAMPLE 5.13: TRANSIENT THREE-DIMENSIONAL FLOW

We present here examples of 3-D FS seepage by using the FEM, France [31]. For predicting the time-dependent phreatic or FS, an iterative technique was adopted based on the assumption that the time-variant or transient flow problem can be approximated as a series of steady-state problems, each having a slightly varying shape satisfying the governing equation, and each separated by a small interval of time [31]. The change in shape of the flow domain is caused by the continuous movement of the phreatic surface, which is represented in a stepwise fashion. At the beginning of each time interval, the surface configuration and boundary conditions are assumed to be known. Nodal heads and seepage velocity components are calculated by using the minimization of energy functional (e.g., Equation 5.21).

The normal velocity (v_n) is determined using computed seepage velocities. Then, a typical point P on the FS is moved in the normal direction to a new location P′ at time $(t + \Delta t)$ (Figure 5.56), where the movement of the point is computed as

$$v_n \, \Delta t = (v_{nx} + v_{ny} + v_{nz})\Delta t \tag{5.63}$$

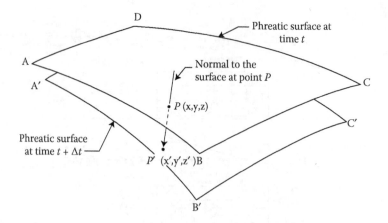

FIGURE 5.56 Movement of point P on the phreatic surface during time Δt. (From France, P.W., *Journal of Hydrology*, 21, 1974, 381–398. With permission.)

Here, Δt represents time increment, and v_{nx}, v_{ny}, and v_{nz} represent the normal components of the actual velocities v_x, v_y, and v_z, respectively. By repeating this process for all nodal points on the phreatic surface, a new configuration, A′, B′, C′, and D′ (Figure 5.56), and a new set of boundary conditions are obtained. The new surface is defined by fitting a polynomial through the temporary points (Figure 5.57);

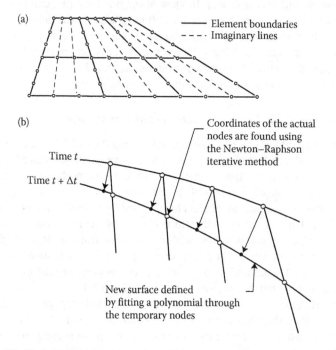

FIGURE 5.57 Defining the phreatic surface. (From France, P.W., *Journal of Hydrology*, 21, 1974, 381–398. With permission.)

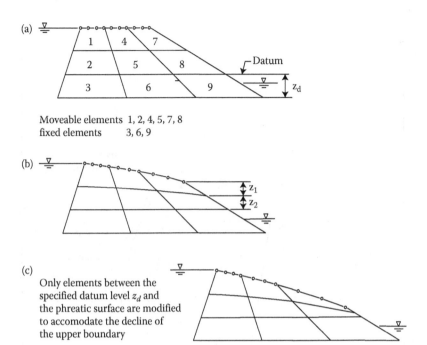

FIGURE 5.58 Modification of elements due to movement of phreatic surface: (a) $t = 0$; (b) $t = t_1$; and (c) $t = t_2$. (From France, P.W., *Journal of Hydrology*, 21, 1974, 381–398. With permission.)

the actual nodal coordinates can be found using the Newton–Raphson method, as reported by France [31]. For convenience and clarity, a 2-D situation is illustrated in Figure 5.57. The elements in the mesh are modified due to the movement of the phreatic surface (Figure 5.58); only the elements between the datum level and the phreatic surface are modified.

The above process can be repeated for the selected time intervals until the full range of profiles have been considered or until a steady-state condition is achieved. When the steady-state condition is achieved, the flow normal to the phreatic surface would become zero, and then the analysis is stopped. If not, the analysis is repeated for the next time step.

Numerical Example: The example considered here represents part of a curved embankment along which a river flows, initially at a constant depth of 100 ft (30.5 m) [31]. The steady-state phreatic surface pertaining to this constant water level is first evaluated. The water level of the river is then assumed to vary in depth along the embankment from 100 ft (30.5 m) at section B–B′ to 70 ft (21.3 m) at section A–A′ (Figure 5.59) and is allowed to fall at the rate of 5 ft/h (1.52 m/h) until it reaches 70 ft (21.3 m) at section B–B′ and 40 ft (12.2 m) at section A–A′. The flow properties, the coefficients of permeability, and the specific storage used in the analysis are shown in Figure 5.59.

Along face A′–B′, the initial phreatic surface was assumed to be 150 ft above the horizontal datum and it was assumed to remain unchanged as its variation due to fluctuations in the river would be negligible. The flow domain was represented by using six elements, and the datum was assumed to coincide with the impervious

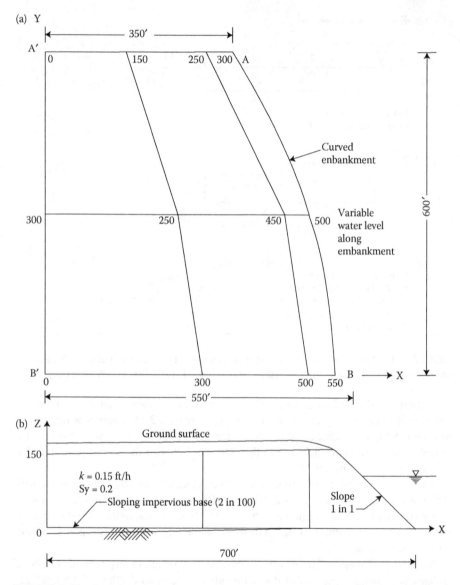

FIGURE 5.59 3-D flow problem. (From France, P.W., *Journal of Hydrology*, 21, 1974, 381–398. With permission.)

base [31]. For the constant water depth, the computed steady FSs for sections A–A′ and B–B′ are presented in Figures 5.60 and 5.61, respectively. These steady FSs were then used as the initial position for the transient problem. Twenty time increments, each having a duration of 5 h, were adopted, but the steady-state condition (due to drawdown) was achieved after 13 increments (65 h). The predicted transient phreatic surfaces for sections A–A′ and B–B′ are shown in Figures 5.62 and 5.63, respectively. The differences in curvature of the profile are clearly seen from these

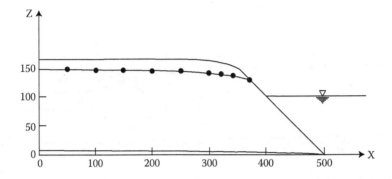

FIGURE 5.60 Steady state flow at section A–A′; downstream water level 100 ft. (From France, P.W., *Journal of Hydrology*, 21, 1974, 381–398. With permission.)

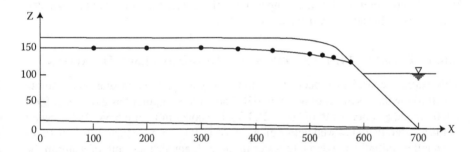

FIGURE 5.61 Steady state flow at section B–B′; downstream water level 100 ft. (From France, P.W., *Journal of Hydrology*, 21, 1974, 381–398. With permission.)

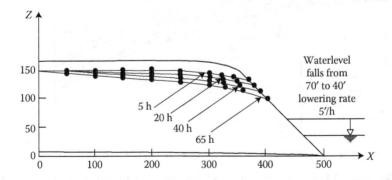

FIGURE 5.62 Unsteady flow at section A–A′ due to drawdown. (From France, P.W., *Journal of Hydrology*, 21, 1974, 381–398. With permission.)

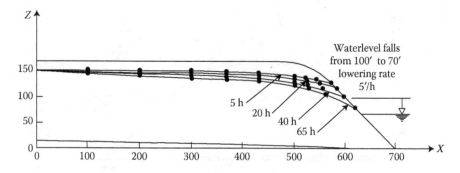

FIGURE 5.63 Unsteady flow at section B–B′ due to drawdown. (From France, P.W., *Journal of Hydrology*, 21, 1974, 381–398. With permission.)

figures. Such differences cannot be captured by a 2-D analysis. This example considers a number of conditions: an irregular flow region, a sloping downstream face, an inclined impervious bed, and a complex phreatic surface that could not be evaluated with a reasonable accuracy using a 2-D analysis.

5.6.8　Example 5.14: Three-Dimensional Flow under Rapid Drawdown

This example considers a special (axisymmetric) case of 3-D flow of groundwater to a well following a sudden drawdown [65]. The circular aquifer has a radius of 99 ft (30.0 m) and a water depth of 70 ft (21.3 m), as shown in Figure 5.64. At the center of the aquifer, there is a well with a diameter of 11 ft (3.35 m), which fully penetrates the aquifer. Initially, the depth of water in the well equals the depth of water in the aquifer, and then it is suddenly lowered to 35 ft (Figure 5.64). The 3-D FE technique, similar to the one used in the preceding example, is used here to determine the flow of groundwater to the well. For steady-state flow, Hall [66] developed an empirical

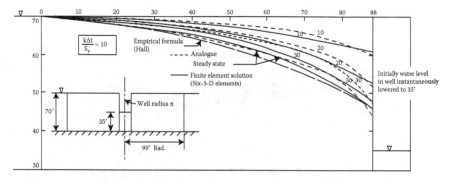

FIGURE 5.64 3-D unsteady flow to well following rapid drawdown. (From France, P.W. et al, *Journal of the Irrigation and Drainage Division, ASCE*, 97, 1971, 165–179. With permission.)

solution for the height of the phreatic surface, measured from the bottom of the aquifer, as given below:

$$h_s = h_w + \frac{(h_e - h_w)[1 - (h_w/h_e)^{2.4}]}{[1 + (1/50)\ln(r_e/r_w)](5/(h_e/h_w))}$$ (5.64)

where h_s is the seepage height, h_w the height of water in the well, r_w the radius of the well, r_e the radius of the aquifer, and h_e the height of water in the aquifer. According to Hall [66], the height (h) of the FS at a distance r (radius) from the center of the well (Figure 5.65) for steady-state flow can be determined from the following equation:

$$h = h_s + (h_e - h_s)\left[2.5\left(\frac{r - r_w}{r_e - r_w}\right) - 1.5\left(\frac{r - r_w}{r_e - r_w}\right)^{1.5}\right]$$ (5.65)

Both transient and steady-state flows, obtained from the 3-D FE analysis, are shown in Figure 5.64 along with the steady-state solutions obtained from Equation 5.64. A 30° sector of the flow domain was represented by six 3-D hexahedral isoparametric elements, each element consisting of 32 nodes, with four nodes along each curved side. As in the previous example, the transient flow was approximated as a series of steady-state problems each having a slightly varying shape satisfying the governing equation, and each separated by a small interval of time (Δt). The FS profiles shown in Figure 5.64 are evaluated at equal increments of $(k(\Delta t/S_y)) = 10$, where k is the permeability coefficient, Δt the time increment, and S_y is the volume of water drained/bulk volume of the medium [65]. The steady-state profile obtained from the FE analysis compares favorably with the empirical solution (Equation 5.65). Although six elements are used in the simulation, they seem to adequately capture the FS profiles.

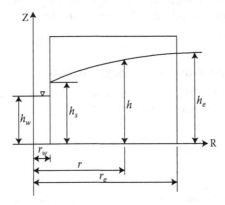

FIGURE 5.65 Schematic of flow toward well. (From France, P.W. et al, *Journal of the Irrigation and Drainage Division, ASCE*, 97, 1971, 165–179. With permission.)

5.6.9 Example 5.15: Saturated–Unsaturated Seepage

This example uses a 3-D saturated–unsaturated seepage theory to analyze the stability of a rockfill dam. The 3-D simulation results are compared with those from 2-D analyses. It was observed that seepage water flows faster and the hydraulic gradients are greater near the abutment in 3-D simulations than in 2-D simulations, meaning a 2-D analyses would underestimate the risk of seepage failure, particularly near the abutment [67].

As noted by Fredlund and Rahardjo [68], Darcy's law is considered valid for seepage in both saturated and unsaturated soils. The primary differences between saturated and unsaturated flows are (i) the coefficient of permeability is not a constant but a function of the degree of saturation or matric suction in case of unsaturated flow and (ii) the volumetric water content in unsaturated soils can vary with time. As shown by Chen and Zhang [67], the GDE for 3-D seepage in a saturated–unsaturated soil can be expressed in terms of the total head h and pore water pressure as follows:

$$\frac{\partial}{\partial x}\left(k_x \frac{\partial h}{\partial x}\right) + \frac{\partial}{\partial y}\left(k_y \frac{\partial h}{\partial y}\right) + \frac{\partial}{\partial z}\left(k_z \frac{\partial h}{\partial z}\right) = \gamma_w \frac{\partial \theta_w}{\partial \psi}\frac{\partial h}{\partial t} \tag{5.66}$$

where k_x, k_y, and k_z are the coefficients of permeability in the x-, y-, and z-directions, respectively, θ_w is the volumetric water content, which is related to the matric suction ψ by $(\partial \theta_w/\partial t) = (-\partial \theta_w/\partial \psi)(\partial u_w/\partial t)$ in which $\psi = u_a - u_w$, u_a being the pore air pressure and u_w being the pore water pressure. It is evident from Equation 5.66 that the soil–water characteristic curves (relationship between θ_w and ψ and permeability values in different directions k_x, k_y, and k_z) are required in the analysis of 3-D seepage in a saturated–unsaturated medium.

Chen and Zhang [67] used a FE program (SVFlux from SoilVision Systems) and a partial differential equation solver (FlexPDE) for the 3-D analysis of the Gouhou dam, using the saturated–unsaturated seepage concept (Equation 5.65). A summary of the input parameters (Table 5.6) and the results are given here, details can be found in Ref. [67].

The Gouhou dam was built in a steep canyon directly above a 10-m-thick gravel layer overlying bedrock (Figure 5.66). The crest was 265 m long and 7 m wide, with

TABLE 5.6
Index Properties Used in the 3-D Seepage Analysis

Material	Specific Gravity, G_s	Initial Water Content, w (%)	Porosity, n (%)	Saturated Permeability, k_{sat} ($\times 10^{-5}$ m/s)
Upper limit of rockfill	2.68	3.5	0.21	2.2
Average of rockfill	2.68	3.5	0.21	11.6
Lower limit of rockfill	2.68	3.5	0.21	231.0
Riverbed gravel	2.71	12.3	0.27	7.4

Source: From Chen, Q. and Zhang, L.M., *Canadian Geotechnical Journal*, 43, 2006, 449–461. With permission.

(a)

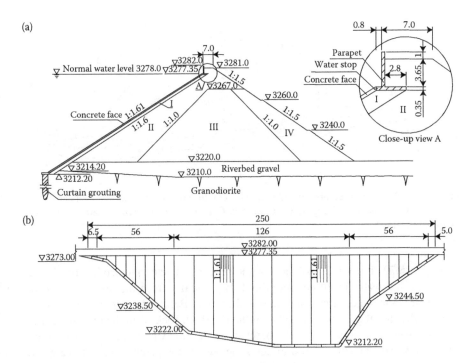

(b)

FIGURE 5.66 Gouhou dam: (a) maximum cross-section profile; (b) upstream elevation view. (From Chen, Q. and Zhang, L.M., *Canadian Geotechnical Journal*, 43, 2006, 449–461. With permission.)

an elevation of 3281 m, as shown. The rockfill dam was divided into four zones, Zone I served as the transition material supporting the concrete face, and zones II through IV constituted the main rockfill. According to forensic investigations, the reservoir water level rose continuously from 3261 m on July 14 to 3277 m on August 27, 1993 around noon time. The seepage water exited from the downstream slope at an elevation of 3260 m, triggering a seepage failure.

The simplified 3-D profile of the dam and boundary conditions used in the FE analysis are shown in Figure 5.67 [67]. Two different cases were considered in the simulation. In Case 1, the rockfill zones were considered uniform, with the concrete face below elevation 3260 m being impervious. The upper limit of the grain-size distribution curve in Figure 5.68 was considered for Zones I and II. For Case 2, an average grain-size distribution was considered for Zone I, while the lower limit gradation was used for Zone II. Zone III was considered to be riverbed gravel in all cases. Also, Case 2 included a 10-m-thick sandwich layer (Zone II) between elevations 3260 and 3270 m, which was ignored in Case 1. The materials for the sandwich layer were assumed to be of lower limit in the gradation curve (Figure 5.68). A summary of the material properties used is given in Table 5.6 [67], while the soil–water characteristic curves and the coefficient of permeability, as a function of matric suction, are shown in Figure 5.69. Only limited results for Case I are presented here. Additional results can be found in Ref. [67].

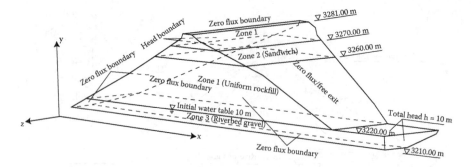

FIGURE 5.67 Simplified profile and boundary conditions for 3-D transient seepage analysis. (From Chen, Q. and Zhang, L.M., *Canadian Geotechnical Journal*, 43, 2006, 449–461. With permission.)

Figure 5.70 shows the evolution of the phreatic surface pictorially for Case I. It is seen that water infiltrates into the dam gradually from the upper part of the upstream slope, where the concrete was defective. Before the perched water table reaches the initial groundwater table in the riverbed, rockfill in the dam is divided into two zones: (a) zone within the perched water table where the pore water pressure is positive, and (b) the unsaturated zone outside the perched water table where pore water pressures are negative [67]. Pore water pressure contours in the cross section at z = 150 m are shown in Figure 5.71, the contours with zero pore water pressure representing the phreatic surfaces. It is seen that the pore water pressures decrease gradually from a maximum at elevation 3260 m on the upstream surface

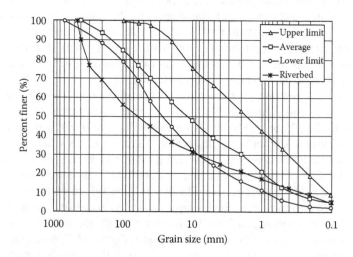

FIGURE 5.68 Average, lower and upper limits, and riverbed grain-size distribution curves. (From Chen, Q. and Zhang, L.M., *Canadian Geotechnical Journal*, 43, 2006, 449–461. With permission.)

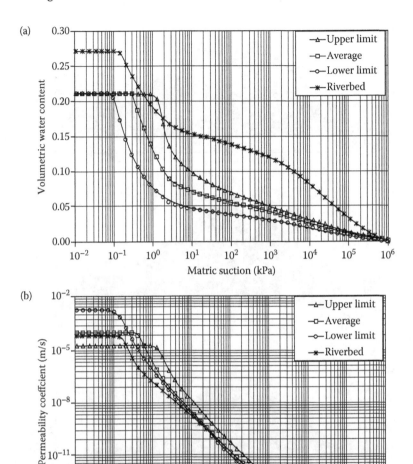

FIGURE 5.69   Soil-water characteristic curves used in the analysis. (a) Volumetric water content; (b) permeability coefficient. (From Chen, Q. and Zhang, L.M., *Canadian Geotechnical Journal*, 43, 2006, 449–461. With permission.)

to zero at the perched water table. As noted by Chen and Zhang [67], after 4.5 days, the bottom of the perched water table reaches the initial groundwater table in the riverbed, forming a seepage channel connecting the upper portion of the upstream slope to the riverbed. Comparatively, 2-D FE analyses by Chen et al. [69] show that the perched water table reaches the riverbed in 5 days, indicating that 3-D seepage occurs more rapidly than the 2-D seepage. Additional results on total head contours, velocity vectors, material anisotropy, and stratification of rockfill are given by Chen and Zhang [67].

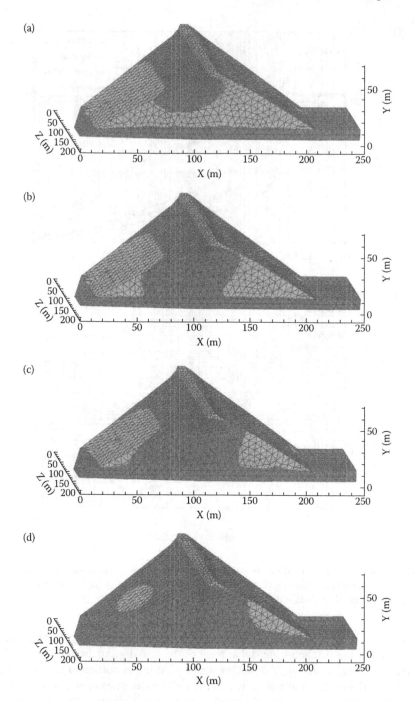

FIGURE 5.70   Evolution of phreatic surface for Case I: (a) $t = 2$ days; (b) $t = 4.5$ days; (c) $t = 6.5$ days; and (d) $t = 9$ days. (From Chen, Q. and Zhang, L.M., *Canadian Geotechnical Journal*, 43, 2006, 449–461. With permission.)

(a)

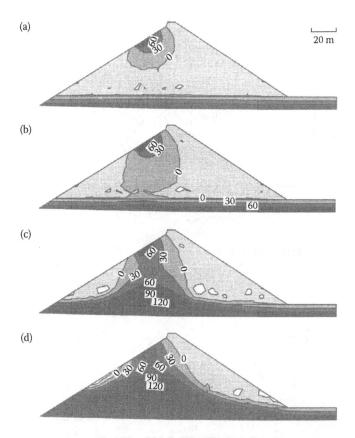

20 m

(b)

(c)

(d)

FIGURE 5.71 Pore water pressure contours (kPa) within the cross-section for Case I at z = 150 m. (a) $t = 2$ days; (b) $t = 4.5$ days; (c) $t = 6.5$ days; and (d) $t = 9$ days. (From Chen, Q. and Zhang, L.M., *Canadian Geotechnical Journal*, 43, 2006, 449–461. With permission.)

PROBLEMS

Problem 5.1

Figure P5.1 shows a sheet pile wall driven in the soil. Adopt a mesh layout for the zone extending to 25 m on both sides of point "a" on the ground surface. Apply a head of 12 m on the nodes along the left vertical and horizontal boundaries, and zero along the right vertical and horizontal boundaries. Assume the bottom boundary to be impervious. Adopt a FE mesh, and solve for fluid heads at nodes using computer code such as STRESEEP-2DFE and calculate the flow across section a–b.

Problem 5.2

Adopt a suitable mesh for the nonhomogeneous dam with two layers, and $k_1 = 0.50$ ft/year and $k_2 = 0.25$ ft/year (Figure P5.2). The upstream and downstream heads are 20 and 5 m, respectively. Solve for fluid heads and the steady FS. You may use suitable code, for example, STRESEEP-2DFE, and solve for nodal heads and the FS by using the IM and VM, and compare the results from the two.

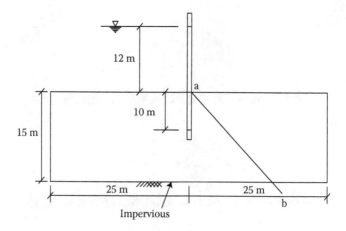

FIGURE P5.1   Sheet pile wall, confined seepage.

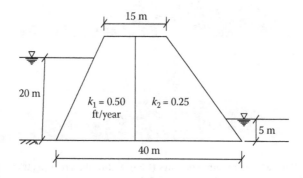

FIGURE P5.2   Free surface seepage in dam.

APPENDIX A

ONE-DIMENSIONAL UNCONFINED SEEPAGE

Certain conditions of seepage in a soil bank can be approximated by 1-D idealization. Flow through earth banks with vertical faces like in parallel drains or ditches, sheet-pile walls, and quay walls are examples of such approximation (Figure A.1). Since the approximation needs only 1-D elements, it can provide significant savings in time for formulation and computational efforts, compared to the 2-D and 3-D models. We present here a FE formulation for the 1-D seepage.

FINITE ELEMENT METHOD

Seepage through a porous material with the vertical faces ($\alpha = 90°$) is assumed to be 1-D, which is based on the Darcy's law and Dupuit assumption [3,8,15,70,71]. We can use the linearized differential equations (Equation 5.9b) for 1-D specialization as follows:

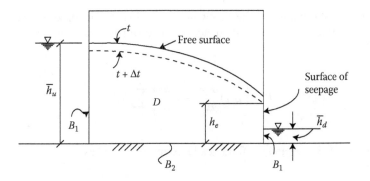

FIGURE A.1   Rectangular flow domain.

$$k_x \bar{h}(x,t)\,\frac{\partial^2 h}{\partial x^2} = n\frac{\partial h}{\partial t} \tag{A.1}$$

For the case of rise in fluid head, the mean fluid head $\bar{h}$ can be expressed as

$$\bar{h}(x,t) = \frac{h(x,t) + h(x,t+\Delta t)}{2} \tag{A.2a}$$

For simplicity, the mean head for the rise can be adopted as

$$\bar{h}(0,t) = \frac{h(0,t) + h(0,t+\Delta t)}{2} \tag{A.2b}$$

The mean head for the case of drawdown is assumed as

$$\bar{h}(0,t) = \frac{h_e(t) + h_e(t+\Delta t)}{2} \tag{A.2c}$$

where h_e is the exit head (Figure A.1).

The boundary conditions are given by

$$h = \bar{h}(0,t) \quad \text{on } B_1 \tag{A.3a}$$

$$h = \bar{h}_d \quad \text{on } B_1 \tag{A.3b}$$

and

$$k_x \frac{\partial h}{\partial x} = \bar{q}(x,t) \quad \text{on } B_2 \tag{A.3c}$$

where B_1 is the part of the boundary such as entrance face in the reservoir (Figure A.1), B_2 is the part of the boundary such as base (rock) and impervious vertical core, and D is the 1-D domain of flow. The initial condition is expressed as

$$h = \bar{h}_o(x,0) \tag{A.3d}$$

Because of the nature of the differential equation that involves $\bar{h}$, we use the Galerkin's weighted residual method, according to which the residual, R, is given by [24,25,72]

$$R = \left[B(t) \frac{\partial^2 h}{\partial x^2} - n \frac{\partial h}{\partial t} \right] \sum_i^n N_i h_i \tag{A.4}$$

where $B(t) = k_x \bar{h}(o,t)$ and the fluid head h is approximated as

$$h(x,t) = \sum_{i=1}^n N_i(x) h_i(t) \tag{A.5}$$

where $N_i(x)$ are interpolation functions, $N_1 = 1/2(1 - L)$, $N_2 = 1/2(1 + L)$ in which L is the local coordinate (Figure A.2), h_i are the nodal heads for the 1-D element (Figure A.2), and n denotes the number of degrees of freedom. Higher-order approximation, for example, quadratic and cubic, can also be used [24,25].

Now, according to the weighted residual (Galerkin) method

$$\int_D RN_m \, dD = 0 \tag{A.6a}$$

where the interpolation functions, N_m, are used as weighting functions, W. The substitution of Equation A.4 into Equation A.6a leads to

$$\int_D \left[B(t) \frac{\partial^2 h}{\partial t} - n \frac{\partial h}{\partial t} \right] \sum_1^n (N_i h_i) N_m \, dD = 0 \tag{A.6b}$$

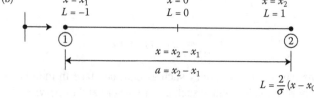

FIGURE A.2 1-D idealization, (a) 1-D idealization of flow domain and finite element mesh; (b) typical line element.

By applying Green's theorem, Equation A.6b leads to FE equations for the domain divided into M number of elements:

$$[k]\{h_i\} + [P]\{\dot{h}_i\} = \{Q\} \tag{A.7a}$$

where

$$k_{mi} = \sum_{1}^{M} \int \left[B(t) \cdot \frac{\partial N_m}{\partial x} \cdot \frac{\partial N_i}{\partial x} \right] dx \tag{A.7b}$$

$$P_{mi} = \sum_{1}^{M} \int n N_m N_i \, dx \tag{A.7c}$$

and

$$\dot{Q}_m = -\sum_{i}^{M} \int N_m \bar{q} \, dB \tag{A.7d}$$

where $[k]$, $[P]$, and $\{Q\}$ are material property (permeability) matrix, porosity (specific yield) matrix, and "load" vector, respectively, B denotes boundary, and M denotes the number of elements.

Time integration is needed to solve Equation A.7a. By using forward difference (Figure A.3), Equation A.7a can be written as [5,24,25,65,72]

$$[k]\{h_{t+\Delta t}\} + [P]\left(\frac{\{h_{t+\Delta t}\} - \{h_t\}}{\Delta t} \right) = \{Q_t\} \tag{A.8a}$$

Equation A.8a reduces to

$$\left([k] + \frac{[P]}{\Delta t} \right)\{h_{t+\Delta t}\} = \{Q_t\} + \frac{[P]}{\Delta t}\{h_t\} \tag{A.8b}$$

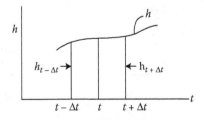

FIGURE A.3 Finite difference approximation for first derivative.

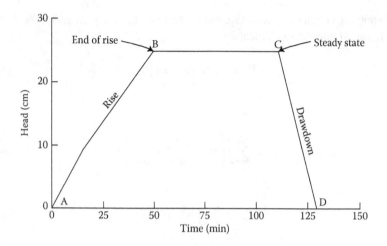

FIGURE A.4   Variation fo head at inflow boundary.

Equation A.8b can be solved for the fluid head at time $t + \Delta t$ using known material properties, prescribed boundary heads at $x = 0$ and $x = \ell$ (length), and initial condition at time $t = 0$. Equation A.8b is applicable during rise and steady state, until point C (Figure A.4).

During the drawdown condition, C to D in Figure A.4, the FS lags behind the level of water in the reservoir. Hence, it is necessary to modify the time-dependent entrance head to account for the surface of seepage, A–E and the exit head, $h_e(t + \Delta t)$ (Figure A.5). The entrance boundary nodal head can be adopted as the exit head $h_e(t + \Delta t)$. The evaluation of the exit head requires a special procedure, based on the

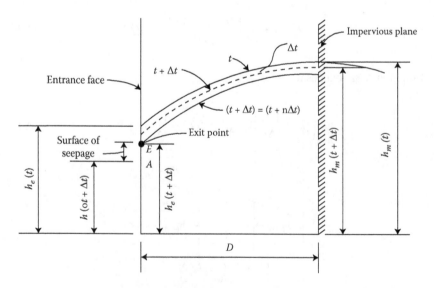

FIGURE A.5   Location of exit point and surface of seepage.

Pavlovsky's method of fragments [1,3]. This procedure has been described previously in this chapter for the 2-D case.

EXAMPLE

A number of problems have been solved by using the 1-D procedure [71]. For example, the predictions from the 1-D procedure compares well with the test data from the Hele–Shaw model with an entrance angle equal to 90° [10,11]. Here, an example for seepage between drains after sudden drawdown is presented.

Figure A.6 shows a series of parallel drains. Szabo and McCaig [58] proposed an analog solution for configurations of FS after sudden drawdown in parallel drains. The initial water level in the model simulating the drains was 19.7 in (50.0 cm) and was lowered suddenly to zero head (Figure A.6). The permeability coefficient equal to 0.0674 in/s (0.171 cm/s) and the porosity equal to 0.886 were adopted [58]. A total of six elements with seven nodes (Figure A.7) were used for the 1-D computer predictions. The first five elements were 2.0 in (5.08 cm) of length each and the last element was 2.5 in (16.35 cm) length. The time step $\Delta t = 10$ s was used. Only a half of the region between two drains was discretized due to the symmetry; the gradient across the center line was assumed to be zero.

France et al. [73] solved the problem of parallel drains using 2-D and 3-D FE procedures, and compared their predictions with the analog solutions developed by Szabo and McCaig [58]. The approximate 1-D procedure described above was used for the same problem, and in Figure A.7, the predictions are compared with the analog [58] and 3-D [65] solutions. The 1-D solution that used six 1-D elements shows satisfactory comparisons with the analog solution. Also, it can be considered as good as the 3-D solution.

Hence, if the geometry and properties of a problem permit 1-D idealization, it can provide a satisfactory solution for the analysis and design with considerable economy.

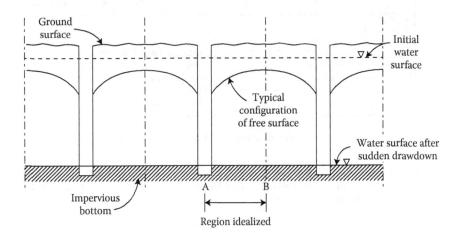

FIGURE A.6   Drawdown in parallel drains.

(a) Finite element mesh; Six 1–D elements

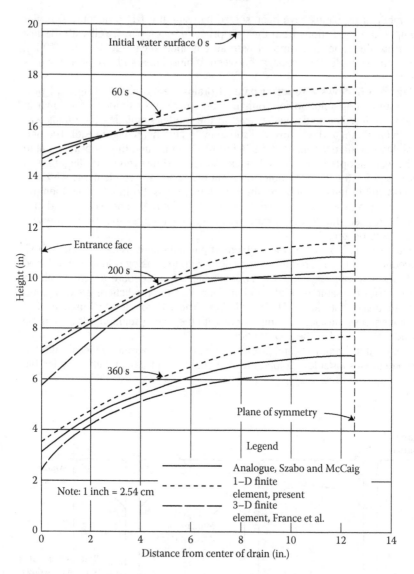

FIGURE A.7 Finite element mesh and comparisons of finite element and analogue solutions for sudden drawdown in drains. (Adapted from Desai, C.S., *Journal of the Irrigation and Drainage, ASCE*, 99(IR1), 1973, 71–86.)

REFERENCES

1. Polubarinova-Kochina P. Ya., *Theory of Ground Water Movement*, translated by De Wiest, R.J.M., Princeton University Press, Princeton, NJ, 1962.
2. Todd, D.K., *Ground Water Hydrology*, John Wiley & Sons, New York, 1960.
3. Harr, M.E., *Groundwater and Seepage*, McGraw-Hill Book Co., New York, 1962.
4. De Wiest, R.J.M. (Ed), *Flow through Porous Media*, Academic Press, New York, 1963.
5. Desai, C.S. and Abel, J.F., *Introduction to the Finite Element Method*, Van Nostrand Reinhold Co., New York, 1972.
6. Desai, C.S., Finite element procedures for seepage analysis using an isoparametric element, *Proceedings of the Symposium on Applied Finite Element Methods in Geotechnical Engineering*, U.S. Army Eng. Waterways Expt. Station, Vicksburg, MS, 1972.
7. Desai, C.S., Finite element methods for flow in porous media, Chap. 8 in Gallagher, R.H. et al., (Eds), *Finite Element in Fluids*, John Wiley & Sons, London, 1975.
8. Desai, C.S. and Sherman, W.C., Unconfined transient seepage in sloping banks, *Journal of Soil Mechanics and Foundations Division, ASCE*, 97(SM2), 357–373, 1971.
9. Brahma, S.P. and Harr, M.E., Transient development of free surface in a homogeneous earth dam, *Geotechnique*, 12, Dec. 1962, 283–302.
10. Desai, C.S., Analysis of Transient Seepage Using Viscous Flow Model and Numerical Methods, Report No. S-70-3, U.S. Army Engineer Waterways Experiment Station, U.S. Corps of Engineers, Vicksburg, MS, 1970.
11. Desai, C.S., Analysis of Transient Seepage Using a Viscous-Flow Model and the Finite Difference and Finite Element Methods, Report S-73-5, U.S. Army Engineer Waterways Experiment Station, U.S. Corps of Engineers, Vicksburg, MS, 1973.
12. Allada, S.R. and Quon, D., Stable, explicit numerical solution of the conduction equation for multidimensional nonhomogeneous media, *Heat Transfer Chemical Engineering Progress Symposium Series 64*, 62, 1966, 151–156.
13. Larkin, B.K., Some stable explicit difference approximations to the diffusion equation, *Journal of Mathematical Computation*, 18(86), Apr. 1964.
14. Saul'ev, V.K., A method of numerical solution for the diffusion equation, *Doklady Akademii Nauk SSSR (NS)*, 115, 1077–1079, 1957.
15. Dvinoff, A. and Harr, M.E., Phreatic surface location after drawdown, *Journal of Soil Mechanics and Foundations Division, ASCE*, 97(SM1), Jan. 1971, 47–58.
16. Irmay, S., On the meaning of the Dupuit and Pavlovskii approximations in aquifer flow, *Journal of Water Resources Research*, 5(2), 2nd Quarter, 1967.
17. Desai, C.S., Finite element residual schemes for unconfined seepage, Technical Note, *International Journal for Numerical Methods in Engineering*, 10, 1976, 1415–1418.
18. Desai, C.S. and Li, G.C., A residual flow procedure and application for free surface flow in porous media, *Advances in Water Resources*, 6, 1983, 27–35.
19. Li, G.C. and Desai, C.S., Stress and seepage analysis of earth dams, *Journal of Geotechnical Engineering, ASCE*, 109(7), 1983, 946–960.
20. Desai, C.S. and Baseghi, B., Theory and verification of residual flow procedure for 3-D free surface seepage, *International Journal of Advances in Water Resources*, 11(4), 1988, 192–203.
21. Baseghi, B. and Desai, C.S., Laboratory verification of the residual flow procedure for three-dimensional free surface flow, *Water Resources Research*, 26(2), 1990, 259–272.
22. Newlin, C.W. and Rossier, S.C., Embankment drainage after instantaneous drawdown, *Journal of Soil Mechanics and Foundations Division, ASCE*, 93(SM6), Nov. 1967, 79–95.
23. Hele-Shaw, H.S., Streamline motion of a viscous film, *Proceedings, 68th Meeting of the British Association of the Advancement of Science*, London, 1899.

24. Desai, C.S., *Elementary Finite Element Method*, Prentice-Hall, Englewood Cliffs, N.J., USA, 1979; Revised as Desai, C.S. and Kundu, T., *Introductory Finite Element Method*, CRC Press, Boca Raton, FL, USA, 2001.

25. Zienkiewicz, O.C. and Taylor, R.L., *The Finite Element Method*, McGraw-Hill Book Company, New York, 1989.

26. Gong, M. and Desai, C.S., Some Applications of Computer Code for Seepage Analysis, Report, Dept. of Civil Eng. and Eng. Mechs., University of Arizona, Tucson, AZ, 1983.

27. Desai, C.S., Codes for seepage analysis using variable and invariant meshes, SEEP-2DFE, and STRESEEP-2DFE, Tucson, AZ, 1983, 2000.

28. Taylor, R.L. and Brown, C.B., Darcy flow solutions with a free surface, *Journal of Hydraulics Divisions, ASCE*, 93(HY2), 1967, 25–33.

29. Finn, W.D.L. Finite element analysis of seepage through dams, *Journal of Soil Mechanics and Foundations Division, ASCE*, 93(SM6), 1967, 41–48.

30. Desai, C.S., Lightner, J.G., and Somasundaram, S., A numerical procedure for three-dimensional transient free surface seepage, *Advances in Water Resources*, 6, Sep. 1983, 175–181.

31. France, P.W., Finite element analysis of three-dimensional ground water flow problems, *Journal of Hydrology*, 21, 1974, 381–398.

32. Desai, C.S., Flow through porous media, in *Numerical Methods in Geotechnical Engineering,* C.S. Desai and J.T. Christian (Eds), McGraw-Hill Book Co., New York, 1977.

33. Terzaghi, K. and Peck, R.B., *Soil Mechanics in Engineering Practice*, John Wiley and Sons, New York, 1967.

34. Desai, C.S. and Kuppusamy, T., Development of Phreatic Surfaces in Earth Embankments, Report VPI-E-80.22, to Water and Power Resources Service, Dept. of Civil Eng., Virginia Tech, Blacksburg, VA, USA, 1980.

35. Desai, C.S., Free surface seepage through foundation and berm of cofferdams, *Journal of Indian Geotechnical Society*, V(1), 1975, 1–10.

36. Baiocchi, C., Free boundary problems in the theory of fluid flow through porous media, *Proceedings, International Congress on Mathematics*, Vancouver, Canada, Vol. II, 1974, 237–243.

37. Baiocchi, C. and Friedman, A., A filtration problem in a porous medium with variable permeability, *Annali di Matematica Pura ed Applicata*, 114(4), 1977, 377–393.

38. Alt, H.W., The fluid flow through porous media. Regularity of the free surface, *Manuscripta Mathematica*, 21, 1977, 255–272.

39. Alt, H.W., Numerical solution of steady-state porous flow free boundary problems, *Numerical Mathematics*, 36, 1980, 73–98.

40. Duvat, G. and Lions, J.L., Inequalities in mechanics and physics, *Grundlagen der Math. Wiss.*, 219, Springer, Berlin, 1976.

41. Lions, J.L. and Stampacchia, G., Variational inequalities, *Communications on Pure and Applied Mathematics*, 20, 1967, 493–519.

42. Oden, J.T. and Kikuchi, N., Theory of variational inequalities with application to problems of flow through porous media, *International Journal of Engineering Science*, 18(10), 1980, 1173–1284.

43. Bruch, J.C., Fixed domain methods for free and moving boundary flows in porous media, *Transport in Porous Media*, 6, 1991, 627–649.

44. Caffrey, J. and Bruch, J.C., Three-dimensional seepage through homogeneous dam, *Advances in Water Resources*, 12, 1979, 167–176.

45. Bouwer, H., Unsaturated flow in ground-water hydraulics, *Journal of Hydraulics Division, ASCE*, 90(HY5), 1964, 121.

46. Freeze, R.A., Influence of unsaturated flow domain on seepage through earth dams, *Water Resources Research*, 7(4), 1971, 929.

47. Neuman, S.P., Saturated-unsaturated seepage by finite elements, *Journal of Hydraulics Division, ASCE*, 99(HY12), 1973, 2233–2251.
48. Cathie, D.N. and Dungar, R., The influence of the pressure permeability relationship on the stability of a rock-filled dam, *Proceedings, Conference on Criteria and Assumptions in the Numerical Analysis of Dams*, Swansea, UK, 1975, 30–845.
49. Westbrook, D.R., Analysis of inequality and residual flow procedures and an iterative scheme for free surface seepage, *International Journal for Numerical Methods in Engineering*, 21, 1985, 1791–1802.
50. Rubin, J., Theoretical analysis of two-dimensional transient flow of water in unsaturated and partly saturated soils, *Journal of Soil Science Society of America*, 32(5), 1968, 607–615.
51. Bathe, K.J. and Khoshguftar, M.R., Finite element free surface seepage analysis without mesh iteration, *International Journal for Numerical and Analytical Methods in Geomechanics*, 3, 1979, 13–22.
52. Freeze, R.A., Three-dimensional transient, saturated-unsaturated flow in a ground water basin, *Journal of Water Resources Research*, 7(2), 1971, 347.
53. Bromhead, E.N., Discussion of finite element residual schemes for unconfined flow, *International Journal of Numerical Methods in Engineering*, 11(5), 1977, 80.
54. Desai, C.S., Free surface flow through porous media using a residual procedure, *Proceedings, Finite Elements in Fluids*, Vol. 5, Gallagher, R.H. et al. (Eds), John Wiley and Sons Limited, 1984.
55. Hele-Shaw, H.S., Experiments on the nature of surface resistance of water and of streamline motion under certain experimental conditions, *Transactions, Institution of Naval Architecture*, Vol. 40, 1898.
56. Desai, C.S., Seepage analysis of earth banks under drawdown, *Journal Soil Mechanics and Foundations Engineering Division, ASCE*, 98(SM1), 1972, 98.
57. Baseghi, B. and Desai, C.S., Three-Dimensional Seepage through Porous Media with the Residual Flow Procedure, Report, Dept. of Civil Eng. and Eng. Mechs., The University of Arizona, Tucson, AZ, 1987.
58. Szabo, B.A. and McCaig, I.W., A mathematical model for transient free surface flow in nonhomogeneous or anisotropic porous media, *Water Resources Bulletin*, 4(3), 1968, 5–18.
59. Duncan, J.M. and Chang, C.Y., Nonlinear analysis of stress and strain in soils, *Journal of Soil Mechanics and Foundations Divsion, ASCE*, 96(SM5), 1970, 1629–1653.
60. Drucker, D.C. and Prager, W., Soil mechanics and plasticity analysis of limit design, *Quarterly of Applied Mathematics*, 10(2), 1952, 157–165.
61. Nobari, E.S. and Duncan, J.M., Effects of Reservoir Filling on Stresses and Movements in Earth and Rockfill Dams, Report No. S-72-2, U.S. Army Engineers Waterways Experiment Station, Vicksburg, MS, 1972.
62. Clough, G.W., Ground Water Level in Silty and Sandy Mississippi River Banks, Report, Mississippi River Commission, Corps of Engineers, Vicksburg, MS, Aug. 1966.
63. Hvorslev, M.J., Time Log and Soil Permeability in Ground Water Observations, Bulletin No. 36, U.S. Army Engineers Waterways Experiment Station, Vicksburg, MS, Apr. 1951.
64. Krinitzsky, E.L. and Wire, J.C., Ground Water in Alluvium of the Lower Mississippi Valley (Upper and Central Areas), Technical Report No. 3-658, Vol. 1, U.S. Army Engineers Waterway Experimental Station, Vicksburg, MS, Sep. 1964.
65. France, P.W., Parekh, C.J., Peters, J.C., and Taylor, C., Numerical analysis of free surface seepage problems, *Journal of the Irrigation and Drainage Division, ASCE*, 97, 1971, 165–179.
66. Hall, H.P., An investigation of steady flow towards a gravity well, *La Houille Blanche*, 10, 1955, 8–35.
67. Chen, Q. and Zhang, L.M., Three-dimensional analysis of water infiltration into the Gouhou rockfill dam using saturated-unsaturated seepage theory, *Canadian Geotechnical Journal*, 43, 2006, 449–461.

68. Fredlund, D.G. and Rahardjo, H., *Soil Mechanics for Unsaturated Soils*, John Wiley & Sons, Inc., New York, 1993.

69. Chen, Q., Zhang, L.M., and Zhu, F.Q., Effects of material stratification on the seepage field in a rockfill dam, *Proceedings of the 2nd Chinese National Symposium on Unsaturated Soils,* Hangzhou, China 23–24, Apr. 2005, 508–515.

70. Dicker, D., Transient free-surface flow in porous media, in *Flow through Porous Media*, Dewiest, R.J.M. (Ed), Academic Press, New York, NY, 1969.

71. Desai, C.S., Approximate solution for unconfined seepage, *Journal of the Irrigation and Drainage, ASCE*, 99(IR1), 1973, 71–86.

72. Wilson, E.L. and Nickell, R.E., Application of the finite element method to heat conduction analysis, *Nuclear Engineering and Design*, 4, 1966, 276–286.

6 Flow through Porous Deformable Media
One-Dimensional Consolidation

6.1 INTRODUCTION

We considered in Chapter 5 the flow of fluid (water) through porous geologic materials, commonly called seepage, when the skeleton of the material is assumed to be rigid. In practice, however, it is common that the deformations in the soil skeleton take place simultaneously with fluid flow through the pores, and there occurs relative motion between fluid and solid (soil). In other words, the deformation and flow are coupled and influence each other. One of the specializations of water flow through porous and deforming geologic medium is called *consolidation* in the geotechnical literature [1–3].

In this chapter, we consider the 1-D consolidation in which the relative motions between the soil particles (skeleton) and water is not considered. In Chapter 7, we will consider general 3-D coupled system including the relative motions. One- and 2-D consolidation equations can be derived as special cases of the general 3-D theory; the application for the latter and dynamic problems will also be included in Chapter 7.

6.2 ONE-DIMENSIONAL CONSOLIDATION

6.2.1 Review of One-Dimensional Consolidation

The 1-D consolidation theory with the related effective stress principle was proposed by Terzaghi in various publications [2,3]. It has been described in many publications, for example, by Taylor [4,5] and Suklje [6]. Rendulic [7] extended the 1-D theory for multidimensional consolidation. Subsequently, various investigations have modified and applied the 1-D theory with linear material behavior (e.g., [8–19]). The 1-D theory has been modified to include nonlinear soil behavior [20–24]. Over the years, the 1-D theory has been applied for the solution of problems, often simulated in the laboratory and/or in the field (e.g., [25–28]).

Computation of 1-D or vertical settlement can be performed using the 1-D theory, with various assumptions such as (a) the structure is small compared to the size of the foundation (Figure 6.1), (b) the load from the structure is applied essentially in the vertical direction, (c) significant displacement is caused only in the vertical direction,

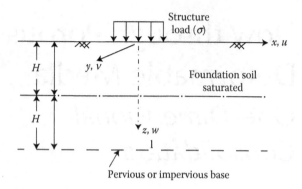

FIGURE 6.1 Consolidation of soil.

while that in other directions can be neglected, (d) the soil is fully saturated and homogeneous, (e) the mechanical behavior of the soil can be simulated using linear elasticity, (f) the flow is laminar, (g) the soil solids and water in the pores are incompressible, (h) the soil permeability is constant, and (i) the small strain theory is applicable.

6.2.2 GOVERNING DIFFERENTIAL EQUATIONS

By using the principle of continuity, we can obtain the following differential equation governing the 1-D flow [2,23,27]:

$$\frac{\partial}{\partial z}\left(\frac{k}{\gamma_w}\cdot\frac{\partial p}{\partial z}\right) = \frac{\partial \varepsilon_v}{\partial t} = -m_v \frac{\partial \sigma'}{\partial t} \tag{6.1}$$

where k is the coefficient of permeability, p is the pore water pressure, ε_v is the volumetric strains, m_v is the coefficient of volume change, z is the vertical coordinate, t is the time, and σ' is the effective (vertical) stress proposed in Terzaghi theory [1–3]:

$$\sigma = \sigma' + p \tag{6.2}$$

where σ is the total stress and p is the pore water pressure. From this equation, we can write

$$\frac{\partial \sigma}{\partial t} = \frac{\partial \sigma'}{\partial t} + \frac{\partial p}{\partial t} \tag{6.3a}$$

Since the total stress (applied load) does not change, we can write

$$\frac{\partial \sigma'}{\partial t} = -\frac{\partial p}{\partial t} \tag{6.3b}$$

The introduction of Equation 6.3b in Equation 6.1 leads to

$$\frac{k}{\gamma_w m_v} \frac{\partial^2 p}{\partial z^2} = \frac{\partial p}{\partial t} \tag{6.4a}$$

or

$$c_v \frac{\partial^2 p}{\partial z^2} = \frac{\partial p}{\partial t} \tag{6.4b}$$

where $c_v = k/(\gamma_w\, m_v)$, is the coefficient of consolidation. Equation 6.4 represents the Terzaghi 1-D consolidation equation.

6.2.2.1 Boundary Conditions

The differential equation (Equation 6.4) can be solved for specific and known boundary conditions. Two of the common boundary conditions are depicted in Figures 6.2a and 6.2b. The surface (top) and bottom are pervious in Figure 6.2a, while the bottom is impervious in Figure 6.2b. Mathematically, at a pervious boundary the pressure is atmospheric; hence, the pore water pressure is zero. Thus, we have

$$p(0, t) = 0 \tag{6.5a}$$

$$p(2H, t) = 0 \tag{6.5b}$$

At an impervious boundary, the fluid flow is zero (Figure 6.2b). This boundary condition can be expressed as

$$\frac{\partial p}{\partial z}(2H, t) = 0 \tag{6.5c}$$

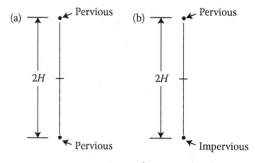

FIGURE 6.2 Boundary conditions in 1-D idealization. (a) Pervious top and bottom; (b) pervious top and impervious bottom.

The initial condition can be expressed as

$$p(z, 0) = \bar{p}(z) \tag{6.5d}$$

where the pore water pressures at time $t = 0$ are specified (overbar) in the 1-D domain.

It may be mentioned that although the problem involves coupled phenomenon, in the 1-D theory, it can be uncoupled in two equations (Equations 6.2 and 6.4). The 1-D consolidation theory by Terzaghi [1–3] can be obtained as a special case of the general formulation, as discussed in Chapter 7.

An alternative for consolidation theory can be derived in terms of strains in the consolidating layer [22]. We can obtain, from the effective stress principle (Equation 6.2)

$$\frac{\partial p}{\partial z} = \frac{\partial \sigma}{\partial z} - \frac{\partial \sigma'}{\partial z} \tag{6.6}$$

Now, insertion of Equation 6.6 in Equation 6.1 yields

$$\frac{\partial}{\partial z}\left[\frac{k}{\gamma_w}\left(\frac{\partial \sigma}{\partial z} - \frac{\partial \sigma'}{\partial z}\right)\right] = \frac{\partial \varepsilon_v}{\partial t} \tag{6.7a}$$

Assuming that the total stress (increment) σ does not change and $\varepsilon_v = \varepsilon$, strain exists in the vertical direction only (i.e., $\varepsilon_x = \varepsilon_y = 0$), Equation 6.7a reduces to the following linear equation:

$$\frac{\partial}{\partial z}\left(-\frac{k}{\gamma_w}\frac{\partial \sigma'}{\partial z}\right) = \frac{\partial \varepsilon}{\partial t} \tag{6.7b}$$

If $(\partial \sigma/\partial z)$ changes, Equation 6.7b will result into a nonlinear differential equation.

6.2.3 STRESS–STRAIN BEHAVIOR

We assume that the linear strain–stress law for the soil is given by

$$\sigma' = -\frac{\varepsilon}{m_v} \tag{6.8a}$$

from which we obtain

$$\frac{\partial \sigma'}{\partial z} = -\frac{1}{m_v} \cdot \frac{\partial \varepsilon}{\partial z} \tag{6.8b}$$

The use of this equation in Equation 6.7b gives

$$\frac{\partial}{\partial z}\left[-\frac{k}{\gamma_w}\left(-\frac{1}{m_v} \cdot \frac{\partial \varepsilon}{\partial z}\right)\right] = \frac{\partial \varepsilon}{\partial t} \tag{6.9a}$$

or

$$\frac{\partial}{\partial z}\left(\frac{k}{\gamma_w m_v}\frac{\partial \varepsilon}{\partial z}\right) = \frac{\partial \varepsilon}{\partial t} \qquad (6.9b)$$

If k, m_v, and γ_w are constants, this equation reduces to

$$c_v \frac{\partial^2 \varepsilon}{\partial z^2} = \frac{\partial \varepsilon}{\partial t} \qquad (6.9c)$$

where $c_v = k/(\gamma_w m_v)$. This equation has the same mathematical form as Equation 6.4b.

6.2.3.1 Boundary Conditions

The boundary conditions for Equation 6.9c are different than those for Equation 6.4b. For the conditions of pervious top and bottom, as $p = 0$, boundary conditions can be written as follows:

$$\varepsilon(0, t) = -\sigma' m_v \qquad (6.10a)$$

and

$$\varepsilon(2H, t) = -\sigma' m_v \qquad (6.10b)$$

and the initial condition is given by

$$\varepsilon(z, o) = \bar{\varepsilon}(z) \qquad (6.10c)$$

where $\bar{\varepsilon}(z)$ will be zero, if we assume the initial strain in the soil is zero. It can be equal to the computed strain for the next increment of loading. Equation 6.9c is based on the following assumptions: (i) the total stress or incremental stress is constant with depth, (ii) m_v is constant with depth, but can be considered to depend on the stress, and (iii) c_v is constant with depth. From a practical viewpoint, these assumptions can have significant influence on the solutions from Equation 6.9c. However, use of a numerical procedure like the FEM, the above quantities can spatially vary, that is, from element to element, although it is usually assumed to be constant within an element.

In the linear Equations 6.4b and 6.9c, the value of c_v is essentially constant. Its average values can be used both in the normally consolidated (NC) and overconsolidated (OC) regions [25,29]; this is considered to be supported by test data (Figure 6.3), which shows the variation of ε_v and c_v versus log (σ'). The value of c_v is usually greater for the OC region compared to that in the NC region. Such a difference in c_v can have significant effect on the rate of consolidation, particularly when an OC soil is subjected to sufficiently large stress such that it enters into the NC state. This can be modeled by using the FEM, as described subsequently.

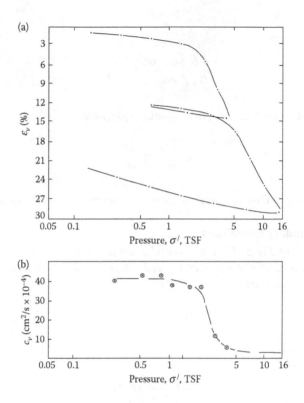

FIGURE 6.3 Variation of ε_v and c_v with pressure (1 TSF = 4.58 N/cm²). (a) Volumetric strain versus pressure; (b) coefficient of consolidation versus pressure. (From Lamb, T.W. et al., The Performance of a Foundation Under a High Embankment, Research Report R71-22, Soil Mech. Div., Dept. of Civil Eng., MIT, Cambridge, MA, 1972; Lamb, T.W. and Whitman, R.V., *Soil Mechanics*, John Wiley & Sons, New York, 1969. With permission.)

6.3 NONLINEAR STRESS–STRAIN BEHAVIOR

The consolidation equation is often solved assuming that the stress–strain behavior is linear (Equation 6.8a). However, many soils exhibit nonlinear stress–strain–volume change behavior. Various investigators have considered nonlinear behavior for the (numerical) solution of consolidation assuming nonlinear elastic, elastoplastic, elastoviscoplastic, and other behavior [e.g., 20–22]. Koutsoftas and Desai [23] and Desai et al. [27] have developed one linear and two nonlinear procedures for consolidation by using Equations 6.4b and 6.9c. Some of the descriptions herein are adopted from Koutsoftas and Desai [23], in which the first author performed research on consolidation during his stay at Virginia Tech, Blacksburg, VA, USA.

6.3.1 PROCEDURE 1: NONLINEAR ANALYSIS

In this procedure, Equation 6.4b is first solved with the assumption of linear behavior according to Equation 6.8a between effective stress and strain. The solution provides the values of pore water pressures at different time levels. The effective stress at a time level is then computed using Equation 6.2.

The nonlinear relation between effective stress and strain (Figure 6.4) is used to evaluate the strain under NC and OC conditions:

$$\varepsilon_{nc}(t) = CR\log\left(\frac{\sigma'_f}{\sigma'_o}\right) \quad \text{when } \sigma'_f > \sigma'_o = \sigma'_{max} \tag{6.11a}$$

$$\varepsilon_{oc}(t) = RR\log\left(\frac{\sigma'_f}{\sigma'_o}\right) \quad \text{when } \sigma'_f < \sigma'_{max} \tag{6.11b}$$

When an OC soil is subjected to a stress increment sufficiently large so that the soil enters the NC regime, the final strain is given by

$$\varepsilon_f = RR\log\left(\frac{\sigma'_{max}}{\sigma'_o}\right) + CR\log\left(\frac{\sigma'_f}{\sigma'_{max}}\right) \tag{6.11c}$$

where ε_{nc}, ε_{oc}, and ε_f are the NC, OC, and final strains, respectively, σ'_o is the initial effective stress, σ'_f is the final effective stress, σ'_{max} is the maximum effective stress, CR is the slope of the NC behavior (line), and RR is the slope of the OC behavior (line) (Figure 6.4). Now, the strains at any time $\varepsilon(t)$ can be computed by using Equations 6.11 in which σ'_f is replaced by the computed effective stress at any time, $\sigma'(t)$.

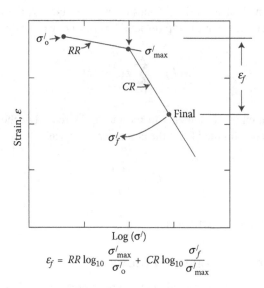

$$\varepsilon_f = RR\log_{10}\frac{\sigma'_{max}}{\sigma'_o} + CR\log_{10}\frac{\sigma'_f}{\sigma'_{max}}$$

FIGURE 6.4 Nonlinear stress–strain behavior in 1-D consolidation. (From Koutsoftas, D.C. and Desai, C.S., One-dimensional consolidation by finite elements: Solutions of some practical problem, Report No. VPI-E-76-17, Dept. of Civil Eng., VPI&SU, Blacksburg, VA, 1976. With permission.)

In the aforementioned procedure, the pore water pressure, that is, effective stress is computed based on the linear stress–strain relation (Equation 6.8a), while the nonlinear behavior is accounted for by use of Equations 6.11. Hence, the procedure may be referred to as "pseudo-nonlinear" [23,27].

6.3.2   PROCEDURE 2: NONLINEAR ANALYSIS

Equation 6.9c in terms of strain involves intrinsically the nonlinear behavior. Its solution yields strains at any depth and time. Then, the settlement can be computed by using the following equation:

$$\varepsilon(z,t) = U(z,t) \times \varepsilon_f \qquad (6.12)$$

where $U(z,t)$ is the degree of consolidation at a depth z and time t, and ε_f is the final strain (Figure 6.4). It was shown [22] that the degree of consolidation from solution of Equations 6.9c and 6.4b are the same; hence, $U(z,t)$ can be obtained from Equation 6.4b or Equation 6.9c.

6.3.2.1   Settlement

Once $\varepsilon(z,t)$ is computed, we can evaluate the (vertical) displacement or settlement $w(z,t)$ at a point or in an element i as

$$w(z,t)_i = \varepsilon(z,t)_i \times (\ell)_i \qquad (6.13a)$$

where i denotes an element in the FE mesh (see later) and ℓ is the length of an element. The total settlement at the top (ground level), $w(T)$ can be then found as

$$w(T) = \sum_{i=1}^{M} \varepsilon(z,t)_i (\ell)_i \qquad (6.13b)$$

where M is the number of elements in the mesh. As required, the total degree of consolidation can be obtained from the computer results as

$$\bar{U} = \frac{w(t)}{w_f} \qquad (6.13c)$$

where w_f is the final (total) settlement.

6.3.3   ALTERNATIVE CONSOLIDATION EQUATION

An alternative equation for consolidation can be expressed as [22]

$$c_v \frac{\partial^2 \sigma_r'}{\partial z^2} = \frac{\partial \sigma_r'}{\partial t} \qquad (6.14a)$$

where σ'_r can be called (relative or effective) stress, which is given by

$$\sigma'_r(t) = \log \frac{\sigma'(t)}{\sigma'_f} = \frac{\varepsilon(t) - \varepsilon_f}{CR} \quad (6.14b)$$

The boundary conditions for Equation 6.14 can be expressed as follows.

6.3.3.1 Pervious Boundary
For a previous boundary, we have

$$\sigma'_r(o,t) = \sigma'_r(2H,t)$$

$$= \log \left(\frac{\sigma}{\sigma'_f} \right) \quad (6.15a)$$

6.3.3.2 Impervious Boundary at 2H
The impervious boundary at $2H$ can be expressed as

$$\frac{d\sigma'_r(2H,t)}{dz} = \frac{d}{dz}\left[\log\left(\frac{\sigma'(z)}{\sigma'_f}\right)\right]$$

$$= 0 \quad (6.15b)$$

Hence, the solution in terms of the effective stress at any time from Equation 6.14a using a numerical method can be substituted in Equation 6.14b to evaluate the strain at any time, $\varepsilon(t)$. Also, the pore water pressure at any time can be computed using Equation 6.2 once $\sigma'(t)$ is found from Equation 6.14a. Thus, it is not necessary to solve Equation 6.4b and use Equations 6.11.

Nonlinear behavior can occur also due to non-Darcy flow condition. In the above equations, we have assumed the linear Darcy law is valid. It can be expressed as

$$v = -k\frac{\partial \phi}{\partial z} \quad (6.16a)$$

where ϕ is the hydraulic head ($= z + p/\gamma_w$). We can incorporate a non-Darcy law in the formula. One such law presented by Schmidt and Westman [30] is given by

$$v = -k\left(\frac{\partial \phi}{\partial z}\right)^b \quad (6.16b)$$

where b is the parameter determined from laboratory tests and found to be in the range of 1–2.38 [30–32].

6.4   NUMERICAL METHODS

Equations 6.4b, 6.9c, and 6.14a can be solved by analytical (closed form) methods [1–3], for simplified material properties, loading, and boundary conditions. However, for many practical problems involving nonlinear behavior, nonhomogeneous materials, and complex boundary conditions, numerical methods such as FD [33–36], FE [37–39], and BE [40] procedures can be used. Here, we present descriptions of the FD and FE methods for 1-D consolidation [24,26].

6.4.1   FINITE DIFFERENCE METHOD

We present FD approximations to the consolidation Equation 6.4b by using the following schemes for specific representation of the derivatives involved:

6.4.1.1   FD Scheme No. 1: Simple Explicit

The FD form of Equation 6.4b can be expressed as follows (Figure 6.5):

$$c_v \left(\frac{p_{i-1}^t - 2p_i^t + p_{i+1}^t}{\ell^2} \right) = \frac{p_i^{t+1} - p_i^t}{\Delta t} \tag{6.17a}$$

or

$$\Delta T \left(p_{i-1}^t - 2p_i^t + p_{i+1}^t \right) = p_i^{t+1} - p_i^t \tag{6.17b}$$

or

$$p_i^{t+1} = \Delta T \left[p_{i-1}^t - p_i^t \left(2 - \frac{1}{\Delta T} \right) + p_{i+1}^t \right] \tag{6.17c}$$

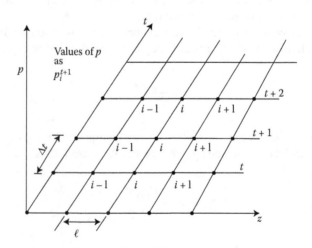

FIGURE 6.5   Finite difference discretization.

where $\Delta T = (c_v \Delta t)/\ell^2$ is the incremental time factor; the time factor is defined as $(c_v t/H^2)$. Since the initial condition, that is, pore water pressure at time $= 0$ is known, we solve Equation 6.17c from one point to the next, from $t = 0 + \Delta t$, $t = \Delta t + \Delta t$, and so on. Since the value of p at $t + 1$ can be computed by knowing its values at the (known) previous time increment, this scheme is called an *explicit*; it is also known as the Euler scheme [23,26,33–36].

6.4.1.2 FD Scheme No. 2: Implicit, Crank–Nicholson Scheme

Here, Equation 6.4b is expressed in the FD form as

$$\frac{\Delta T}{2}\left[(p_{i-1}^{t+1} - 2p_i^{t+1} + p_{i+1}^{t+1}) + (p_{i-1}^t - 2p_i^t + p_{i+1}^t)\right] = p_i^{t+1} - p_i^t \qquad (6.18a)$$

or

$$\frac{\Delta T}{2}p_{i+1}^{t+1} - p_i^{t+1}(\Delta T + 1) + \frac{\Delta T}{2}p_{i+1}^{t+1} = -\frac{\Delta T}{2}p_{i-1}^t + p_i^t(\Delta T + 1) + p_{i+1}^t \qquad (6.18b)$$

Thus, the values of p at $t + 1$ for nodes i are unknowns in the set of simultaneous equations that result from applying Equation 6.18 to all nodes in the problem. This scheme involves solution of simultaneous equations, and is called an *implicit* (Crank–Nicholson) scheme [23,26,33–36].

6.4.1.3 FD Scheme No. 3: Another Implicit Scheme

In this scheme, the FD from Equation 6.4b is expressed as

$$\Delta T(p_{i-1}^{t+1} - 2p_i^{t+1} + p_{i+1}^{t+1}) = p_i^{t+1} - p_i^t \qquad (6.19a)$$

or

$$\Delta T p_{i-1}^{t+1} - (2\Delta T + 1)p_i^{t+1} + \Delta T p_{i+1}^{t+1} = -p_i^t \qquad (6.19b)$$

This *implicit* scheme also involves a set of simultaneous equations when Equation 6.19 is used for all points in the problem [33–36].

6.4.1.4 FD Scheme No. 4A: Special Explicit Scheme

According to this scheme, which is referred to as the Saul'ev procedure, the FD approximation is given by [26,35,40]

$$\Delta T(p_{i-1}^{t+1} - p_i^{t+1} - p_i^t + p_{i+1}^t) = p_i^{t+1} - p_i^t \qquad (6.20a)$$

or

$$-(\Delta T + 1)p_i^{t+1} = -\Delta T p_{i-1}^{t+1} + (\Delta T - 1)p_i^t - p_{i+1}^t \qquad (6.20b)$$

The first derivative from $i-1$ to i is written at $t+1$, and at t from i to $i+1$ (Figure 6.5). The term p_{i-1}^{t+1} is known since it has been computed before, at $i-1$. Thus, this scheme is called *special explicit* because we can solve for p, point by point, which does not involve solutions of simultaneous equations.

6.4.1.5 FD Scheme No. 4B: Special Explicit

This scheme involves FD expressions in alternating direction (i.e., in one direction for a given time interval and then in the opposite direction for the next time interval); hence, it is called the ADEP [40–42]. The FD approximation is given by Equation 6.20 for a given direction during $(t+1)$, and the following equation in the reverse direction during $(t+2)$:

$$\Delta T(p_{i+1}^{t+2} - p_i^{t+2} - p_i^{t+1} + p_{i-1}^{t+1}) = p_i^{t+2} - p_i^{t+1} \qquad (6.21a)$$

or

$$-(\Delta T + 1)p_i^{t+2} = -\Delta T p_{i+1}^{t+2} + (\Delta T - 1)p_i^{t+1} - p_{i-1}^{t+1} \qquad (6.21b)$$

Solution of these equations involves known values of p_{i+1} at $t+2$, which were just computed following the reverse direction. Hence, by solving Equations 6.20 and 6.21, we obtain the values of p at $t+1$ and $t+2$ in an *explicit* manner.

6.4.2 FINITE ELEMENT METHOD

We present a brief description of the 1-D consolidation; details are given in Refs. [18,23,26,39].

Figure 6.6b shows the 1-D FE idealization of consolidation for the shaded zone (Figure 6.6a). A generic element and nodes are shown in Figure 6.6c. Assuming linear variation for the unknown, that is, pore water pressure, p, strain, ε, and relative stress, σ_r', corresponding to Equations 6.4b, 6.9c, and 6.14a, respectively, can be expressed as follows:

$$p(z,t) = N_1(z)p_1(t) + N_2(z) \cdot p_2(t) \qquad (6.22a)$$

$$\varepsilon(z,t) = N_1(z)\varepsilon_1(t) + N_2(z)\varepsilon_2(t) \qquad (6.22b)$$

$$\sigma_r'(z,t) = N_1(z)\sigma_{r1}'(t) + N_2(t)\sigma_{r2}'(t) \qquad (6.22c)$$

These equations can be expressed in a matrix form. For example, for p, Equation 6.22a, we can write

$$p(z,t) = [N]\{p_n\} \qquad (6.23a)$$

where $[N]$ is the matrix of interpolation functions, N_1 and N_2, and $\{p_n\}^T = [p_1 \; p_2]$ is the vector of nodal pore water pressures. The interpolation functions are expressed as

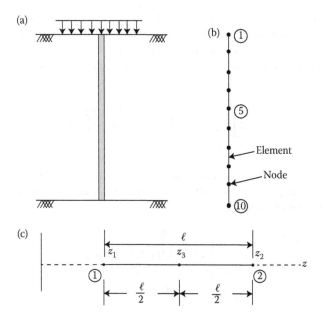

FIGURE 6.6 One-dimensional FE consolidation. (a) 1-D consolidation model; (b) finite element discretization; and (c) generic 1-D element.

$$N_1 = \frac{1}{2}(1 - L) \tag{6.23b}$$

and

$$N_2 = \frac{1}{2}(1 + L) \tag{6.23c}$$

where L is the local coordinate given by

$$L = \frac{z - z_3}{\ell/2} \tag{6.23d}$$

Here, z is the coordinate of a point within the element, z_3 is the coordinate of the midpoint of the element, and ℓ is the length of an element.

The above linear approximation for the unknown is similar to that for (static) problems in Chapter 2. However, since the problem is now time-dependent, the nodal values of the unknowns are time-dependent. Hence, a different approach compared to the static case is needed. For instance, in Equation 6.22, the interpolation function is dependent on spatial coordinate, z, whereas the nodal values of p are treated as time dependent. The solution is obtained in two steps. The first step is almost identical to the solution for static problems wherein the time dependence of nodal

p is temporarily suppressed. Then, as will be seen subsequently, the solution will result in matrix difference equations in time, which are solved by using appropriate numerical integration schemes like the FD method.

The variation (energy) function, Ω, for an unknown, say p, can be expressed as [33,39]

$$\Omega = A\int_{z_1}^{z_2}\left[\frac{1}{2}c_v\left(\frac{\partial p}{\partial z}\right)^2 + \frac{\partial p}{\partial t}p\right]dz - \int_{z_1}^{z_2}\bar{q}pdz \qquad (6.24)$$

where A is the area of the element, z_1 and z_2 are nodal coordinates of the element (Figure 6.6c), and $\bar{q}$ is the intensity of the fluid flux. Now, the derivative of p with respect to z can be expressed as

$$\frac{\partial p}{\partial z} = \frac{\partial}{\partial z}\left[\frac{1}{2}(1-L)p_1 + \frac{1}{2}(1+2)p_2\right] \qquad (6.25a)$$

$$= \frac{1}{\ell}[-1 \quad 1]\begin{Bmatrix}p_1\\p_2\end{Bmatrix}$$

$$= [B]\{p_n\} \qquad (6.25b)$$

where $[B]$ is the transformation matrix.

The derivative with respect to time can be expressed as

$$\frac{\partial p}{\partial t} = \frac{\partial}{\partial t}\left[\frac{1}{2}(1-L)p_1(t) + \frac{1}{2}(1+L)p_2(t)\right]$$

$$= [N]\begin{Bmatrix}\dfrac{\partial p_1}{\partial t}\\\dfrac{\partial p_2}{\partial t}\end{Bmatrix} = [N]\begin{Bmatrix}\dot{p}_1\\\dot{p}_2\end{Bmatrix}$$

$$= [N]\{\dot{p}_n\} \qquad (6.26)$$

where $\{\dot{p}_n\}^T = [\dot{p}_1\ \dot{p}_2]$, and the overdot denotes the derivative with respect to time.

By substituting p, $(\partial p/\partial z)$, and $(\partial p/\partial t)$ in Equation 6.24, and equating to zero the variations $(\partial\Omega/\partial p_1)$ and $(\partial\Omega/\partial p_2)$, we obtain the following element equations [39]:

$$\frac{Ac_v}{\ell}\begin{bmatrix}1 & -1\\1 & -1\end{bmatrix}\begin{Bmatrix}p_1\\p_2\end{Bmatrix} + \frac{A\ell}{6}\begin{bmatrix}2 & 1\\1 & 2\end{bmatrix}\begin{Bmatrix}\partial p_1/\partial t\\\partial p_2/\partial t\end{Bmatrix} = \frac{\bar{q}\ell}{2}\begin{Bmatrix}1\\1\end{Bmatrix} \qquad (6.27a)$$

or

$$[k_\alpha]\{p_n\} + [k_t]\{\dot{p}_n\} = \{Q(t)\} \qquad (6.27b)$$

where $[k_\alpha]$ and $[k_t]$ are the material property matrices, and $\{Q(t)\}$ denotes the vector of nodal applied flux.

Equation 6.27 is often used for a homogeneous layer, in which c_v is constant. For layered systems, c_v, can vary with z, and it is useful to consider the following equations in which k and m_v may vary:

$$\frac{k}{\gamma_w} \frac{\partial^2 p}{\partial z^2} = m_v \frac{\partial p}{\partial t} \tag{6.28a}$$

Then, the element equations can be derived as [39]

$$a_1 \begin{bmatrix} 1 & -1 \\ -1 & 1 \end{bmatrix} \begin{Bmatrix} p_1 \\ p_2 \end{Bmatrix} + a_2 \begin{bmatrix} 2 & 1 \\ 1 & 2 \end{bmatrix} \begin{Bmatrix} \dot{p}_1 \\ \dot{p}_2 \end{Bmatrix} = \frac{\bar{q}(t)\ell}{2} \begin{Bmatrix} 1 \\ 1 \end{Bmatrix} \tag{6.28b}$$

where

$$a_1 = \frac{Ak}{\gamma_w \ell} \tag{6.28c}$$

and

$$a_2 = \frac{Am_v \ell}{6} \tag{6.28d}$$

Then, the values of k and m_v can be considered to vary with z.

6.4.2.1 Solution in Time
We adopt two different time integration schemes described as follows.

6.4.2.1.1 FE—Scheme 1
Equation 6.27b is a matrix differential equation in which we have the time-dependent term $(\dot{p}_n)$. We can use numerical time integration to express the equation in terms of p at nodal points in time. We illustrate the time integration using the simple Euler scheme, by expressing approximately the time derivative $(\dot{p})$ in the FD form as (Figure 6.7)

$$\left(\frac{\partial p}{\partial t}\right)_i^{t+1} \approx \frac{p_i^{t+1} - p_i^t}{\Delta t} \tag{6.29}$$

where Δt denotes the chosen time increment.

By substituting $\dot{p}_1$ and $\dot{p}_2$ at the element node points 1 and 2 by using Equation 6.29, we can write the element equations (Equation 6.27b) as follows:

$$\left([k_\alpha] + \frac{1}{\Delta t}[k_t]\right) \begin{Bmatrix} p_1^{t+1} \\ p_2^{t+1} \end{Bmatrix} = \{Q^{t+1}\} + \frac{1}{\Delta t}[k_t] \begin{Bmatrix} p_1^t \\ p_2^t \end{Bmatrix} \tag{6.30a}$$

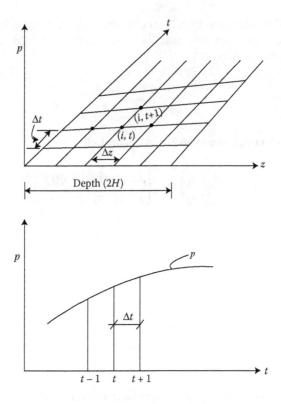

FIGURE 6.7   Time integration for FE analysis.

or

$$[\bar{k}] \cdot \begin{Bmatrix} p_1^{t+1} \\ p_2^{t+1} \end{Bmatrix} = \{\bar{Q}\} \tag{6.30b}$$

where

$$[\bar{k}] = [k_\alpha] + \frac{1}{\Delta t}[k_t]$$

and

$$\{\bar{Q}\} = \{Q^{t+1}\} + \frac{1}{\Delta t}[k_t]\begin{Bmatrix} p_1^t \\ p_2^t \end{Bmatrix}$$

The above Euler integration procedure is referred to as *FE—Scheme 1.*

6.4.2.1.2   FE—Scheme 2

In *FE—Scheme 2*, we use another time integration scheme in which p is expressed as the average at the midpoint of the time increment, Δt:

$$p^{t+\frac{1}{2}} = \frac{1}{2}(p^{t+1} + p^t) \tag{6.31}$$

which leads to the element equations as

$$\left([k_\alpha] + \frac{2}{\Delta t}[k_t]\right)\begin{Bmatrix} p_1^{t+\frac{1}{2}} \\ p_2^{t+\frac{1}{2}} \end{Bmatrix} = \{Q\} + \frac{2}{\Delta t}[k_t]\begin{Bmatrix} p_1^t \\ p_2^t \end{Bmatrix} \tag{6.32}$$

6.4.2.2   Assemblage Equations

The element Equations 6.30a and 6.32 can be assembled by satisfying the inter-element compatibility of p at common nodes between two adjacent elements. The assemblage or global equations can be expressed as

$$[K]\{r\} = \{R\} \tag{6.33}$$

where $[K]$ is the global property matrix composed of $[k_\alpha]$ and $[k_t]$, $\{r\}$ is the global vector of nodal p, and $\{R\}$ is the global vector of nodal applied forces (fluxes). For Equations 6.30a and 6.32, the vector $\{r\}$ is related to the time level $t+1$ or $t+1/2$, respectively.

6.4.2.3   Boundary Conditions or Constraints

There are two types of constraints or boundary conditions that occur in a time-dependent problem: (1) constraints on the boundary of the problem, and (2) initial conditions at the starting time, which often is assumed to be zero ($t = 0$):

$$p(o,t) = \bar{p}_o(t) \quad \text{for } t > 0 \tag{6.34a}$$

$$p(h,t) = \bar{p}_h(t) \quad \text{for } t > 0 \tag{6.34b}$$

where $h = 2H$ is the thickness of the consolidating layer (Figure 6.1) and the overbar denotes the known or specified value of p. The initial condition is expressed as

$$p(z,o) = \bar{p}(z), \quad o \leq z \leq 2H, \text{ and } t \leq 0 \tag{6.34c}$$

6.4.2.4 Solution in Time

Now, the boundary constraints are introduced in Equation 6.33. Then, the time integration is performed by starting at time $0 + \Delta t$ or $(0 + 1)$. Since the values of pore pressures at time $= 0$ are known from the initial conditions, we can solve Equation 6.33 for pore water pressures p at time $0 + \Delta t$. The solution is propagated for various times, that is, $t + \Delta t$, since the values at t are available from the computations at the previous time level and so on.

6.4.2.5 Material Parameters

The material parameters can be obtained from laboratory and/or field tests. The coefficient of permeability can be derived from laboratory permeameter test. Parameters m_v, a_v, λ, and κ can be obtained from triaxial and consolidation tests.

6.5 EXAMPLES

A number of example problems for settlements using the 1-D theory and FE or FD methods are presented in this section. The FE computer code CONS-1DFE [43] is mainly used for solution of problems presented; D.C. Koutsoftas contributed actively toward obtaining these solutions [23]. The code possesses the following capabilities:

1. Single- and multi-layered systems.
2. The parameter c_v can be a function of time. Also, values of c_v can depend upon whether the effective stress at any depth and time is smaller or greater than the maximum past pressure. In other words, different values of c_v can be used for the NC and OC regions.
3. The loading (stress) can be applied instantaneously and maintained constantly with time. It can also be applied in a finite number of steps over specified time intervals.

6.5.1 EXAMPLE 6.1: LAYERED SOIL—NUMERICAL SOLUTIONS BY VARIOUS SCHEMES

Desai and Johnson [24,26] have used the five foregoing FD schemes, and two FE schemes presented before, for 1-D consolidation. This study contained comparisons for convergence, numerical stability, and computational time for the FD and FE schemes. Here, we present an example solved by using various schemes.

Figure 6.8 shows a three-layered saturated system with material properties c_v and k, which were presented by Barden and Younan [28]; the properties were adopted for pressure increment 20–40 psi (138–276 kPa). The number of nodes and elements used are also shown in Figure 6.8. Figure 6.9 shows dissipation of pore water pressure versus time factor for the bottom layer, from closed-form solution, measurements, and numerical solutions, at the interface between the second and third layer; the measurements were obtained by using transducers (at the interface). The results from the FD schemes 2 and 3 were very close; hence, only those from schemes 3,

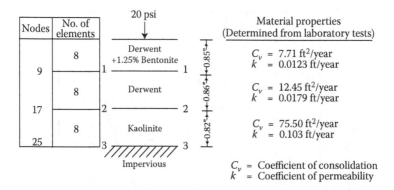

FIGURE 6.8 Three-layered soil systems. (From Desai, C.S. and Johnson, L.D., *International Journal of Numerical Methods in Engineering*, 7, 1973, 243–254. With permission.)

4A, and 4B are included. FD explicit scheme No. 1 required very small Δt, and in general was not stable; hence, it is not shown in Figure 6.9. All numerical schemes seem to yield almost the same correlation with test results, except during the initial time period.

The times taken for numerical predictions by various schemes for the solution of the one-layered system, depicted in Figure 6.6, are plotted in Figure 6.10. As can be seen, the *explicit* schemes (1, 4A, and 4B) take the least time, while the FE schemes take the maximum time. The time taken by the FD *implicit* scheme is between the

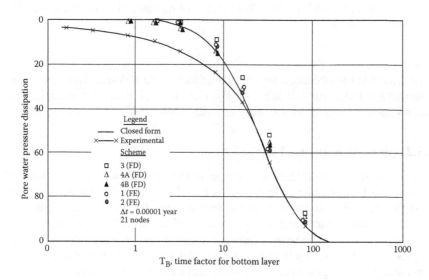

FIGURE 6.9 Pore water pressure dissipation at interface between middle and bottom layer: comparisons between numerical, closed form and test results. (From Desai, C.S. and Johnson, L.D., *International Journal of Numerical Methods in Engineering*, 7, 1973, 243–254. With permission.)

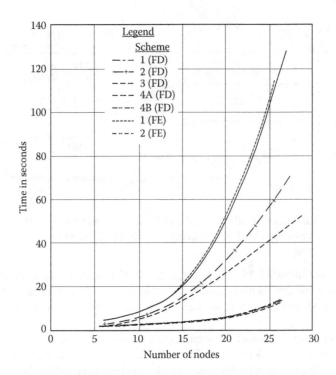

FIGURE 6.10 Comparison of total computational times ($\Delta T = 0.125$). (From Koutsoftas, D.C. and Desai, C.S., One-dimensional consolidation by finite elements: Solutions of some practical problem, Report No. VPI-E-76-17, Dept. of Civil Eng., VPI&SU, Blacksburg, VA, 1976. With permission.)

FD *explicit* and FE schemes. It may be noted that these results are for a simple homogeneous case and linear soil behavior. For more realistic cases, for example, layer systems and nonlinear behavior, the results can be different.

6.5.2 EXAMPLE 6.2: TWO-LAYERED SYSTEM

Figure 6.11a shows a two-layered system analyzed by Boehmer and Christian [44]. We used the FE computer code-cons-1D FE [43] to obtain the numerical solution using the value of c_v for the first layer to be four times that for the second (bottom) layer c_{v2}; the value of the latter was adopted as 0.07 ft²/day (65.1 cm²/day).

Figure 6.12 shows comparisons between results using 2-D consolidation [44] and the present 1-D FE computations for normalized depth z/H versus normalized excess pore water pressure $\Delta p/q$, where $q = \sigma$ is the applied stress. The results are plotted for various values of the time factor, $T = (c_v t/H^2)$, in which $c_v = 0.07$ ft²/day (65.1 cm²/ day) was used. Figure 6.13 shows the isochrones of excess pore water pressures for the two-layered system in which the c_v of the bottom layer is four times that of the top layer (Figure 6.11b). The results from the 2-D FE for typical time factors are

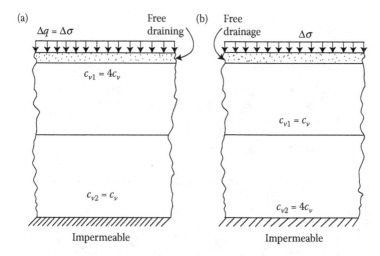

FIGURE 6.11 Consolidation in two-layered system. (From Koutsoftas, D.C. and Desai, C.S., One-dimensional consolidation by finite elements: Solutions of some practical problem, Report No. VPI-E-76-17, Dept. of Civil Eng., VPI&SU, Blacksburg, VA, 1976; Boehmer, J.W. and Christian, J.T., *Journal of the Soil Mechanics and Foundations Engineering, ASCE*, 96(SM4), 1970. With permission.)

compared with those from the present 1-D FE analyses. The correlation between the two results is considered very good. Thus, for problems with certain loading, geometry, and boundary conditions (Figure 6.11) the 1-D solution can also yield acceptable results.

6.5.3 EXAMPLE 6.3: TEST EMBANKMENT ON SOFT CLAY

Figure 6.14 shows a test embankment on a soft to medium Boston Blue clay, about 100 ft (30.50 m) thick presented by Lamb et al. [25]; the following details of soil tests and field data are adopted from Ref. [25]; the soil properties are also shown in the figure. Figure 6.15 shows various physical properties of the soil from results of oedometer tests; it shows that the top half of the layer is overconsolidated, whereas the bottom half is normally consolidated. Field measurements for settlement of the surface and excess pore water pressures with depth were reported over a period of 10 years [25].

Lamb et al. [25] performed comprehensive tests to determine the compressibility of the soil; Figure 6.16 shows the coefficient of consolidation and coefficient of volume change using the oedometer and triaxial stress path tests. Lamb et al. [25] used such data to predict the performance of the embankment by using the analytical solution for the 1-D consolidation equation. Two analyses as shown in Figure 6.17 were performed: (1) with average values of c_v and m_v, and (2) with distributed values of c_v and m_v.

In the 1-D FE procedure presented here, the average curve for the co-efficient of consolidation versus depth (shown in Figure 6.16) was used. The surface settlements,

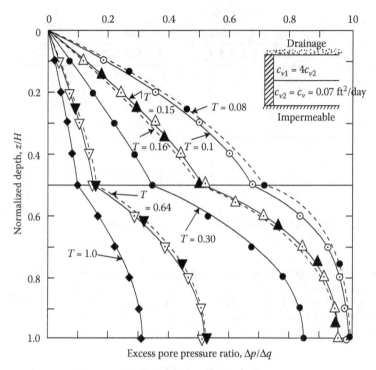

Note: Time factor T computed using $c_v = 0.07$ ft^2/day

Legend: ● ▲▼   Results from Boehmer and Christian [44]

All other symbols: Results based on 1-D F.E. solution

FIGURE 6.12 Isochrones during consolidation for two-layered system with $c_v = 4c_v$. (From Koutsoftas, D.C. and Desai, C.S., One-dimensional consolidation by finite elements: Solutions of some practical problem, Report No. VPI-E-76-17, Dept. of Civil Eng., VPI&SU, Blacksburg, VA, 1976; Boehmer, J.W. and Christian, J.T., *Journal of the Soil Mechanics and Foundations Engineering, ASCE,* 96(SM4), 1970. With permission.)

$w(t)$, were evaluated by using the nonlinear stress–strain model and the following equation:

$$w(t) = \sum_{i=1}^{M} \varepsilon_i \ell_i = \sum_{i=1}^{M} \left[\lambda \log\left(\frac{\sigma'}{\sigma_o'} \right) \right] \ell_i \qquad (6.35)$$

The average values of $\lambda = 0.25$ were used for the settlement calculations.

The measured excess pore water pressures after 10 years are compared with those from the 1-D theory [25], and from the present 1-D FE analysis (Figure 6.17a). The two analyses do not compare well with the measured values. The results from 1-D theory with distribution of c_v and m_v [25], and those from the 1-DFE are essentially the same.

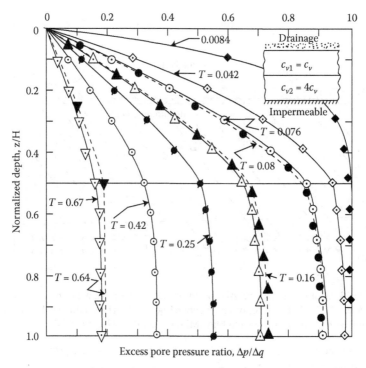

FIGURE 6.13 Isochrones during consolidation for two-layered system with $c_{v2} = 4c_v$. (Adapted from Koutsoftas, D.C. and Desai, C.S., One-dimensional consolidation by finite elements: Solutions of some practical problem, Report No. VPI-E-76-17, Dept. of Civil Eng., VPI&SU, Blacksburg, VA, 1976; Boehmer, J.W. and Christian, J.T., *Journal of the Soil Mechanics and Foundations Engineering, ASCE,* 96(SM4), 1970.)

The difference in predictions and test data can be due to a number of factors such as the Terzaghi 1-D theory does not account for multidimensional effects; hence, the pore water pressure predictions are higher than the field data. Another factor may be that the laboratory test data are affected by the sample disturbance. Hence, we performed additional analyses by adopting c_v of overconsolidated soil two times (chosen arbitrarily) that from laboratory tests, and c_v for normally consolidated case as three times (chosen arbitrarily) that from the laboratory tests. Figure 6.17b shows good comparisons between measured data and predictions from 1-DFE procedure with such modified parameters.

Figure 6.18 shows the measured displacements at a (surface) location. Lamb et al. [25] reported that the data from this location were probably representative of the section analyzed. A number of parametric calculations for settlement predictions were made such as by varying the values of k, c_v, and m_v. The predictions obtained

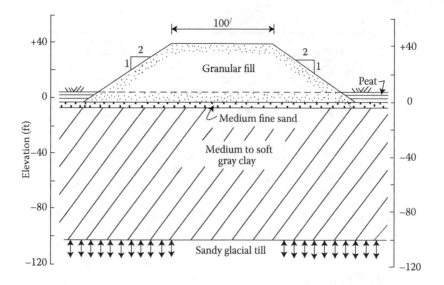

FIGURE 6.14 Consolidation analysis: northeast test embankment (1 ft = 0.305 m). (From Lamb, T.W. et al., The Performance of a Foundation Under a High Embankment, Research Report R71-22, Soil Mech. Div., Dept. of Civil Eng., MIT, Cambridge, MA, 1972. With permission.)

by adjusting the value of c_v as three times the laboratory value, provided good correlation with the measurements.

6.5.4 EXAMPLE 6.4: CONSOLIDATION FOR LAYER THICKNESS INCREASES WITH TIME

The thickness of the consolidating layer increases with time in situations like construction of dam with a clay core when the sediment (soil) deposited during construction increases with time, and deposition of sediments in the ocean by rivers which carry great amounts of sediment.

A closed-form solution for the problem was presented by Gibson [10] for the clay layer thickness increases linearly with time, with the layer lying on impermeable base, and also on permeable base. Hence, two analyses were performed by using the CONS-1-DFE procedure in which the consolidating layer lies on impermeable and permeable bases; in both cases, the top of the layer was assumed to be permeable.

Figures 6.19a and 6.19b show excess pore pressure distributions with depth for the two cases, in comparison with the closed-form solution reported by Gibson [10]. The results are shown for three values of time factor (ft/year) defined as $T = m_v^2 t / c_v$. The 1-DFE predictions compare very well with the closed form solutions, and the results show that the excess pore water pressures increase with time continually, and therefore, the degree of consolidation decreases with time.

6.5.5 EXAMPLE 6.5: NONLINEAR ANALYSIS

We solve the consolidation problem involving a thick deposit of Boston Blue clay (Figure 6.14), which contains approximately the top half of OC clay and the bottom

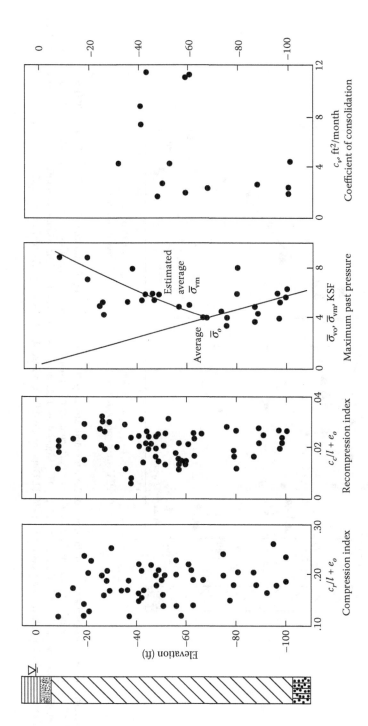

FIGURE 6.15 Quantities from oedometer test: northeast test embankment (1 KSF = 4.79 N/cm²; 1 ft²/month = 929 cm²/month, 1 ft = 30.48 cm). (From Lamb, T.W. et al., The Performance of a Foundation Under a High Embankment, Research Report R71-22, Soil Mech. Div., Dept. of Civil Eng., MIT, Cambridge, MA, 1972; Koutsoftas, D.C. and Desai, C.S., One-dimensional consolidation by finite elements: Solutions of some practical problem, Report No. VPI-E-76-17, Dept. of Civil Eng, VPI&SU, Blacksburg, VA, 1976. With permission.)

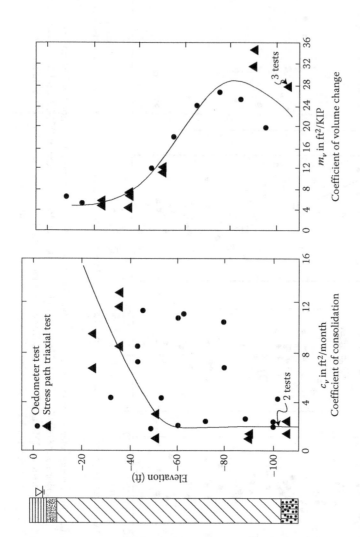

FIGURE 6.16 Comparisons of data from oedometer and stress path triaxial tests (1 ft²/month = 929 cm²/month; 1 ft²/KIP = 0.21 cm²/N, 1 ft = 30.48 cm). (Adapted from Lamb, T.W. et al., The Performance of a Foundation Under a High Embankment, Research Report R71-22, Soil Mech. Div., Dept. of Civil Eng., MIT, Cambridge, MA, 1972; Koutsoftas, D.C. and Desai, C.S., One-dimensional consolidation by finite elements: Solutions of some practical problem, Report No. VPI-E-76-17, Dept. of Civil Eng., VPI&SU, Blacksburg, VA, 1976.)

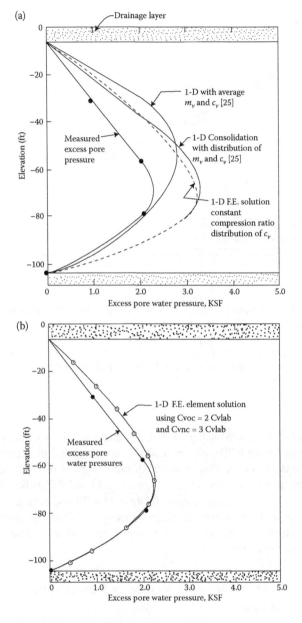

FIGURE 6.17 Comparisons of predicted and measured excess pore water pressure 10 years after construction (1 KSF = 4.79 N/cm²; 1 ft = 30.48 cm). (a) Excess pore water pressure versus elevation for variations for c_v and 1-D FE solution; (b) excess pore water pressure versus elevation for 1-D FE solution and measured values. (Adapted from Koutsoftas, D.C. and Desai, C.S., One-dimensional consolidation by finite elements: Solutions of some practical problem, Report No. VPI-E-76-17, Dept. of Civil Eng., VPI&SU, Blacksburg, VA, 1976; Lamb, T.W. et al., The Performance of a Foundation Under a High Embankment, Research Report R71-22, Soil Mech. Div., Dept. of Civil Eng., MIT, Cambridge, MA, 1972.)

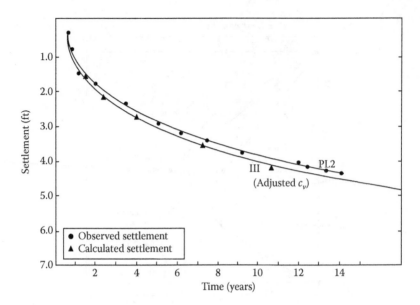

FIGURE 6.18 Comparisons between FE predictions and measured settlements. (Adapted from Koutsoftas, D.C. and Desai, C.S., One-dimensional consolidation by finite elements: Solutions of some practical problem, Report No. VPI-E-76-17, Dept. of Civil Eng., VPI&SU, Blacksburg, VA, 1976; Lamb, T.W. et al., The Performance of a Foundation Under a High Embankment, Research Report R71-22, Soil Mech. Div., Dept. of Civil Eng., MIT, Cambridge, MA, 1972.)

half of NC clay (Figure 6.15) using a nonlinear analysis. The nonlinear distribution of c_v is shown in Figure 6.20a. Figure 6.20b shows computed distributions of axial strains over the depth of the clay at three different degrees of consolidation based on the foregoing nonlinear Procedure 1. Figures 6.21a and 6.21b show distributions of strains at two typical degrees of consolidation, $\overline{U} = 33$ and 88% (approximately), respectively. The predictions were obtained by using non linear procedures 1 and 2, described previously. The predictions for the nonlinear analyses can be considered to be satisfactory and realistic.

6.5.6 EXAMPLE 6.6: STRAIN-BASED ANALYSIS OF CONSOLIDATION IN LAYERED CLAY

In this example, we analyze 1-D consolidation in layered clay using the FD approach. The analysis method, proposed by Mikasa [45] and extended by Kim and Mission [46], is based on compressive strain instead of excess pore pressure used in the formula by Terzaghi [2]. The numerical solutions consider interface boundary conditions in terms of infinitesimal strains, and the results are compared with those obtained from the consolidation theory by Terzaghi [2].

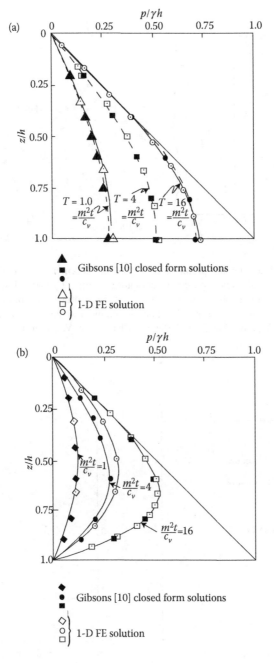

FIGURE 6.19 Excess pore water pressures versus depth for layer with impermeable base: increasing thickness with time. (a) Impermeable base; (b) permeable top and base. (Adapted from Gibson, R.E., *Geotechnique*, 8(4), 1958, 171–182; Koutsoftas, D.C. and Desai, C.S., One-dimensional consolidation by finite elements: Solutions of some practical problem, Report No. VPI-E-76-17, Dept. of Civil Eng., VPI&SU, Blacksburg, VA, 1976.)

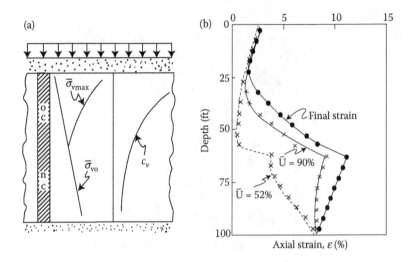

FIGURE 6.20 Finite element predications for test embankment on clay (1 ft = 30.48 cm). (a) Variations of initial stress and c_v; (b) computed strains at various levels of consolidation (U) (U, degree of consolidation). (Adapted from Koutsoftas, D.C. and Desai, C.S., One-dimensional consolidation by finite elements: Solutions of some practical problem, Report No. VPI-E-76-17, Dept. of Civil Eng., VPI&SU, Blacksburg, VA, 1976.)

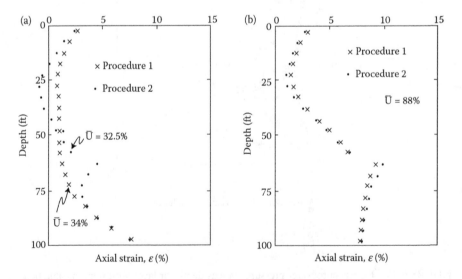

FIGURE 6.21 Predicted axial strains versus depth for test embankment on clay (1 ft = 30.48 cm). (a) $U \simeq 33\%$; (b) $U = 88\%$. (Adapted from Koutsoftas, D.C. and Desai, C.S., One-dimensional consolidation by finite elements: Solutions of some practical problem, Report No. VPI-E-76-17, Dept. of Civil Eng., VPI&SU, Blacksburg, VA, 1976.)

Mikasa [45] derived the following generalized 1-D consolidation equation in terms of strains for a clay layer having homogenous consolidation properties

$$c_v \frac{\partial^2 \varepsilon}{\partial z^2} = \frac{\partial \varepsilon}{\partial t} \tag{6.36}$$

where ε is the compressive strain, c_v is the coefficient of consolidation, z is the depth, and t is the consolidation time (Figure 6.22). The relationship between strain (ε) and the change in effective vertical stress $\Delta\sigma'$ at a given depth and time can be expressed as

$$\varepsilon = m_v \Delta\sigma_{v'} = m_v (\Delta\sigma_v - p) \tag{6.37}$$

where m_v is the volume compressibility coefficient and p is the excess pore pressure. The interface boundary condition requires that there be a single excess pore pressure at the interface and that the flow q from one layer into the other layer be equal (Figure 6.23). Using Darcy's law, these conditions can be expressed as follows:

$$q = \frac{k_1}{\gamma_w} \left(\frac{\partial p}{\partial z} \right)_1 A = \frac{k_2}{\gamma_w} \left(\frac{\partial p}{\partial z} \right)_2 A \tag{6.38a}$$

or

$$q = k_1 \left(\frac{\partial p}{\partial z} \right)_1 = k_2 \left(\frac{\partial p}{\partial z} \right)_2 \tag{6.38b}$$

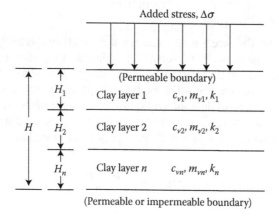

FIGURE 6.22 Consolidation in layered clay: geometry and properties. (From Kim, H.-J. and Mission, J.L., *International Journal of Geomechanics, ASCE*, 11(1), 2011, 72–77. With permission.)

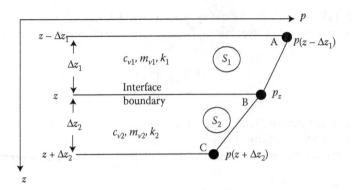

FIGURE 6.23 Variation of excess pore-water pressure at sublayers. (From Kim, H.-J. and Mission, J.L., *International Journal of Geomechanics, ASCE*, 11(1), 2011, 72–77. With permission.)

where k is the coefficient of permeability, γ_w is the unit weight of water, A is the cross-sectional area, and subscripts 1 and 2 indicate the upper layer and lower layer, respectively (Figure 6.23). Equation 6.38b can be written in a FD form as

$$k_1 \frac{p_z - p_{(z-\Delta z_1)}}{\Delta z_1} = k_2 \frac{p_{(z+\Delta z_2)} - p}{\Delta z_2} \tag{6.39}$$

Using the nodal notations in Figure 6.23, Equation 6.39 can be written in terms of nodal subscripts A, B, and C and solved for the excess pore water pressure at the interface, p_B, as

$$p_B = \frac{\alpha_1 p_A + \alpha_2 p_C}{\alpha_1 + \alpha_2} \tag{6.40}$$

where p_A and p_C are the excess pore pressures at A and C, respectively, $\alpha_1 = (k_1/\Delta z_1)$ and $\alpha_2 = (k_2/\Delta z_2)$ (Figure 6.23). The continuity relation in Equation 6.40 shows that the excess pore pressure at the interface is unique and produces the settlements S_1 and S_2 in the sublayer above and below the interface, respectively. As shown by Kim and Mission [46], for a given $\Delta\sigma$ the sublayer consolidation settlements, S_1 and S_2, can be approximated by integrating numerically the excess pore pressure distribution. Using the trapezoidal method of integration, S_1 and S_2 can be expressed as follows:

$$S_i = m_{vi}\Delta\sigma\Delta z_i - m_{vi}\left(\frac{p_i + p_{i+1}}{2}\right)\Delta z_i \tag{6.41}$$

Similarly, if the strain at the interface is defined by two adjacent strains ε_{B1} and ε_{B2} above and below the interface, respectively (Figure 6.24), the sublayer settlements S_1 and S_2 can be calculated in terms of strain ε, using the trapezoidal rule, as

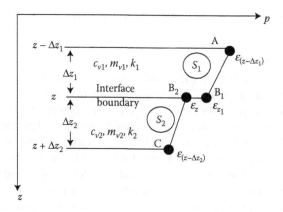

FIGURE 6.24 Variation of strain at sublayers. (From Kim, H.-J. and Mission, J.L., *International Journal of Geomechanics, ASCE*, 11(1), 2011, 72–77. With permission.)

$$S_i = \left(\frac{\varepsilon_i + \varepsilon_{i+1}}{2} \right) \Delta z_i \qquad (6.42)$$

From Equation 6.37, the interface strains ε_{B1} and ε_{B2} can be written as

$$\varepsilon_{B1} = m_{vi}(\Delta\sigma - p_B) \qquad (6.43a)$$

$$\varepsilon_{B2} = m_{v2}(\Delta\sigma - p_B) \qquad (6.43b)$$

Finally, using Equation 6.40, Equations 6.43a and 6.43b can be rewritten in the following form:

$$\varepsilon_{B1} = \varepsilon_A \left[\frac{\alpha_1}{(\alpha_1 + \alpha_2)} \right] + \varepsilon_C \left[\frac{\alpha_2}{(\alpha_1 + \alpha_2)} \left(\frac{m_{v1}}{m_{v2}} \right) \right] \qquad (6.44a)$$

$$\varepsilon_{B2} = \varepsilon_C \left[\frac{\alpha_2}{(\alpha_1 + \alpha_2)} \right] + \varepsilon_C \left[\frac{\alpha_1}{(\alpha_1 + \alpha_2)} \left(\frac{m_{v2}}{m_{v1}} \right) \right] \qquad (6.44b)$$

To use Equations 6.44a and 6.44b in the FD form of the governing differential equation (Equation 6.36), two adjacent nodes are defined at the interface (Figure 6.24). These nodes are assumed to have the same elevation, but different strains for the same excess pore pressure. This treatment of interface conditions allows more accurate estimate of settlements in the sublayers above and below the interface. The solution involves determining the initial and boundary conditions, applying Equation 6.36 to the FD grid in each layer except at the interface. Equations 6.44a and 6.44b are used to determine the interface strains at any time step. Additional details can be found in Refs. [45,46].

6.5.6.1 Numerical Example

A two-layered saturated clay profile is subjected to a constant and uniform surcharge load of 18 ton/m², as shown in Figure 6.25. Layer thicknesses and consolidation properties are also shown in the figure. Two different cases are considered. For Case 1, the bottom boundary is considered impermeable, which implies one-way drainage (from the top). For Case 2, the same boundary is considered permeable, allowing drainage from both top and bottom. FD solutions were obtained using $\Delta z = 0.5$ m and time step, $\Delta t = 1$ day, which satisfies the stability criterion [45]. Excess pore water pressure profiles at $t = 800$ days for both cases are shown in Figure 6.26. The solutions adequately capture the different boundary conditions (impermeable and permeable) at the bottom. The corresponding strain profiles at $t = 800$ days are shown in Figure 6.27. Equivalent solutions obtained from the Terzaghi's consolidation theory are superimposed on the same figure for comparison. The results obtained from the strain-based approach considered in this example compare well with those from the Terzaghi's consolidation theory [2]. The time (t) versus settlement (S) curve in Figure 6.28 shows that the strain-based method gives almost identical results as those from the Terzaghi's consolidation theory [2].

6.5.7 EXAMPLE 6.7: COMPARISON OF UNCOUPLED AND COUPLED SOLUTIONS

In this example, we compare uncoupled, coupled, and the 1-D consolidation theories by Terzaghi [2] using the FEM. It is seen that the Terzaghi FE solutions may not satisfy the flow continuity conditions at the interfaces between soil layers. Also, the

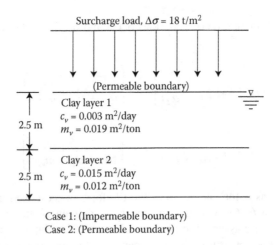

FIGURE 6.25 Two-layer clay including consolidation properties. (From Kim, H.-J. and Mission, J.L., *International Journal of Geomechanics, ASCE*, 11(1), 2011, 72–77. With permission.)

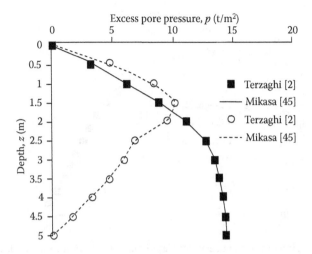

FIGURE 6.26 Excess pore-water profiles for Case 1 and Case 2. (From Kim, H.-J. and Mission, J.L., *International Journal of Geomechanics, ASCE*, 11(1), 2011, 72–77. With permission.)

average degree of consolidation, as defined by settlement and excess pore pressure, is different for layered systems [47].

6.5.7.1   Uncoupled Solution

As noted earlier, for layered systems, coefficient of consolidation, c_v, can vary with depth, z. The governing differential equation for 1-D consolidation (Equation 6.28a) can be written in an uncoupled form as

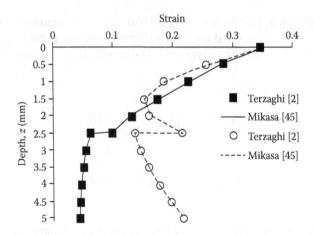

FIGURE 6.27 Strain profiles for Case 1 and Case 2. (From Kim, H.-J. and Mission, J.L., *International Journal of Geomechanics, ASCE*, 11(1), 2011, 72–77. With permission.)

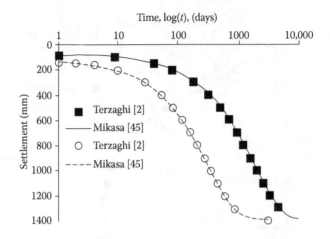

FIGURE 6.28 Time–settlement curve for Case 1 and Case 2. (From Kim, H.-J. and Mission, J.L., *International Journal of Geomechanics, ASCE*, 11(1), 2011, 72–77. With permission.)

$$\frac{1}{\gamma_w}\frac{\partial}{\partial z}k\frac{\partial p}{\partial z} = m_v\frac{\partial p}{\partial t} \qquad (6.45)$$

After solution by the Galerkin-weighted residual method [48], Equation 6.45 leads to the element equation of the form

$$[k_C]\{p\} + [m_m]\left\{\frac{dp}{dt}\right\} = \{0\} \qquad (6.46)$$

where $[k_C]$ and $[m_m]$ are the fluid conductivity and mass matrices, respectively. Using linear interpolations and fixed time steps, the ordinary (matrix) differential equation (Equation 6.46) can be written at two consecutive time steps "0" and "1" as follows:

$$[k_C]\{p\}_0 + [m_m]\left\{\frac{dp}{dt}\right\}_0 = \{0\} \qquad (6.47a)$$

$$[k_C]\{p\}_1 + [m_m]\left\{\frac{dp}{dt}\right\}_1 = \{0\} \qquad (6.47b)$$

Using a weighted average of the gradients at the beginning and end of the time interval Δt, we can write from Equations 6.47a and 6.47b

$$\{p\}_1 = \{p\}_0 + \Delta t \left((1-\theta)\left\{\frac{dp}{dt}\right\}_0 + \theta\left\{\frac{dp}{dt}\right\}_1 \right)$$ (6.47c)

where $0 \le \theta \le 1$. Elimination of $\{dp/dt\}_0$ and $\{dp/dt\}_1$ from Equations 6.47a and 6.47c leads to the following recurrence equation, after assembly between steps "0" and "1"

$$([M_m] + \theta\Delta t[K_C])\{p\}_1 = ([M_m] - (1-\theta)\Delta t[K_C])\{p\}_0$$ (6.48)

The solution of Equation 6.48 gives the distribution of excess pore pressure. The corresponding distribution of settlement, s, can be obtained from

$$s = \int_z^{D-z} m_v(\sigma - p)dz$$ (6.49)

where σ is the total stress. It is shown subsequently that by applying this solution to Terzaghi's 1-D consolidation equation (Equation 6.4b) will give wrong answers because it is unable to explicitly represent changes in the permeability k, and is therefore unable to enforce the interface flow continuity Equation 6.47.

6.5.7.2 Coupled Solution

For this case, the governing differential and continuity equations, respectively, are given by

$$\frac{\partial p}{\partial z} + \frac{\partial}{\partial z}\left(\frac{1}{m_v}\frac{\partial s}{\partial z}\right) = 0$$ (6.50a)

$$\frac{\partial}{\partial z}\left(\frac{k}{\gamma_w}\frac{\partial p}{\partial z}\right) + \frac{\partial}{\partial t}\frac{\partial s}{\partial z} = 0$$ (6.50b)

where s represents settlement at depth z. Following the Galerkin-weighted residual method [48], Equations 6.50a and 6.50b lead to element matrix equations of the form

$$[k_m]\{s\} + [c]\{p\} = \{f\}$$ (6.51a)

$$[c]^T\left\{\frac{ds}{dt}\right\} - [k_C]\{p\} = \{0\}$$ (6.51b)

where $[k_m]$ and $[k_C]$ are the solid stiffness and fluid conductivity matrices, respectively, $[c]$ is the connectivity matrix and $\{f\}$ is total applied force vector. Denoting $\{\Delta f\}$ as the change in load between successive times and $\{\Delta s\}$ as the resulting change in displacements, and using linear interpolation in time in terms of θ, Equation 6.51a leads to

$$\{\Delta s\} = \Delta t \left((1 - \theta) \left\{ \frac{ds}{dt} \right\}_0 + \theta \left\{ \frac{ds}{dt} \right\}_1 \right) \qquad (6.52a)$$

Likewise, Equation 6.51b can be written at the two consecutive time levels to obtain the derivatives which can then be eliminated to yield the following recurrence equations in incremental form:

$$\begin{bmatrix} [k_m] & [c] \\ [c]^T & -\theta \Delta t [k_C] \end{bmatrix} \begin{Bmatrix} \{\Delta s\} \\ \{\Delta p\} \end{Bmatrix} = \begin{Bmatrix} \{\Delta f\} \\ \Delta t [k_C]\{p\}_0 \end{Bmatrix} \qquad (6.52b)$$

Finally, the displacements and pore pressure vectors are updated at each time increment using the following equations:

$$\{s\}_1 = \{s\}_0 + \{\Delta s\} \qquad (6.53a)$$

$$\{p\}_1 = \{p\}_0 + \{\Delta p\} \qquad (6.53b)$$

The average degree of consolidation can then be expressed either in terms of excess pore pressure, $U_{av} = 1 - (1/D)\int_0^D (p/p_0)dz$ or settlement $U_{av} = (s_t/s_u)$, where s_t and s_u represent settlements at time t and ultimate settlement U, respectively.

6.5.7.3 Numerical Example

A two-layer system, shown in Figure 6.29, is analyzed here using the aforementioned method (after Huang and Griffiths [47]) and the FE formula of Terzaghi's 1-D consolidation theory [2], for comparison. The geometric and material properties are shown in Table 6.1. Both elements have the same c_v, but different k/γ_w and m_v.

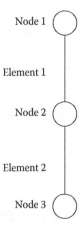

FIGURE 6.29 Schematic of a two-layer system. (From Huang, J. and Griffiths, D.V., *Geotechnique*, 60(9), 2010, 709–713. With permission.)

TABLE 6.1

Geometric and Material Parameters Used

Element	Length	k/γ_w	m_v	C_v
1	1	10	10	1
2	1	1	1	1

For the conventional 1-D consolidation theory by Terzaghi [2], the element matrices are given by

$$[k_C]_1 = [k_C]_2 = \begin{bmatrix} 1 & -1 \\ -1 & 1 \end{bmatrix}$$ (6.54a)

$$[m_m]_1 = [m_m]_2 = \begin{bmatrix} 1/3 & 1/6 \\ 1/6 & 1/3 \end{bmatrix}$$ (6.54b)

In view of Equations 6.54a and 6.54b, the global fluid conductivity matrix, $[K_C]$, and global mass matrix, $[M_m]$, can be written as

$$[K_C] = \begin{bmatrix} 1 & -1 & 0 \\ -1 & 2 & -1 \\ 0 & -1 & 1 \end{bmatrix}$$ (6.55a)

$$[M_m] = \begin{bmatrix} 1/3 & 1/6 & 0 \\ 1/6 & 2/3 & 1/6 \\ 0 & 1/6 & 1/3 \end{bmatrix}$$ (6.55b)

For the uncoupled case, element fluid conductivity and mass matrices are given as

$$[k_C]_1 = \begin{bmatrix} 10 & -10 \\ -10 & 10 \end{bmatrix}; \quad [k_C]_2 = \begin{bmatrix} 1 & -1 \\ -1 & 1 \end{bmatrix}$$ (6.56a)

$$[m_m]_1 = \begin{bmatrix} 10/3 & 10/6 \\ 10/6 & 10/3 \end{bmatrix}; \quad [m_m]_2 = \begin{bmatrix} 1/3 & 1/6 \\ 1/6 & 1/3 \end{bmatrix}$$ (6.56b)

In view of Equations 6.56a and 6.56b, the global fluid conductivity and mass matrices can be expressed as

$$[K_C] = \begin{bmatrix} 10 & -10 & 0 \\ -10 & 11 & -1 \\ 0 & -1 & 1 \end{bmatrix} \tag{6.57a}$$

$$[M_m] = \begin{bmatrix} 10/3 & 10/6 & 0 \\ 10/6 & 11/3 & 1/6 \\ 0 & 1/6 & 1/3 \end{bmatrix} \tag{6.57b}$$

As noted by Huang and Griffiths [47], from the comparison of Equation 6.55a with Equation 6.57a and Equation 6.55b with Equation 6.57b, it is evident that the global fluid conductivity and mass matrices are different that will lead to different solutions for pore pressure distributions and settlements for layered soils. For layered soils, the solutions reported by Huang and Griffiths [47] are considered to be correct.

REFERENCES

1. Terzaghi, K., *Erdbaumechanik auf Bodenphysikalischer Gundlage*, F. Deuticke, Vienna, 1925.
2. Terzaghi, K., *Theoretical Soil Mechanics*, John Wiley & Sons, New York, 1943.
3. Terzaghi, K. and Peck, R.B., *Soil Mechanics in Engineering Practice*, John Wiley & Sons, New York, 1955.
4. Taylor, D.W. and Merchant, W., A theory of clay consolidation accounting for secondary compression, *Journal of Mathematical Physics*, 9(3), 1940, 67–185.
5. Taylor, D.W., *Fundamentals of Soil Mechanics*, John Wiley & Sons, New York, 1955.
6. Suklje, L., *Rheological Aspects of Soil Mechanics*, Wiley Interscience, London, 1969.
7. Rendulic, L., Porenziffer und Poren Wasserdruck in Tonen, *Der Bauingenieur*, 17, 1936, 559–564.
8. Mandel, J., Consolidation des Sols, *Geotechnique*, 3, 1953, 287–299.
9. Skempton, A.W. and Bjerrum, L., A contribution to settlement analysis of foundations on clay, *Geotechnique*, 7, 1957, 168–178.
10. Gibson, R.E., The progress of consolidation in a clay layer increasing in thickness with time, *Geotechnique*, 8(4), 1958, 171–182.
11. Lo, K.Y., Secondary compression of clays, *Journal of the Soil Mechanics and Foundations Engineering, ASCE*, 87(4), 1961, 61–87.
12. Schiffman, R.L. and Gibson, R.E., Consolidation of nonhomogeneous clay layers, *Journal of the Soil Mechanics and Foundations Engineering, ASCE*, 90(SM5), 1964, 1–30.
13. Gibson, R.E., England, G.L., and Hussey, M.J.L., The theory of one-dimensional consolidation of saturated clays, *Geotechnique*, 17, 1967, 261–273.
14. Desai, C.S., A rheological model for consolidation of layered soils, *Journal of Indian National Society of Soil Mechanics and Foundation Engineering*, 8(4), 1969, 359–374.
15. Schiffman, R.L. and Stein, J.R., One-dimensional consolidation of layered systems, *Journal of the Soil Mechanics and Foundations Engineering, ASCE*, 96(SM4), 1970, 1495–1504.
16. Davis, R.O., Numerical approximation of one-dimensional consolidation, *International Journal of Numerical Methods in Engineering*, 4, 1972, 279–287.

17. Mesri, G. and Rokhsar, A., Theory of consolidation of clays, *Journal of the Geotechnical. Engineering Division, ASCE*, 100, 1974, 889–904.
18. Desai, C.S. and Christian, J.T., *Numerical Methods in Geotechnical Engineering*, McGraw-Hill Book Co., New York, 1977.
19. Suklje, L. and Kovačič, I., Consolidation of drained multilayer viscous soils, in *Evaluation and Prediction of Subsidence*, (Saxena, S.K., Editor), Int. Conference, Pensacola Beach, FL, 1978.
20. Barden, L., Consolidation of clay with non-linear viscosity, *Geotechnique*, 15, 1965, 345–362.
21. Kuppusamy, T. and Anandkrishnan, M., Nonlinear consolidation characteristics of thick clay layer, *Journal of Indian National Society of Soil Mechanics and Foundation Engineering*, 1(3), 1971, 237–248.
22. Davis, E.H. and Raymond, G.P., A nonlinear theory of consolidation, *Geotechnique*, 15, 1965, 161–173.
23. Koutsoftas, D.C. and Desai, C.S., One-dimensional consolidation by finite elements: Solutions of some practical problem, Report No. VPI-E-76-17, Dept. of Civil Eng., VPI&SU, Blacksburg, VA, 1976.
24. Desai, C.S. and Johnson, L.D., Evaluation of some numerical schemes for consolidation, *International Journal of Numerical Methods in Engineering*, 7, 1973, 243–254.
25. Lamb, T.W., D'Appolonia, D.J., Karlsud, K., and Kirby, R.C., The Performance of a Foundation Under a High Embankment, Research Report R71-22, Soil Mech. Div., Dept. of Civil Eng., MIT, Cambridge, MA, 1972.
26. Desai, C.S. and Johnson, L.D., Evaluation of two finite element schemes for one-dimensional consolidation, *International Journal of Computers and Structures*, 2(4), 1972, 469–486.
27. Desai, C.S., Kuppusamy, T., Koutsoftas, D.C., and Janardharam, R., A one-dimensional finite element procedure for nonlinear consolidation, *Proceedings of the 3rd International Conference on Num. Methods in Geomechanics*, Aachen, Germany, April 1979.
28. Barden, L. and Younan, N.A., Consolidation of layered clays, *Can. Geotech. J.*, 6(4), 1969, 413–429.
29. Lamb, T.W. and Whitman, R.V., *Soil Mechanics*, John Wiley & Sons, New York, 1969.
30. Schmidt, J.D. and Westmann, R.A., Consolidation of porous media with non-Darcy flow, *Journal of Engineering Mechanics Division, ASCE*, 99(EM6), 1973, 1201–1216.
31. Hansbo, S., Consolidation of clay, with special reference to influence of vertical sand drains, *Proceedings, Swedish Geotech. Institute*, 18, 1960, 1–159.
32. Parkin, A.K., Field solutions for turbulent seepage flow, *J. of Soil Mech. and Found. Div., ASCE*, 97(SM1), 1971, 209–219.
33. Courant, R. and Hilbert, D., *Methods of Mathematical Physics*, Interscience Publishers Ltd., Vol. 1, London, 1953.
34. Crandall, S.H., *Engineering Analysis*, McGraw-Hill, New York, 1956.
35. Richtmeyer, R.D. and Morton, K.W., *Difference Methods for Initial Value Problems*, Interscience, New York, 1957.
36. Forsythe, G.E. and Wasow, W.R., *Finite Difference Methods for Partial Differential Equations*, John Wiley, New York, 1960.
37. Zienkiewicz, O.C., *The Finite Element Method in Structural and Continuum Mechanics*, McGraw-Hill Publ. Co., London, 1967.
38. Desai, C.S. and Abel, J.F., *Introduction to the Finite Element Method*, Van Nostrand Reinhold Co., New York, 1972.
39. Desai, C.S., *Elementary Finite Element Method*, Prentice Hall, Englewood Cliffs, NJ, 1979; revised as *Introductory Finite Element Methods* by Desai, C.S. and Kundu, T., CRC Press, Boca Raton, FL, 2001.

40. Saul'ev, V.K., *On a Method of Numerical Integration of the Equation of Diffusion*, Doklady Akad. Nauk., USSR, Vol. 115, 1957, pp. 1077–1079.
41. Larkin, B.K., Some stable explicit difference approximations to the diffusion equation, *Journal of Mathematical Computing*, 18, 1964, 196–202.
42. Desai, C.S. and Sherman, W.C., Unconfined transient seepage in sloping banks, *Journal of the Soil Mechanics and Foundations Engineering, ASCE*, 97, 1971, 357–373.
43. Desai, C.S., *CONS-1DFE: Computer Code for One-Dimensional Coordination*, Report, Virginia Tech, Blacksburg, VA, 1975.
44. Boehmer, J.W. and Christian, J.T., Plane strain consolidation by finite elements, *Journal of the Soil Mechanics and Foundations Engineering, ASCE*, 96(SM4), 1970.
45. Mikasa, M., *The Consolidation of Soft Clay—A New Consolidation Theory and Its Application*, Kajima Institution, Tokyo, 1963 (in Japanese).
46. Kim, H.-J. and Mission, J.L., Numerical analysis of one-dimensional consolidation in layered clay using interface boundary relations in terms of infinitesimal strain, *International Journal of Geomechanics, ASCE*, 11(1), 2011, 72–77.
47. Huang, J. and Griffiths, D.V., One-dimensional consolidation theories for layered soil and coupled and uncoupled solutions by the finite element method, *Geotechnique*, 60(9), 2010, 709–713.
48. Sandhu, R.S. and Wilson, E.L., Finite element analysis of seepage in elastic media, *Journal of Engineering Mechanics, ASCE*, 95(EM3), 1969, 641–652.

7 Coupled Flow through Porous Media

Dynamics and Consolidation

7.1 INTRODUCTION

The behavior of (saturated) porous soil–structure systems subjected to dynamic or static loads should be defined by taking into consideration the coupling between flow and deformation. Figure 7.1 shows the schematic of such a soil–structure system.

The behavior of the mixture of soil and water is affected by the deformation of the solid particles (skeleton), the relative motion (sliding) between particles and water, the deformation of pore water, and the movement of pore water through pores. We present in this chapter a general formulation for such an interacting system, from which seepage, consolidation, dynamic behavior, and behavior of dry geologic media can be obtained as special cases.

7.2 GOVERNING DIFFERENTIAL EQUATIONS

The formulations for linear elastic materials presented by Biot [1–3] have been used very often, with the FEM [4–8]. Since the soil and rock behavior is usually nonlinear, modifications have been introduced in Biot's formulation. In this chapter, we present Biot's theories for nonlinear materials with elastoplastic and DSC models [9]. Before we present the equations, we define various terms relevant to saturated porous materials.

7.2.1 POROSITY

Figure 7.2a shows the schematic of a porous material element consisting of solid (particles) and fluid (water). The porosity diagram is shown in Figure 7.2b. Then, the porosity $\bar{n}$ is defined as

$$\bar{n} = \frac{V_v}{V_v + V_s} \tag{7.1}$$

where V_v is the volume of pores equal to the volume of fluid ($= V_f$) and V_s is the volume of solids. The density, ρ, of the mixture is expressed as

$$\rho = (1 - \bar{n})\,\rho_s + \bar{n}\rho_f \tag{7.2}$$

where ρ_s and ρ_f are the densities of solids and fluids, respectively.

Advanced Geotechnical Engineering

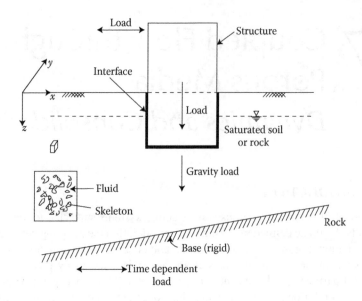

FIGURE 7.1 Schematic of structure-foundation system.

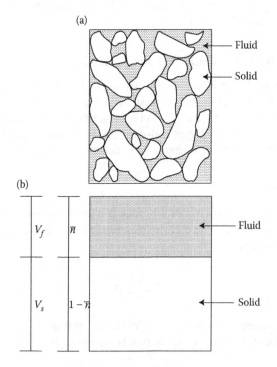

FIGURE 7.2 Soil-fluid element and porosity. (a) Soil element with solids and fluid; (b) representation of porosity. (Adapted from Wathugala, W. and Desai, C.S., Nonlinear and Dynamic Analysis of Porous Media and Applications, Report to National Science Foundation, Washington, DC, Dept. of Civil Eng. and Eng. Mech., Univ. of Arizona, Tucson, AZ, USA, 1990.)

The symbolic deformations of solids and fluids are shown in Figure 7.3. The terms in this figure are as follows: u_i $(i = 1, 2, 3)$ are the displacement components for the solid in 1 (x), 2 (y), and 3 (z) directions, and U_i $(i = 1, 2, 3)$ denotes displacement components of the fluid. Relative displacements can occur between the solid and fluid for loadings such as dynamic. Then, w_i $(i = 1, 2, 3)$ denotes such relative displacements between solid and fluid, averaged over the face of the solid skeleton, given by

$$w_i = \frac{Q_i}{A_i} = \bar{n} \, (U_i - u_i) \tag{7.3}$$

where A_i is the area normal to the ith direction. The volume of fluid moving through an area of the skeleton normal to the ith direction, Q_i, is expressed as

$$Q_i = A_i \, \bar{n} \, (U_i - u_i) \tag{7.4}$$

Figure 7.3b shows the relative displacement, w_i.

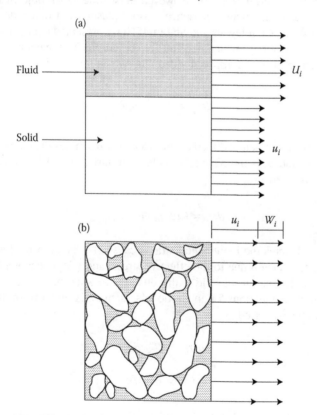

FIGURE 7.3 Displacements in element with two phases. (a) Displacement of different phases; (b) relative displacement of fluid. (Adapted from Wathugala, W. and Desai, C.S., Nonlinear and Dynamic Analysis of Porous Media and Applications, Report to National Science Foundation, Washington, DC, Dept. of Civil Eng. and Eng. Mech., Univ. of Arizona, Tucson, AZ, USA, 1990.)

For small strains, the strain tensor, ε_{ij}, is given by

$$\varepsilon_{ij} = \frac{1}{2}\,(u_{i,j} + u_{j,i}) \tag{7.5}$$

where $u_{i,j}$ denotes derivative of solid displacement u_i ($i = 1, 2, 3$) and so on. The change in the volume of fluid in a unit volume of the skeleton, ξ, is given by

$$\xi = w_{i,i} \tag{7.6}$$

where (i, i) denote the summation, that is, $w_{i,i} = w_{1,1} + w_{2,2} + w_{3,3}$.

7.2.2 CONSTITUTIVE LAWS

Biot [1–3] has developed the formulation for coupled, solid–fluid behavior by assuming linear elastic material behavior. However, most solid (soil)–fluid (water) media behave nonlinearly and may experience elastic, plastic, and creep deformations. Zienkiewicz [4] and Zienkiewicz and Shiomi [5] have presented equations by assuming incremental plasticity, which account for the nonlinear behavior. Accordingly, the total incremental stress, $d\sigma_{ij}$, is divided into two components as

$$d\sigma_{ij} = d\sigma'_{ij} + dp\delta_{ij} \tag{7.7}$$

where $d\sigma'_{ij}$ is the incremental effective stress tensor, dp denotes the incremental pore water pressure, and δ_{ij} is the Kronecker delta. Similarly, the total strain increment, $d\varepsilon_{ij}$, can be expressed as

$$d\varepsilon_{ij} = (d\varepsilon_{ij})_{\sigma'} + (d\varepsilon_{ij})_p \tag{7.8}$$

where $(d\varepsilon_{ij})_{\sigma'}$ denotes the incremental strains caused by the deformation of soil or rock grains and skeleton due to the effective stress, and $(d\varepsilon_{ij})_p$ denotes the strains caused by the deformations of solid grains due to pore water pressure.

The constitutive equations for an elastic–plastic material in terms of the effective quantities can be expressed as

$$d\sigma'_{ij} = C^{ep}_{ijk\ell}\, d\varepsilon'_{k\ell} \tag{7.9a}$$

$$d\sigma' = C^{ep}\, d\varepsilon \tag{7.9b}$$

The first equation, Equation 7.9a, is expressed in tensor rotation, while the second equation, Equation 7.9b, is expressed in matrix rotation. Here, $C^{ep}_{ijk\ell} = C^e_{ijk\ell} - C^p_{ijk\ell}$, where the first term relates to elastic behavior while the second term arises from inelastic or plastic behavior. The latter is derived based on the particular yield criterion or function chosen (e.g., conventional plasticity: von Mises, Drucker–Prager, and Mohr–Coulomb) or continuous yielding (e.g., critical state and cap), HISS plasticity, and related flow rule [9].

The bulk elastic behavior of solid grains (skeleton) can be expressed as

$$(d\varepsilon_{ij})_p = \frac{dp}{3K_s}\delta_{ij} \tag{7.10}$$

where K_s denotes the bulk modulus of the solid grains. The substitution of Equations 7.8 through 7.10 in Equation 7.7 leads to

$$d\sigma_{ij} = C_{ijk\ell}^{ep} d\varepsilon_{k\ell} - \left(\frac{dp}{3K_s}\right)C_{ijk\ell} + dp\delta_{ij} \tag{7.11}$$

The common terms without dp in the last two terms can be expressed as

$$\delta_{ij} - \frac{C_{ijk\ell}^{ep}}{3K_s}\delta_{k\ell} = \alpha\delta_{ij} + \beta_{ij} \tag{7.12}$$

where α is a scalar term. Assuming that the deviatoric part, β_{ij}, can be neglected [5], Equation 7.11 reduces to

$$d\sigma_{ij} = C_{ijk\ell}^{ep} d\varepsilon_{k\ell} + \alpha dp\delta_{ij} \tag{7.13}$$

For elastic materials, $C_{ijk\ell}^{ep}$ reduces to $C_{ijk\ell}^{e}$ as

$$C_{ijk\ell}^{e} = 2\mu\delta_{ik}\delta_{j\ell} + \lambda\delta_{ij}\delta_{k\ell} \tag{7.14}$$

where μ and λ are Lame's constants. Now, by using Equations 7.12 and 7.14, α can be deduced as

$$\alpha = 1 - \frac{\lambda + (2/3)\mu}{K_s} = 1 - \frac{K}{K_s} \tag{7.15}$$

where K is the bulk modulus of the soil skeleton, which for elastoplastic material can be derived as

$$K = \frac{\delta_{ij}\, C_{ijk\ell}^{ep}\, \delta_{k\ell}}{9} \tag{7.16}$$

7.2.2.1 Volumetric Behavior

The following four items can influence the volume change behavior of the mixture:

1. The volume of fluid flowing out $d\xi$; this causes decrease of the volume of the mixture.
2. The change in compressive strain in the fluid due to the pore pressure change, dp, equal to $(dp\,\bar{n})/K_f$; this causes an increase in the volume of the mixture.

3. The change in the compressive strain of solid grains due to the change in pore water pressure, dp, equal to $dp(1 - \bar{n})/K_s$; this causes an increase in the volume of the mixture.
4. The change in compressive strain in solid grains due to the change in effective stress, $d\sigma'_{ij}$, equal to $-d\sigma'_{ii}/3K_s$; this causes an increase in the volume of the mixture.

In view of the above four items, the volume change of the mixture, the continuity condition, can be expressed as

$$d\xi = d\varepsilon_{ii} + \frac{dp}{K_s}(1 - \bar{n}) + \frac{dp\bar{n}}{K_f} - \frac{d\sigma'_{ii}}{3K_s} \tag{7.17}$$

The substitution of Equations 7.7 and 7.13 into Equation 7.17 leads to [5]:

$$dp = M(\alpha d\varepsilon_{kk} + d\xi) \tag{7.18}$$

where α is defined in Equation 7.15 and M is expressed as

$$M = \frac{K_f \cdot K_s}{K_s\bar{n} + K_f(\alpha - \bar{n})} \tag{7.19}$$

Equations 7.13 and 7.18 can be applied for both elastic and elastic–plastic material behavior assuming that the bulk modulus for solid grains and fluid are invariant.

If the bulk modulus for the solid grains is much higher than that for the soil skeleton, that is, if $K_s \gg K$, the values of α in Equation 7.15 tend to unity. Then, Equation 7.13 can be expressed as

$$d\sigma'_{ij} = d\sigma_{ij} - dp\delta_{ij}$$
$$= C_{ijk\ell}d\varepsilon_{k\ell} \tag{7.20}$$

Equation 7.20 denotes the effective stress concept [10], implying that the deformation of the solid skeleton is affected by the effective stress.

7.3 DYNAMIC EQUATIONS OF EQUILIBRIUM

As noted before, the solid and fluid components of the mixture are coupled. Thus, there will be two governing equations for the mixture in terms of the displacement and fluid movement or pressure. According to Biot [3], Zienkiewicz [4], and Desai and coworkers [6–9], those equations are given by

$$\sigma_{ij,j} + (1 - \bar{n})\rho_s b_i + \bar{n}\rho b_i$$
$$- (1 - \bar{n})\rho_s \ddot{u}_i - \bar{n}\rho_f \ddot{U}_i = 0 \tag{7.21a}$$

and

$$h_i = p_{,i} + \rho_f b_i - \rho_f \ddot{U}_i \tag{7.21b}$$

where b_i denotes the components of body force per unit mass, u_i and U_i are the displacements of the skeleton and the fluid, respectively, ρ_s and ρ_f are the densities of solid grains and fluid, respectively, h is the fluid head (in unit of pressure), the comma denotes the first derivative with respect to spatial coordinates (x_i, $i = 1, 2, 3$), and the over dot denotes time derivative.

If we substitute Equations 7.2, 7.3, and 7.20 into Equation 7.21a, it simplifies to

$$\sigma_{ij} + \rho b_i - \rho \ddot{u}_i - \rho_f \ddot{w}_i = 0 \tag{7.22}$$

The equations governing the flow of fluid through the pores of the mixture are, according to the Darcy's law, given by

$$\dot{w}_i = k_{ij} h_{,j} \tag{7.23}$$

where k_{ij} is the permeability tensor. The substitution of Equations 7.4 and 7.23 into Equation 7.21b gives

$$p_{,i} + \rho_f b_i = \rho_f \ddot{u}_i - (\rho_f / \overline{n}) \, \ddot{w}_i - k_{ij}^{-1} \dot{w}_j = 0 \tag{7.24}$$

Equations 7.22 and 7.24 are the dynamic equations of equilibrium for the primary variables (unknown) u and w. This is often called the u–w formulation. If we assume that the relative velocity, $\dot{w}$, is very small, and can be neglected, the procedure is called u–p formulation [4], which can be suitable for quasi-static problems such as consolidation, which do not involve dynamic effects.

7.4 FINITE ELEMENT FORMULATION

The FEM is suitable for the solution of the coupled problems involving a mixture of solid and fluid (water). A schematic of the FE discretization is shown in Figure 7.4, together with two nodal unknowns, displacements, u_i ({u}) and w_i ({w}).

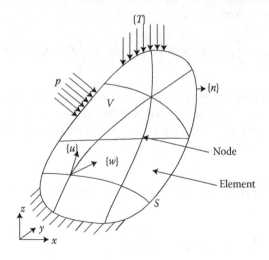

FIGURE 7.4 Finite element discretization.

We express displacements u_i and w_i at any point in an element as

$$u_i = N_u^a \bar{U}_i^a \ (i = 1 \text{ to } 3; a = 1 \text{ to } N_{ue}) \tag{7.25}$$

and

$$w_i = N_w^b \cdot W_i^b \ (i = 1 \text{ to } 3; b = 1 \text{ to } N_{we}) \tag{7.26}$$

where N_u and N_w denote interpolation or shape functions for u and w, respectively [11,12], N_{ue} and N_{we} are node numbers per element for u and w and $\bar{U}_i$ and W_i denote nodal values of u_i and w_i, respectively. Now, we can write variations, δu_i and δw_i, and second time derivatives, $\ddot{u}_i$ and $\ddot{w}_i$, required for the later Equations 7.30 and 7.31 as

$$\left.\begin{aligned} \delta u_i &= N_u^a \delta \bar{U}_i^a \\ \delta w_i &= N_w^b \delta W_i^b \\ \ddot{u}_i &= N_u^a \ddot{\bar{U}}_i^a \\ \ddot{w}_i &= N_w^b W_i^b \end{aligned}\right\} \ (i = 1, 2, 3; a = 1, 2,..., N_{ue}; b = 1,2,..., N_{we}) \tag{7.27a,b,c,d}$$

The FE equations are often derived by using the virtual work principle. It requires that for arbitrary compatible (virtual) displacements, δu_i and δw_i, with relevant Equations 7.22 and 7.24, respectively, the work done over the system must vanish. First, applying virtual displacement δu_i to Equation 7.22, we have

$$\int_V (\sigma_{ij,j} + \rho b_i - \rho \ddot{u}_i - \rho_f \ddot{w}_i) \, \delta u_i dV = 0 \tag{7.28}$$

where V is the volume of the domain of the system. By using the Gauss theorem, Equation 7.28 can be expressed as

$$\int_V \rho \ddot{u}_i \delta u_i dV + \int_V \rho_f \ddot{w}_i \delta u_i dV + \int_V \sigma_{ij} \delta u_{i,j} dV = \int_S T_i \delta u_i dS + \int_V \rho b_i \delta u_i dV \tag{7.29}$$

where S denotes the surface (boundary) of the domain on which the fraction loading, $T_i = \sigma_{ij} n_j$, is applied and n_j is the unit vector normal to the surface, S.

Substituting Equation 7.27 into Equation 7.29 and eliminating the arbitrary nodal displacement $\delta \bar{U}_i$, we have

$$\sum_V \left\{ \int_{V_e} \sigma_{ij} N_{u,j}^a dV + \ddot{\bar{U}}_i^c \int_{V_e} \rho N_u^c N_u^a dV + \ddot{W}_i^b \int_{V_e} \rho_f N_u^a N_u^b dV \right\}$$

$$= \sum_{V_e} \left\{ \int_{S_e} N_u^a T_i dS + \int_{V_e} \rho b_i N_u^a dV \right\} \tag{7.30}$$

where V_e denotes the volume of the element and c varies from 1 to N_u, the number of nodes in the entire domain.

Now, applying the virtual displacement, δw_i to Equation 7.24, we obtain

$$\sum_V \left\{ \int_{V_e} pN_{w,i}^b dV + \ddot{U}_i^a \int_{V_e} \rho_f N_w^b N_u^a dV + W_i^d \int_{V_e} \frac{\rho_f}{n} N_w^b N_w^d dV + \dot{W}_i^d \int_{V_e} k_{ij}^{-1} N_w^b N_w^d dV \right\}$$

$$= \sum_V \left\{ \int_{S_e} pN_w^b n_i dS + \int_{V_e} \rho_f b_i N_w^b dV \right\} \tag{7.31}$$

where d varies from 1 to N_w, and where $N_u = N_w$ are the number of nodes in the entire domain.

Equations 7.30 and 7.31 can be expressed in the following form:

$$\begin{bmatrix} M_{uuij}^{ac} & M_{uwij}^{ad} \\ M_{wuij}^{bc} & M_{wwij}^{bd} \end{bmatrix} \begin{Bmatrix} \ddot{U}_j^c \\ \ddot{W}_j^d \end{Bmatrix} + \begin{bmatrix} O & O \\ O & C_{wwij}^{bd} \end{bmatrix} \begin{Bmatrix} \dot{U}_j^c \\ \dot{W}_j^d \end{Bmatrix}$$

$$+ \begin{Bmatrix} \displaystyle\int_V N_{u,i}^a \sigma_{ij} dV \\ \displaystyle\int_V N_{w,i}^b p\, dV \end{Bmatrix} = \begin{Bmatrix} f_{ui}^a \\ f_{wi}^b \end{Bmatrix} \tag{7.32a}$$

or in matrix notation as follows:

$$\begin{bmatrix} [M_{uu}] & [M_{uw}] \\ [M_{wu}] & [M_{ww}] \end{bmatrix} \begin{Bmatrix} \{\ddot{U}\} \\ \{\ddot{W}\} \end{Bmatrix} + \begin{bmatrix} O & O \\ O & C_{ww} \end{bmatrix} \begin{Bmatrix} \{\dot{U}\} \\ \{\dot{W}\} \end{Bmatrix}$$

$$+ \begin{bmatrix} [K_{uu}] & [K_{uw}] \\ [K_{uw}] & [K_{ww}] \end{bmatrix} \begin{Bmatrix} \{U\} \\ \{W\} \end{Bmatrix} = \begin{Bmatrix} \{f_u\} \\ \{f_w\} \end{Bmatrix} \tag{7.32b}$$

The third term in Equation 7.32b relates to the constitution model, Equation 7.9 and results, after substitutions of u_i, w_i and their derivatives, into the stiffness equations. Other terms in Equation 7.32 are expressed as

$$\text{ADD } K_{uu} \text{ and so on.}$$

$$M_{uuij}^{ac} = \delta_{ij} \int_V pN_u^a N_u^a dV \tag{7.33a}$$

$$M_{uwij}^{ad} = \delta_{ij} \int_V \rho_f N_u^a N_w^d dV \tag{7.33b}$$

$$M^{bc}_{wuij} = \delta_{ij} \int_V \rho_f N^b_w N^c_u dV \tag{7.33c}$$

$$M^{bd}_{wwij} = \delta_{ij} \int_V \frac{\rho_f}{n} N^b_w N^d_w dV \tag{7.33d}$$

$$C^{bd}_{wwij} = \int_V k^{-1}_{ij} N^b_w N^d_w dV \tag{7.33e}$$

$$f^a_{ui} = \int_S N^a_u T_i dS + \int_V \rho b_i N^a_u dV \tag{7.33f}$$

$$f^b_{wi} = \int_S p N^b_w n_i dS + \int_V \rho_f b_i N^b_w dV \tag{7.33g}$$

In Equation 7.32, the summation is performed over the entire volume, V, and boundary, S.

7.4.1 TIME INTEGRATION: DYNAMIC ANALYSIS

We can express Equation 7.32 in a general form as

$$M_{ij}\ddot{x}_j + C_{ij}\dot{x}_j + K_{ij}x_j = f_i$$
$$i, j = 1 \text{ to } N_D \tag{7.34a}$$

where N_D is the number of degrees of freedom, M_{ij}, C_{ij}, and K_{ij} represent the components of mass, damping, and stiffness matrices, respectively, f_i represents the force function or loads, and x_j, $\dot{x}_j$, and $\ddot{x}_j$ denote the displacement, velocity, and acceleration (related to $\bar{U}$ and W), respectively. We can express Equation 7.34a in matrix notation as

$$[M]\{\ddot{x}\} + [C]\{\dot{x}\} + [K]\{x\} = \{f\} \tag{7.34b}$$

where $\{x\}$ contains both $\{\bar{U}\}$ and $\{W\}$

Equations 7.34 represent an initial value problem that needs to be solved for values of x_j at a given time t, knowing the values of x_j ($t = 0$) and $\dot{x}_j$ ($t = 0$). A number of procedures for time integration of Equation 7.34 are available. The Newmark method is often used, which is described below [13]:

7.4.1.1 Newmark Method

For the integration, the time domain is discretized into (equal) time steps (Figure 7.5). Now, we can express the current time at which the solution is desired as time

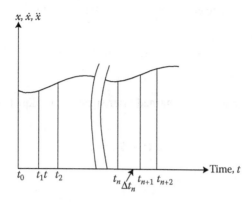

FIGURE 7.5 Time Integration and steps in dynamic FEM.

level, $t + \Delta t$; t denotes previous time level. By using the Taylor's series expansion for $x_i(t + \Delta t)$, where Δt is the time increment, it can be expressed as

$$x_i(t + \Delta t) = x(t) + \dot{x}_i(t) \cdot \Delta t + \ddot{x}_i(t + \alpha\Delta t) \cdot \frac{\Delta t^2}{2} \ldots$$

$$t \leq (t + \alpha\Delta t) \leq (t + \Delta t)$$

(7.35)

where α is a fraction, $0 < \alpha \leq 1$. The acceleration $\ddot{x}_i(t + \alpha\Delta t)$ is expressed as

$$\ddot{x}(t + \alpha\Delta t) = (1 - 2\beta)\,\ddot{x}_i(t) + 2\beta\,\ddot{x}_i(t + \Delta t)$$

(7.36)

where β is a parameter. Using Equations 7.35 and 7.36, we have

$$\ddot{x}_i(t + \Delta t) = \frac{1}{\beta\Delta t^2}\{x_i(t + \Delta t) - x(t)$$

$$- \Delta t\,\dot{x}_i(t)\} - \frac{1 - 2\beta}{2\beta}\,\ddot{x}_i(t)$$

(7.37)

The velocity is now expressed as

$$\dot{x}_i(t + \Delta t) = \dot{x}_i(t) + \Delta t\ddot{x}_i(t + \alpha\Delta t);$$

$$t \leq (t + \bar{\alpha}\Delta t) \leq (t + \Delta t)$$

(7.38)

where $\bar{\alpha}$ is a fraction, $0 \leq \bar{\alpha} \leq 1$.
The acceleration $\ddot{x}_i(t + \bar{\alpha}\Delta t)$ is now expressed as

$$\ddot{x}_i(t + \bar{\alpha}\Delta t) = (1 - \gamma)\,\ddot{x}(t) + \gamma\,\ddot{x}_i(t + \Delta t)$$

(7.39)

where γ is a parameter. The equation for $\dot{x}_i(t + \Delta t)$ is obtained from Equations 7.38 and 7.39:

$$\dot{x}_i(t + \Delta t) = \dot{x}_i(t) + \Delta t\{(1 - \gamma)\,\ddot{x}_i(t)$$
$$+ \gamma\,\ddot{x}_i(t + \Delta t)\} \tag{7.40}$$

The following equations are obtained after substituting Equations 7.37 and 7.40 into Equation 7.34:

$$\bar{K}_{ij}x_j(t + \Delta t) = \bar{f}_i \tag{7.41}$$

where

$$\bar{K}_{ij} = \frac{1}{\beta\Delta t^2}\cdot M_{ij} + \frac{\gamma}{\beta\Delta t}C_{ij} + K_{ij} \tag{7.42}$$

$$\bar{f}_i = f_i(t + \Delta t) + M_{ij}\left\{\frac{x_j(t)}{\beta\Delta t^2} + \frac{\dot{x}_j(t)}{\beta\Delta t} + \left(\frac{1}{2\beta} - 1\right)\ddot{x}_j(t)\right\}$$
$$+ C_{ij}\left\{\frac{\gamma}{\beta\Delta t}x_j(t) + \left(\frac{\gamma}{\beta} - 1\right)\dot{x}_j(t) + \left(\frac{\gamma}{2\beta} - 1\right)\Delta t\,\ddot{x}_j(t)\right\} \tag{7.43}$$

Usually, the mass matrix M_{ij} and damping matrix C_{ij} are constant during the time variation. If the material is assumed to be linearly elastic, the stiffness matrix K_{ij} is also constant; hence, $x_i(t + \Delta t)$, at time $(t + \Delta t)$, can be obtained by solving Equation 7.41 only once. However, for nonlinear materials characterized, say, by an elastoplastic model, K_{ij} is a function of x_i (or strain and stress) and varies during each increment or over increments of loading, and during iterations under an increment; often, iterative procedures such as the Newton–Raphson method [14] is used to solve for $x_i(t + \Delta t)$. Once the revised value of K_{ij} is used, we can solve Equation 7.41 for x_j at time step $(t + \Delta t)$. Once $x_j(t + \Delta t)$ is found, the velocity $\dot{x}(t + \Delta t)$ and $\ddot{x}(t + \Delta t)$ can be found from Equations 7.40 and 7.37, respectively.

The accuracy and stability of the computer solution for a given problem may depend on factors such as discretization of space and time, material properties, and boundary conditions. For linear problems, the stability for the Newmark scheme has been investigated by various researchers [15,16] and is expressed for unconditional stability as

$$2\beta \ge \gamma \ge 0.5\beta \tag{7.44}$$

where β and γ are parameters in the time integration scheme. The following criterion governs conditional stability in which the equation for the selection of the time step, Δt, is given by

$$\omega\,\Delta t\,\Omega_c = \xi\left(\gamma - \frac{1}{2}\right) + \frac{\left[(\gamma/2) - \beta + \xi^2\,(\gamma - (1/2))^2\right]^{(1/2)}}{(\gamma/2) - \beta} \tag{7.45}$$

where ξ is the damping ratio, and ω is the maximum natural frequency of the system. The recommended values are $\alpha = 0.5$ and $\beta = 0.25$.

The above stability consideration with the use of allowable Δt may yield results that are stable, that is, the solution remains in bounds, but they may not necessarily provide acceptable accuracy. Hence, accuracy should also be considered for realistic solutions.

7.4.2 CYCLIC UNLOADING AND RELOADING

Many soil–structure systems are subjected to cycles of loading, unloading, and reloading (Figure 7.6). Constitutive models are often valid for the continuously hardening (or virgin) behavior, which tends, asymptotically, to the ultimate condition. After unloading and reloading, the behavior usually reverts to the virgin curve. Unloading and reloading can involve accumulation of irreversible or plastic deformations. However, for simplicity, the unloading and reloading are often assumed to be linear (elastic), often, with an average of unloading and reloading moduli, following the same line (Figure 7.6), which may not be realistic for some materials.

It is difficult to use the theory of plasticity to simulate unloading by using a contracting yield surface for unloading. This is because the theory of plasticity requires that during loading and also unloading, the yield surface should be convex, which may not be the case. Sometimes, procedures are used in which plastic behavior is accommodated by using ad hoc schemes. One such procedure, developed in Refs. [8,17] is described below.

We consider two cases: (1) one-way and (2) two-way cyclic loading. The former is shown schematically in Figures 7.6 and 7.7. The initial part of the loading up to A and its continuation after the end of reloading is called the virgin response. The

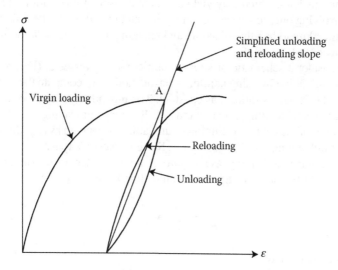

FIGURE 7.6 Stress-Strain curves during virgin, unloading and reloading. (Adapted from Shao, C. and Desai, C.S., Implementation of DSC Model for Nonlinear and Dynamic Analysis of Soil-Structure Interaction, Report to National Science Foundation, Washington, DC, Dept. of Civil Eng. and Eng. Mech., Univ. of Arizona, Tucson, AZ, USA, 1997.)

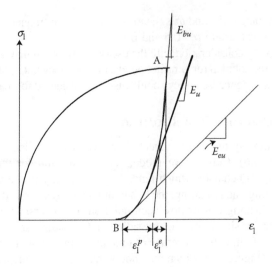

FIGURE 7.7 Schematic representation of unloading model. (Adapted from Shao, C. and Desai, C.S., Implementation of DSC Model for Nonlinear and Dynamic Analysis of Soil-Structure Interaction, Report to National Science Foundation, Washington, DC, Dept. of Civil Eng. and Eng. Mech., Univ. of Arizona, Tucson, AZ, USA, 1997.)

unloading and reloading phases are called non-virgin loading. The two-way cyclic loading is depicted in Figure 7.8b, which often occurs in dynamic loading.

The virgin phase of the behavior is modeled by using nonlinear elastic, elastoplastic, or other suitable models. The non-virgin part is often modeled by using linear or nonlinear elasticity, which may yield acceptable results for certain problems, for example, involving only few cycles of unloading and reloading. In the following, we present a procedure to handle unloading and reloading using interpolation functions with nonlinear elasticity [8,17].

Let us consider a schematic of stress–strain curve, σ_1 versus ε_1 (Figure 7.7). To simulate the unloading by using nonlinear elastic model, we compute the variable E_u and assume the Poisson's ratio, v, as constant. The unloading takes place at the end of given incremental loading, point A; thus, A–B denotes unloading.

The elastic modulus at the beginning of unloading is denoted as E_{bu}, whereas that at the end of unloading, it is denoted by E_{eu}. Since irreversible (plastic) deformations can take place during unloading; as an approximation, we denote an effective modulus as E_p. The elastic modulus, E_u, during unloading is given by

$$\frac{1}{E_u} = \frac{1}{E_{bu}} + \frac{1}{E_p} \tag{7.46}$$

where E_p is computed as

$$E_p = p_a K_1 \left(\frac{p_a}{\sqrt{J_{2D}^b} - \sqrt{J_{2D}}} \right)^{K_2} \tag{7.47}$$

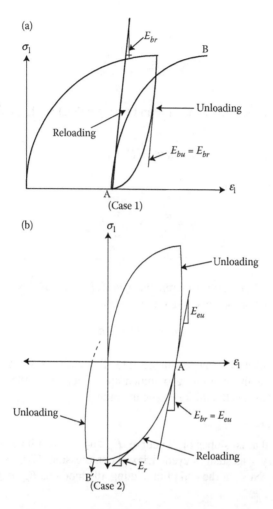

FIGURE 7.8 Reloading cases for one-way and two-way loadings. (a) Reloading case 1: A to B; (b) reloading case 2: A to B′. (Adapted from Shao, C. and Desai, C.S., Implementation of DSC Model for Nonlinear and Dynamic Analysis of Soil-Structure Interaction, Report to National Science Foundation, Washington, DC, Dept. of Civil Eng. and Eng. Mech., Univ. of Arizona, Tucson, AZ, USA, 1997.)

where p_a is the atmospheric pressure constant used to nondimensionalize K_1 and K_2, which are constants, J_{2D}^b and J_{2D} are the second invariants of the deviatoric stress tensors at the beginning of unloading, point A, and at the current state during unloading, respectively.

Consider a special case under CTC test, when $\sigma_1 > \sigma_2 = \sigma_3$. Then

$$\sqrt{J_{2D}} = \frac{1}{\sqrt{3}} (\sigma_1 - \sigma_3) \qquad (7.48a)$$

and

$$E_p = p_a K_1 \left(\frac{\sqrt{3} p_a}{\sigma_1^b - \sigma_1} \right)^{K_2} \tag{7.48b}$$

where σ_1^b and σ_1 are stresses at point A and during unloading, respectively. Also, we can write

$$d\sigma_1 = E_u d\varepsilon_1 \tag{7.48c}$$

Then, the axial strain is

$$d\varepsilon_1 = \frac{d\sigma_1}{E_u} = \frac{d\sigma_1}{E_{bu}} + \frac{d\sigma_1}{E_p}$$

$$= d\varepsilon_1^e + d\varepsilon_1^p \tag{7.49}$$

where $d\varepsilon_1^e$ is the elastic strain increment, and $d\varepsilon_1^p = d\varepsilon_1/E_p$, is the "irreversible" strain (Figure 7.7). For the general case, we can write

$$\{d\sigma\} = [C_u]\{d\varepsilon\} \tag{7.50}$$

where $[C_u]$ is a constitutive matrix during unloading with varying E_u and constant v.

We note that at the beginning of unloading (Equation 7.48b), $E_p = \infty$; hence, $d\varepsilon_1^p = 0$, which ensures the initial elastic unloading.

7.4.2.1 Parameters

The parameters, that is, slopes of the curve, E_{bu} and E_{eu}, can be determined on the basis of laboratory tests such as cyclic triaxial compression (CTC) and simple shear (SS), (Figure 7.7). Now, at the end of unloading, the modulus E_{eu} for the CTC test is given by

$$\frac{1}{E_u} = \frac{1}{E_{bu}} + \frac{1}{K_1 p_a} \left(\frac{\sigma_1^b - \sigma_1^{eu}}{\sqrt{3} p_a} \right)^{K_2} \tag{7.51}$$

For the plastic strain for CTC test, we have

$$\varepsilon_1^p = \int d\varepsilon_1^p = \int \frac{d\sigma_1}{E_p} = \int_{\sigma_1^{eu}}^{\sigma_1^b} \frac{1}{K_1 p_a} \left(\frac{\sigma_1^b - \sigma_1}{\sqrt{3} p_a} \right)^{K_2} d\sigma_1 \tag{7.52}$$

where K_1 and K_2 are material parameters. The solution of Equations 7.51 and 7.52 yields K_1 and K_2:

$$K_2 = \frac{\sigma_1^b - \sigma_1^{eu}}{\varepsilon_1^p} \left(\frac{1}{E_{eu}} - \frac{1}{E_{bu}} \right) \tag{7.53a}$$

$$K_1 = \frac{\sqrt{3}}{(K_2 + 1)\,\varepsilon_1^p}\left(\frac{\sigma_1^b - \sigma_1^{eu}}{\sqrt{3}p_a}\right)^{K_2+1}$$ (7.53b)

For the 3-D stress and for isotropic material, they can be derived as

$$K_2 = \frac{\sqrt{3}\left(\sqrt{J_{2D}^b} - \sqrt{J_{2D}^{eu}}\right)}{\varepsilon_1^p}\left(\frac{1}{E_{eu}} - \frac{1}{E_{bu}}\right) - 1$$ (7.53c)

$$K_1 = \frac{\sqrt{3}}{\left(K_2 + 1\right)\,\varepsilon_1^p}\left(\frac{\sqrt{J_{2D}^b} - \sqrt{J_{2D}^{eu}}}{p_a}\right)^{K_2+1}$$ (7.53d)

The parameters can also be found from shear $(\tau - \gamma)$ tests, where τ and γ are shear stress and strain, respectively [8,17].

7.4.2.2 Reloading

The stress–strain equations for reloading can be expressed as

$$\{d\sigma^a\} = R\,[C^{DSC}]\{d\varepsilon\} + (1 - R)\,[C^e]\{d\varepsilon\}$$ (7.54)

where $[C^{DSC}]$ is the constitutive matrix for the DSC model, $[C^e]$ is the elastic matrix at the beginning of reloading, and R is the interpolation variable, with $R = 0$ at the beginning of reloading and $R = 1$ at the end of reloading.

There are two cases of reloading: one-way and two-way (Figure 7.8). The modulus, E_r, used to compute $[C^{DSC}]$ and $[C^e]$ for reloading is different. For the one-way case, the modulus at the beginning of reloading, E_{br}, can be assumed as the modulus, E_{bu}, at the beginning of unloading (Figure 7.8a). For the two-way loading, the modulus, E_{br}, for reloading, can be assumed to be equal to the modulus at the end of unloading, E_{eu} (Figure 7.8b).

The value of the modulus, E_r, during reloading is found from the following procedure. First, a parameter, S, is defined as

$$S = \frac{(\sigma_{bu} - \sigma)d\sigma}{\|\sigma_{bu} - \sigma\|\,\|d\sigma\|}; \quad -1 \le S \le 1$$ (7.55)

where σ_{bu}, σ, and $d\sigma$ are stresses at the beginning of unloading, the current stress during reloading, and the stress increment, respectively. The parameter $S = -1$ indicates one-way loading and $S = 1$ indicates two-way loading. The modulus, E_{br}, at the beginning of reloading, is interpolated between E_{bu} and E_{eu}:

$$\frac{1}{E_{br}} = \frac{1 - S}{2E_{bu}} + \frac{1 + S}{2E_{eu}}$$ (7.56a)

Advanced Geotechnical Engineering

and the modulus during reloading, E_r, is computed as

$$\frac{1}{E_r} = \frac{1-R}{E_{br}} + \frac{R}{E}$$ (7.56b)

where E is the elastic modulus of the material. Thus, at the beginning of reloading ($R=0$), $E_r = E_{br}$, which ensures a smooth transition or continuity from unloading to reloading in the two-way case. At the end of reloading, $E_r = E$, which ensures a smooth transition from reloading to virgin loading. The parameter R is often defined as

$$R = \frac{\sqrt{J_{2D}}}{\sqrt{J_{2D}^c}}$$ (7.57)

where J_{2D} and J_{2D}^c are the second invariants of the deviatoric stress tensors at the beginning of the last unloading, and at current state during reloading, respectively.

7.5 SPECIAL CASES: CONSOLIDATION AND DYNAMICS-DRY PROBLEM

7.5.1 CONSOLIDATION

In Chapter 6, we considered 1-D consolidation based on the Terzaghi theory. Now, we can derive multidimensional consolidation as the special case of Equation 7.32.

In the absence of inertia and with no relative displacements between solid and fluid, the governing equations can be expressed as follows [1,18,19]:

$$\sigma'_{ij,j} + \delta_{ij}p_{,j} + \rho F_i = o$$ (7.58a)

The conditions of continuity are

$$k_{ij}(p_{,j} + \rho_w F_j) + \dot{u}_{i,i} = o$$ (7.58b)

By following a similar procedure as for the foregoing dynamic case, for the fully coupled behavior, we can derive the FE equations as [18–21]

$$[K_{uu}]\{\bar{U}(t)\} + [K_{up}]\{p_n(t)\} = -\{M_1\} + \{M_2\} + \{P_1\}$$ (7.59a)

$$[K_{up}]\{\bar{U}(t)\} - g * [K_{pp}]\{p_n(t)\} = g * \{M_3\} - g * \{P_2\}$$ (7.59b)

where $*$ denotes the convolution product and is defined as

$$g * f_1(t) = \int_o^t F(\tau) \cdot g(t - \tau)\, d\tau$$

$$= \int_o^t F(\tau)\, d\tau \tag{7.60}$$

where $g = 1$ and τ denotes time within the time increment from t to $t + \Delta t$.

Note that in Equation 7.59, we have replaced $\{w\}$ by $\{p\}$ because, in the case of consolidation, the relative displacement is assumed to be zero. Hence, we use the following expression for p_i instead of w_i in Equation 7.26:

$$p_i = N_p^b p_{ni}^b; (i = 1 \text{ to } 3); b = 1 \text{ to } N_{pe}) \tag{7.61}$$

where p_i denotes the pore water pressure in an element, N_p^b is the interpolation function, and p_{ni} denotes nodal pore water pressures.

Now, the various terms in Equation 7.59 are expressed as

$$[K_{uu}] = \int_V [B_e]^T [C][B_e]\, dV \tag{7.62a}$$

$$[K_{up}] = \int_V [B_\Delta]\{N_p\}^T\, dV \tag{7.62b}$$

$$[K_{pp}] = \int_V [B_q]^T [R][B_q]\, dV \tag{7.62c}$$

$$\{M_1\} = \int_V [B_e]^T \{\sigma_o\}\, dV \tag{7.62d}$$

$$\{M_2\} = \int_V [N_u]\{\rho F\}\, dV \tag{7.62e}$$

$$\{M_3\} = \int_V [B_q]^T [R]\, \{\rho_w F\}\, dV \tag{7.62f}$$

$$\{P_1\} = \int_{S_3} [N_u]^T [N_u]\, \{\bar{T}\}\, dS \tag{7.62g}$$

$$\{P_2\} = \int_{S_4} [N_p]^T [N_p]\, \{\bar{Q}\} \tag{7.62h}$$

where various related terms are defined as

$$\{\varepsilon\} = [B_e]\{\bar{U}\}$$
$$\{\sigma'\} = [C][B_e]\{\bar{U}\} + \{\sigma_o\}$$
$$\{\varepsilon_v\} = [B_\Delta]\{\bar{U}\}$$
$$\left\{\frac{\partial p}{\partial x_i}\right\} = [B_q]\{p_n\}$$

where $[R]$ denotes the coefficient of permeability matrix, $[N_u]$ and $[N_p]$ are interpolation matrices for the displacement of solids and pore water pressures, respectively, $\{Q\}$ is the fluid flux vector, and $\{\sigma_o\}$ denotes initial stress.

Now, with $g(t - \tau) = 1$, and choosing the time interval for t to $t + t\Delta$ and using Equation 7.59, we obtain

$$[K_{uu}]\{\bar{U}\}_{t+\Delta t} + [K_{up}]\{p_n\}_{t+\Delta t} = -\{M_1\}_{t+\Delta t} + \{M_2\}_{t+\Delta t} + \{P_1\}_{t+\Delta t} \quad (7.63a)$$

$$[K_{up}]^T\{\bar{U}\}_{t+\Delta t} - \frac{\Delta t}{2}[K_{pp}]\{P_n\}_{t+\Delta t} = [K_{up}]^T\{\bar{U}\}_t + \frac{\Delta t}{2}\cdot[K_{up}]\{p_n\} + \frac{\Delta t}{2}[\{M_3\}_t$$

$$+ \{M_3\}_{t+\Delta t}] - \frac{\Delta t}{2}[\{P_2\}_t + \{P_2\}_{t+\Delta t}] \quad (7.63b)$$

which can be written as

$$\begin{bmatrix} [K_{uu}] & [K_{up}] \\ K_{up} & -\dfrac{\Delta t}{2}[K_{pp}] \end{bmatrix} \left\{ \begin{matrix} \bar{U} \\ p_n \end{matrix} \right\}_{t+\Delta t} = \left\{ \begin{matrix} \{R_u\} \\ \{R_p\} \end{matrix} \right\} = \{R\} \quad (7.64)$$

where $\{R\}$ is the global load vector that is composed of the right-hand side of Equation 7.59.

Equation 7.64 can be integrated in time to obtain results for displacements, and pore pressures at $t + \Delta t$ based on their values at previous step t. The values at the first interval 0 to Δt is found based on the initial and boundary condition at $t = 0$. Stresses, strains, velocities, and so on can be found from computed values of displacements and pore water pressures.

7.5.1.1 Dynamics-Dry Problem

The FE equations for the general coupled problem involve solid displacements, u_i, and relative displacements, w_i. The special case that does not involve relative displacement and inertia is consolidation, as described above. In that case, the relative displacement is replaced by pore water pressures.

Another special case involving no water, that is, dry materials, constitutes the dynamics of dry systems. The equations that can be specialized from Equation 7.32 in simple form, without suffix and prefix, can be derived as

$$[M]\{\ddot{U}\} + [K]\{\bar{U}\} = \{f\} \tag{7.65a}$$

where $[M]$ and $[K]$ are mass and stiffness matrices, respectively, and are defined as

$$[M] = \int_V \rho[N]^T[N]dV \tag{7.65b}$$

$$[K] = \int_V [B]^T[C][B]dV \tag{7.65c}$$

where $[B]$ is the strain–displacement transformation matrix related to solids. The damping matrix is not included in Equation 7.65. The damping matrix is often derived in terms of stiffness and mass matrices [22], and a part of damping may be taken care of through $[K]$, if hysteretic damping is incorporated in the constitutive model for the soil.

The load vector is expressed as

$$\{f\} = \int_V [N]^T\{\bar{X}\}dV + \int_S [N]^T\{\bar{T}\}dS \tag{7.65d}$$

where $\{\bar{X}\}$ is the time-dependent body force vector, $\{\bar{T}\}$ is the time-dependent surface loading vector, and the over bar denotes known quantities.

7.5.2 Liquefaction

Liquefaction in saturated soils, often under dynamic (earthquake) loading, is an important factor in the analysis and design of geotechnical structures.

A number of procedures and criteria have been proposed to identify liquefaction, often based on ad hoc and empirical considerations; they are usually based on index properties such as blow count, and critical stress and strain criteria, from laboratory and/or field observations [23–29].

However, there is an increased recognition that the role of basic mechanism in the deforming material can provide enhanced understanding and modeling of liquefaction. Basic approaches are often derived from on energy consideration, and the DSC. The former is presented in Refs. [30–37] and the use of the DSC for liquefaction is presented in Refs. [9,38–42].

Liquefaction generally occurs as a consequence of modification of material's microstructure during deformation, and represents one of the unstable states in the microstructure. During deformation, the microstructure can assume various threshold or unstable states such as transition from compressive to dilative volume change, peak (stress) condition, softening or degradation, and critical state at the end of the residual phase and ultimate failure. Initiation of liquefaction is considered to be the unstable state near the critical condition.

In the DSC, the disturbance, D, that represents the coupling between RI and FA state (Appendix 1) and relates to the microstructural changes and to threshold or unstable states, can be defined by using measured behavior such as stress–strain, volumetric, pore water pressure, and nondestructive P- and S-wave velocities. For example, D from stress–strain and S-wave velocity can be expressed as

$$D_\sigma = (\sigma^i - \sigma^a)/(\sigma^i - \sigma^c) \qquad\qquad (7.66a)$$

$$D_v = (V^i - V^a)/(V^i - V^c) \qquad\qquad (7.66b)$$

where σ and V denote measured stress and velocity, and i, a, and c denote RI, observed, and FA states, respectively. Figure 7.9a shows the static and cyclic stress–strain behavior from which disturbance D_σ can be derived (Figure 7.9b). Figure 7.10a shows measured shear wave velocity [32,41] during the 1995 Port Island, Kobe, Japan earthquake, from which disturbance D_v can be obtained (Figure 7.10b).

It has been found that the initial liquefaction occurs at the point on the disturbance curve, denoted by critical disturbance D_c, at which the curvature (second derivative) of D (Equation AI.40a, Appendix 1), is the minimum. Such point can also be located at the intersection of the tangents to the initial and later parts of the curve (Figure 7.9b). The disturbance denoted by D_f (Figures 7.9b and 7.10b) can denote the

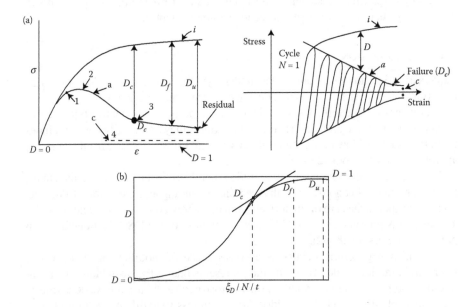

FIGURE 7.9 Static and cyclic behavior in DSC and distribution of disturbance. (a) Disturbance in static and cyclic responses; (b) variation versus ξ_D, N or time and critical disturbance.

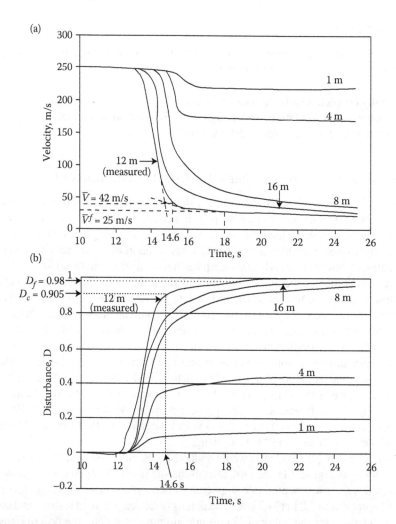

FIGURE 7.10 Shear wave velocity and disturbance at different depths. (a) Shear wave velocity versus time at different depths: measured at 12 m and interpolated at 1, 4, 8, and 16 m; (b) disturbance versus time at different depths. (From Desai, C.S. *Journal of Geotechnical and Geoenvironmental Engineering*, ASCE, 126(7), 2000, 618–631. With permission.)

final liquefaction at which the material can be considered to have failed; however, it still possesses certain strength (see Figure 7.10a). The disturbance D_u (Figure 7.9b) denotes the ultimate disturbance close to unity to which D approaches. Thus, the identification of liquefaction in the DSC is based on a fundamental mechanism in the deforming material and does not depend on index properties.

It is believed that the DSC provides a general and simpler approach for identification and analysis of liquefaction. We have used the DSC method for analysis and prediction of liquefaction in some of the example problems below.

7.6 APPLICATIONS

We present below applications involving validations and comparisons between predictions and measurements for a number of problems:

1. Fully coupled, involving soils and fluid
2. Coupled (consolidation), involving no relative motion and no inertia
3. Dynamics for structures in dry materials

7.6.1 Example 7.1: Dynamic Pile Load Tests: Coupled Behavior

The behavior of a pile founded in nonlinear saturated soil (clay) is affected by important factors such as *in situ* conditions, pile driving, consolidation, and one-way and two-way dynamic (cyclic) loading. Hence, the analysis and design of such piles, often used in the offshore environment, require consideration of these factors as well as the nonlinear behavior of soils and interfaces between structure (pile) and soil. The advances in computer methods have reached the state when their use for analysis and design can be beneficial to the profession. However, it is desirable to validate the computer procedures by comparing predictions with laboratory and/or field measurements. An example that includes a comprehensive treatment of the above factors and considers validations is presented below.

The work involving applied research and applications was conducted under a project supported by National Science Foundation (NSF), Washington, DC with collaboration between the University of Arizona, Tucson, AZ and Earth Technology Corporation, Long Beach, CA. The personnel engaged were Professors Hudson Matlock and Chandrakant Desai, Dr. G.W. Wathugala, Dr. Po Lam, Mr Dwayne Bogard, Dr. J. Audibert, and Mr L. Cheang.

Comprehensive pile load tests were conducted at Sabine Pass near the coast of the Gulf of Mexico; Figure 7.11 shows details of the site [8,20,43,44]. Six field tests were performed with instrumented pile segments (probes) of diameters 1.72 in (4.37 cm) called X-probe, and 3.0 in (7.62 cm). The details of laboratory, field testing, and computer validations for the 3.0 in (7.62 cm) pile are presented here; the descriptions of field testing are adopted from Ref. [44].

A schematic of the loading system used for the pile test is shown in Figure 7.12. The load is applied to the probe through the N-rods. The shear is transferred from the pile segment (probe) to the surrounding soil. It is measured by monitoring the difference between axial loads at two adjacent load cells; Figure 7.13 shows details of the instrumentation on the probe, including strain gages and load cells, porous element for fluid pressure, and linear voltage differential transducers (LVDTs). Two load cells (A) in the pile segments were placed at a distance of 31.6 in (80.30 cm), connected in a single Wheatstone bridge such that the bridge output is directly proportional to the shear transfer between two load cells.

Lateral earth pressures and pore water pressures are measured by using two pressure transducers (B) installed between two load cells (Figure 7.13). The total pressure transducer is affected only by forces normal to the outer face of the measurement unit; hence, the total radial pressure on the surface of the probe is measured. In Figure 7.13,

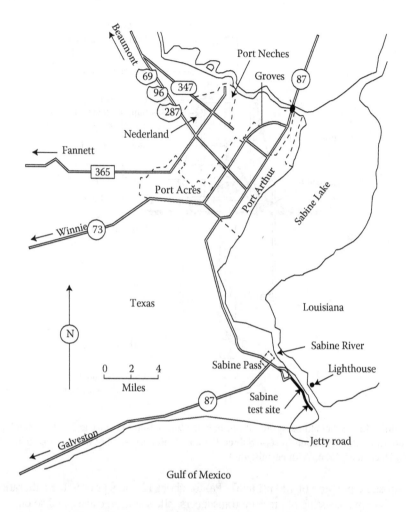

Sabine test site

FIGURE 7.11 Site of instrumented pile segment tests at Sabine, Texas. (Adapted from Wathugala, W. and Desai, C.S., Nonlinear and Dynamic Analysis of Porous Media and Applications, Report to National Science Foundation, Washington, DC, Dept. of Civil Eng. and Eng. Mech., Univ. of Arizona, Tucson, AZ, USA, 1990; The Earth Technology Corporation, Pile Segments Tests—Sabine Pass, ETC Report No. 85-007, Long Beach, CA, USA, December, 1986.)

a slip joint is located between the pile segment and the cutting shoe to measure relative displacement. The direct current LVDT is mounted on the pile segment and the core of the LVDT is attached to the cutting shoe. Consequently, during the driving of the probe, the cutting shoe and the pile segment move together. However, during a load test, only the pile segment moves because of the provision of the slip joint (Figure 7.13). To ensure the water tightness of the probe, the system was subjected to

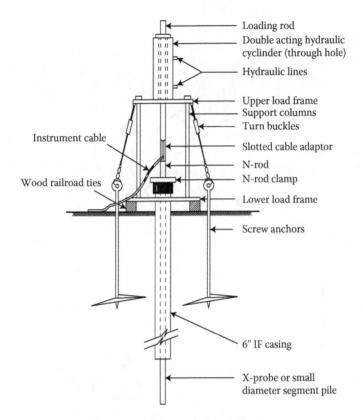

Loading rod
Double acting hydraulic
cyclinder (through hole)
Hydraulic lines

Upper load frame
Support columns
Turn buckles

Instrument cable

Slotted cable adaptor
N-rod
N-rod clamp

Wood railroad ties

Lower load frame

Screw anchors

6" IF casing

X-probe or small
diameter segment pile

FIGURE 7.12 Schematic diagram of portable loading set-up. (From The Earth Technology Corporation, Pile Segments Tests—Sabine Pass, ETC Report No. 85-007, Long Beach, CA, USA, December, 1986. With permission.)

a hydrostatic pressure of 100 psi (690 kPa) for about 12 hours prior to the calibration of the system consisting of pressure transducers, pile segment, cables, and so on.

The 3.0 in (7.62 cm) probe was driven in the soil by using a 300 lb (1334 N) casing hammer with a drop of 3 ft (91 cm) for each blow. The field test data in terms of shear stress transfer versus displacement and time, pore water pressures versus time, and total and effective horizontal stress versus time are analyzed here and compared with the FE predictions, as presented below.

Field test results for 3.0 in (7.62 cm) pile: The pore water pressure with respect to the initial pressure in the soil around the pile increases during driving, which reduces the effective stresses. Then, before loading the structure, the excess pore water pressure is reduced due to its dissipation, and the soil becomes stronger due to consolidation.

Tension tests to failure at different consolidation levels were performed to study the increase in pile capacity during consolidation. Figure 7.14 shows the measured shear transfer versus displacement at different consolidation levels (degree of consolidation, U), which indicates that the pile capacity increases with consolidation after pile driving.

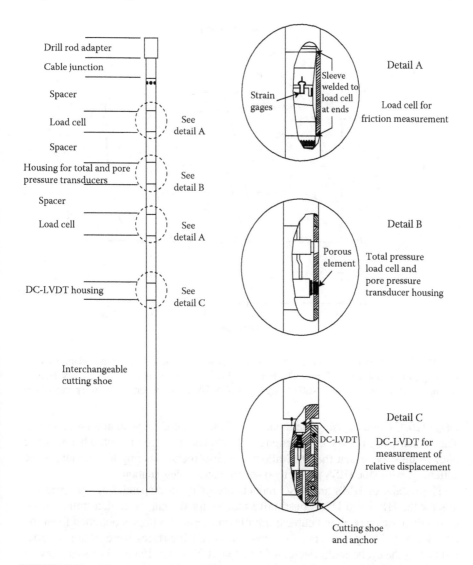

FIGURE 7.13 Pile segment, 3.0 in (7.62 cm) diameter and instrumentation. (From The Earth Technology Corporation, Pile Segments Tests—Sabine Pass, ETC Report No. 85-007, Long Beach, CA, USA, December, 1986. With permission.)

To simulate practical loading, one-way and two-way cyclic load tests were performed near the end of consolidation. In *one-way* loading, the reloading is in the same direction as the original loading, while if the reloading is in the opposite direction, it is called *two-way* loading. A number of loading (to failure)–unloading (to zero shear transfer) and reloading cycles were performed.

Field load tests—Simulation: The FE procedure used was based on the generalized Biot's theory, as described earlier in this chapter. The field tests considered were analyzed using the FE procedure with the HISS plasticity model by Wathugala and

3 in (7.62 mm) pile

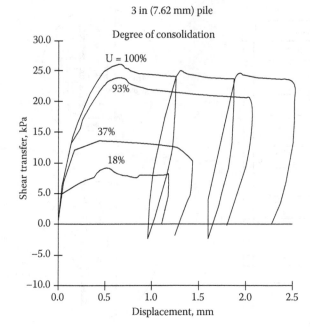

FIGURE 7.14 Measured shear transfer versus displacements at different consolidation levels, U, degree of consolidation. (From The Earth Technology Corporation, Pile Segments Tests—Sabine Pass, ETC Report No. 85-007, Long Beach, CA, USA, December, 1986. With permission.)

Desai [20,43], and by using the model based on the DSC by Shao and Desai [8,17]. The comparisons between predictions and field test data are presented below using both models. Note that the DSC allows for disturbance leading to degradation or softening, while the HISS plasticity does not include degradation.

The details of HISS and DSC models are given in Appendix 1. The parameters for the HISS and DSC constitutive models for the clay were determined from comprehensive triaxial and cubical (multiaxial) tests on samples obtained from the field [45,46]. The parameters for clay–pile (steel) interfaces were obtained from tests using the cyclic-multi-degree-of-freedom (CYMDOF-P) interface shear device [47,48]. The material parameters are shown in Table 7.1.

7.6.1.1 Simulation of Phases

The computer code simulated various phases from *in situ* to dynamic (cyclic) loading. Figure 7.15 depicts the phases simulated in the analysis presented herein. The FE mesh used is shown in Figure 7.16, together with the boundary conditions [17,20]. One of the differences is that the interface elements were used in the study presented in Refs. [8,17], while they were not used in the previous study [20,43]. Also, the HISS plasticity model was used in Refs. [20,43], while the DSC model with the HISS plasticity for the RI behavior was used for the study in Refs. [8,17].

The mesh contains a total of 225 nodes and 192 elements. The elements in contact with the pile were assigned interface properties (Table 7.1); the thickness of the

TABLE 7.1
Material Parameters for Clay and Interface

	Clay	Interface
Relative Intact (RI) State		
Elastic		
E	10,350 kPa	4300 kPa
v	0.35	0.42
Plasticity		
γ	0.047	0.077
β	0	0
n	2.8	2.6
$3R$	0	0
h_1	0.0001	0.000408
h_2	0.78	2.95
h_3	0	0.0203
h_4	0	0.0767
Fully Adjusted (FA) State Critical State		
$\bar{m}$	0.0694	0.123
λ	0.1692	0.298
e_o^c	0.9033	1.359
Disturbance Function		
D_u	0.75	1.0
A	1.73	0.816
Z	0.3092	0.418
Unloading and Reloading		
E_{bu}	34,500 kPa	4300 kPa
E_{eu}	3450 kPa	400 kPa
ε_1^p	0.005	0.0305
Others		
Permeability	2.39×10^{-10} m/s	2.39×10^{-10} m/s
Density of soils (ρ_s)	2.65 mg/m^3	2.65 mg/m^3
Bulk modulus (K_s) of soil grain	10^9 kPa	
Bulk modulus (K_f) of water	10^8 kPa	
Density of water (ρ_f)	1.0 mg/m^3	

Note: h_i ($i = 1$–4) are parameters in a special yield function [17,43].

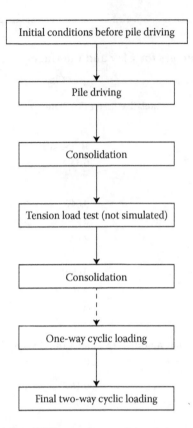

FIGURE 7.15   Representation of different phases of simulation.

interface element was assumed to be 1.40 mm thick. The pile was assumed to be rigid so that pile motion is simulated as prescribed displacements of nodes in contact with the pile. The various phases (Figure 7.15) are briefly described below.

In situ conditions before pile driving: The initial (*in situ*) stresses and pore water pressures for horizontal ground level were derived as

$$\sigma'_v = \gamma_s h \tag{7.67a}$$

$$\sigma'_h = K_o \sigma'_v \tag{7.67b}$$

$$p = \gamma_w h \tag{7.67c}$$

where σ'_v and σ'_h are the vertical and horizontal effective stresses at a point below the surface at depth h, respectively, γ_s is the submerged unit weight of soil, γ_w is the unit weight of water, p is the pore water pressure, and K_o is the coefficient of lateral earth pressure at rest. The value of K_o for normally consolidated clays can be approximated using Jaky's empirical formula [49]:

$$K_o = (1 - \sin \varphi') \tag{7.68}$$

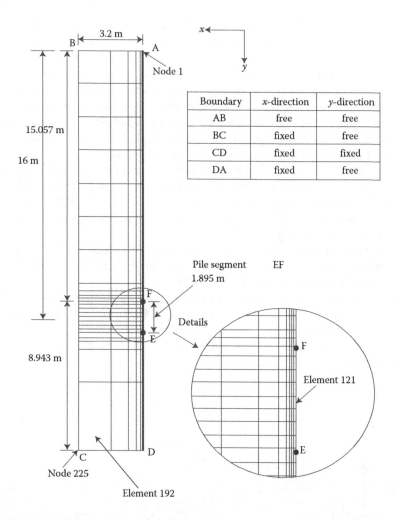

FIGURE 7.16 Finite element mesh and boundary conditions. (Adapted from Wathugala, W. and Desai, C.S., Nonlinear and Dynamic Analysis of Porous Media and Applications, Report to National Science Foundation, Washington, DC, Dept. of Civil Eng. and Eng. Mech., Univ. of Arizona, Tucson, AZ, USA, 1990; Shao, C. and Desai, C.S., Implementation of DSC Model for Nonlinear and Dynamic Analysis of Soil-Structure Interaction, Report to National Science Foundation, Washington, DC, Dept. of Civil Eng. and Eng. Mech., Univ. of Arizona, Tucson, AZ, USA, 1997.)

where ϕ' is the effective angle of friction. Using the value of $\phi' = 12.50°$ given in Ref. [17], K_o was found to be 0.784.

The initial values of the hardening parameters α_o (Equation A1.29a, Appendix 1) was found using the foregoing values of stresses in yield function (Equation A1.28, Appendix 1). For simplicity, it was assumed that the initial hardening was caused only by volumetric plastic strain; hence, the disturbance, D, was adopted as zero, because it was assumed to be dependent on the accumulated deviatoric plastic strain only.

Pile driving: When the pile is driven in the soil, it causes a change in the *in situ* stresses and pore water pressures. It also causes the soil around and below the pile to get pushed away to accommodate the pile. The strain path method (SPM) proposed by Baligh [50] and presented in Refs. [8,17] is used to model the effect of pile driving; it was developed on the basis of measurements. The computed stresses are used to find the stress path for each point. Then, effective stresses were calculated by using the adopted constitutive model. The total stresses and pore water pressures are found using the equilibrium equations. Thus, the SPM yields modified stresses and pore water pressures due to the driving of the pile segments.

Consolidation: Next, the consolidation phase modifies stresses and pore water pressures developed at the end of pile driving. The computer code DSC-DYN2D [51] was used to compute the stresses and pore water pressures at the end of consolidation, which were required as the initial condition for the subsequent one-way and two-way cyclic loading (Figure 7.15).

Simulation of loading: The code DSC-DYN2D [51] was used to simulate various steps, including one- and two-way loadings. The comparisons between predictions and field data for typical phases are given below.

Consolidation: Comparisons between predictions by using the HISS [20,43] and DSC [8,17] models and measurements at the end of consolidation are shown in Figure 7.17, for the 3 in (7.62 cm) probe at the center of the element 121 (Figure 7.16), 53 ft (16 m) below the ground surface. The normalized excess pore water pressure, p_n, was computed as

$$p_n = \frac{p - p_c}{p_o - p_c} \qquad (7.69)$$

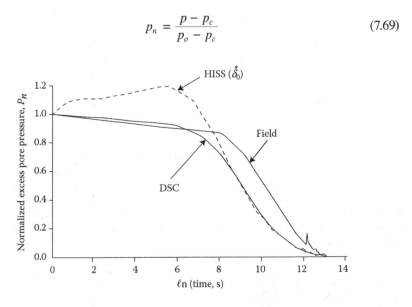

FIGURE 7.17 Comparisons between predictions from HISS and DSC models, and field consolidation behavior. (Adapted from Shao, C. and Desai, C.S., Implementation of DSC Model for Nonlinear and Dynamic Analysis of Soil-Structure Interaction, Report to National Science Foundation, Washington, DC, Dept. of Civil Eng. and Eng. Mech., Univ. of Arizona, Tucson, AZ, USA, 1997.)

where p, p_o, and p_c are the current, initial, and end of consolidation pore water pressures, respectively. It is noted that the DSC model yields improved correlation with field data compared to the prediction using the HISS model.

One-way cyclic loading: It was reported by Earth Technology Corporation [44] that the tension tests conducted after the consolidation did not significantly influence the settlement curve. Hence, the tension tests were not simulated. The simulation of the one-way cyclic loading at the end of consolidation is described below.

The vertical displacements measured in the one-way loading were applied to the nodes in contact with the pile segment (from E to F in Figure 7.16) in about 135 steps. Figures 7.18a through 7.18c compare the predictions using the HISS and DSC models with the measurements. The shear transfer was calculated on the basis of the accumulated computed vertical stresses at the nodes (E to F, Figure 7.16). Overall, the DSC model predictions and the field data compare very well, better than those by the HISS model. One of the reasons could be that the DSC accounts for degradation and softening. The predictions for the unloading and reloading loops are considered to be satisfactory.

Two-way cyclic loading: Five cycles of loading–unloading–reloading were simulated in two-way cyclic loading for the 3 in (7.62 cm) probe. The vertical displacements measured in the field were applied to the nodes from E to F (Figure 7.16). The comparisons between predictions (DSC and HISS models) and field data are presented as follows:

Figure 7.19a: Shear transfer versus pile displacement
Figure 7.19b: Shear transfer versus time
Figure 7.20: Pore water pressure versus time
Figure 7.21: Effective horizontal stress versus time

It can be seen that predictions compare very well with the field data, and, in general, the DSC model provides improved correlations compared to that by the HISS model.

Finally, we can conclude that the FEM with the DSC model can provide highly satisfactory predictions for challenging geotechnical problems such as dynamic behavior of driven piles in saturated soils.

7.6.2 EXAMPLE 7.2: DYNAMIC ANALYSIS OF PILE-CENTRIFUGE TEST INCLUDING LIQUEFACTION

A pile load test under dynamic loading in a centrifuge is analyzed here using the code DSC-DYN2D [51] with the DSC constitutive model for the sand and interface between pile and sand. The National Geotechnical Centrifuge facility at the University of California, Davis, with a radius of 9.0 m and a shaking table was used for dynamic tests on piles. The centrifuge facility has a maximum model mass of about 2500 kg with an available bucket area of 4.0 m^2, and a maximum centrifugal acceleration of 50 g; the details of the facility are presented in Refs. [52,53].

A number of pile tests with single, four and nine pile groups were performed using the centrifuge facility. Here, we have considered and simulated event J in the model, referred to as CSP3 [53]. The foundation soil consisted of two layers. The upper 9.3 m contained medium-dense Nevada sand with $D_r = 55\%$. The lower

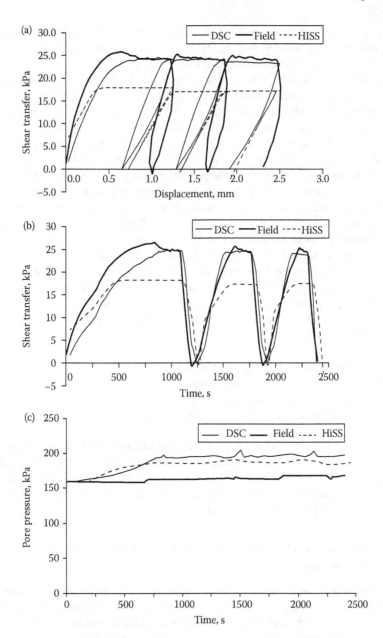

FIGURE 7.18 Comparisons between predictions from HISS and DSC models, and measurements for one-cyclic loading tests. (a) Shear transfer versus pile displacements; (b) shear transfer versus time; and (c) pore water pressures versus time. (Adapted from Shao, C. and Desai, C.S., Implementation of DSC Model for Nonlinear and Dynamic Analysis of Soil-Structure Interaction, Report to National Science Foundation, Washington, DC, Dept. of Civil Eng. and Eng. Mech., Univ. of Arizona, Tucson, AZ, USA, 1997.)

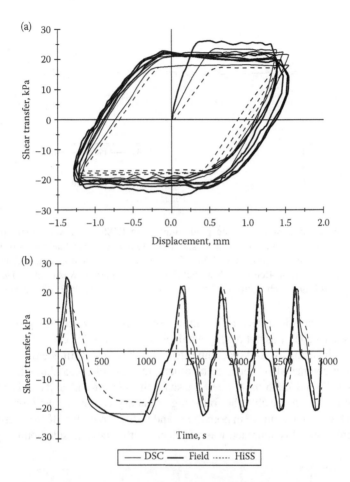

FIGURE 7.19 Comparisons between HISS and DSC predictions and field measurements for two-way cyclic loading. (a) Shear transfer versus pile displacement; (b) shear transfer versus time. (From Shao, C. and Desai, C.S., Implementation of DSC Model for Nonlinear and Dynamic Analysis of Soil-Structure Interaction, Report to National Science Foundation, Washington, DC, Dept. of Civil Eng. and Eng. Mech., Univ. of Arizona, Tucson, AZ, USA, 1997. With permission.)

layer, which was 11.4 m thick contained dense Nevada sand ($D_r = 80\%$) (Figure 7.22). Further details about the structure, pile, and soil can be found in Refs. [52,53].

In this study, we have analyzed the single pile test. The pile was made of aluminum with a diameter of 0.67 m, wall thickness of 72 mm, and embedment depths of 20.7 and 16.8 m; the former was considered herein. The linear elastic properties of the pile material was adopted as $E = 70.0$ GPa and $v = 0.33$. The DSC model was used to characterize the sand and interface; details of the DSC model are given in Appendix 1. The parameters for the sand and interface are shown in Table 7.2 [54,55]. The parameters for the interface between the aluminum pile and soil were obtained by using an artificial neural network (ANN) procedure [54,55].

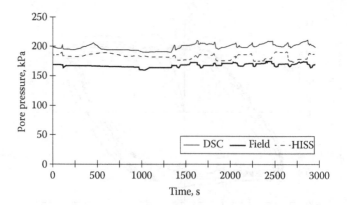

FIGURE 7.20 Comparisons between predictions from HISS and DSC models and field measurements for pore water pressure versus time: Two-way cyclic loading. (From Shao, C. and Desai, C.S., Implementation of DSC Model for Nonlinear and Dynamic Analysis of Soil-Structure Interaction, Report to National Science Foundation, Washington, DC, Dept. of Civil Eng. and Eng. Mech., Univ. of Arizona, Tucson, AZ, USA, 1997. With permission.)

It would be realistic to use a 3-D simulation. However, when this problem was analyzed, such a code was not readily available with the DSC. Hence, the single pile was modeled by using the 2-D procedure [55] with the plane strain idealization; such an approximation has also been used by others, for example, Anandarajah [56].

Figures 7.23a and 7.23b show the FE mesh for the pile and soil, and details of mesh around the pile, respectively. The following boundary conditions were introduced: AB and CD were free to move in both the x- and y-directions, and BC was restrained in the y-direction. This approach was consistent with repeating side boundary [57];

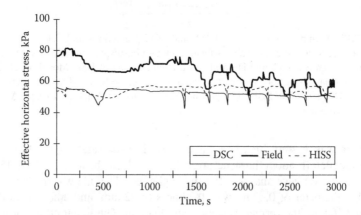

FIGURE 7.21 Comparisons between predictions from HISS and DSC modes and field measurements for effective horizontal stress versus time: two-way cyclic loading. (From Shao, C. and Desai, C.S., Implementation of DSC Model for Nonlinear and Dynamic Analysis of Soil-Structure Interaction, Report to National Science Foundation, Washington, DC, Dept. of Civil Eng. and Eng. Mech., Univ. of Arizona, Tucson, AZ, USA, 1997. With permission.)

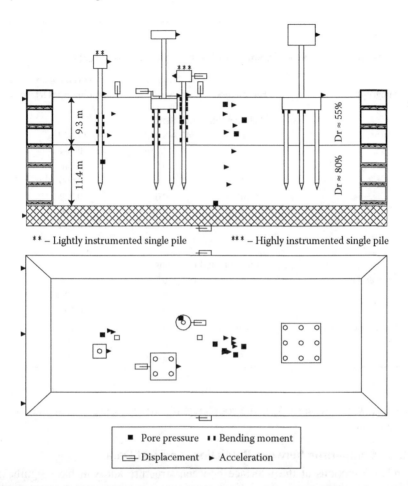

** – Lightly instrumented single pile   *** – Highly instrumented single pile

■ Pore pressure ▮▮ Bending moment
▭ Displacement ► Acceleration

FIGURE 7.22 Details of centrifuge test CSP3. (Adapted from Kutter, B.L. et al., Design of large earthquake simulator at U.C. Davis, *Proceedings, Centrifuge 94*, Balkema, Rotterdam, pp. 169–175, 1994; Wilson, D.W., Boulanger, R.W., and Kutter, B.L., Soil-Pile-Superstructure Interaction at Soft or Liquefiable Soil Sites. Centrifuge Data for CSP1 and CSP5, Report Nos. 97/02, 97/06, Center for Geotechnical Modeling, Dept. of Civil and Environ. Engng., Univ. of California, Davis, CA, USA, 1997.)

here, the displacements of the nodes on the side boundary on the same horizontal plane were assumed to be the same.

The base of the mesh was subjected to the acceleration–time history shown in Figure 7.24 [52,53]. Before such a load was applied, the *in situ* stresses and pore water pressures were introduced at the center of an element by using the following equations:

$$\sigma'_v = \gamma_s h \; ; \; \sigma'_h = K_o \sigma'_v$$
$$K_o = v/(1 - v); \; p_o = \gamma_w h \tag{7.70}$$

where σ'_v and σ'_h are effective vertical and horizontal stresses at depth h, respectively, γ_s is the submerged unit weight of soil, γ_w is the unit weight of water, p_o is the

TABLE 7.2

Parameters for Nevada Sand and Sand–Aluminum Interface

Subgroup	Parameters	Nevada Sand	Nevada Sand Aluminum Interface
Relative Intact (RI) State			
Elasticity	E	40,848.8 kPa	14.6 MPa
	ν	0.316	0.384
Plasticity	γ	0.0675	0.246
	β	0.0	0.000
	3R	0.0	0.0
	n^*	4.1	3.350
	a_1	0.1245	0.620
	η_1	0.0725	0.570
Fully Adjusted (FA) State			
Critical state	$\bar{m}$	0.22	0.304
	λ	0.02	0.0278
	e_o^c	0.712	0.791
Disturbance function	Du	0.99	0.99
	A	5.02	0.595
	Z	0.411	1.195

initial pore water pressure, K_o is the coefficient of earth pressure at rest, and ν is the Poisson's ratio.

7.6.2.1 Comparison between Predictions and Test Data

The relative motions at the interface between structure and soil have significant influence on the behavior of the soil-structure system. To identify such an effect, the FE analyses were conducted with and without interface. It was found that the predictions with the interface provided improved correlations with measurements.

Figures 7.25a and 7.25b show the comparisons between pore water pressures with time in two elements, No. 139 near the pile, and No. 9 away from the pile (Figure 7.23). The predictions for element 139 (near the pile, Figure 7.23b) with provision of interface element show highly satisfactory correlations with the observed data. Figure 7.25a also shows that the liquefaction can occur after about 13 s. It can also be seen that the predictions from the analysis without interface do not show satisfactory correlation with the test data. Figure 7.25b shows comparisions between predictions and test data for Element 9, with and without interface. The closer correlation for both can be due to the far distance of the element from the pile.

Figure 7.26 shows typical disturbance versus time plots. It was obtained on the basis of laboratory cyclic triaxial tests on Nevada sand [53,55] with a confining pressure of 80 kPa. The critical disturbance, D_c, at which liquefaction (microstructural instability) may occur, is noted as 0.87 in the figure; the average value from various tests was found to be 0.86. Table 7.3 shows the times taken for various elements for

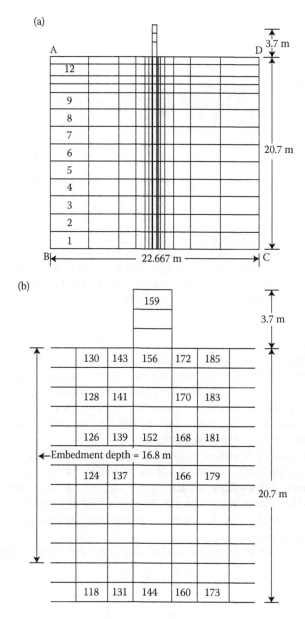

FIGURE 7.23 Finite element mesh for pile and soil. (a) Overall FE mesh; (b) FE mesh near pile. (Adapted from Pradhan, S.K. and Desai, C.S., Dynamic Soil-Structure Interaction Using Disturbed State Concept, Report to National Science Foundation, Washington, DC, Dept. of Civil Eng. and Eng. Mech., Univ. of Arizona, Tucson, AZ, USA, 2002.)

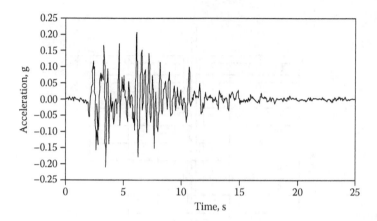

FIGURE 7.24 Base motion acceleration. (Adapted from Kutter, B.L. et al., Design of large earthquake simulator at U.C. Davis, *Proceedings, Centrifuge 94*, Balkema, Rotterdam, pp. 169–175, 1994; Wilson, D.W., Boulanger, R.W., and Kutter, B.L., Soil-Pile-Superstructure Interaction at Soft or Liquefiable Soil Sites. Centrifuge Data for CSP1 and CSP5, Report Nos. 97/02, 97/06, Center for Geotechnical Modeling, Dept. of Civil and Environ. Engng., Univ. of California, Davis, CA, USA, 1997.)

the excess pore water pressure (U_w) to reach the initial effective vertical stress, and $D = D_c$. The former defines the conventional empirical procedure for liquefaction [58]. It can be seen that according to the conventional procedure, times taken for liquefaction are consistently higher than those for the critical disturbance procedure. It may be noted that the critical disturbance procedure is based on microstructural modifications in the soil. It implies that the microstructural instability can occur earlier than liquefaction by the conventional procedure.

Figure 7.27 shows a variation of D with time for Element 139 in the interface between pile and sand, and for element 126 away from the interface, in the sand. The results for element 139 show the tendency to reach the initiation of the liquefaction state ($D_c = 0.86$) earlier than element 126. Hence, it may be noted that the liquefaction for this problem can begin in and near the interface elements, earlier than in the surrounding elements in soil.

Figure 7.28 shows variations of R_1 and R_2 with time; these terms are defined as

$$R_1 = \frac{V_\ell}{V_s} \tag{7.71a}$$

and

$$R_2 = \frac{V_\ell}{V_p} \tag{7.71b}$$

where V_ℓ, V_s, and V_p are the liquefied volume (in which $D_c = 0.86$ or greater is reached), total soil volume, and pile volume, respectively. It can be seen that both reach stable values of about 0.32 and 10.0 for R_1 and R_2, respectively, at the time of liquefaction in about 12 s. Such quantities can be used for analysis and design.

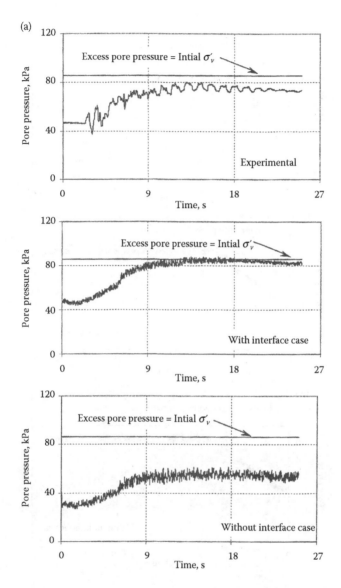

FIGURE 7.25 Comparisons for predicted and measured pore water pressures. (a) Element 139 near pile; (b) Element 9 away from pile. (Adapted from Pradhan, S.K. and Desai, C.S., Dynamic Soil-Structure Interaction Using Disturbed State Concept, Report to National Science Foundation, Washington, DC, Dept. of Civil Eng. and Eng. Mech., Univ. of Arizona, Tucson, AZ, USA, 2002.)

7.6.3 EXAMPLE 7.3: STRUCTURE–SOIL PROBLEM TESTED USING CENTRIFUGE

Popescu and Prevost [59] have presented an analysis of a number of problems tested in the centrifuge facilities at Princeton and Cambridge Universities. We present below one of the problems involving structure–soil system tested at the Princeton facility (Figure 7.29a). The geometry of the tested model corresponded to the Niigata

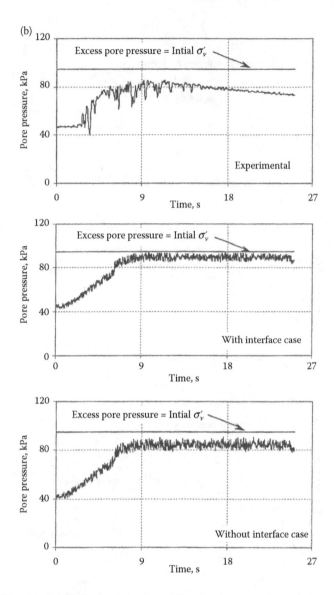

FIGURE 7.25 (continued) Comparisons for predicted and measured pore water pressures. (a) Element 139 near pile; (b) Element 9 away from pile. (Adapted from Pradhan, S.K. and Desai, C.S., Dynamic Soil-Structure Interaction Using Disturbed State Concept, Report to National Science Foundation, Washington, DC, Dept. of Civil Eng. and Eng. Mech., Univ. of Arizona, Tucson, AZ, USA, 2002.)

apartment, which was damaged due to liquefaction in the 1964 Niigata earthquake. For the test, the structure was placed in Nevada sand at the relative density $D_r = 60\%$. Two centrifuge tests were performed at $100\ g$ centrifuge acceleration [60].

The structure and foundation soil are shown in Figure 7.29a, including locations of measuring devices such as accelerometers and pore water pressure transducers.

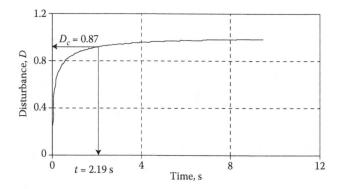

FIGURE 7.26 Critical disturbance D_c, $\sigma_3 = 80$ kPa. (Adapted from Pradhan, S.K. and Desai, C.S., Dynamic Soil-Structure Interaction Using Disturbed State Concept, Report to National Science Foundation, Washington, DC, Dept. of Civil Eng. and Eng. Mech., Univ. of Arizona, Tucson, AZ, USA, 2002.)

The FE mesh is shown in Figure 7.29b, which consists of 119 elements and 154 nodes. The plane strain idealization was adopted because the structure extends over the whole width of the centrifuge box. The boundary conditions for the mesh in Figure 7.29b were as follows:

1. Prescribed acceleration to the degree of freedom of solid phase at base and lateral nodes
2. Impervious base and side boundaries
3. Restrained vertical motion at the base for both solid and fluid phases
4. Impervious interface at the sand–structure interface
5. Full friction at the base
6. No friction at the sides of the structure

7.6.3.1 Material Properties

Material parameters at $D_r = 60\%$ are given in Table 7.4 [59].

TABLE 7.3

Times to Liquefaction Based on Conventional and Disturbance Methods

Element	Conventional $U_w = \sigma_v'$	DSC $D = D_c$
143	1.74 s	1.23 s
130	1.83	1.29
104	2.55	1.92
78	3.36	2.67
52	3.81	2.94
26	9.03	8.22

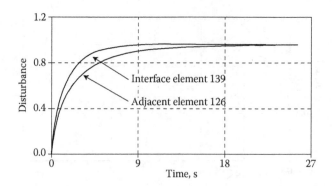

FIGURE 7.27 Disturbance versus time in elements 126 and 139. (Adapted from Pradhan, S.K. and Desai, C.S., Dynamic Soil-Structure Interaction Using Disturbed State Concept, Report to National Science Foundation, Washington, DC, Dept. of Civil Eng. and Eng. Mech., Univ. of Arizona, Tucson, AZ, USA, 2002.)

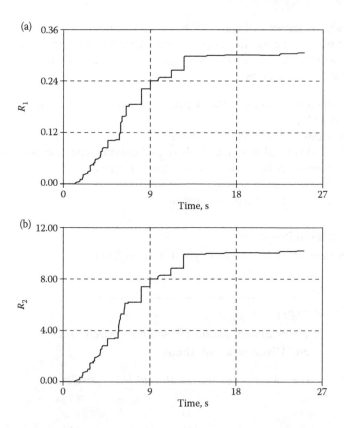

FIGURE 7.28 Variation of disturbance ratio, R_1 and R_2. (a) R_1 versus time; (b) R_2 versus time. (Adapted from Pradhan, S.K. and Desai, C.S., Dynamic Soil-Structure Interaction Using Disturbed State Concept, Report to National Science Foundation, Washington, DC, Dept. of Civil Eng. and Eng. Mech., Univ. of Arizona, Tucson, AZ, USA, 2002.)

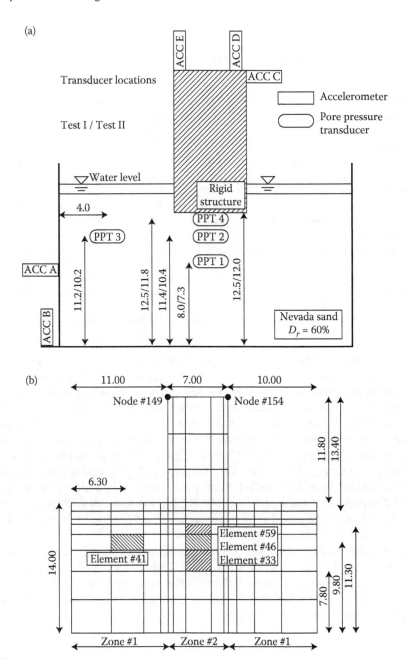

FIGURE 7.29 (a) Structure in centrifuge facility at Princeton university; (b) finite element mesh of structure and foundation soil. (Adapted from Popescu, R. and Prevost, J.H., *Soil Dynamics and Earthquake Engineering,* 12, 1993, 73–90.)

TABLE 7.4

Material Parameters for Nevada Sand

Property	Nevada Sand		
	Dr = 40%	Dr = 60%	Dr = 70%
Mass density—solid (kg/m³)	2670.0	2670.0	2670.0
Porosity	0.424	0.398	0.384
Low-strain shear modulus (MPa)	25.0	30.0	35.0
Low-strain bulk modulus (MPa)	54.2	65.0	75.8
Reference mean effective normal stress (kPa)	100.0	100.0	100.0
Power exponent	0.7	0.7	0.7
Fluid bulk modulus (MPa)	2000.0	2000.0	2000.0
Friction angle at failure (compression and extension)	33°	35°	37°
Cohesion (kPa)	0.0	0.0	0.0
Maximum deviatoric strain compression/extension (%)	8.0/7.0	6.0/5.0	4.0/4.0
Dilation angle	30°	30°	33°
Dilation parameter	0.15	0.13	0.085
Permeability (m/s)	6.6×10^{-5}	5.6×10^{-5}	4.7×10^{-5}

Note: For details of the constitutive model and parameters, consult Refs. [59,62,63].

State parameters: The mass density, ρ_s, from routine soil classification tests, for the Nevada sand was adopted from Ref. [61]. The porosity n^w was derived from the relative density, maximum and minimum void ratios. The permeability of the sand was obtained from constant-head permeability tests reported in Ref. [61], at $D_r = 40\%$, 60%, and 90%. The underlying material in Zone 2 (Figure 7.29b) was much denser due to the heavy weight of the structure, for which the bearing pressure was about 200 kPa. Hence, parameters for $D_r = 70\%$ were used for Zone 2, whereas for Zone 1, the parameters for $D_r = 60\%$ were used. The permeability of sand in the centrifuge model was found from the recorded excess pore water pressure–time histories during the diffusion phase in the centrifuge test; the value of $k_m = 2.8 \times 10^{-5}$ m/s resulted from the tests. It can be seen that this is lower than $k_m = 5.6 \times 10^{-5}$ m/s from the laboratory constant-head permeability test (Table 7.4).

Constitutive model: A kinematic hardening elastoplastic model [62,63] was used to define the mechanical behavior of the sand (Nevada); a nonassociative plasticity rule was used. Various parameters in the model are described below.

The low-strain shear modulus, G, was determined from isotropic consolidated compression tests and was expressed as

$$G = G_o \left(\frac{\sigma_3}{\sigma_{30}} \right)^n \tag{7.72a}$$

where G_o is found from $(\sigma_1 - \sigma_3)$ and deviatoric strain (ε_d) at 0.10%:

$$G_0 = \frac{\sigma_1 - \sigma_3}{2\varepsilon_d} \tag{7.72b}$$

in which $\varepsilon_d = ((3\varepsilon_1 - \varepsilon_v)/2)$, and ε_1 and ε_v are axial and volumetric strains, respectively. In Equation 7.72a, σ_3 is the (initial) confining (mean) pressure, n is the exponent, and $\sigma_{30} = 100$ kPa is the reference pressure. The value of G_o and n were found from plots of Equation 7.72a, based on laboratory tests [59].

The bulk modulus, K_o, for the soil was found from:

$$K_o = \frac{2G_o(1 + v)}{3(1 - 2v)}$$

(7.73)

The plasticity part of the model contains the following parameters:
Friction angle in compression (and extension), φ_c, is found as

$$\varphi_c = \arcsin\left(\frac{3\eta}{6 + \eta}\right)$$

$$\eta = \frac{q}{p}\bigg|_{\text{at failure}}$$

(7.74)

The value of the maximum deviatoric strain, ε_d^m, was adopted according to prior experience [59].

The dilation angle, $\bar{\phi}$, was determined based on particle characteristics using the procedure in Ref. [64]. The value of 30° was found for lower densities, while at high density $(D_r = 70\%)$, it was adopted as 33° [59]. The dilation parameter, X_p, was determined based on liquefaction strength analysis from data of und-rained cyclic laboratory triaxial and simple shear tests; details are given in Ref. [61]. Table 7.4 includes the parameters for the elastic–plastic model for Nevada sand [59].

The structural material was considered as linear elastic with parameters given below:

$E = 1000$ MPa, $v = 0.30$, $\rho_s = 1520$ kg/m³

7.6.3.2 Results

Figure 7.30 shows the input acceleration to the box, and comparisons between computed and measured horizontal and vertical accelerations [60]. Figure 7.30a shows the input acceleration to the test box. The comparisons between computed and measured horizontal accelerations at the Node 154 at the top of the structure (Figure 7.29b) and recorded during test #2-ACC C are shown in Figures 7.30b and 7.30c, respectively. Similar comparisons for the vertical accelerations at Node 149 are shown in Figures 7.30d and 7.30e. The computed and measured pore water pressure from test nos. 1 and 2 for various elements are shown in Figure 7.31. It can be seen that the computed values are in good agreement with the measurements.

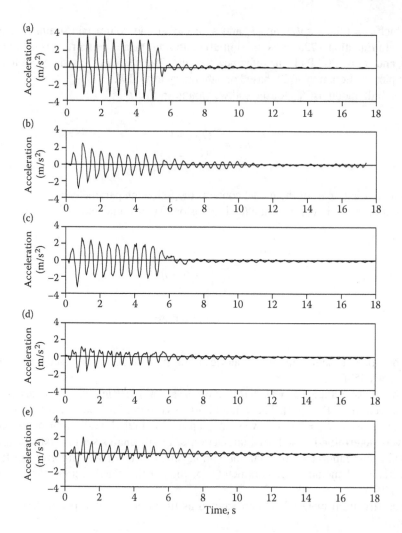

FIGURE 7.30 Computer predictions and measurements of horizontal and vertical accelerations. (Adapted from Popescu, R. and Prevost, J.H., *Soil Dynamics and Earthquake Engineering*, 12, 1993, 73–90.)

7.6.4 EXAMPLE 7.4: CYCLIC AND LIQUEFACTION RESPONSE IN SHAKE TABLE TEST

The computer code (DSC-DYN2D) with the DSC model [51] was employed to predict the cyclic and liquefaction behavior of sand using a shake table test [65–67].

Figure 7.32a shows the shake table test equipment reported by Akiyoshi et al. [65]. The test involved the use of saturated Fuji river sand [65,68]. However, since the parameters for the DSC model were not available for that sand, for the computer analysis, we used the properties of the Ottawa sand. Based on the similar grain size behavior for both sands (Figure 7.32b) and similar elastic moduli, index properties, and cycles to liquefaction (Table 7.5), such an assumption is considered to be appropriate.

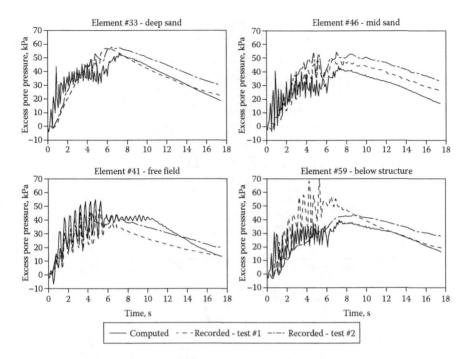

FIGURE 7.31 Computer predictions and measurements for excess pore water pressures in various elements. (Adapted from Popescu, R. and Prevost, J.H., *Soil Dynamics and Earthquake Engineering,* 12, 1993, 73–90.)

The DSC parameters for the Ottawa sand were determined from a series of multiaxial (true axial) tests for relative density $D_r = 60\%$ and under confining pressures, $\sigma_o' = 69$, 138, and 207 kPa [67]. Cyclic tests were performed on saturated cubical specimens ($10 \times 10 \times 10$ cm) at a value of B close to unity. Cyclic deviatoric stresses of amplitude $\sigma_d = (\sigma_1 - \sigma_3)_d, = 35$, 70, and 100 kPa for initial effective stresses, $\sigma_o' = 69$, 138, and 207 kPa were applied. Typical test data for $\sigma_o' = 138°$ kP are given in Figures 7.33a, for applied stress, measured axial strains, and pore water pressures. Figure 7.33b shows measured time-dependent σ_d for three confining pressures. The DSC parameters were found using the above test data by following the procedure presented in Appendix 1. They are listed in Table 7.6.

Finite element analysis: Figure 7.34 shows the FE mesh for the shake table test, which contained 160 elements (120 for soil and 40 for steel box) and 190 nodes. The material of the box was assumed to be linear elastic with $E = 200 \times 10^6$ kPa and $v = 0.30$. The idea of repeating side boundaries [6,57,69] was employed. In this approach, displacements of the nodes on the side boundary on the same horizontal plane were assumed to be the same. The bottom boundary was restrained in the vertical direction, but it was free to move in the horizontal direction. The applied load involves horizontal displacements, X, on the bottom nodes given by the following equation:

$$X = u \sin(2\pi ft) \qquad (7.75)$$

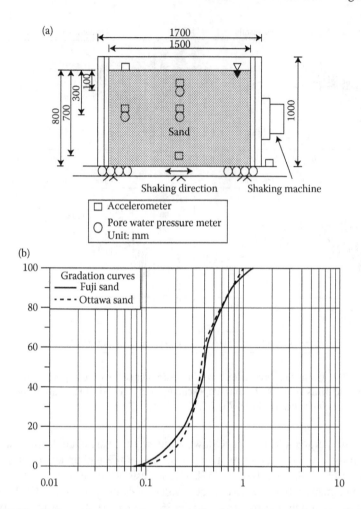

FIGURE 7.32 Shake table setup and grain size distributions for sands. (a) Shake table test set-up; (b) grain size distribution curves of Ottawa sand and Fuji river sand. (Adapted from Park, I.J. and Desai, C.S. Analysis of Liquefaction in Pile Foundations and Shake Table Tests using the Disturbed State Concept, Report, Dept. of Civil Eng. and Eng. Mechanics, University of Arizona, Tucson, AZ, USA, 1997; Akiyoshi, T. et al. *International Journal for Numerical and Analytical Methods in Geomechanics,* 20(5), 1996, 307–329.)

where u is the amplitude (= 0.0013 m), f the frequency (= 5 Hz), and t the time. The FE analysis was performed for 50 cycles with time steps, $\Delta t = 0.001$ s from time 0.0 to 2.0 s, and $\Delta t = 0.05$ s, from time 2.0 to 10.0 s.

7.6.4.1 Results

The computed and measured excess pore water pressure with time at the point (depth = 300 mm) shown as a solid dot (Figure 7.34) are presented in Figure 7.35. The test data indicate that liquefaction occurred after about 2.0 s when the pore

TABLE 7.5
Properties of Ottawa and Fuji River Sands

Properties	Ottawa Sand [67]		Fuji River Sand [65,68]
Deformation and strength			
Elastic modulus, E	193,000 kPa		170,600 kPa
Poisson's ratio, v	0.38		0.35
Angle of friction, φ'	38.0°		37.0°
	Index		
Total density, γ_t	19.63 kN/m³		19.80 kN/m³
Specific gravity, G_s	2.64		2.68
Maximum void ratio, e_{max}	0.77		1.08
Minimum void ratio, e_{min}	0.46		0.53
Coefficient of uniformity, C_u	2.00		2.21
Mean particle size, D_{50}	0.38 mm		0.40 mm
Cycles to Liquefaction, N_f	D_r	$\bar{\sigma}_0'$	N_f
Fuji river sand	47%	98 kPa	8
	75%	98 kPa	10
Ottawa sand	60%	69 kPa	5
	60%	138 kPa	7
	60%	207 kPa	9

water pressure equaled the initial effective stress. It can be seen that the FE with DSC predictions compare very well with the measurements.

Figures 7.36a through 7.36d show the growth of disturbance in different elements of the mesh for typical times = 0.50, 1.00, 2.00, and 10.00 s. The computed plots of the growth of disturbance at the above point (depth = 300 mm) are shown in Figure 7.36e. The laboratory tests on the Ottawa sand showed that liquefaction initiates at an average value of the critical disturbance $D_c = 0.84$ (Figure 7.37) [40,66]. At times below 2.0 s, the value of disturbance is below the critical value. However, at time equal to 2.0 s and in the vicinity, disturbance has reached the value equal to about 0.84 or higher; this can be considered to represent the initiation of liquefaction. It is also evident from Figure 7.36 that the disturbance continues to grow after time = 2.0 s, and when it occurs equal to and beyond the critical disturbance in the major part of the box, the sand may be considered to experience complete failure.

7.6.5 EXAMPLE 7.5: DYNAMIC AND CONSOLIDATION RESPONSE OF MINE TAILING DAM

Conventional design methods may not allow integrated analysis for consolidation and dynamic loading because of the effects of factors such as geometry and non-linear behavior of the materials. The FE method can account for such factors and

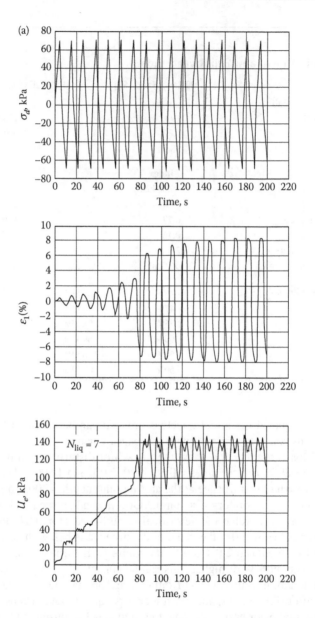

FIGURE 7.33 Laboratory test results for sand for $D_r = 60\%$. (a) Applied deviatoric stress, measured strain, and excess pore water pressure versus. time for typical initial stress $\sigma_o = 138$ kPa; (b) deviatoric stress versus. strain, $D_r = 60\%$. (Adapted from Gyi, M.M. and Desai, C.S., Multiaxial Cyclic Testing of Saturated Ottawa Sand, Report, Dept. of Civil Eng. and Eng. Mech., Univ. of Arizona, Tucson, AZ, USA, 1996.)

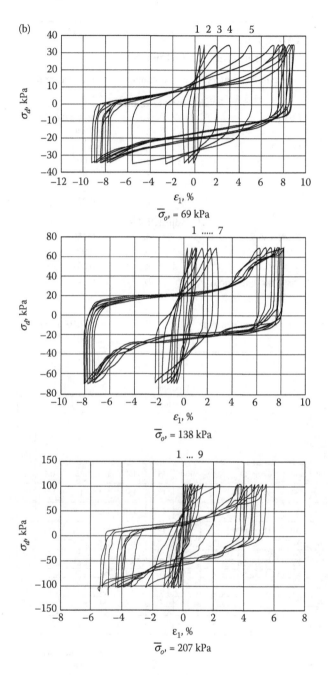

FIGURE 7.33 (continued) Laboratory test results for sand for $D_r = 60\%$. (a) Applied deviatoric stress, measured strain, and excess pore water pressure versus. time for typical initial stress $\sigma_o = 138$ kPa; (b) deviatoric stress versus. strain, $D_r = 60\%$. (Adapted from Gyi, M.M. and Desai, C.S., Multiaxial Cyclic Testing of Saturated Ottawa Sand, Report, Dept. of Civil Eng. and Eng. Mech., Univ. of Arizona, Tucson, AZ, USA, 1996.)

TABLE 7.6
DSC Parameters for Ottawa Sand

Group of Parameters	Parameters	Ottawa Sand
Relative Intact (RI) State		
Elasticity	E	193,000 (kPa)
	ν	0.380
Plasticity	γ	0.123
	β	0.000
	n	2.450
	a_l	0.8450
	η_1	0.0215
Fully Adjusted (FA)		
Critical state	$\bar{m}$	0.150
	λ	0.02
	e_o^c	0.601
Disturbance function	D_u	0.99
	A	4.22
	Z	0.43
Unloading and reloading	E_{eu}	177,600 (kPa)
	ε_1^p	0.0013

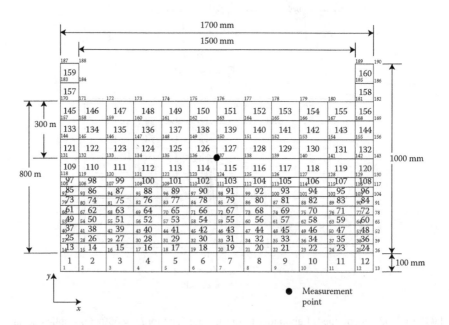

FIGURE 7.34 Finite element mesh for shake table test. (Adapted from Park, I.J. and Desai, C.S. Analysis of Liquefaction in Pile Foundations and Shake Table Tests using the Disturbed State Concept, Report, Dept. of Civil Eng. and Eng. Mechanics, University of Arizona, Tucson, AZ, USA, 1997.)

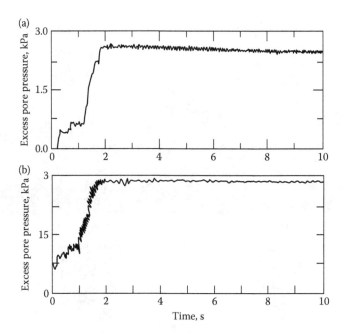

FIGURE 7.35 Excess pore water pressure at depth = 300 mm. (a) Measured (Adapted from Akiyoshi, T. et al. *International Journal for Numerical and Analytical Methods in Geomechanics*, 20(5), 1996, 307–329.); (b) computed by using DSC model. (Adapted from Park, I.J. and Desai, C.S. Analysis of Liquefaction in Pile Foundations and Shake Table Tests using the Disturbed State Concept, Report, Dept. of Civil Eng. and Eng. Mechanics, University of Arizona, Tucson, AZ, USA, 1997.)

provide detailed results for displacements, stresses, pore water pressures, factors of safety, liquefaction, and quantity of seepage. Such data can lead to improved and informed design and construction decisions. The following example presents FE analysis with appropriate constitutive models for design for additional construction of a mine tailing dam known as Reservation Canyon Tailing Dam (Figure 7.38) at Barrick Mercur Gold Mines, Utah [70–75].

The crest of the original dam was at an elevation of 7260 ft (2114 m) (Figure 7.39). The buttress impoundment part of the dam was at an elevation of 7330 ft (2236 m). It was proposed to increase the height of the dam to about 7360 ft (2245 m). The increase in height warranted detailed analysis of the safety and stability of the enlarged structure. We could use the conventional slip circle (and other) methods for the safety and stability analysis. However, we believed that for realistic predictions, it was necessary to perform integrated computer (FE) analysis that can take into account the effects of sequential construction, nonlinear material properties, consolidation, and potential earthquake loading.

Dam details: The dam is composed of the outer shell (Zone II) made of bulk fill material, clay core (Zone V), chimney drain (Zone IV), and the inner backside shell (Zone VI). The latter is made of the run-of-mine (ROM) material (Figure 7.38). The lower part of the tails, called the bulk discharge tails (maximum of about 110 ft (34 m)

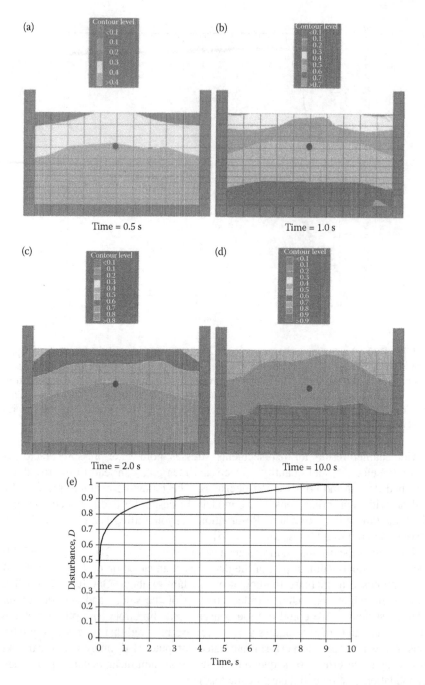

FIGURE 7.36 Growth of disturbance at various times: (a) 0.50 s; (b) 1.0 s; (c) 2.0 s; (d) 10 s; and (e) disturbance versus time at 300 mm. (Adapted from Park, I.J. and Desai, C.S. Analysis of Liquefaction in Pile Foundations and Shake Table Tests using the Disturbed State Concept, Report, Dept. of Civil Eng. and Eng. Mechanics, University of Arizona, Tucson, AZ, USA, 1997.)

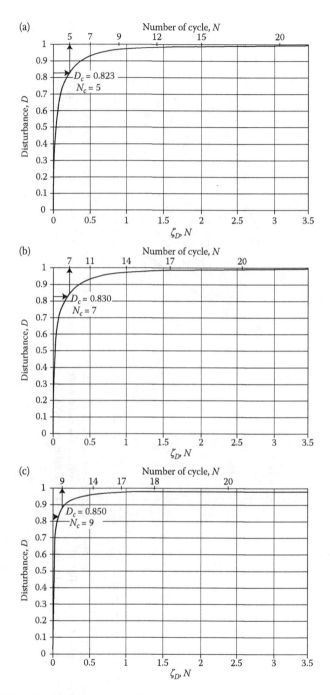

FIGURE 7.37 Disturbance versus $\zeta_D(N)$ and number of cycles at D_c. (a) $\sigma'_o = 69$ kPa; (b) $\sigma'_o = 138$ kPa; and (c) $\sigma'_o = 207$ kPa. (Adapted from Park, I.J. and Desai, C.S. Analysis of Liquefaction in Pile Foundations and Shake Table Tests using the Disturbed State Concept, Report, Dept. of Civil Eng. and Eng. Mechanics, University of Arizona, Tucson, AZ, USA, 1997.)

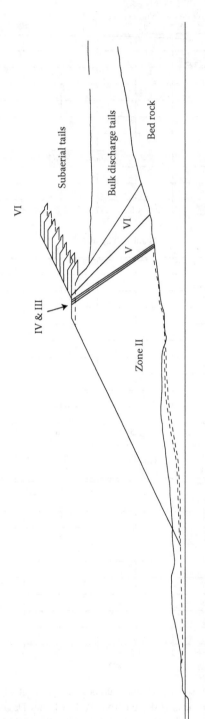

FIGURE 7.38 Tailing dam with buttress. (Adapted from Desai, C.S., Settlement and Seismic Analyses of Reservation Canyon Tailing Dam and Impoundment, Report, C. Desai, Tucson, AZ, 1995; Cross-sections and Material Properties Assumed in Previous Models, Parts of Previous Reports, provided by Physical Resources Eng., Inc. (White, D.), Tucson, AZ,USA, 1993.)

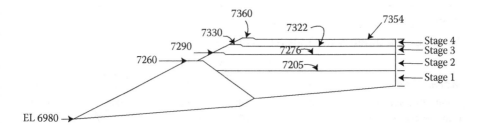

FIGURE 7.39 Elevations and stages of consolidation. (Adapted from Desai, C.S., Settlement and Seismic Analyses of Reservation Canyon Tailing Dam and Impoundment, Report, C. Desai, Tucson, AZ, 1995.)

thickness) are overlain by the subaerial tails of recent deposits. The details of the original dam design, buttress, construction procedures, and methods of deposition of bulk discharge and subaerial tails are given in various reports, for example, Refs. [71,72,74].

7.6.5.1 Material Properties

Using linear elastic models for the materials did not provide satisfactory predictions for the trends observed in the field. Hence, HISS (see Appendix 1) plasticity model was used for bulk discharge and subaerial tails. The parameters for different materials are given in Table 7.7a [71–74]. The linear elastic and plasticity parameters were determined from available triaxial tests. For bulk discharge tails, test data reported by Knight-Piesold were used [73]. Triaxial undrained and drained tests conducted by Geotest Express Testing [74] were also used.

TABLE 7.7a
Material Properties and Elastic Parameters

					Property				
	γ_b	γ_s			$E \times 10^6$		c	φ	
Material	pcf	pcf	w	n	psf	v	psf	(deg)	k (ft/s)
1. Bulk fill (II) (outer shell)	125	125	0.32	0.464	1.00	0.40	576	35	4.29×10^{-6a}
2. Clay core (V)	120	123	0.37	0.522	0.75	0.48	288	30	1.3×10^{-9}
3. Chimney drain (IV)	125	125	0.27	0.422	1.65	0.30	0	35	1.08×10^{-4}
4. Bulk fill (ROM-VI) (inner shell and buttresses)	130	135	0.30	0.486	1.00	0.40	850	37	7.2×10^{-5}
5. Bulk discharge tails[b]	110	115	0.37	0.539	1.30	0.48	0	31	2.5×10^{-7}
6. Subaerial tails[b]	100	121	0.42	0.660	1.20	0.48	450	32	5.9×10^{-7}

Note: γ_b = bulk density, γ_s = saturated density, w = water constant, n = porosity, k = hydraulic conductivity, E = elastic modulus, v = Poisson s ratio, c = cohesion, φ = angle of friction.

[a] This pertains only to the bottom horizontal layer, a–b–c–d (Figure 7.40).

[b] Nonlinear parameters are given in Table 7.7b.

TABLE 7.7b
Nonlinear Parameters for Tails (HISS Plasticity Model)

Material	γ	β	N	a_1	η_1
			Property		
5. Bulk discharge tails	0.086	0.0	2.1	0.00099	0.803
6. Subaerial tails	0.110	0.0	2.1	0.000277	1.283

Parameters: g = gravitational constant = 32 ft/s², G_a = specific gravity = 2.7, γ_w = density of water = 62.4 pcf, ρ = mass density water = 1.95 lb-s²/ft⁴, γ_s = density of solids = 168.48 pcf, ρ_s = mass density of solids = 5.265 lb-s²/ft⁴. (1 pcf = 157 N/m³, 1 ft/s = 0.3048 m/s, 1 ft/s² = 0.3048 m/s², 1 lb-s²/ft⁴ = 515 N-s²/m⁴).

7.6.5.2 Finite Element Analysis

The dam-soils system was idealized using the plane strain assumption. The overall (dynamic) behavior of the dam-soils system is influenced by *in situ* stress, nonlinear material behavior, construction sequences, and the consolidation in the bulk discharge and subaerial tails. Initially, the stresses in various zones in the dam were computed as follows:

$$\sigma_y = \gamma_s y$$
$$\sigma_x = K_o \sigma_y$$
$$\tau_{xy} = 0 \tag{7.76}$$
$$K_o = \frac{\nu}{1-\nu}$$
$$p_o = \gamma_w y$$

where σ_x and σ_y are horizontal and vertical stresses (at the element center), respectively, τ_{xy} is the shear stress, K_o is the coefficient of the lateral earth pressure at rest, ν is the Poisson s ratio, p_o is the initial pore water pressure, and γ_s and γ_w are the densities of the soil and water, respectively.

The consolidation in subaerial and bulk discharge tails was simulated by dividing them into four stages, 1 through 4 in Figure 7.39. The first layer constituted the bulk discharge tails, and the other three layers represented the subaerial tails. The consolidation was simulated over 5 years; during the consolidation phase, the time step Δt was varied over different times:

Stage	Variable Δt (s)	Total Time (Days)
1	100 to 5×10^5	102.87
2	500 to 1×10^5	218.60
3	500 to 2×10^6	450.10
4	1000 to 5×10^6	1028.70
	Total time	1800.27 days
		$\cong$ 5.0 years

In the FE analysis, stresses and pore water pressures at the end of the previous stage were adopted as the initial conditions for the subsequent stage. The FE mesh after the fourth stage is shown in Figure 7.40.

7.6.5.3 Dynamic Analysis

The FE mesh used for the dynamic analysis is shown in Figure 7.40, which shows the mesh after the consolidation stages. The initial stresses and pore water pressures for the dynamic analysis were those at the end of the consolidation stages. The initial displacement in the dam-soil system were set equal to zero before the application of earthquake load, which is shown in Figure 7.41; it represents the displacement record, integrated from the acceleration record for the El Centro earthquake. The possibility of such an earthquake of amplitude 7.5 on the Richter scale was identified in Ref. [75]. The record (Figure 7.41) was applied at the bottom nodes.

Boundary conditions: The nodes on the bottom, that is, the top of the bed rock were constrained against both horizontal (u) and vertical (v) displacements for the consolidation phase. For the dynamic analysis, the bottom nodes were subjected to the horizontal earthquake motion (Figure 7.41).

The right-hand boundary (Figure 7.40) was placed far away (730 ft = 223 m) from the top of the buttress; the following two boundary conditions were investigated in the parametric study:

a. $u = 0$, $v =$ free, no flow
b. $u =$ earthquake record (Figure 7.41), $v =$ free, no flow

It was found from the parametric analysis that predictions from the above two boundary conditions were not significantly different, particularly near the dam and buttress. Hence, the boundary conditions in case (a) were used. At the top boundary, u and v and flow were assumed to be free.

A nonwoven geotextile with an opening of the size of 100 meshes was installed along the inclined junction between the buttress and tails (Figure 7.40). Seepage (partial) through the horizontal parts of the junction (not crossed) was permitted. To simulate such a partial flow condition, alternate nodes marked X were fixed against seepage, whereas the other nodes, marked O, were allowed free seepage.

It was assumed that partial flow can occur through the dam. This was simulated by assigning high permeability to the bottom layer of the mesh for the dam, a–b–c–d (Figure 7.40). Thus, seepage from the upstream of the dam, through the core and sand drain, occurs predominantly through the bottom layer (a–b–c–d).

7.6.5.4 Earthquake Analysis

As stated earlier, the possibility of an earthquake of magnitude 7.5 on the Richter scale was considered [75]. Because the El Centro earthquake had similar properties, it was proposed to apply the El Centro earthquake record (acceleration, Figure 7.41a; it was integrated to compute the displacement–time record (Figure 7.41b), which was applied at the bottom nodes (Figure 7.40)). The vertical component was not included for this example.

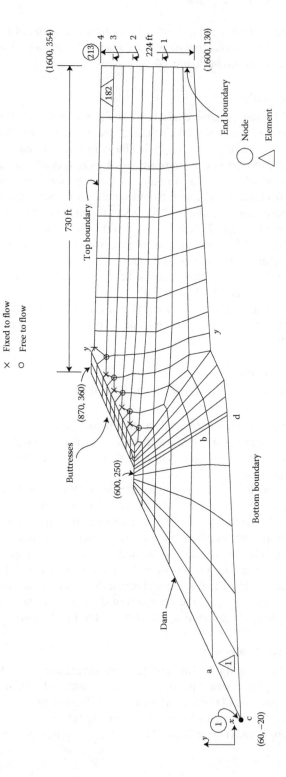

FIGURE 7.40 Finite element mesh and dimensions (1 ft = 0.305 m). (Adapted from Desai, C.S., Settlement and Seismic Analyses of Reservation Canyon Tailing Dam and Impoundment, Report, C. Desai, Tucson, AZ, 1995.)

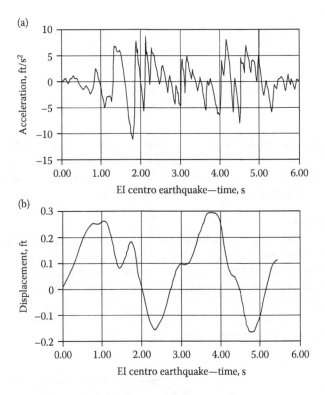

FIGURE 7.41 El centro earthquake record (1 ft = 0.305 m, 1 ft/s² = 0.305 m/s²) (a) Acceleration; (b) displacements. (Adapted from Desai, C.S., Settlement and Seismic Analyses of Reservation Canyon Tailing Dam and Impoundment, Report, C. Desai, Tucson, AZ, 1995; Preliminary Assessment of the Seismic Potential of the Oquirrh Fault Zone, Utah, 1982: Mercur Gold Project, Report, by Woodward-Clyde Conf., San Francisco, CA, for Davy McKee Corp., 1982.)

7.6.5.5 Design Quantities

Factor of safety: The factor of safety, F_s, was expressed as

$$F_s = \frac{\tau_a}{\tau_m} \tag{7.77a}$$

where τ_a and τ_m are the allowable and maximum computed shear stresses, respectively. In terms of the strength parameters, c and φ, and principal stresses, σ_1 and σ_3, the above equation can be expressed as

$$F_s = \frac{2c\,(\cos\phi) + (\sigma_1 + \sigma_3)\sin\phi}{\sigma_1 - \sigma_3} \tag{7.77b}$$

7.6.5.6 Liquefaction

The following liquefaction factor, L_f, was expressed based on empirical procedures [76,77]:

$$L_f = \frac{p_e}{\sigma'_{vo}} \qquad (7.78)$$

where p_e is computed excess pore water pressure, and σ'_{vo} is the initial effective vertical stress. If $L_f \geq 1$, liquefaction can occur. However, as reported in Ref. [78], a sand layer may liquefy when L_f is in the range of 0.50–0.60. Hence, in this study, we have considered that liquefaction can occur when $L_f \geq 0.50$.

7.6.5.7 Results

The code DSC-SSTDYN [51], which allows static, consolidation, and dynamic analyses, was used for consolidation and then dynamic analysis. It is based on the generalized Biot's theory as presented before and includes coupled analysis in which both displacements and pore water pressures are treated as unknowns.

The FE predictions were obtained in terms of principal stress vectors, displacement vectors, and contours of pore water pressures and of liquefaction factors. The liquefaction potential was computed only for the tailings.

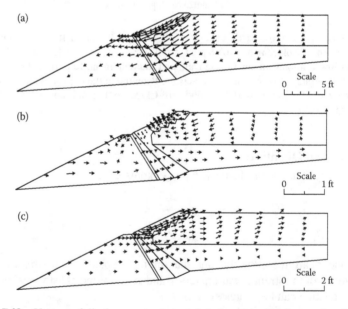

FIGURE 7.42 Vectors of displacements at stage 4 during consolidation, and various times during earthquake (1 ft = 0.305 m). (a) After stage 4 consolidation; (b) after $t = 4.0$ s during earthquake; and (c) after $t = 5.0$ s during earthquake. (Adapted from Desai, C.S., Settlement and Seismic Analyses of Reservation Canyon Tailing Dam and Impoundment, Report, C. Desai, Tucson, AZ, 1995.)

Figures 7.42a through 7.42c show the vectors of displacement at the end of consolidation (Stage 4), and at times $t = 4.0$ and 5.0 s during the earthquake, respectively. The factors of safety at the end of consolidation, and at $t = 2.5$ and 5.0 s during the earthquake are shown in Figures 7.43a through 7.43c, respectively. Figures 7.44a and 7.44b show liquefaction factors at $t = 2.5$ and 5.0 s during the earthquake, respectively. It can be seen that there were small zones where the computed factor of safety was less than one, and the liquefaction factor was greater than 0.50. Although they are not severe, it was thought advisable to consider the results in the decision process for the design and the construction (extension) of the tailing dam.

7.6.5.8 Validation for Flow Quantity

The trends and magnitudes of the computed results in Figures 7.42 through 7.44 are considered to be realistic. However, no field data were available to validate them. One field data that was available was field observations for quantity of flow through the buttress for the height before the proposed extension.

The quantity of flow through the buttress from the FE predictions is computed below, only for horizontal flow, Q_x, for height including the extension.

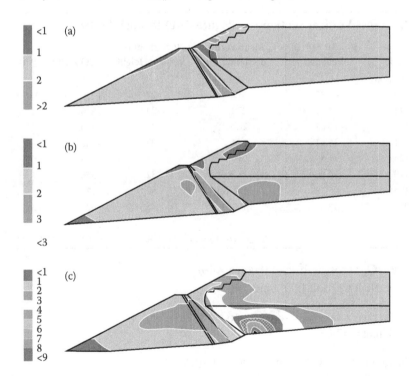

FIGURE 7.43 Contours of factors of safety at various times. (a) Factors of safety after stage 4 consolidation; (b) factors of safety after 2.5 s during earthquake; and (c) factors of safety after 5.0 s during earthquake. (Adapted from Desai, C.S., Settlement and Seismic Analyses of Reservation Canyon Tailing Dam and Impoundment, Report, C. Desai, Tucson, AZ, 1995.)

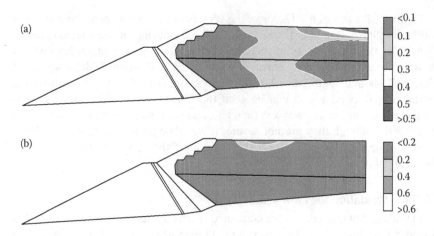

FIGURE 7.44 Liquefaction factors at various times during earthquake (a) After 2.5 s during earthquake; (b) after 5.0 s during earthquake. (Adapted from Desai, C.S., Settlement and Seismic Analyses of Reservation Canyon Tailing Dam and Impoundment, Report, C. Desai, Tucson, AZ, 1995.)

Q_x across Vertical Section y–y (Figure 7.40) through Buttress

Node	$V_x \times 10^{-8}$ (ft/s) Predicted	y_m Mean Vertical Coordinate (ft)	Increment of Vertical Height	$\Delta Q \times 10^{-8}$ ft³/s/ft
167 (top)	0	339	—	
149	−8.85	312	27	−239 (−8.85 × 27)
144	−15.01	290	22	−330
143	−10.02	267	23	−230
141	−16.63	244	23	−382
126	−11.57	219	25	−289
119	−10.80	187	32	−346
106	−5.38	145	42	−226
93	−13.52	100	45	−608
92	0			
		Total Q_x across y–y, Figure 7.40		−2650 × 10⁻⁸ (ft³/s)/ft

7.6.5.9 Q_x across a–b–c–d (Figure 7.40)

Between Nodes 17 and 18 (Figure 7.40)

- Average x-velocity = -0.761×10^{-8} ft/s
- Height = 47 ft

Therefore, $Q_x = -0.761 \times 47 = -36 \times 10^{-8}$ (ft³/s)/ft
Therefore, the net flow through the buttress is

$$= Q_x(y-y) - Q_x(a-b-c+d)$$

$$= -[2650 - (-36)] \times 10^{-8} = -2614 \times 10^{-8} \text{ (ft}^3\text{/s)/ft}$$

Now, the total length of the buttress is approximately 1200 ft (366 m). Therefore, the total flow through the buttress, Q_B, is

$$Q_B = (-2614 \times 1200) \times 10^{-8} \text{ ft}^3/\text{s}$$

$$= -3.14 \times 10^{-2} \text{ ft}^3/\text{s} = -0.053 \text{ m}^3/\text{min}$$

$$= -13.98 \text{ gallons/min for the entire height.}$$

Note: 1 ft = 0.305 m and 1 gallon = 0.00379 m³.

The flow across the buttress at the present time (before the extension) can be computed proportional to the height:

$$Q_B \text{ (present)} = 13.98 \times 70 = 9.8 \text{ gallons/min}$$

The measured flow across the buttress at the present time before extension was about 5–10 gallons/min [72,73,75], which compares well with the computed value, 9.8 gallons/min.

7.6.6   EXAMPLE 7.6: SOIL–STRUCTURE INTERACTION: EFFECT OF INTERFACE RESPONSE

The relative motions between structure and foundation (soil or rock) influence significantly the behavior of the system. Hence, it is advisable to consider for such motions at the interface. The dynamic analysis of a concrete pile foundation system with and without interface model is considered herein [39]. Figure 7.45 shows the model steel pile, 0.076 m diameter and 1.85 m length, founded in fully saturated Ottawa sand, $D_r = 60\%$. As shown in the figure, an interface zone was included between the pile segment and the soil. The FE mesh in Figure 7.16 was adopted for this analysis; it also shows the details of the FE mesh near the segment, and the boundary conditions. Figures 7.46a and 7.46b show the meshes with and without the provision of interfaces, respectively. Figure 7.46c shows the timewise displacements applied to the nodes on the pile segment.

The system was idealized as axisymmetric and the FE mesh contained 192 elements and 225 nodes. The four-node isoparametric elements were used for soil and interface; eight elements (from Nodes 100–109, 109–118, etc.) were used for the interface zone. The thickness of the interface was assumed to be 0.012 m. The sand and the interface were modeled by using the DSC-HISS model, which is described in Appendix 1. The material parameters were determined from triaxial and multiaxial tests for Ottawa sand, and interface shear device for the Ottawa sand–concrete interface; they are shown in Table 7.8.

The loading involved prescribed displacements in the following form:

$$v = \bar{v} \sin(2\pi ft) \tag{7.79}$$

where v is the applied displacement, $\bar{v}$ is its amplitude, f is the frequency, and t is time. The amplitude was adopted as 0.01 m with a frequency of 0.50 Hz (Figure 7.46c).

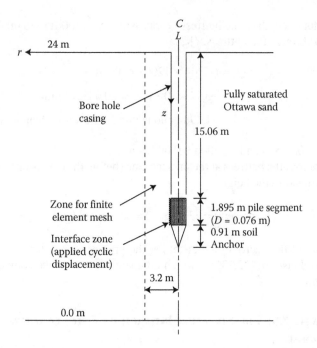

FIGURE 7.45 Details of pile segment, interface and soil. (Adapted from Wathugala, W. and Desai, C.S., Nonlinear and Dynamic Analysis of Porous Media and Applications, Report to National Science Foundation, Washington, DC, Dept. of Civil Eng. and Eng. Mech., Univ. of Arizona, Tucson, AZ, USA, 1990; Park, I.J. and Desai, C.S. Analysis of Liquefaction in Pile Foundations and Shake Table Tests using the Disturbed State Concept, Report, Dept. of Civil Eng. and Eng. Mechanics, University of Arizona, Tucson, AZ, USA, 1997; The Earth Technology Corporation, Pile Segments Tests—Sabine Pass, ETC Report No. 85-007, Long Beach, CA, USA, December, 1986.)

To investigate the influence of the interface behavior, FE analyses using the DSC-DYN 2D [51] were conducted for two cases:

Case 1, Without Interface: Here, it was assumed that no relative motion between the structure and soil can occur. Results are presented to compare displacements at Nodes 136 and 137; the former node was on the pile whereas the latter was in soil adjacent to Node 136. The pore water pressures were plotted in Elements 121 and 122 (Figure 7.46c).

Case 2, With Interface: Here, it was assumed that relative motion would occur between the pile and soil. Then, Element 121 becomes the interface element, and so on.

7.6.6.1 Comparisons

Figures 7.47a and 7.47b show displacements in the y-direction at Nodes 136 and 137 for the cases with and without interface, respectively. It can be seen that Node 136 experiences almost the same magnitudes of displacement in both cases. However,

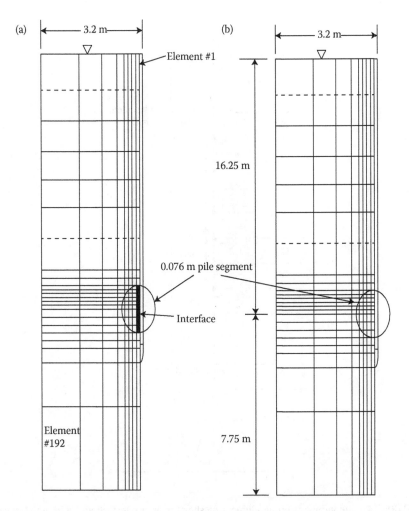

FIGURE 7.46 Finite element meshes for with interface and without interface and loading. (a) With interface; (b) without interface; and (c) detailed mesh around interface and loading. (Adapted from Wathugala, W. and Desai, C.S., Nonlinear and Dynamic Analysis of Porous Media and Applications, Report to National Science Foundation, Washington, DC, Dept. of Civil Eng. and Eng. Mech., Univ. of Arizona, Tucson, AZ, USA, 1990; Park, I.J., and Desai, C.S. Analysis of Liquefaction in Pile Foundations and Shake Table Tests using the Disturbed State Concept, Report, Dept. of Civil Eng. and Eng. Mechanics, University of Arizona, Tucson, AZ, USA, 1997.)

the magnitudes of displacements for Case 2 show significantly smaller values for node 137 in the soil. Thus, considerable relative motions at the interface can occur and influence the overall response.

Figures 7.48a and 7.48b show the variations of pore water pressures with time for with and without interface. The patterns of computed pore water pressures appear to be similar in both cases. However, the values between the upper and lower peaks show

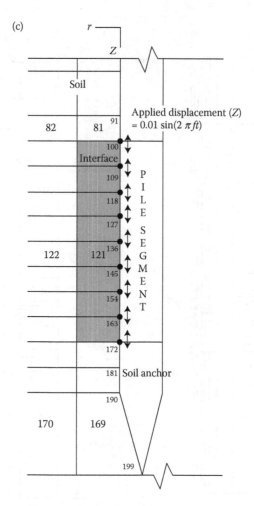

FIGURE 7.46 (continued) Finite element meshes for with interface and without interface and loading. (a) With interface; (b) without interface; and (c) detailed mesh around interface and loading. (Adapted from Wathugala, W. and Desai, C.S., Nonlinear and Dynamic Analysis of Porous Media and Applications, Report to National Science Foundation, Washington, DC, Dept. of Civil Eng. and Eng. Mech., Univ. of Arizona, Tucson, AZ, USA, 1990; Park, I.J. and Desai, C.S. Analysis of Liquefaction in Pile Foundations and Shake Table Tests using the Disturbed State Concept, Report, Dept. of Civil Eng. and Eng. Mechanics, University of Arizona, Tucson, AZ, USA, 1997.)

a significant difference; it is about 70 kPa for the case without interface whereas it is about 30 kPa for the case with interface. There is a difference between the two shapes; in case 2, the pore water pressure shows rounded peaks compared to relatively sharper peaks in Case 1.

It can be said that the relative motions at the interface influence the response of the soil–structure system, and need to be considered in the analysis and design.

TABLE 7.8

Model Parameters for Sand and Interface

Parameters		Ottawa Sand–Concrete Interface	Ottawa Sand
		Relative Intact (RI) State	
Elasticity	E	3183.0 kPa	193,000 kPa
	v	0.42	0.380
Plasticity	γ	0.109	0.123
	n	3.12	2.45
	a_1	0.289	0.8450
	η_1	0.470	0.0215
		Fully Adjusted (FA) State	
Critical state	$\bar{m}$	0.22	0.15
	λ	0.0131	0.02
	e_o^c	0.598	0.601
Disturbance function	Du	0.99	0.99
	A	0.595	4.22
	Z	0.665	0.43

7.6.7 EXAMPLE 7.7: DYNAMIC ANALYSIS OF SIMPLE BLOCK

Figure 7.49 shows a block material (like a footing on smooth rigid base) subjected to a uniform harmonic loading with an amplitude of 250 N/m²; force $F(t) = F_0 \sin \omega t$, where F_0 is the amplitude and ω is the frequency. The physical and mechanical properties are given below:

Width = 1.0 m, height = 1.0 m
Thickness = 0.2 m, modulus of elasticity = 207×10^6 kPa, $v = 0.0$ (for effective axial load only)
Mass density $\rho_s = 1923$ Ns²/m⁴
Amplitude of applied load $F_0 = 50$ N

The amplitude of 50 N is applied in terms of normal pressure, which equals the distributed pressure of 250 N/m². The FE predictions were obtained by using DSC-DYN2D [51]; the time step $\Delta t = 0.002$ s was used.

To compare the prediction by the FE elements, the following closed-form equations were used [22]:

$$\text{Displacement}, v(t) = \frac{F_o}{k}\left[\frac{1}{1-\beta^2}\right]\sin\bar{\omega}t \qquad (7.80a)$$

$$\text{Velocity}, \dot{v}(t) = \frac{F_o\bar{\omega}}{k}\left[\frac{1}{1-\beta^2}\right]\cos\bar{\omega}t \qquad (7.80b)$$

where $\beta = (\bar{\omega}/w)$ and ω is the undamped natural frequency of the system and is equal to

$$w = \sqrt{\frac{k}{m}} \qquad (7.80c)$$

where k and m are the stiffness and mass of the block, respectively.

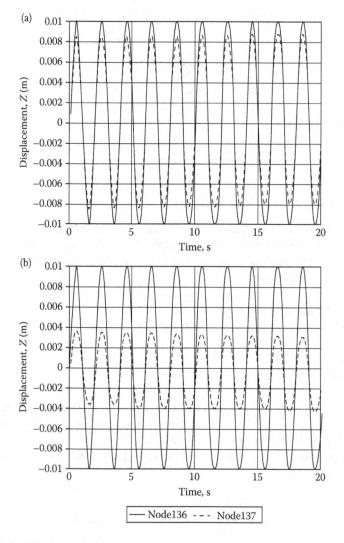

FIGURE 7.47 Computed displacements with time. (a) Without interface; (b) with interface. (Adapted from Park, I.J. and Desai, C.S. Analysis of Liquefaction in Pile Foundations and Shake Table Tests using the Disturbed State Concept, Report, Dept. of Civil Eng. and Eng. Mechanics, University of Arizona, Tucson, AZ, USA, 1997.)

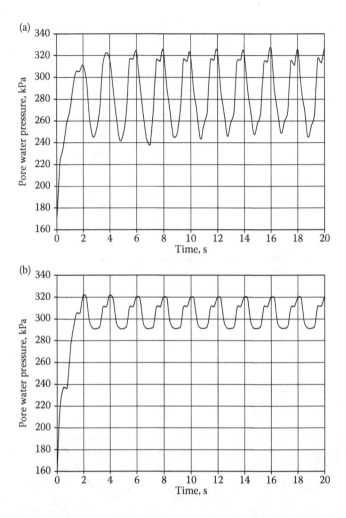

FIGURE 7.48 Computed pore water pressures in element 121. (a) Without interface; (b) with interface. (Adapted from Park, I.J. and Desai, C.S. Analysis of Liquefaction in Pile Foundations and Shake Table Tests using the Disturbed State Concept, Report, Dept. of Civil Eng. and Eng. Mechanics, University of Arizona, Tucson, AZ, USA, 1997.)

The predicted and closed-form solutions are shown in Figures 7.50a and 7.50b for displacement versus time, and velocity versus time, respectively. It is observed that the predicted displacement and velocity values match, almost exactly, with the closed-form solutions (Equation 7.80).

7.6.8 EXAMPLE 7.8: DYNAMIC STRUCTURE–FOUNDATION ANALYSIS

The structure–foundation system tested under the SIMQUAKE program conducted by the University of New Mexico under a research project supported by the Electric

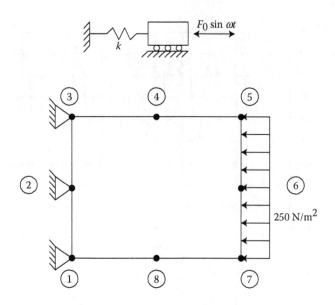

FIGURE 7.49 Block-single element.

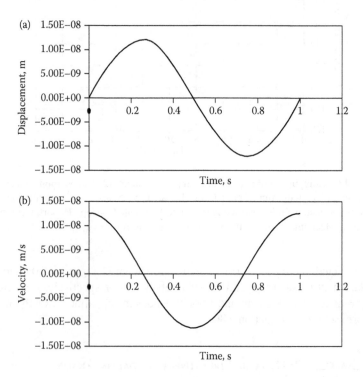

FIGURE 7.50 Comparisons between predictions and theory at Node No. 6. (a) Displacements; (b) velocities.

Power Research Institute (EPRI) was considered in this example [78–83]. It involved the response of a 1/8-scale model nuclear containment structure subjected to a strong ground motion generated by blasts (Figure 7.51a). The details of testing and field measurement have been reported by Vaughan and Isenberg [78,89].

Mesh and material properties: The FE mesh is shown in Figure 7.51b, which involved an approximate plane strain idealization. The concrete structure rested on a soil foundation consisting of two soil layers and one backfill, which were included in the mesh (Figure 7.51b). A cap-type yield plasticity model [80,81] was used for soils, and the linear elastic model was used for concrete. The parameters determined from laboratory (triaxial) tests were adopted from Refs. [78,79] and are presented in

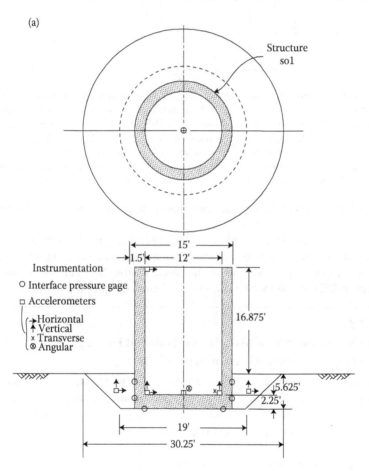

(a)

FIGURE 7.51 Scale model of simquake structure, instrumentation and finite element mesh (1 ft = 0.305 m). (a) SIMQUAKE structure; (b) finite element mesh. (Adapted from Vaughan, D.K. and Isenberg, J., Nonlinear Soil-Structure Analysis of SIMQUAKE II, Final Report, Research Project 8102, Electric Power Res. Inst. (EPRI), Weidlinger Associates, Menlo Park, CA, USA, 1982; Zaman, M.M. and Desai, C.S., Soil-Structure Interaction Behavior for Nonlinear Dynamic Problems, Report to National Science Foundation, Washington, DC, Dept. of Civil Eng. and Eng. Mech., Univ. of Arizona, Tucson, AZ, USA, 1982.)

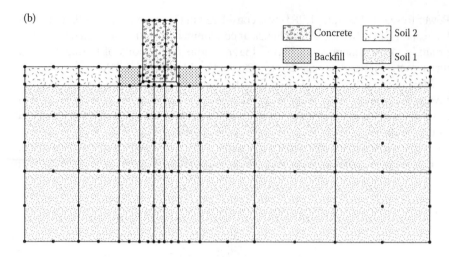

FIGURE 7.51 (continued) Scale model of SIMQUAKE structure, instrumentation and finite element mesh (1 ft = 0.305 m). (a) SIMQUAKE structure; (b) finite element mesh. (Adapted from Vaughan, D.K. and Isenberg, J., Nonlinear Soil-Structure Analysis of SIMQUAKE II, Final Report, Research Project 8102, Electric Power Res. Inst. (EPRI), Weidlinger Associates, Menlo Park, CA, USA, 1982; Zaman, M.M. and Desai, C.S., Soil-Structure Interaction Behavior for Nonlinear Dynamic Problems, Report to National Science Foundation, Washington, DC, Dept. of Civil Eng. and Eng. Mech., Univ. of Arizona, Tucson, AZ, USA, 1982.)

Table 7.9. The details of the cap plasticity model are given in Appendix 1, including the definition of the parameters.

Interface: The thin-layer interface element [84], between the structure and soil, described in Appendix 1 was used in the FE simulation. The Ramberg–Osgood (RO) [85] model can be used to simulate the nonlinear elastic shear behavior of the interface (Figure 7.52); the RO model is also described in Chapter 2.

TABLE 7.9

Cap Model Parameters for Materials for Soils and Elastic Parameters for Concrete (Definitions Given in Appendix 1)

Parameters (1)	Soil 1 (2)	Soil 2 (3)	Backfill (4)	Concrete (5)
E psi (kPa)	46,435 (3.2×10^5)	22,113 (1.5×10^5)	13,100 (9.0×10^4)	4×10^6 (2.7×10^7)
v	0.3	0.3	0.26	0.2
α psi (kPa)	470.0 (3240)	470.0 (3240)	470.0 (3240)	
θ	0.0	0.0	0.0	
β 1/psi (1/kPa)	0.165 (0.024)	0.165 (0.024)	0.165 (0.024)	
γ psi (kPa)	390.0 (2,689)	390.0 (2,689)	390.0 (2,689)	
D 1/psi (1/kPa)	0.0007 (0.000102)	0.0007 (0.000102)	0.00065 (0.000094)	
W	0.06	0.06	0.06	
Z	0.0	0.0	0.0	
R	2.5	2.5	2.5	
γ_W lb/in³ (kg/cm³)	0.0637 (0.00176)	0.0637 (0.00176)	0.0637 (0.00176)	0.58 (0.0016)

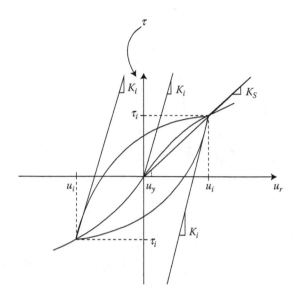

FIGURE 7.52 Ramberg–Osgood Model. (Adapted from Zaman, M.M. and Desai, C.S., Soil-Structure Interaction Behavior for Nonlinear Dynamic Problems, Report to National Science Foundation, Washington, DC, Dept. of Civil Eng. and Eng. Mech., Univ. of Arizona, Tucson, AZ, USA, 1982.)

The shear behavior of the interface was expressed by using the RO model:

$$u_r = u_y \left(\frac{\tau}{K_i u_y} \right) (1 + \alpha) \left| \frac{\tau}{K_i u_y} \right|^{R-1} \tag{7.81a}$$

The unloading–reloading responses were simulated using the following function:

$$u_r - u_i = u_y \left(\frac{\tau - \tau_i}{K_i u_y} \right) \left(1 + \frac{2\alpha}{2^R} \right) \left| \frac{\tau - \tau_i}{K_i u_y} \right|^{R-1} \tag{7.81b}$$

where u_r is the relative (shear) displacement, τ is the interface shear stress, u_i and τ_i are relative displacement and shear stress, respectively, at the point of loading reversal (Figure 7.52), K_i is the initial shear stiffness that was assumed at the point of load reversal, u_y is the reference displacement, and α and R are parameters. The normal behavior was expressed as

$$\sigma_n = a_0 (e^{a_1 \, v_r} - 1) \tag{7.82}$$

where σ_n and v_r are the normal stress and normal relative displacements, respectively, and a_0 and a_1 are constants.

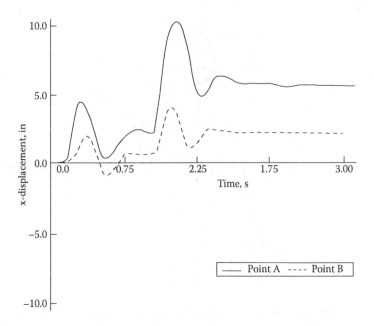

FIGURE 7.53 Horizontal displacement-time history input (1 inch = 2.54 cm). Below ground level, coordinates (50,5) and (100,5) from center line of structure. (Adapted from Zaman, M.M. and Desai, C.S., Soil-Structure Interaction Behavior for Nonlinear Dynamic Problems, Report to National Science Foundation, Washington, DC, Dept. of Civil Eng. and Eng. Mech., Univ. of Arizona, Tucson, AZ, USA, 1982.)

Since no laboratory test data were available for interfaces between the concrete and the backfill materials used in the EPRI project, pertinent properties of the surrounding soils (i.e., backfill and Soil 1) were used for representing the normal behavior of the interface elements. Shearing properties of the interface elements in the analysis were estimated from the cyclic tests involving sand–concrete interfaces [82].

Loading: The input history of displacement versus time applied on the boundary is shown in Figure 7.53.

7.6.8.1 Results

Typical comparisons between predicted and observed velocities and contact pressure are shown in Figures 7.54a and 7.54b, and 7.55, respectively. Comparisons between computed acceleration versus time for two types of interfaces, bonded (no interface elements) and frictional (allows relative motion), are shown in Figure 7.56.

It can be seen from the above figures that the FE predictions provide very good correlations with the field measurements for about 1.0 s, but thereafter they are not as good. In the case of bonded and frictional interface, the computed results show that the frictional interface, which allows relative motion, yields higher acceleration than the bonded one. It can be stated that the provision of interface with relative motions can modify the response, that is, displacements, velocities, and accelerations, depending on factors such as location, frequency related to loading, and physical properties. In view of the complexity of the problem, the computer predictions are considered to be satisfactory.

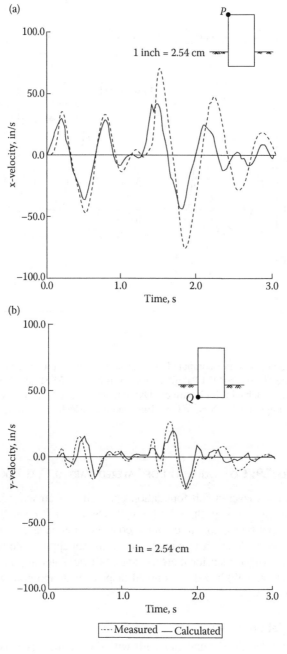

FIGURE 7.54 Comparison of computed and measured velocities (1 in/s = 2.54 cm/s). (a) x-velocity at point *P*; (b) y-velocity at point *Q*. (Adapted from Zaman, M.M. and Desai, C.S., Soil-Structure Interaction Behavior for Nonlinear Dynamic Problems, Report to National Science Foundation, Washington, DC, Dept. of Civil Eng. and Eng. Mech., Univ. of Arizona, Tucson, AZ, USA, 1982.)

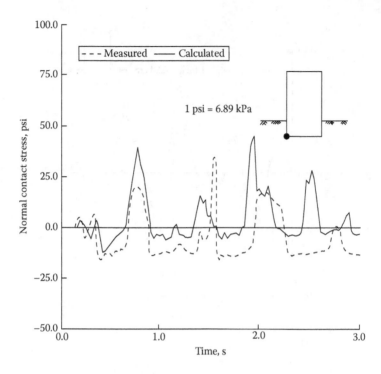

FIGURE 7.55 Comparison of computed and measured contact pressure beneath upstream corner of structure (1 psi = 6.895 kPa). (Adapted from Zaman, M.M. and Desai, C.S., Soil-Structure Interaction Behavior for Nonlinear Dynamic Problems, Report to National Science Foundation, Washington, DC, Dept. of Civil Eng. and Eng. Mech., Univ. of Arizona, Tucson, AZ, USA, 1982.)

7.6.9 EXAMPLE 7.9: CONSOLIDATION OF LAYERED VARVED CLAY FOUNDATION

Figure 7.57 shows a layered soil foundation consisting of anisotropic varved clay subjected to loadings through three buildings [18,86,87]. 1-D consolidation theory (Chapter 6) may not be suitable to compute consolidation behavior in such layered and anisotropic soils. Hence, it was necessary to use an appropriate procedure that allows for multi-(two-) dimensional effects and realistic material properties. An FE procedure (Equation 7.59) based on coupled displacement and pore water pressure was employed herein.

7.6.9.1 Material Properties

Figure 7.57 shows details of foundation soils with three buildings constructed in the Hackensack Meadows region of New Jersey [86]. The top soil was very stiff varved clay underlain by a soft to firm varved clay deposit. A number of laboratory tests were performed [86] for the determination of the values of past pressures, over-consolidation ratio, compression ratio, swell ratio, recompression ratio (RR), index

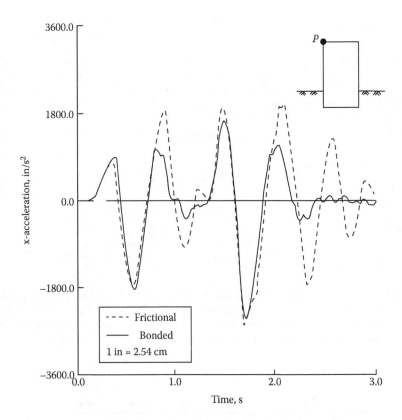

FIGURE 7.56 Influence of interface condition on horizontal acceleration-time history at top of structure in SIMQUAKE II, Point P (1 in/sec² = 2.54 cm/s²). (Adapted from Zaman, M.M. and Desai, C.S., Soil-Structure Interaction Behavior for Nonlinear Dynamic Problems, Report to National Science Foundation, Washington, DC, Dept. of Civil Eng. and Eng. Mech., Univ. of Arizona, Tucson, AZ, USA, 1982.)

properties, moduli, E and v, and coefficient of permeabilities of soils. Figure 7.58 shows the distribution of initial overburden pressure, $\bar{\sigma}_{vo}$, and the increase in pressure, $\Delta\bar{\sigma}_v$, due to the fill and the building load. The applied load increment leads to the final value of the total load, $\bar{\sigma}_v$, which is lower than the maximum pressure, $\bar{\sigma}_{v\max}$. Figure 7.59 shows the results for swell ratio versus RR [18,86].

The increase in the applied stress in Layer 1 (Figure 7.57) was assumed to be about 600 lb/ft² (29 kPa). The average value of the RR is 0.033, (Figure 7.59). By assuming $v = 0.4$ for Layer 1, we can find E as follows:

$$D = \frac{\bar{\sigma}}{0.435 \times RR} = \frac{600}{0.435 \times 0.033}$$

$$= 41,800\,\text{lb/ft}^2\ (2010\,\text{kPa})$$

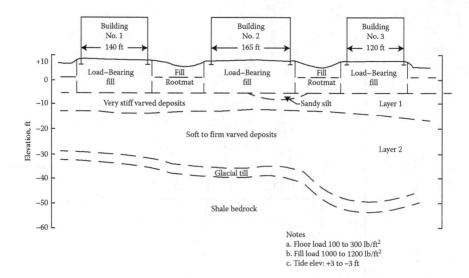

FIGURE 7.57 Building-foundation on varved clay (1 ft = 0.305 m). (Adapted from Baker, G.L. and Marr, W.A., *Proceedings, Annual Meeting*, Transportation Research Board, Washington, DC, 1976; Siriwardane, H.J. and Desai, C.S., Nonlinear Two-Dimensional Consolidation Analysis, Technical Report, VA Tech., Blacksburg, VA, USA, 1980.)

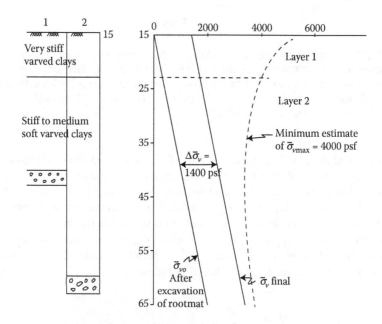

FIGURE 7.58 Foundation soils, initial and applied pressures (1 psf = 47.88 pa, 1 ft = 0.305 m). (Adapted from Baker, G.L. and Marr, W.A., *Proceedings, Annual Meeting,* Transportation Research Board, Washington, DC, 1976.)

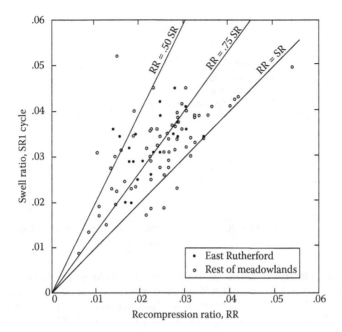

FIGURE 7.59 Properties of varved clays from experiments. (Adapted from Baker, G.L. and Marr, W.A., *Proceedings, Annual Meeting*, Transportation Research Board, Washington, DC, 1976.)

where D is the constrained modulus. Then,

$$E = \frac{D\,(1 - 2v)\,(1 + v)}{1 - v}$$
$$= \ 19,000 \ \text{lb/ft}^2 \ (910 \ \text{kPa})$$

Similarly, the value of E for Layer 2, assuming an increase in load of about 1250 lb/ft² (59 kPa) gave a value of about 40,000 lb/ft² (1910 kN/m²). The soil was assumed to be isotropic for the mechanical behavior, that is, $E_x = E_y$. However, it was assumed anisotropic for fluid flow (consolidation).

The coefficient of permeability for the soil was estimated from the laboratory tests as

$$\text{Layer 1}: k_x \ = 2.5 \times 10^{-4} \ \text{ft/day} \ (0.77 \ \text{m/day})$$
$$\text{Layer 2}: k_x \ = 1.5 \times 1.0^{-4} \ \text{ft/day} \ (0.40 \ \text{m/day})$$

Because of the varved nature of the soil, the results are expected to be affected considerably with the anisotropy defined by the ratio k_x/k_y. A parametric study was performed by varying this ratio. First, the values of k_x for both layers were assumed,

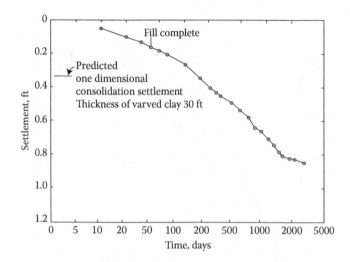

FIGURE 7.60 Observed settlements versus time for building No. 2 (1 Ft = 0.305 m). (Adapted from Baker, G.L. and Marr, W.A., *Proceedings, Annual Meeting*, Transportation Research Board, Washington, DC, 1976; Siriwardane, H.J. and Desai, C.S., Nonlinear Two-Dimensional Consolidation Analysis, Technical Report, VA Tech., Blacksburg, VA, USA, 1980.)

and then according to the assumed ratio (k_x/k_y), the values of k_y were computed for FE analysis. The parametric variations are shown in Figure 7.63.

7.6.9.2 Field Measurements

Consolidation settlements were measured in the field under buildings Nos. 1, 2, and 3 (Figure 7.57). A typical measurement for settlement under building No. 2 is shown in Figure 7.60. It can be seen that the observed settlement was far greater (about 2.5 times) than that obtained using the 1-D Terzaghi theory [10,12]. The discrepancy between the observed data and the 1-D theory is considered to be significant. The reasons for the discrepancy can be due to factors such as sample disturbance, multi-dimensional effects, and laboratory and field conditions. However, it is believed that the anisotropic permeability can have significant influence.

7.6.9.3 Finite Element Analysis

The structure–foundation system was assumed to be approximately symmetrical. Hence, the right-hand half of the system was discretized into FE (Figure 7.61). Some of the main factors considered in the FE computation were rate of loading, multi-dimensional effects, which included movements of foundation, irregular geometry, and the anisotropy in permeability of varved clay.

The variable loading (Figure 7.62) was applied to top nodes in the mesh (Figure 7.61). As stated before, a number of FE analyses were performed by varying the ratio k_x/k_y (Figures 7.63a and 7.63b). These figures show comparisons between predictions for various values of k_x/k_y and field observations for building No. 2, under which the thickness of the varved clays was about 30 ft (9.0 m). Only six results are shown

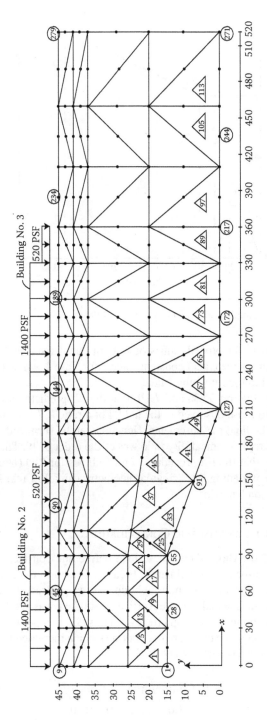

FIGURE 7.61 Finite element mesh and loading (1 ft = 0.305 m). (Adapted from Siriwardane, H.J. and Desai, C.S., Nonlinear Two-Dimensional Consolidation Analysis, Technical Report, VA Tech., Blacksburg, VA, USA, 1980.)

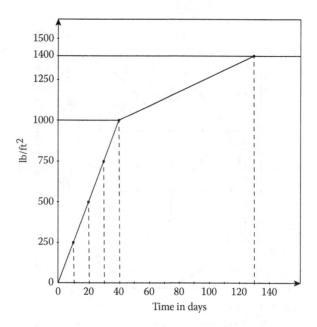

FIGURE 7.62 History of applied loading (1 lb/ft^2 = 47.88 Pa).

(Figures 7.63a and 7.63b) for typical values of k_x/k_y in comparison with field data. Here, the values of E and v were kept constant, except for Case 1, where the value of E was twice compared to the foregoing value.

It can be seen from Figures 7.63a and 7.63b that the FE predictions yield trends of settlement similar to those observed in the field. For a ratio k_x/k_y, of about 10, the results were close to the field measurements. Hence, it can be surmised that for the varved clay, the anisotropic permeability ratio was of the order of 10. The important finding of the study is that the FE procedure, which allows for 2-D (lateral) movement, anisotropy, rate of loading, and irregular geometry can allow realistic predictions of the consolidation behavior.

7.6.10 EXAMPLE 7.10: AXISYMMETRIC CONSOLIDATION

Yokoo et al. [88] used the Biot's theory described before and considered an application involving axisymmetric consolidation. Figure 7.64 shows the axisymmetric problem including the FE mesh. The bottom boundary was placed at a distance of 20 m from the ground surface and the side (vertical) boundary was at 60 m from the center line. The bottom boundary on rigid rock was considered impervious. Uniform load, p, was applied through the circular area (footing) which increased linearly from zero (very small value) to 5.0 ton/m^2 as [88]:

$$p = \begin{cases} 1.0 \times 10^{-6}\ t\ (\text{ton/m}^2)(t \le 0.5 \times 10^7 \text{s}) \\ 5.0 \qquad (\text{ton/m}^2)(t > 0.5 \times 10^7 \text{s}) \end{cases} \tag{7.83}$$

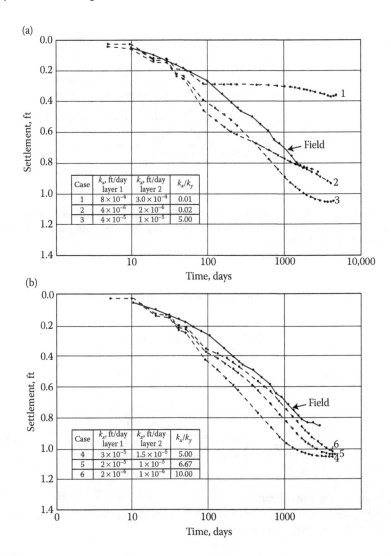

FIGURE 7.63 Comparison between predictions and field data, and results for various k_x/k_y. (a) Settlement versus time for Cases 1–3; (b) settlement versus time for Cases 4–6. (Adapted from Siriwardane, H.J. and Desai, C.S., Nonlinear Two-Dimensional Consolidation Analysis, Technical Report, VA Tech., Blacksburg, VA, USA, 1980.)

7.6.10.1 Details of Boundary Conditions

Axis of symmetry:

$$u_r = 0, v_r = 0 (r = 0 \text{ m}) \tag{7.84a}$$

Surface of bed rock:

$$u_r = u_z = 0, \quad v_z = 0 \, (z = 20 \text{m}) \tag{7.84b}$$

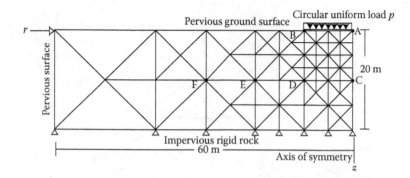

FIGURE 7.64 Finite element mesh for Axisymmetric Idealization (Adapted from Yokoo, Y., Yamagata, K., and Nagoka, H., *Japanese Society of Soil Mechanics and Foundation Engineering,* 11(1), 1971, 29–46.)

Vertical boundary (60 m) away from the center line:

$$u_r = 0, \sigma_{rz} = 0, h = 0 \ (r = 60\,\text{m}) \tag{7.84c}$$

Ground surface:

$$h = 0 \ (z = 0 \text{ m})$$
$$\sigma_{zz} = \begin{cases} p \ (z = 0\,\text{m}, r \le 10\,\text{m}) \\ o \ (z = 0\,\text{m}, r > 10\,\text{m}) \end{cases} \tag{7.84d}$$

where h denotes the water head.

The clay layer was assumed to be transversely isotropic with five elasticity coefficients [89]; the parameters were adopted as

$$E_2 = \begin{cases} 2.0 \times 10^2 & (t/\text{m}^2)\,(z \le 10\,\text{m}) \\ 2.0 \times 10^2\,(0.5 + z/20)(t/\text{m}^2)\,(z \ge 10\,\text{m}) \end{cases}$$
$$E_1 = 2.5\,E_2; \ G_2 = 0.5\,E_1,$$
$$v_1 = 0.30, \ v_2 = 0.2,$$
$$k_z = \begin{cases} 1.0 \times 10^{-8} & (\text{m/s})\,(z \le 10\,\text{m}) \\ 1.0 \times 10^{-8}\,(1.2 - 0.02z)\ (\text{m/s})\ (z \ge 10\,\text{m}) \end{cases} \tag{7.85}$$
$$k_r = 5.0\,K_z$$
$$k_{rz} = 0.0$$
$$\gamma_w = 1.0\,\text{ton/m}^2, \text{unit weight of water}$$

Various terms in the transversely isotropic medium are shown below [89]:

$$\varepsilon_z = \sigma_z/E_2 - v_2\sigma_r/E_2 - v_2\sigma_\theta/E_2$$
$$\varepsilon_r = -v_2\sigma_z/E_2 + \sigma_r/E_1 - v_1\sigma_\theta/E_1 \qquad (7.86)$$
$$\varepsilon_\theta = -v_2\sigma_z/E_2 - v_1\sigma_r/E_1 + \sigma_\theta|E_1$$
$$2\varepsilon_{zr} = \tau_{zr}/G_2$$

7.6.10.2 Results

The clay foundation was divided into 93 triangular elements with 59 nodes. Figure 7.65 shows the computed variation of consolidation settlement versus time for the 2-D problem, for points A and B (Figure 7.64). It also shows the variation of uniform load, p, over time. The settlement of the ground surface for the 1-D solution (Chapter 6) is also shown for comparison. It can be seen that the settlement is more rapid from the 2-D case compared to that from the 1-D case; a reason can be that the horizontal permeability for the 2-D case was higher.

The variation of water head with time for points C, D, E, F (Figure 7.64) are shown in Figure 7.66. The water heads at points C and D reach their maximum values when the load p reaches the maximum. The water heads at point E and F, which are away from the axis of symmetry, reach their maximum values at higher times.

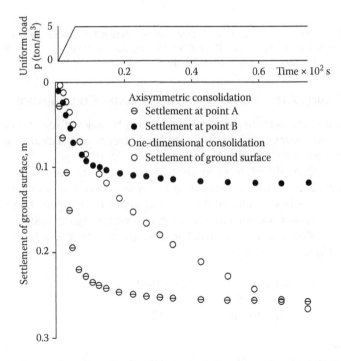

FIGURE 7.65 Settlement of ground surface for points A and B, Fig. 7.64. (Adapted from Yokoo, Y., Yamagata, K., and Nagoka, H., *Japanese Society of Soil Mechanics and Foundation Engineering,* 11(1), 1971, 29–46.)

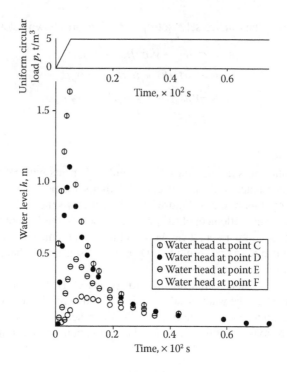

FIGURE 7.66 Variation of water head during consolidation for various points, Fig. 7.64. (Adapted from Yokoo, Y., Yamagata, K., and Nagoka, H., *Japanese Society of Soil Mechanics and Foundation Engineering*, 11(1), 1971, 29–46.)

7.6.11 EXAMPLE 7.11: TWO-DIMENSIONAL NONLINEAR CONSOLIDATION

Sometimes, it is necessary to consider the nonlinear behavior of soils for consolidation. Such a procedure based on foregoing Biot's theory and critical state for the constitutive model for soils was presented in Refs. [87,90,91].

Figure 7.67 shows the FE mesh for the 2-D problem; it involves six-noded triangular element. The width of the loaded area, B, was assumed to be equal to 10.0 ft (3.05 m). For a realistic condition, the load was applied linearly for up to 25 days ($T_v = 0.07$), and then was assumed to be invariant (inset in Figure 7.68).

The details of the critical state model used are given in Appendix 1. The material parameters adopted are given below:

E (initial) = 13,000 psf (623.0 kN/m²), $v + 0.40$
$M = 1.05$; $\lambda = 0.14$; $\kappa = 0.05$
$e_o = 0.90$; $k_x = k_y = 4 \times 10^{-5}$ ft/day (1.22 × 10⁻⁵ m/day)

7.6.11.1 Results

Figures 7.68 and 7.69 show the timewise variations of surface settlements and the nondimensional term $(U \times 100)/B$ versus T_v at node 29, where U is the time-dependent displacement, B is the width of the loaded zone, $T_v = (C_v t)/H^2$, C_v is the coefficient

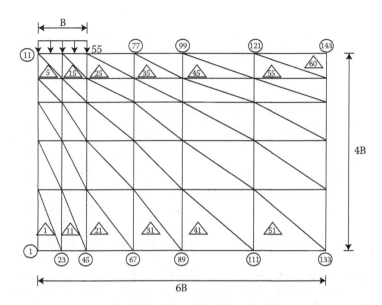

FIGURE 7.67 Finite element mesh for Two-dimensional consolidation. (Adapted from Siriwardane, H.J. and Desai, C.S., Nonlinear Two-Dimensional Consolidation Analysis, Technical Report, VA Tech., Blacksburg, VA, USA, 1980.)

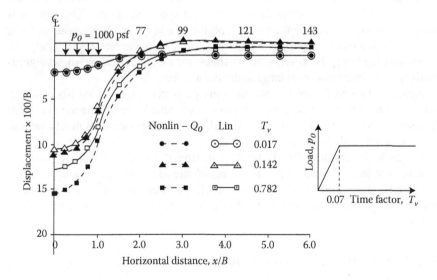

FIGURE 7.68 Surface settlements at various times: Nonlinear and linear analyses (1 psf = 47.88 Pa). (Adapted from Siriwardane, H.J. and Desai, C.S., Nonlinear Two-Dimensional Consolidation Analysis, Technical Report, VA Tech., Blacksburg, VA, USA, 1980.)

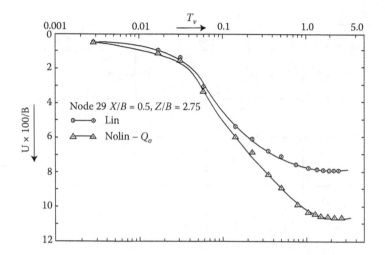

FIGURE 7.69   Settlement predictions with time [T_v] by linear and nonlinear procedures. (Adapted from Siriwardane, H.J. and Desai, C.S., Nonlinear Two-Dimensional Consolidation Analysis, Technical Report, VA Tech., Blacksburg, VA, USA, 1980.)

of consolidation, and H is the drainage length. The value C_v was found by using E. The predicted pore water pressure, p, was nondimensionalized with respect to p_o (p/p_o), where p_o is the applied surface loading (Figure 7.68). The applied load, p_o, was adopted as 1000 psf (48.0 kPa).

The above results show predictions for both linear elastic and nonlinear (critical state model) analyses using the residual load approach [87,90]. The latter was computed on the basis of the incremental plastic strain; its details are given in Refs. [11,87]. The settlements from linear and nonlinear schemes are not significantly different at earlier times. However, at larger times, the nonlinear scheme shows significantly higher settlements compared to the linear scheme.

Figure 7.70 shows dissipation of pore water pressures at the section Nodes 45–55 in Figure 7.67. The nonlinear approach shows somewhat lower dissipation than from the linerar analysis, during earlier and later time periods. However, at middle period, it shows significantly lower dissipation.

The above results show that the predictions from the nonlinear approach in terms of settlement and pore water pressure (effective stress) can be significantly different compared to those from the linear approach. Hence, if the soil behavior is nonlinear, it is advisable to use an appropriate nonlinear constitutive model for realistic analysis and design.

7.6.12   EXAMPLE 7.12: SUBSIDENCE DUE TO CONSOLIDATION

Settlement due to consolidation can occur due to factors such as drainage and removal of water from the subsoil; in addition, due to external loads. We now present an example of subsidence due to fluid flow in a cavity in the foundation soil [91]. The three-building system on layered foundation in Example 7.9 (Figure 7.57) was adopted for

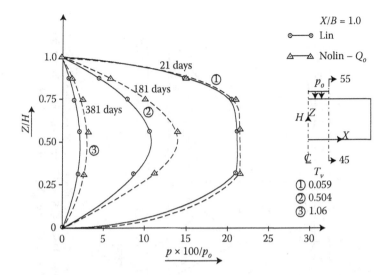

FIGURE 7.70 Distributions of pore water pressures at section 45–55 (Figure 7.67) for linear and nonlinear analyses. (Adapted from Siriwardane, H.J. and Desai, C.S., Nonlinear Two-Dimensional Consolidation Analysis, Technical Report, VA Tech., Blacksburg, VA, USA, 1980.)

this analysis. The soil was modeled by using the continuous yield critical state model, and the material in the cavity was assumed to be fractured; hence, free drainage was allowed in the cavity. The FE mesh with the cavity is shown in Figure 7.71.

The FE incremental nonlinear procedure used is the same as in Example 7.11 [11,91]. Two sets of predictions are presented here. In the first set, the material properties are assumed to be linear, and a parametric study is performed to analyze the influence of simulating the cavity flow in the foundation. In the second set, predictions are presented to show the effect of the nonlinear behavior of soil represented by using the plasticity (critical state) model. As in Example 7.9, the loads were applied at the top of stiff varved clay.

7.6.12.1 Linear Analysis: Set 1

The material properties assumed for layers 1 and 2 and the cavity are shown below:

	E	v	k
	lb/ft² (kN/m²)		ft/day (m/day)
Layer 1	13,000 (618)	0.4	2×10^{-5} (0.6×10^{-5})
Layer 2	40,000 (1900)	0.4	2×10^{-5} (0.3×10^{-5})
Cavity	40,000 (1900)	0.4	Variable

The cavity was assumed to contain highly permeable (fractured) material with high permeability ($k_x = k_y$), which varied as $k_x = k_y = 1 \times 10^{-4}$, 1×10^{-3}, and 1×10^{-2} ft/day (0.3×10^{-4}, 0.3×10^{-3}, 0.3×10^{-2} m/day). The free drainage in the cavity was simulated by introducing zero pore water pressure at the nodes on the walls of the cavity.

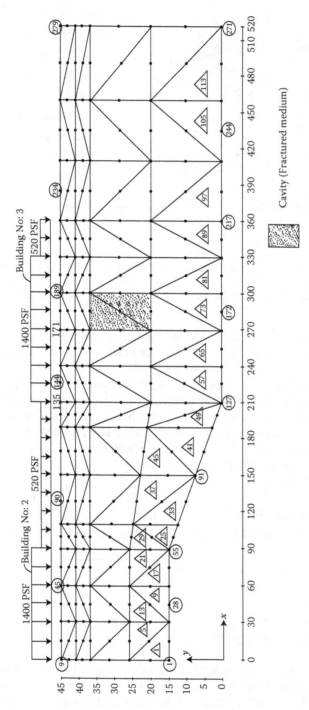

FIGURE 7.71 Finite element mesh, loading and cavity (1 ft = 0.305 m, 1 psf = 47.88 pa). (Adapted from Siriwardane, H.J. and Desai, C.S., Nonlinear Two-Dimensional Consolidation Analysis, Technical Report, VA Tech., Blacksburg, VA, USA, 1980; Desai, C.S. and Siriwardane, H.J., *Proceedings of the International Conference on Evaluation and Prediction of Subsidence*, Pensacola, Florida, USA, ASCE, New York, pp. 500–515, 1978.)

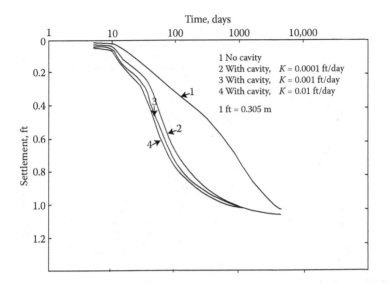

FIGURE 7.72   Settlements of Node 171 with and without cavity and influence of permeability of fractured medium in cavity. (Adapted from Siriwardane, H.J. and Desai, C.S., Nonlinear Two-Dimensional Consolidation Analysis, Technical Report, VA Tech., Blacksburg, VA, USA, 1980; Desai, C.S. and Siriwardane, H.J., *Proceedings of the International Conference on Evaluation and Prediction of Subsidence,* Pensacola, Florida, USA, ASCE, New York, pp. 500–515, 1978.)

The variations of predicted settlements with time at node 171 (Figure 7.71) are shown in Figure 7.72. The settlements along the ground surface at typical times 5, 130, and 1050 days are shown in Figure 7.73. It is assumed that free drainage occurred in the cavity from time $t = 0$. These figures indicate that the occurrence of drainage significantly influences the rate of subsidence (settlement). However, the magnitudes of permeability influence the rate of settlement to a lesser extent. The final subsidence is practically the same for all permeabilities.

7.6.12.2   Nonlinear Analysis

As stated before, the soil was considered to be elastoplastic with critical state parameters given in Example 7.11. The elastic properties are given under Section 7.6.12.1 above. The cavity was assumed to be linear elastic. Figure 7.74 shows predicted settlements with time at Node 171. It is evident from Figure 7.74 that the rate and total settlement are significantly greater for the nonlinear case than for the linear case.

7.6.13   EXAMPLE 7.13: THREE-DIMENSIONAL CONSOLIDATION

In certain practical situations, it can be necessary to analyze the problem as fully 3-D. Such cases may involve irregular geometrics, nonsymmetrical loading conditions, and junction of components in a structure.

A 3-D computer procedure was developed based on the generalized Biot's theory described earlier; one of the special cases of this procedure is consolidation. A simple example that was simulated by the 3-D code [9,92] is presented herein.

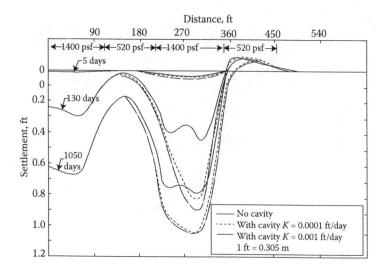

FIGURE 7.73 Surface subsidence with and without cavity. (Adapted from Siriwardane, H.J. and Desai, C.S., *Nonlinear Two-Dimensional Consolidation Analysis*, Technical Report, VA Tech., Blacksburg, VA, USA, 1980; Desai, C.S. and Siriwardane, H.J., *Proceedings of the International Conference on Evaluation and Prediction of Subsidence*, Pensacola, Florida, USA, ASCE, New York, pp. 500–515, 1978.)

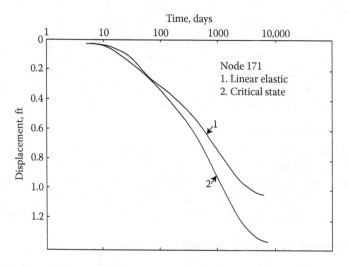

FIGURE 7.74 Comparison between settlements at node 171 for linear and nonlinear analyses (1 ft = 0.305 m). (Adapted from Siriwardane, H.J. and Desai, C.S., *Nonlinear Two-Dimensional Consolidation Analysis*, Technical Report, VA Tech., Blacksburg, VA, USA, 1980; Desai, C.S. and Siriwardane, H.J., *Proceedings of the International Conference on Evaluation and Prediction of Subsidence*, Pensacola, Florida, USA, ASCE, New York, pp. 500–515, 1978.)

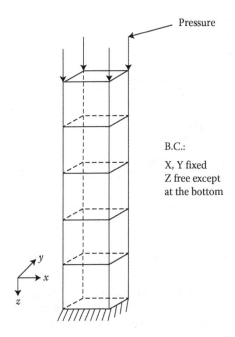

FIGURE 7.75 Finite element 3-D mesh for consolidation.

Figure 7.75 shows a saturated column undergoing consolidation due to applied pressure on the top, and the mesh involving five elements. The material properties used for the 3D simulation are given below:

$E = 1000$ kPa, $v = 0.0$, porosity, $n = 0.50$ permeability $k_x = 0.001$ m/s

The boundary conditions were adopted as follows:

- Displacements in the x- and y-directions are fixed at all nodes.
- Displacements in vertical z-direction are free at all nodes except at the nodes on the bottom.
- Gradients of pore waters were zero on the side boundaries and the bottom.

It may be noted that the problem considered is essentially 1-D. Hence, the 1-D (Terzaghi) theory can be used for theoretical predictions [10]. Figure 7.76 shows comparisons between the theoretical and FE predictions for displacement versus time at the top. The correlation is considered to be very good.

7.6.14 EXAMPLE 7.14: THREE-DIMENSIONAL CONSOLIDATION WITH VACUUM PRELOADING

In this example, we use 3-D coupled FEM to evaluate the performance of a ground improvement project involving vacuum preloading and prefabricated vertical drains (PVDs). The project is located on a reclaimed land at Nansha Port in Guangzhou,

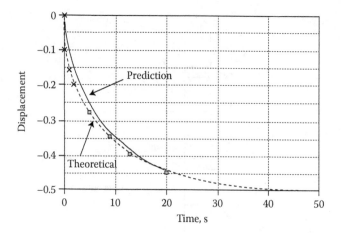

FIGURE 7.76   Comparison between theoretical and 3-D FE displacements with time.

China [93]. The site was reclaimed using clay slurry dredged from the seabed that formed the first 3–4 m of the soil deposit. This reclaimed soil was underlain by muddy and weak silty clay and clay layers (Table 7.10). Because the soft deposits had very low bearing capacity and the possibility of undergoing excessive settlements, PVDs were installed and the site was subjected to vacuum preloading. Figure 7.77a shows a plan view of the site, including the quadrant where instrumentations were installed for monitoring [93]. The instrumentation included settlement gauges, pyrometers, multilevel gauges, and inclinometers [93]. Figure 7.77b shows a plan view of the area considered in the 3-D simulation.

The zone where PVDs were installed measured 180 m × 190 m in plan dimensions (Figure 7.78b). The PVDs were installed in a square pattern with a spacing of 1 m and a depth of 22 m. A 2-m-thick sand blanket served as a platform for the installation of PDVs and horizontal perforated pipes required for applying vacuum pressure (85 kPa for 90 days).

TABLE 7.10
Soil Parameters Used in Finite Element Analysis

Soil Layer	Thickness (m)	Density (g/cm³)	Permeability (cm/s) k_h	k_h	k_v	Poison's Ratio, υ	Comp. Modulus (MPa)
Backfill sand	2	1.8	1×10^{-3}	9.62×10^{-4}	1×10^{-3}	0.3	11.24
Dredged slurry	4	1.75	5×10^{-6}	4.94×10^{-6}	1×10^{-7}	0.32	1.24
Muddy soil	4	1.6	1×10^{-8}	6.88×10^{-9}	1×10^{-8}	0.40	0.95
Silty clay	3	1.7	4×10^{-6}	3.83×10^{-6}	4×10^{-6}	0.35	5.6
Mucky soil	3	1.75	5×10^{-6}	4.81×10^{-6}	5×10^{-7}	0.38	2.1
Silty clay	6	1.7	4×10^{-6}	3.92×10^{-6}	1×10^{-6}	0.35	5.6
Clay	18	1.8	1×10^{-7}	1×10^{-7}	1×10^{-7}	0.32	3.7
Mixed slurry wall	7	1.5	1×10^{-6}	1×10^{-6}	1×10^{-6}	0.34	0.63

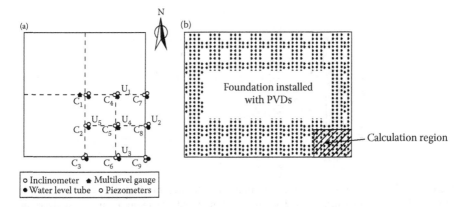

FIGURE 7.77 (a) Plan view of test area and instrumentation area; (b) plan view of modeling area. (Adapted from Chen, P. and Dong, Z., Simplified 3D Finite-Element Analysis of Soft Foundation Improved by Vacuum Preloading, www.seiofbluemountain.com/upload/product/201010/2010ythy09a12.pdf, pp. 830–837, 2010.)

A commercial FE code (ADINA v.8.3), including Biot's coupled 3-D consolidation theory, was used in the analysis. The FE mesh used in the analysis is shown in Figure 7.78a. It consisted of 88,725 20-noded elements and 95,832 nodes. To keep the FE model to a manageable level, only a 17 m × 17 m area with 22 m depth, containing 289 PDVs, was considered. Horizontal boundaries were placed 50 m away from the PVD zone and the bottom boundary was placed at 40 m depth. Thus, the FE mesh was 67 m × 67 m × 40 m in dimensions. A plan view of the model area with the PVD and the mixed slurry wall (MSW) is shown in Figure 7.78b. In the four side boundaries, the lateral displacements and flow were constrained, but they were allowed to move in the vertical direction. All displacements and flow were constrained at the bottom boundary. The top surface was assumed to be drainable, with a constant pore water pressure of 85 kPa. The slurry wall was assumed to have a very low

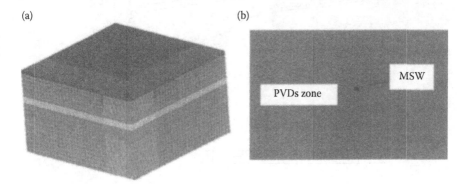

FIGURE 7.78 (a) Finite element mesh used; (b) plan view of PVD zone and mixed slurry wall. (Adapted from Chen, P. and Dong, Z., Simplified 3D Finite-Element Analysis of Soft Foundation Improved by Vacuum Preloading, www.seiofbluemountain.com/upload/product/201010/2010ythy09a12.pdf, pp. 83037, 2010.)

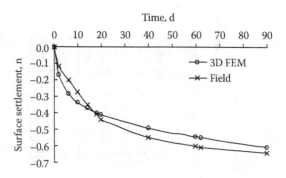

FIGURE 7.79 Variation of surface settlement at center of finite element mesh with time. (Adapted from Chen, P. and Dong, Z., Simplified 3D Finite-Element Analysis of Soft Foundation Improved by Vacuum Preloading, www.seiofbluemountain.com/upload/product/201010/2010ythy09a12.pdf, pp. 830–837, 2010.)

permeability (almost impermeable), while the PVDs were assumed to have a much higher permeability (1.0×10^{-2} cm/s). In the influence zone, smear effects were considered following the approach proposed by Chen et al. [94]. The soils were assumed to behave elastically. Additional details of the FE analysis are given in Ref. [93].

Figure 7.79 shows the increase in surface settlement with time. The FE predictions are also compared with the field measurements. In the beginning stage of vacuum preloading, the predicted settlements are higher than the measured values; a reverse trend is seen after about 20 days. Both the predicted and measured settlements become stable after about 40 days of preloading.

The variation of surface settlement with depth and time of vacuum preloading at the center of the idealized domain is shown in Figure 7.80. About 50% of the total

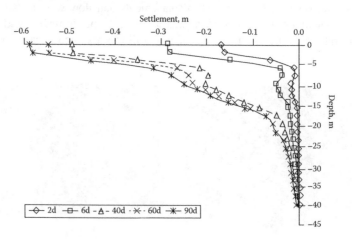

FIGURE 7.80 Variation of settlement at center of FEM mesh with depth and time. (Adapted from Chen, P. and Dong, Z., Simplified 3D Finite-Element Analysis of Soft Foundation Improved by Vacuum Preloading, www.seiofbluemountain.com/upload/product/201010/2010ythy09a12.pdf, pp. 830–837, 2010.)

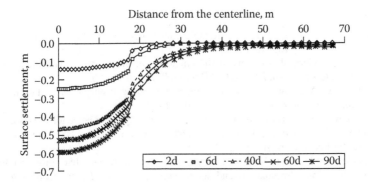

FIGURE 7.81 Surface settlement profiles with time. (Adapted from Chen, P. and Dong, Z., Simplified 3D Finite-Element Analysis of Soft Foundation Improved by Vacuum Preloading, www.seiofbluemountain.com/upload/product/201010/2010ythy09a12.pdf, pp. 830–837, 2010.)

settlement appears to take place within 6 days showing the effectiveness of preloading. Most settlements take place within 20 m depth; some settlements (about 8 cm) are seen in a depth of between 25 m and 30 m.

The variation of surface settlement with distance from the center of the idealized domain is shown in Figure 7.81 for different vacuum preloading periods. Settlements within the PDV region (17 m) are larger than those outside that region, showing the effectiveness of preloading. These results also indicate that much larger consolidation times would be required without the PDVs, as expected.

A comparison of measured and predicted excess pore water pressure is shown in Figure 7.82. Overall, the predicted reduction in pore pressure compared well with the field measurements. The trend in reduction pore water pressure reflects the consolidation mechanism due to vacuum preloading. Overall, the results show that the FE analysis is a useful tool for the simulation of 3-D consolidation.

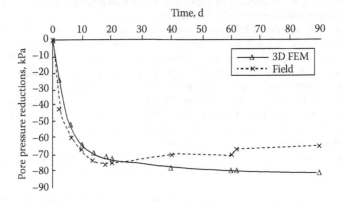

FIGURE 7.82 Comparison of predicted and measured excess pore-water pressure. (Adapted from Chen, P. and Dong, Z., Simplified 3D Finite-Element Analysis of Soft Foundation Improved by Vacuum Preloading, www.seiofbluemountain.com/upload/product/201010/2010ythy09a12.pdf, pp. 830–837, 2010.)

REFERENCES

1. Biot, M.A., General theory of three-dimensional consolidation, *Journal of Applied Physics*, 12, 1941, 155–164.
2. Biot, M.A., General solutions of the equations of elasticity and consolidation for a porous material, *Journal of Applied Physics*, 27, 1956, 91–96.
3. Biot, M.A., Mechanics of deformation and acoustic propagation in Porous media, *Journal of Applied Physics*, 33, 1962, 1482–1498.
4. Zienkiewicz, O.C., Basic formulation of static and dynamic behavior of soil and other bonus media, *Numerical Methods in Geomechanics*, (Martin, J.B., Editor), D. Reidel Pub., Dordrecht, Germany, pp. 3–39, 1981.
5. Zienkiewicz, O.C. and Shiomi, T., Dynamic behavior of saturated Porous media: The generalized biot formulation and its numerical solution, *International Journal for Numerical and Analytical Methods in Geomechanics*, 8, 1984, 71–96.
6. Desai, C.S. and Galagoda, H.M., Earthquake analysis with generalized plasticity models for saturated soils, *International Journal of Earthquake Engineering and Structural Dynamics*, 18(6), 903–919, 1989.
7. Wathugala, W. and Desai, C.S., Nonlinear and Dynamic Analysis of Porous Media and Applications, Report to National Science Foundation, Washington, DC, Dept. of Civil Eng. and Eng. Mech., Univ. of Arizona, Tucson, AZ, USA, 1990.
8. Shao, C. and Desai, C.S., Implementation of DSC Model for Nonlinear and Dynamic Analysis of Soil-Structure Interaction, Report to National Science Foundation, Washington, DC, Dept. of Civil Eng. and Eng. Mech., Univ. of Arizona, Tucson, AZ, USA, 1997.
9. Desai, C.S., *Mechanics of Material and Interfaces: The Disturbed State Concept*, CRC Press, Boca Raton, FL, USA, 2001.
10. Terzaghi, K., *Theoretical Soil Mechanics*, John Wiley and Sons, New York, NY, USA, 1943.
11. Desai, C.S. and Abel, J.F., *Introduction to the Finite Element Method*, Van Nostrand Reinhold Co., New York, NY, USA, 1972.
12. Desai, C.S., *Elementary Finite Element Method*, Prentice-Hall, Englewood Cliffs, NJ, USA, 1979. Revision published (Desai, C.S. and Kundu, T.) by CRC Press, Boca Raton, FL, USA, 2001.
13. Newmark, N.M., A method of computation for structural dynamics, *Journal of the Engineering Mechanics Division, ASCE*, 85, 1959, 67–94.
14. Booth, A.D., *Numerical Methods*, Butterworth, London, UK, 1966.
15. Bathe, K.J. and Wilson, E.L., Stability and accuracy analysis of direct integration method, *Earthquake Engineering and Structural Dynamics*, 1, 1973, 283–291.
16. Hughes, T.J.R., Analysis of transient algorithms with particular reference to stability behavior, Chapter in *Computational Methods for Transient Analysis*, Belytschko, T. and Hughes, T.J.R. (Editors), North-Holland Publishing Co., New York, NY, USA, 1983.
17. Shao, C. and Desai, C.S., Implementation of DSC model and application for analysis of pile tests under cyclic loading, *International Journal for Numerical and Analytical Methods in Geomechanics*, 24(6), 2000, 601–624.
18. Desai, C.S. and Saxena, S.K., Consolidation analysis of layered anisotropic foundations, *International Journal for Numerical and Analytical Methods in Geomechanics*, 1, 1977, 5–23.
19. Sandhu, R.S. and Wilson, E.L., Finite element analysis of flow of saturated porous elastic media, *Journal of the Engineering Mechanics Division, ASCE*, 95(3), 1969, 641–652.
20. Desai, C.S., Wathugala, G.W., and Matlock, H., Constitutive model for cyclic behavior of cohesive soils, II: Applications, *Journal of Geotechnical Engineering, ASCE*, 119(4), 1993, 730–748.

21. Shao, C. and Desai, C.S., Implementation of DSC Model for Dynamic Analysis of Soil-Structure Interaction Problems, Report, Dept. of Civil Eng. and Eng. Mech., Univ. of Arizona, Tucson, AZ, USA, 1998.

22. Clough, R.W. and Penzien, J., *Structural Dynamics*, McGraw-Hill, Inc., New York, NY, USA, 1993.

23. Casagrande, A., *Liquefaction and Cyclic Deformation of Sands- a Critical Review*, Harvard Soil Mechanics Series No. 88, Harvard University, Cambridge, Mass., 1976.

24. Castro, G., and Poulos, SA.J., Factors affecting liquefaction and cyclic mobility, *Journal of Geotechnical Engineering, ASCE*, 103(6), 1977, 502–516.

25. Seed, H.B., Soil liquefaction and cyclic mobility evaluation for level ground during earthquakes, *Journal of Geotechnical Engineering, ASCE*, 105(2), 1979, 201–255.

26. National Research Council(NRC), Liquefaction of Soils during Earthquake, Rep. No. CETS-EE01, National Academic Press, Washington, D.C., 1985.

27. Ishihara, K., Liquefaction and flow failure during earthquakes, *Geotechnique*, 43(3), 1993, 351–415.

28. Arulanandan, K. and Scott, R.F. (Eds), *VELACS-Verification and Numerical Procedures for the Analysis of Soil Liquefaction Problems, Vols. I and II*, Balkema, Rotterdam, The Netherlands, 1994.

29. Shibata, T.,Oka, F., and Ozawa, Y., Characteristics of ground deformation due to liquefaction, *Soils and Foundations*, Tokyo, Vol. Jan, NS (372 p), 1996, 65–79.

30. Nemat-Nasser, S., and Shokooh, A., A unified approach to densification and liquefaction of cohesionless sand, *Canadian Geotechnical Journal*, Ottawa, 16, 1979, 659–678.

31. Davis, R.O. and Berrill, J.B., Energy dissipation and seismic liquefaction in sands, *Earthquake Engineering and Structural Dynamics*, 19, 1982, 59–68.

32. Davis, R.O. and Berrill, J.B., Energy dissipation and liquefaction at port Island, Kobe, *Bulletin of the New Zealand National Society for Earthquake Engineering*, Waikanae, New Zealand, 31, 1998a, 31–50.

33, Davis, R.O., and Berrill, J.B., Site-specific prediction of liquefaction, *Geotechnique*, 48(2), 1998b , 289–293.

34. Berrill, J.B., and Davis, R.O., Energy dissipation and seismic liquefaction in sands-revised model *Soils and Foundations*, Tokyo, 25(2), 1985, 106–118.

35. Figueroa, J.L., Saada, A.S., Ling, L., and Dahisaria, M.N., Evaluation of soil liquefaction by energy principles, *Journal of Geotechnical Engineering, ASCE*, 120(9), 1994, 1554–1569.

36. Liang, L., Figuroa, J.L., and Saada, A.S., Liquefaction under random loading: Unit energy approach, *Journal of Geotechnical Engineering, ASCE*, 121(11), 1995, 776–781.

37. Kayen, R.E., and Mitchell, J.K., Assessment of liquefaction potential during earthquakes by arias intensity, *Journal of Geotechnical and Geoenvironmental Engineering, ASCE*, 123(12), 1997, 1162–1174.

38. Desai, C.S., Shao, C., and Rigby, D., Discussion of 'evaluation of soil liquefaction by energy principles,' by J.L. Figueroa, A.S. Saada, L. Liang, and N.M. Dahisaria, *Journal of Geotechnical Engineering, ASCE*, 122(3), 1996, 241–242.

39. Park, I.J., and Desai, C.S., Analysis of Liquefaction in Pile Foundations and Shake Table Tests using the Disturbed State Concept, Report, Dept. of Civil Eng. and Eng. Mechanics, University of Arizona, Tucson, AZ, USA, 1997.

40. Desai, C.S., Park, I.J., and Shao, C., Fundamental yet simplified model for liquefaction instability, *International Journal for Numerical and Analytical Methods in Geomechanics*, 22, 1998, 721–748.

41. Desai, C.S., Evaluation of liquefaction using disturbed state and energy approaches, *Journal of Geotechnical and Geoenvironmental Engineering, ASCE*, 126(7), 2000, 618–631.

42. Park, I.J., Kim, S. II, and Choi. J.S., Disturbed state modeling of saturated sand under dynamic loads, *Proceedings, 12 World Congress on Earthquake Engineering*, Auckland, New Zealand, 2000.

43. Wathugala, G.W. and Desai, C.S., Constitutive model for cyclic behavior of cohesive soils, I: Theory, *Journal of the Geotechnical Engineering Division, ASCE*, 119(4), 1993, 714–729.

44. The Earth Technology Corporation, Pile Segments Tests—Sabine Pass, ETC Report No. 85-007, Long Beach, CA, USA, December, 1986.

45. Katti, D.R. and Desai, C.S., Modeling Including Associated Testing of Cohesive Soils Using the Disturbed State Concept, Report to National Science Foundation, Washington, DC, Dept. of Civil Eng. and Eng. Mech., Univ. of Arizona, Tucson, AZ, USA, 1991.

46. Katti, D.R. and Desai, C.S., Modeling and testing of cohesive soils using the disturbed state concept, *Journal of Engineering Mechanics, ASCE*, 121(5), 1995, 648.

47. Rigby, D.B. and Desai, C.S., Testing and Constitutive Modeling of Saturated Interfaces in Dynamic Soil-Structure Interaction, Report, National Science Foundation, Washington, DC, Dept. of Civil Eng. and Eng. Mech., Univ. of Arizona, Tucson, AZ, USA, 1996.

48. Desai, C.S. and Rigby, D.B., Cyclic interface and joint shear device including pore pressure effects, *Journal of Geotechnical and Geoenvironmental Engineering, ASCE*, 123(6), 1997, 568–579.

49. Jaky, J., Pressure on Silus, *Proceedings, 2nd International Conference on Soil Mechanics and Foundation Engineering*, Rotterdam, 1, 1948, 103–107.

50. Baligh, M.M., Undrained deep penetration—I: Shear stresses, and II: Pore pressures, *Geotechnique*, 36(4), 1986, 471–485 and 487–501.

51. Desai, C.S., DSC—DYN2D—Dynamic and Static Analysis, Dry and Coupled Porous Saturated Materials, Report, C. Desai, Tucson, AZ, USA, 1999.,

52. Kutter, B.L., Idriss, I.M., Kohnke, T., Lakeland, J., Li, X.S., Sluis, W., Zheng, X., Tauscher, R., Goto, Y., and Kubodera, I., Design of large earthquake simulator at U.C. Davis, *Proceedings, Centrifuge 94*, Balkema, Rotterdam, pp. 169–175, 1994.

53. Wilson, D.W., Boulanger, R.W., and Kutter, B.L., Soil-Pile-Superstructure Interaction at Soft or Liquefiable Soil Sites. Centrifuge Data for CSP1 and CSP5, Report Nos. 97/02, 97/06, Center for Geotechnical Modeling, Dept. of Civil and Environ. Engng., Univ. of California, Davis, CA, USA, 1997.

54. Pradhan, S.K. and Desai, C.S., Dynamic Soil-Structure Interaction Using Disturbed State Concept, Report to National Science Foundation, Washington, DC, Dept. of Civil Eng. and Eng. Mech., Univ. of Arizona, Tucson, AZ, USA, 2002.

55. Pradhan, S.K. and Desai, C.S., DSC model for soil and interface including liquefaction and prediction of centrifuge test, *Journal of Geotechnical and Geoenvironmental Engineering, ASCE*, 132(2), 214–222, 2006.

56. Anandarajah, A., Fully Coupled Analysis of a Single Pile Founded in Liquefiable Sands, Report, Dept. of Civil Eng., The John Hopkins Univ., Baltimore, MD, USA, 1992.

57. Zienkiewicz, O.C., Leung, K.H., and Hinton, E., Earthquake response behavior of soil with damage, *Proceedings of Numerical Methods in Geomechanics*, Z. Eisentein (Editor), Canada, 1982.

58. Seed, H.B., Soil liquefaction and cyclic mobility evaluation for level ground during earthquakes, *Journal of Geotechnical Engineering, ASCE*, 105(2), 1979, 201–255.

59. Popescu, R. and Prevost, J.H., Centrifuge validation of a numerical model for dynamic soil liquefaction, *Soil Dynamics and Earthquake Engineering*, 12, 1993, 73–90.

60. Krstelj, I., Development of an Earthquake Motion Simulator and Its Application in Dynamic Centrifuge Testing, Master of Science Thesis, Princeton University, Princeton, NJ, USA, 1992.

61. Arulmoli, K., Muraleetharan, K.K., and Hossain, M.M., VELACS—Verification of Liquefaction Analyses by Centrifuge Modeling: Laboratory Testing Program: Soil Data Report, The Earth Technology Corporation, Irvine, CA, USA, 1992.

62. Prevost, J.H., A simple plasticity theory for frictional cohesionless soils, *Journal of Soil Dynamics and Earthquake Engineering*, 4(1), 1985, 9–17.

63. Prevost, J.H., Mathematical modeling of monotoric and cyclic undrained clay behavior, *International Journal for Numerical and Analytical Methods in Geomechanics*, 1(2), 1977, 195–216.

64. Koerner, R.M., Effect of particle characteristics on soil strength, *Journal of the Soil Mechanics and Foundations Division, ASCE*, 96(SM4), 1970, 1221–1234.

65. Akiyoshi, T., Fang, H.L., Fuchida, K., and Matsumoto, H., A nonlinear seismic response analysis method for saturated soil-structure system with absorbing boundary, *International Journal for Numerical and Analytical Methods in Geomechanics*, 20(5), 1996, 307–329.

66. Park, I.J. and Desai, C.S., Cyclic behavior and liquefaction of sand using disturbed state concept, *Journal of Geotechnical Engineering, ASCE*, 126(9), 2000, 834–846.

67. Gyi, M.M. and Desai, C.S., Multiaxial Cyclic Testing of Saturated Ottawa Sand, Report, Dept. of Civil Eng. and Eng. Mech., Univ. of Arizona, Tucson, AZ, USA, 1996.

68. Ishihara. K. and Nagase, H., Multidimensional irregular loading tests on sand, *Journal of Soil Dynamics and Earthquake Engineering*, 17(4), 1988, 201–212.

69. Desai, C.S. and Galagoda, H.M., Earthquake analysis with generalized plasticity model for saturated soils, *Earthquake Engineering and Structural Dynamics*, 18(6), 1989, 903–919.

70. Desai, C.S., Settlement and Seismic Analyses of Reservation Canyon Tailing Dam and Impoundment, Report, C. Desai, Tucson, AZ, 1995.

71. Desai, C.S., Shao, C., White, D., and Davis, S., Stability analysis of consolidation and dynamic response of mine tailing dam, *Proceedings of the 5th International Conference on Tailings and Mine Waste*, Fort Collins, CO, USA, January 1998.

72. Cross-sections and Material Properties Assumed in Previous Models, Parts of Previous Reports, provided by Physical Resources Eng., Inc. (White, D.), Tucson, AZ, USA, 1993.

73. Triaxial Tests on Bulk Discharge Tails and Results of Piezocone Measurements, Report, Knight Piesold and Co., Denver, CO, USA, 1988.

74. Triaxial Test Results on Subaerial Tails, Testing Report, Geotest Express, Concord, MA, USA, 1994.

75. Preliminary Assessment of the Seismic Potential of the Oquirrh Fault Zone, Utah, 1982: Mercur Gold Project, Report, by Woodward-Clyde Conf., San Francisco, CA, for Davy McKee Corp., 1982.

76. Seed, H.B., Lee, K.L., Idriss, I.M., and Makdisi, F.I., The slides in the San Fernando dam during the earthquake of February 9, 1971, *Journal of the Geotechnical Engineering Division, ASCE*, 107(GT77), 1975, 651–688.

77. Yamamoto, T., Ohara, S., and Ishikawa, M., Liquefaction of Saturated Sand Deposits Under Non-uniform Vertical Stresses, Technology Report, Faculty of Engineering, Yamaguchi Univ., Tokiwedai, Japan, Vol. 5, No. 2, pp. 71–86, 1993.

78. Vaughan, D.K. and Isenberg, J., Nonlinear rocking response of model containment structures, *Earthquake Engineering and Structural Dynamics*, 11, 1983, 275–296.

79. Vaughan, D.K. and Isenberg, J., Nonlinear Soil-Structure Analysis of SIMQUAKE II, Final Report, Research Project 8102, Electric Power Res. Inst. (EPRI), Weidlinger Associates, Menlo Park, CA, USA, 1982.

80. Zaman, M.M. and Desai, C.S., Soil-Structure Interaction Behavior for Nonlinear Dynamic Problems, Report to National Science Foundation, Washington, DC, Dept. of Civil Eng. and Eng. Mech., Univ. of Arizona, Tucson, AZ, USA, 1982.

81. Zaman, M.M., Desai, C.S., and Drumm, E.C., An interface model for dynamic soil-structure interaction, *Journal of the Geotechnical Engineering Division, ASCE*, 110(9), 1984, 1257–1273.
82. Desai, C.S., Drumm, E.C., and Zaman, M.M., Cyclic testing and modelling of interfaces, *Journal of the Geotechnical Engineering Division, ASCE*, 111(6), 1985, 795–815.
83. Desai, C.S., Dynamic soil-structure with constitutive modelling for soils and interfaces, *Proceedings, Finite Elements for Nonlinear Problems*, Trondhein, Norway, 1985.
84. Desai, C.S., Zaman, M.M., Lightner, J.G., and Siriwardane, H.J., Thin-layer element for interfaces and joints, *International Journal for Numerical and Analytical Methods in Geomechanics*, 8(1), 1984, 19–43.
85. Ramberg, W. and Osgood, W.R., Description of Stress-Strain Curves by Three Parameters, Tech Note 902, National Advisory Comm. Aeronaut., Washington, DC, 1943.
86. Baker, G.L. and Marr, W.A., Consolidation behavior of structural fills on hackensack varved clays, *Proceedings, Annual Meeting*, Transportation Research Board, Washington, DC, 1976.
87. Siriwardane, H.J. and Desai, C.S., Nonlinear Two-Dimensional Consolidation Analysis, Technical Report, VA Tech., Blacksburg, VA, USA, 1980.
88. Yokoo, Y., Yamagata, K., and Nagoka, H., Finite element method applied to Biot's consolidation theory, soils and foundations, *Japanese Society of Soil Mechanics and Foundation Engineering*, 11(1), 1971, 29–46.
89. Lekhniskii, S.G., *Theory of Elasticity of an Anisotropic Body*, Holden Day, 1963 (Translated from Russian by Fern, P.).
90. Siriwardane, H.J. and Desai, C.S., Two numerical schemes for nonlinear consolidation, *International Journal for Numerical and Analytical Methods in Geomechanics*, 17, 1981, 405–426.
91. Desai, C.S. and Siriwardane, H.J., Subsidence due to consolidation including nonlinear behavior, *Proceedings of the International Conference on Evaluation and Prediction of Subsidence*, Pensacola, Florida, USA, ASCE, New York, pp. 500–515, 1978.
92. Desai, C.S., *DSC-DYN3D-Computer Code for Static and Dynamic Analysis: Solid (Porous) Structure and Soil-Structure Problems, Manuals I, II and III*, C, Desai, Tucson, AZ, USA, 2001.
93. Chen, P. and Dong, Z., Simplified 3D Finite-Element Analysis of Soft Foundation Improved by Vacuum Preloading, www.seiofbluemountain.com/upload/product/201010/2010ythy09a12.pdf, pp. 830–837, 2010.
94. Chen, P., Fang, Y., Mo, H., Zhang, G., and Dong, Z., Analysis of 3D FEM for soft foundation improved by vacuum preloading, *Chinese Journal of Geotechnical Engineering*, 31(4), 2009, 564–570.

Appendix 1: Constitutive Models, Parameters, and Determination

A1.1 INTRODUCTION

The subject of constitutive models for geologic materials and interfaces is wide over a large number of publications; for details, books such as Refs. [1–3] can be consulted. Here, we present some brief descriptions of the models, including their parameters and determination, used for various applications in different chapters of this book; the major attention is given to the models related to applications by the authors of this book. For models used in applications by other authors, related references are cited. For the models used in various chapters, references have been made to the details presented in this appendix.

A1.2 ELASTICITY MODELS

For isotropic and elastic material (Figure A1.1a), stress–strain relations in matrix form are given by

$$\{\sigma\} = [C]\{\varepsilon\} \tag{A1.1a}$$

or in the incremental form, they can be expressed as

$$\{d\sigma\} = [C_t]\{d\varepsilon\} \tag{A1.1b}$$

where $\{\sigma\}$ and $\{\varepsilon\}$ are the stress and strain vectors, respectively, d denotes increment, $[C]$ is the constitutive relation matrix, which is generally expressed in terms of two parameters, Young's modulus, E, and Poisson's ratio, v, or shear modulus, G, and bulk modulus, K, and $[C_t]$ the tangent constitutive matrix. For the latter, the tangent values of parameters (E_t, v_t, G_t, K_t) are used in $[C_t]$.

Figures A1.1a through A1.1d show linear elastic, nonlinear elastic, perfectly plastic, and plastic hardening models, respectively. The latter two models are often called conventional plasticity models.

The elastic moduli are determined from the unloading slope (sometimes, the average of unloading and reloading slopes) because only unloading could identify whether the behavior is elastic or inelastic (plastic). Sometimes, in lieu of the unloading data, the moduli are found approximately as slopes at the origin.

For linear elastic assumption, E and v are determined from the slopes of the unloading curve. The parameter v is obtained from the axial strain ε_1 versus

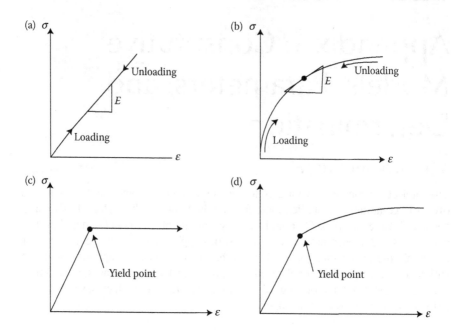

FIGURE A1.1 Elastic and plastic responses. (a) Linear elastic; (b) nonlinear elastic; (c) perfectly plastic response; and (d) hardening response.

volumetric strain $\varepsilon_v(\varepsilon_{ii})$ or ε_{33} curve. The parameters G and K can be determined similarly from tests in terms of shear stress (τ) versus shear strain (γ), and the mean pressure $[p = (J_1/3),\ J_1 = $ first invariant of the stress tensor, $\sigma_{ij}]$ versus volumetric strain $(\varepsilon_{ii} = \varepsilon_v = I_1,$ first invariant of the strain tensor, $\varepsilon_{ij})$, respectively. Figures A1.2a through A1.2d show test curves for finding the parameters, E, v, G, and K, respectively. Various symbols in the figures are defined as $\sigma_1 - \sigma_3 = $ deviatoric stress, $\varepsilon_1 = $ axial strain, $\varepsilon_v = $ volumetric strain, $\tau = $ shear stress, and $\gamma = $ shear strain. A tangent modulus is obtained as the slope at any point on the stress–strain curve; it can also be obtained as the first derivative of the function adopted to simulate the test curve. Table A1.1 shows the relations between E and v and slopes of the stress–strain response under various stress paths, which are shown on the right side of the table. The symbols for stress paths are related to conditions in tests such as compression, extension, simple shear, and proportional loading; for example, CTC, CTE, and SS denote conventional triaxial *compression*, conventional triaxial *extension*, and *simple shear* (on octahedral plane), respectively (see Table A1.1).

A1.2.1 LIMITATIONS

In the linear elastic model, based on the assumption of continuum behavior, the value of the modulus remains constant irrespective of the magnitudes of stress and strain. However, the material, very often, exibits nonlinear relation between stress, strain,

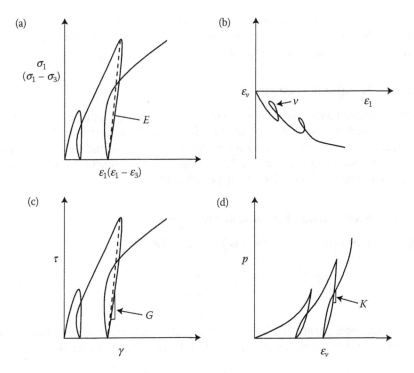

FIGURE A1.2   Elastic parameters from laboratory tests.

TABLE A1.1
Elastic Parameters from Tests under Different Stress Paths

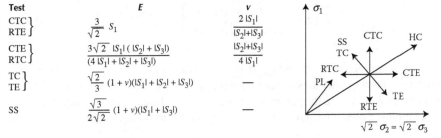

Test	E	v
CTC, RTE	$\dfrac{3}{\sqrt{2}}\,S_1$	$\dfrac{2\,|S_1|}{|S_2|+|S_3|}$
CTE, RTC	$\dfrac{3\sqrt{2}\,|S_1|\,(\,|S_2|+|S_3|)}{(4\,|S_1|+|S_2|+|S_3|)}$	$\dfrac{|S_2|+|S_3|}{4\,|S_1|}$
TC, TE	$\dfrac{\sqrt{2}}{3}\,(1+v)(|S_1|+|S_2|+|S_3|)$	—
SS	$\dfrac{\sqrt{3}}{2\sqrt{2}}\,(1+v)(|S_1|+|S_3|)$	—

Source: Adapted from Desai, C.S., *Mechanics of Materials an Interfaces: The Disturbed State Concept*, CRC Press, Boca Raton, FL, USA, 2001.

Note: S_1 = (average) slope of the unloading/reloading curve, τ_{oct} versus $\varepsilon_i\,(i = 1, 2, 3)$ plots.

and volumetric responses. Moreover, the linear elastic model does not account for factors such as inelastic or plastic deformation, stress paths, full volume dilation, and microcracking leading to fracture, softening, and failure.

A1.2.2 NONLINEAR ELASTICITY

We can represent a nonlinear curve (Figure A1.1b) by using mathematical functions such as hyperbola, parabola, splines, and Ramberg–Osgood (R–O). Then, in the incremental analysis, the behavior is simulated as piecewise linear elastic in which the tangent moduli can be computed as derivatives of the function at given points. We present here two simulations, by hyperbola and R–O function.

A1.2.3 STRESS–STRAIN BEHAVIOR BY HYPERBOLA

The simulation using the hyperbola is given as [1,2,4–7]

$$\sigma = \frac{\varepsilon_1}{a + b\varepsilon_1} \tag{A1.2}$$

where σ can be a quantity such as σ_1 and $(\sigma_2 - \sigma_3)$, and a and b are material parameters or constants, which can be expressed as (Figures A1.3a and A1.3b)

$$a = \frac{1}{E_i} \tag{A1.3a}$$

$$b = \frac{1}{\sigma_{ult}} \tag{A1.3b}$$

where E_i is the initial modulus and σ_{ult} is the ultimate or asymptotic stress (Figure A1.3a). They can be obtained by plotting ε_1/σ versus ε_1 (Figure A1.3b).

The tangent modulus, E_t, can be derived as [1,2,5,6]

$$E_t = \left[1 - \frac{R_f(1 - \sin\varphi)(\sigma)}{2c\cos\varphi + 2\sigma_3\sin\varphi}\right]^2 \bar{K}p_a\left(\frac{\sigma_3}{p_a}\right)^n \tag{A1.4a}$$

where R_f is the failure ratio $= \sigma_f/\sigma_{ult}$, σ_f is the measured failure or peak stress, σ_{ult} is the ultimate (asymptotic) stress, c and φ are the cohesive strength and angle of friction, respectively, $\bar{K}$ and n are the parameters to define dependence of initial modulus, E_i on σ_3, or confining pressure (Figure A1.3c), and p_a is the atmospheric pressure.

A1.2.4 PARAMETER DETERMINATION FOR HYPERBOLIC MODEL

The values of c and φ are found from a set of triaxial test data under different confining pressures by plotting the Mohr diagram. The values of initial modulus,

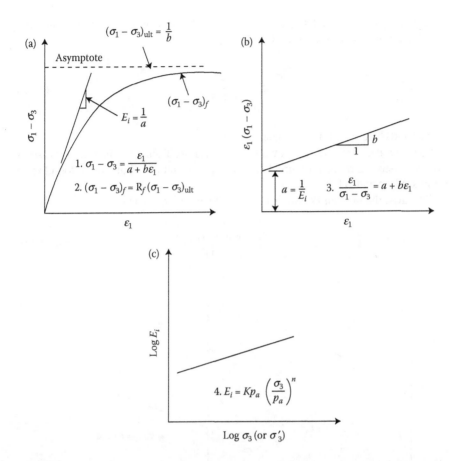

FIGURE A1.3 Hyperbolic model, initial modulus, E_i, and determination of parameters. (a) Stress-strain response by hyperbola; (b) transformation of hyperbola; and (c) initial modulus versus confining stress.

E_i ($= (1/a)$), and ultimate stress, σ_{ult} ($= (1/b)$), are found by plotting the stress–strain data as in Figure A1.3b. The values of E_i, as unloading (or initial) slope, are found from the stress–strain data for different confining pressures. Then, the relation between E_i and σ_3 is expressed as

$$E_i = \bar{K}p_a \left(\frac{\sigma_3}{p_a} \right)^n \tag{A1.4b}$$

The plot of log E_i versus log σ_3 (Figure A1.3c) gives values of $\bar{K}$ and n.

A1.2.4.1 Poisson's Ratio

The procedure for finding the hyperbolic parameters for Poisson's ratio is given below.

First, we plot measured ε_1 and ε_3 (from triaxial) test as in Figure A1.4a; here, the hyperbolic relation between ε_1 and ε_3 is given by

$$\varepsilon_1 = \frac{\varepsilon_3}{f + \varepsilon_3 d} \qquad (A1.5)$$

where f and d are material parameters or constants.

Now, plot $\varepsilon_3/\varepsilon_1$ versus ε_3 as shown in Figure A1.4b. Then, plot the initial Poisson's ratio, v_i versus log σ_3 as in Figure A1.4c. The parameters G, F, and d are found from Figures A1.4b and A1.4c.

Similarly, the tangent Poisson's ratio can be expressed as [1,6]

$$v_t = \frac{G - F \log(\sigma_3/p_a)}{(1 - A)^2} \qquad (A1.6)$$

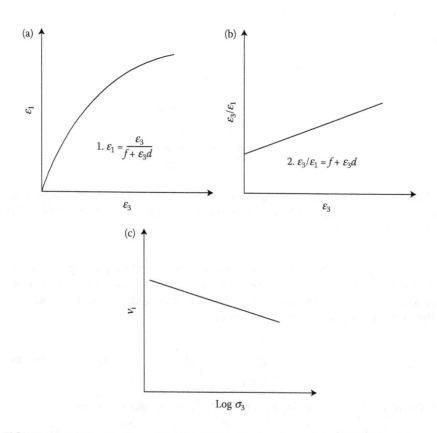

FIGURE A1.4 Hyperbolic model, initial Poisson's ratio v_i, and determination of parameters. (a) ε_1 versus ε_3 response by hyperbola; (b) transformation of hyperbola; and (c) initial Poisson's ratio, v_i versus confining stress.

where

$$A = \frac{\sigma d}{\bar{K} p_a (\sigma_3/p_a)^n \left[1 - ((R_f \sigma(1 - \sin\varphi))/(2c\cos\varphi + 2\sigma_3\sin\varphi))\right]}$$

and G, F, and d are parameters.

The stress–strain curve (Figure A1.3a) can also be expressed in terms of shear stress and strain, and the hyperbolic relation is given by

$$\tau = \frac{\gamma}{A + B\gamma} \tag{A1.7}$$

where A and B are constants and are given by

$$A = \frac{1}{G_i} \quad \text{and} \quad B = \frac{1}{\tau_{\text{ult}}}$$

G_i is the initial shear modulus and τ_{ult} is the asymptotic shear stress.

If E and v are known, K and G can be evaluated from the following expressions:

$$G = \frac{E}{2(1 + v)} \tag{A1.8a}$$

$$K = \frac{E}{3(1 - 2v)} \tag{A1.8b}$$

A1.3   NORMAL BEHAVIOR

The hydrostatic or isotropic curve can be expressed as an exponential function, in terms of p versus ε_v $(= I_1)$ or $J_1/3$ versus I_1 $(= \varepsilon_v)$ as shown in Figure A1.2d

$$p = \left(p_o + p_t\right) e^{\alpha I_1} \tag{A1.9}$$

where p_o and p_t are initial strength and strength in volumetric tension, respectively, and α is a material parameter. Parameters α and p_t can be obtained by plotting p versus $\log I_1$, for a known p_o.

A1.4   HYPERBOLIC MODEL FOR INTERFACES/JOINTS

For a (two-dimensional) interface problem, a natural joint (interface) and its idealized version are depicted in Figures A1.5a and A1.5b, respectively. The natural joint (of unit length in the plane strain assumption) is idealized by assuming a joint of small thickness t with a width equal to that of the natural joint. In the finite element procedure, such an interface is called the *thin-layer element* [7]. The behavior is defined based on shear stiffness, k_{st}, and normal stiffness, k_n; the latter is often

(a)   (b)

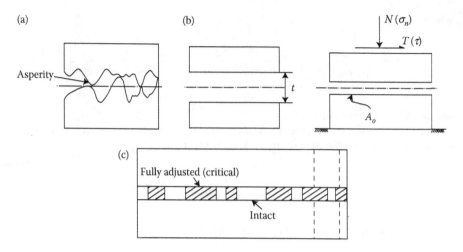

FIGURE A1.5 Idealization of interface or joint and phases in DSC. (a) Natural; (b) idealized; and (c) intact and critical phases.

adopted arbitrarily very high initially, and very low at failure [1,2,7]. The shear stiffness can be expressed in terms of shear stress, τ, and relative shear displacement, u_r (Figure A1.6a) similar to Equation A1.4a as

$$k_{st} = k_j \gamma_w \left(\frac{\sigma_n}{p_a}\right)^n \left(1 - \frac{R_f \tau}{c_a + \sigma_n \tan\delta}\right)^2 \tag{A1.10a}$$

where k_j (initial shear stiffness, k_{si}), n is a material parameter, γ_w is the unit weight of water, σ_n is the normal stress, R_f is the failure ratio, and c_a and δ are the adhesive strength and the interface friction angle, respectively. The parameters can be determined by similar procedures described above for the hyperbolic model for soil or rock.

The normal stiffness can be determined on the basis of normal load tests on an interface. The normal (incremental) load or stress, σ_n, is applied and the resulting relative normal displacement, v_r, is measured (Figure A1.6b). The normal stiffness, k_{nt}, is computed as the slope of the unloading curve or of the curve at the origin. The initial normal stiffness, k_{ni}, can be expressed as

$$k_{ni} = K_j (\sigma_n)^m \tag{A1.10b}$$

where K_j and m are material parameters.

It is desirable to use test data with positive σ_n for evaluating and defining normal stiffness during a slip motion. However, in computer analyses, k_{nt} is often assumed arbitrarily to be of high value, of the order to 10^{10} kPa, under the stipulation that the normal stiffness at the interface is high during the slip motion. For other states of motion in an interface such as separation or debonding, it may be necessary to adopt a smaller value of k_{nt} arbitrarily (of the order of about 10 kPa).

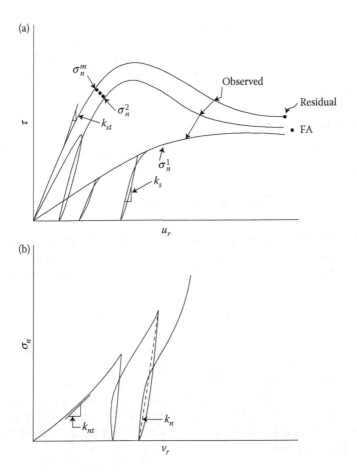

FIGURE A1.6   Shear and normal response of interface. (a) τ versus u_r under different normal stress, σ_n; (b) σ_n versus v_r.

A1.4.1   Unloading and Reloading in Hyperbolic Model

When the material is unloaded from a given load, usually it follows a different path (Figure A1.2a). Then, if it is reloaded from the end of unloading, it also follows a different path. It is essential to model both unloading and reloading for certain problems. In the hyperbolic model, unloading and reloading are often assumed to follow the same path, and the unloading (elastic) modulus, E_{ur}, as a function of the confining stress, σ_3, is given by [1,5]

$$E_{ur} = K_{ur} p_a \left(\frac{\sigma_3}{p_a} \right)^{\bar{n}} \qquad (A1.11)$$

where K_{ur} and $\bar{n}$ are parameters. The procedure for identifying unloading is based on the stress level at the current stress state. If the current stress level is smaller than the maximum previous value reached during (incremental) loading, unloading occurs; then, the modulus, E_{ur}, in Equation A1.11 is used. If the current stress level is greater than the previous maximum value, virgin or primary loading, reloading occurs, and the modulus, E_t (Equation A1.4a), can be used. The parameters K_{ur} and $\bar{n}$ can be determined by plotting log E_{ur} versus log (σ_3/p_a), as in Figure A1.3c.

A1.5 RAMBERG–OSGOOD MODEL

For stress–strain $(\sigma-\varepsilon)$ behavior, the R–O [1,8,9] model is given by (Figure A1.7)

$$\sigma = \frac{(E_i - E_p)\varepsilon}{[1 + \{(E_i - E_p)/(\sigma_y)\varepsilon\}^m]^{1/m}} + E_p\varepsilon \qquad (A1.12a)$$

where E_i and E_p are the initial and final moduli, respectively, σ_y (yield stress in the final region) and m are parameters; the latter defines the order of the curve. In the case of the application of the R–O model for defining p_y-v curves (Chapter 2), various terms are related as

$$p_y = \sigma, \; v = \varepsilon, \; k_o = E_i, \; k_f = E_p, \; p_f = \sigma_y$$

R–O model can be used for other behaviors, for which appropriate parameters can be replaced in Equation A1.12a and Figure A1.7.

The procedure for finding the R–O parameters is described below:

The value of E_i (k_o) is determined as the slope of the stress–strain $(\sigma-\varepsilon)$ or (p_y-v) curve at $\sigma(p)$ equal to zero (at the origin). The slope at a convenient point σ_y (p_f) in the ultimate region can be adopted as E_p (k_f). The value of m can be found by solving the following equation [9]:

$$F(m) = A^m - \frac{1}{R^m}B^m + \left(\frac{1}{R^m} - 1\right) = 0 \qquad (A1.12b)$$

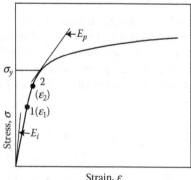

FIGURE A1.7   Ramberg–Osgood representation of stress–strain response.

where R is the ratio between the strains at two selected points on the curve ($R = \varepsilon_2/\varepsilon_1$), $A = (E_i - E_p)/(E_1 - E_p)$, $B = (E_i - E_f)/(E_2 - E_f)$, $E_1 = \sigma_1/\varepsilon_1$, and $E_2 = \sigma_2/\varepsilon_2$; here, 1 and 2 denote the two chosen points (Figure A1.7). Once m is determined from an iterative solution on Equation A1.12b, σ_y can be computed from

$$\sigma_y = \frac{E_i - E_p}{(A^m - 1)^{1/m}} \tag{A1.13}$$

The procedure for finding the R–O parameters is also given in Ref. [9].

A1.6   VARIABLE MODULI MODELS

The stress–strain relations for nonlinear elastic models can be expressed in terms of measured data, e.g., the σ_1 versus ε_1 or $(\sigma_1 - \sigma_3)$ versus ε_1 curve, which can be expressed in the hyperbolic form (Equation A1.4a). Then, the parameter, that is, tangent modulus, can be found as the first derivative of the function (Equation A1.4a). Sometimes, the parameter itself is expressed in terms of stress, strain, stress invariant, and/or strain invariant. They are often called variable moduli (VM) models. Some examples are given below:

$$G = G_0 + \gamma_1(J_1/3) + \gamma_2\sqrt{J_{2D}} \tag{A1.14a}$$

and

$$K = K_0 + K_1 I_1 + K_2 I_1^2 \tag{A1.14b}$$

where G and K are the shear and bulk moduli, and G_0 and K_0 are the initial values, respectively. γ_1 and γ_2, and K_1 and K_2 are other parameters in G and K, respectively. Further details and other VM models are included in Refs. [1,2,10]. The parameters G_0 and K_0 are initial values of the shear and bulk moduli, respectively. The other parameters in the VM model can be determined by solving the equations by using a least squares procedure.

A1.7   CONVENTIONAL PLASTICITY

Models such as von Mises, Mohr–Coulomb, and Drucker–Prager are considered to belong to conventional plasticity; Figure A1.8 shows their plots in $\sigma_1-\sigma_2-\sigma_3$ and $J_{2D}^{1/2}$ versus J_1 stress spaces. In the von Mises and Drucker–Prager criteria, the yield surface plots as a circle in the $\sigma_1-\sigma_2-\sigma_3$ space; this means that the strength of the material is independent of the stress path. Plots of stress paths for common laboratory testing are shown in Figure A1.9 and the dependence of strength on the stress path (e.g., C, E, and S as compression, extension, and simple shear, respectively) is depicted in Figure A1.8c. Here, the material is assumed to behave elastically until a yield stress (σ_y) and plastic or irreversible deformations occur after the yield. In plasticity, the incremental constitutive equations are expressed as

$$\{d\sigma\} = [C^{ep}]\{d\varepsilon\}$$
$$= ([C]^e - [C]^p)\{d\varepsilon\} \tag{A1.15}$$

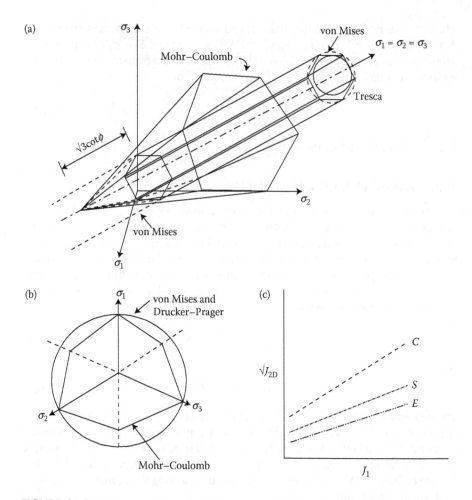

FIGURE A1.8 Plots of various plasticity models. (a) Plots of F for different models in σ_1–σ_2–σ_3 space; (b) plots of various F for different models on Π–plane; and (c) ultimate (failure) envelopes under different stress paths.

where $[C]^e$ is the elastic matrix as in Equation A1.1b and $[C]^p$ is the plasticity matrix derived on the basis of yield condition (function), consistency condition and flow rule [2,3,11], which are stated below for some specific models.

A1.7.1 VON MISES

For this case, the yield function is given by (Figure A1.8a)

$$F = J_{2D} - k^2 = 0 \qquad\qquad\qquad (A1.16a)$$

or

$$= \sqrt{J_{2D}} - k = 0 \qquad\qquad\qquad (A1.16b)$$

where F denotes the yield function, k is a material constant, J_{2D} is the second invariant of the deviatoric stress tensor, $S_{ij} = \sigma_{ij} - p\delta_{ij}$, which is given by

$$J_{2D} = \frac{1}{6}\left[(\sigma_1 - \sigma_2)^2 + (\sigma_2 - \sigma_3)^2 + (\sigma_1 - \sigma_3)^2\right] \qquad (A1.17)$$

The von Mises model requires three constants, E, v, and k; the latter, which is related to the cohesive strength, can be determined by plotting $J_{2D}^{1/2}$ versus J_1 based on tests (triaxial) on the material. Figure A1.9 shows various stress paths used in material testing, in which C denotes compression, S denotes simple shear, and E denotes extension. Note that the behavior can be different under different stress

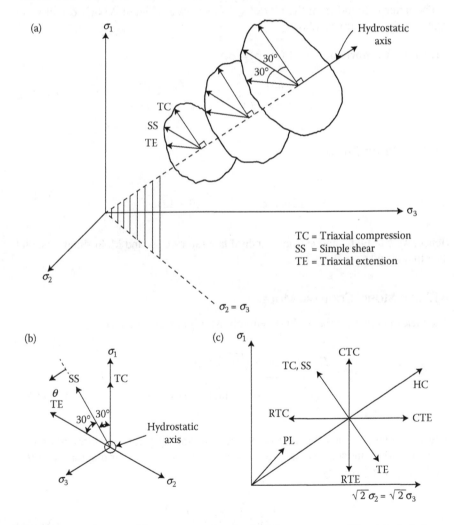

FIGURE A1.9 Common laboratory stress-paths: (a) 3-D stress space; (b) octahedral plane; and (c) triaxial plane (compressive stress is positive).

paths. It may be noted that for saturated cohesive materials, there will be only one envelope in Figure A1.8a, as for the von Mises model. It does not account for the frictional behavior in which the strength depends also on confining stress (angle of friction, φ). The Drucker–Prager and Mohr–Coulomb criteria allow for both cohesion and friction. The yield function for the Drucker–Prager model is given by (Figure A1.8b):

$$\sqrt{J_{2D}} - \alpha J_1 - k = 0 \tag{A1.18}$$

where J_1 is the first invariant of σ_{ij} and is proportional to mean (confining) stress, $p = J_1/3$ and α is another constant. This model requires four parameters, E, v, α, and k. The latter is found from the plot of $\sqrt{J_{2D}}$ versus J_1 (Figure A1.8c). α and k are related to c and φ are as follows.

A1.7.1.1 Compression Test (σ_1, $\sigma_2 = \sigma_3$)

$$\alpha = \frac{2 \sin\varphi}{\sqrt{3}\,(3 - \sin\varphi)}; \quad k = \frac{6c \,\cos\varphi}{\sqrt{3}\,(3 - \sin\varphi)} \tag{A1.19a}$$

A1.7.2 PLANE STRAIN

$$\alpha = \frac{\tan\varphi}{(9 + 12\tan^2\varphi)^{1/2}}; \quad k = \frac{3c}{(9 + 12\tan^2\varphi)^{1/2}} \tag{A1.19b}$$

Hence, we can find c and φ from standard laboratory tests and Mohr diagrams, and then find α and k.

A1.7.3 MOHR–COULOMB MODEL

The yield function for the Mohr–Coulomb (M–C) model is given by

$$F = J_1 \sin\varphi + \sqrt{J_{2D}}\,\cos\theta$$
$$- \frac{\sqrt{J_{2D}}}{3} \sin\varphi \sin\theta - c\cos\varphi = 0 \tag{A1.20}$$

Thus, two elastic (E and v) and two plasticity (c and φ) parameters are required to define the M–C model. The latter are determined from the Mohr diagram, and θ is the Lode angle, given by

$$\theta = \frac{1}{3}\sin^{-1}\left(-\frac{3\sqrt{3}}{2}\,\frac{J_{3D}}{J_{2D}^{3/2}}\right) \tag{A1.21a}$$

and

$$-\frac{\pi}{6} \le \theta \le \frac{\pi}{6} \qquad (A1.21b)$$

A1.8 CONTINUOUS YIELD PLASTICITY: CRITICAL STATE MODEL

Figure A1.10 shows the critical state (CS) concept advanced by Roscoe and coworkers [12,13]. A loose granular material (Figure A1.11a), under loading, experiences increasing compression and ultimately tends to approach the CS, denoted by c (Figure A1.11a), after which the void ratio or the density of the material remains invariant (Figures A1.11a through A1.11c). A dense soil first experiences compression, but usually before the peak stress, it starts to dilate, that is, increases in volume. However, in the CS, it approaches the same void ratio as the loose material. Thus, irrespective of the initial condition, loose or dense, the material approaches the unique (critical) state. The CS can be considered similar to the ultimate or failure condition in conventional plasticity models.

The constitutive model for the CS can be defined by plotting the critical values of $\sqrt{J_{2D}}$ versus J_1 (Figure A1.11d) and the void ratio at the CS, e^c versus J_1 (or $J_1/3p_a$) (Figure A1.11e). Hence, the parameters are $\bar{m}$, λ, and κ, the slope of the critical state line (CSL) in the $\sqrt{J_{2D}}$ versus J_1 space, the slope of the CSL in the e^c versus $\ln (J_1/3p_a)$, and the slope of the swelling curve, κ, respectively (Figures A1.11d and A1.11e). In addition, we need the initial void ratio, e_o, and two elastic constants to define the CS model. The behavior at the CS can be defined by using the following equations [2,12,13]:

$$\sqrt{J_{2D}} = \bar{m}J_1 \qquad (A1.22a)$$

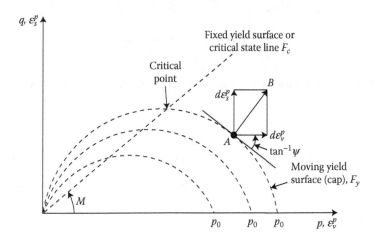

FIGURE A1.10   Critical state model and yield surfaces.

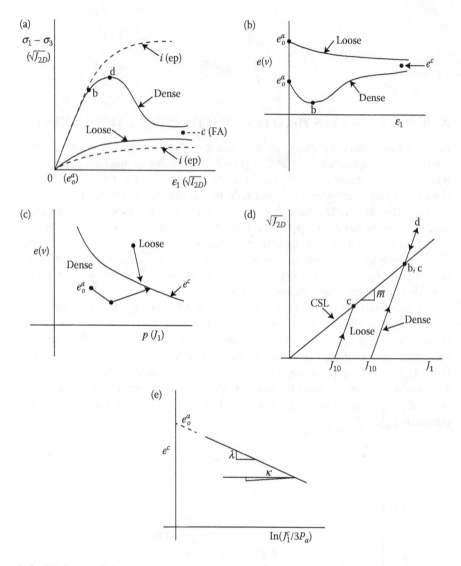

FIGURE A1.11 Behaviour of loose and dense cohension less material and critical state model.

or

$$q = Mp \tag{A1.22b}$$

$$e^c = e_o^c - \lambda \ell n (J_1^c / 3 p_a) \tag{A1.22c}$$

or

$$e^c = e_o^c - \lambda \ell n (p/p_a) \tag{A1.22d}$$

where e_o^c is the initial void ratio, which is the value of e^c corresponding to $J_1 = 3p_a$, and λ is the slope of the CSL in the e-log p plot. The yield surface in the CS concept, F_y (Figure A1.10), defines continuous yield. In the original CS model, terms related to the triaxial tests, $q = \sigma_1 - \sigma_3$ and $p = (\sigma_1 + \sigma_2)/2$, were used to plot, F_y [12,13] (Figure A1.10). However, F_y can also be expressed in terms of three-dimensional stresses, J_1 and J_{2D}. Then, $\bar{m}$ and M are the slopes of the CSL in J_1 versus $\sqrt{J_{2D}}$ (Figure A1.11d), and q ($\sigma_1 - \sigma_3$) versus p (Figure A1.10) spaces, respectively, e^c is the void ratio at CS, and p_a is the atmospheric pressure constant.

The associated relation for incremental volumetric plastic strain, $d\varepsilon_v^p$, is expressed as

$$d\varepsilon_v^p = \frac{de^p}{1+e_o} = \frac{\lambda - \kappa}{1+e_o}\frac{dp_o}{p_o} \tag{A1.23}$$

where κ is the unloading slope in e-log p plot (Figure A1.11e).

The yield surface in the CS concept, F_y, is required to define the continuous yield:

$$F_y = \bar{m}^2 J_1^2 + \bar{m}\, J_1 J_{10} + J_{2D} = 0 \tag{A1.24a}$$

or

$$F_y = M^2 p^2 - M^2 p_0\, p + q^2 = 0 \tag{A1.24b}$$

The above equations are equivalent and the relation between $\bar{m}$ and M is given by $M = 3\sqrt{3\bar{m}}$; note that the former is the slope of the CSL in the $\sqrt{J_{2D}}$ versus J_1 space and the latter is the slope of the CSL in the q–p space.

Thus, the following constants are required to define the CS model:

$$E, v; M(\bar{m}), \lambda(\text{or } \lambda_c), \kappa, e_o$$

The parameters for the CS model can be determined on the basis of triaxial (or multiaxial) and hydrostatic (consolidation) tests. A brief description is given below. Further details are given in Refs. [1,2].

1. The parameters E, v (G, K) are found from the unloading slopes of test curves in different forms such as $(\sigma_1 - \sigma_3)$ versus ε_1 (Figure A1.2).
2. M is found as the slope of the CSL in the q–p space (Figure A1.10) and $\bar{m}$ is found as the slope of CSL in the $\sqrt{J_{2D}} - J_1$ space (Figure A1.11d).
3. λ and κ are found as loading and unloading slopes in the e-log p plot (Figure A1.11e).
4. e_o is the initial void ratio at the starting point of the test.

It may be noted that the above CS model defines the behavior of loose sands and normally consolidated clay, which lie below the CSL (Figure A1.10). Additional considerations are required for the behavior of dense sand and overconsolidated clay in which the stress path crosses the CSL and then approaches the CS; this behavior is defined later.

A1.8.1 CAP MODEL

The continuous yield model for cohesionless material, called the *cap* model, was proposed by DiMaggio and Sandler [1,2,14]. The word "cap" may be considered to refer to yield surfaces that look like *caps*. Two yield surfaces are defined in this model: (a) for continuous yield, F_y, and (b) for failure yield surface or envelope, F_f (Figure A1.12); F_y is expressed as

$$F_y = \left(\frac{J_{10} - J_{1C}}{\sqrt{J_{2DC}}} \right)^2 J_{2D} + (J_1 - J_{1C})^2 = 0 \tag{A1.25}$$

where J_{10} denotes the intersection of F_y and the J_1 axis, J_{1C} is the value of J_1 at the center of the elliptical F_y, and $\sqrt{J_{2DC}}$ is the value of $\sqrt{J_{2D}}$ when $J_1 = J_{1C}$. As in the

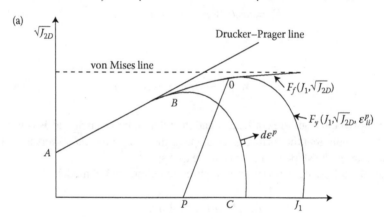

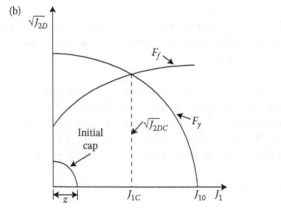

FIGURE A1.12 Cap model. (a) In $\sqrt{J_{2D}} - J_1$ space; (b) intersection of F_f and F_y.

CS model, the term J_{10} defines yielding or hardening, assumed as the function of volumetric plastic strain as

$$J_{10} = -\frac{1}{D}\ell n\left(1 - \frac{\varepsilon_v^p}{W}\right) + Z \qquad\qquad (A1.26)$$

where D, W, and Z are material parameters. The term Z denotes the size of the yield surface due to the initial stress (strain).

The failure or final yield surface is given by

$$F_f = \sqrt{J_{2D}} + \gamma\, e^{-\beta J_1} - \alpha = 0 \qquad\qquad (A1.27)$$

where α, β, and γ are the material parameters.

The number of parameters in the cap model is nine, as listed below:

Elasticity: E, v, or K and G
Plasticity: D, W, Z, R; α, β, γ

where $R = ((J_{10} - J_{1C})/(\sqrt{J_{2DC}}))$ is the first term in the parentheses in Equation A1.25.

Note: The forms of Equations A1.25 through A1.27 may be different in different publications; in that case, the reader could correlate the parameters accordingly.

The parameters can be determined from hydrostatic and triaxial tests under different confining stresses. The plots required are shown in Figures A1.12 and A1.13, and the determination of various parameters is described as follows:

1. α from the intersection of von Mises yield function with $\sqrt{J_{2D}}$ (Figure A1.13a).
2. $\alpha-\gamma$ from the intersection of Drucker–Prager yield function with $\sqrt{J_{2D}}$ (Figure A1.13a).

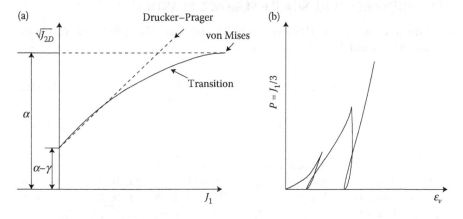

FIGURE A1.13 Parameters in cap model. (a) $\sqrt{J_{2D}}$ versus J_1; (b) p versus ε_v.

3. β from $\beta = -\dfrac{1}{J_1} \ell n \left(\dfrac{\alpha - \sqrt{J_{2D}}}{\gamma} \right)$

4. D from $3pD = -\ell n \left(1 - \dfrac{\varepsilon_v^p}{W} \right)$, where $\varepsilon_v^p = \varepsilon_v - \varepsilon_v^e = W(1 - e^{-3pD})$

Here, p = mean pressure = $J_{10}/3$, ε_v is the total volumetric strain, and ε_v^e is the elastic volumetric strain.

Further details for the determination of parameters are given in Refs. [1,2,14].

A1.8.2 LIMITATIONS OF CRITICAL STATE AND CAP MODELS

Although the CS and cap models account for continuous yield from the beginning of loading, they suffer from the following limitations:

1. They do not account for different strengths along different stress paths, which is common for many geomaterials; in other words, in the CS and cap models, yield surfaces plots circular in the principal stress space, σ_1–σ_2–σ_3.
2. The yielding depends only on the volumetric behavior, that is, ε_v^p. However, for many materials, such as sands, yielding can also depend on plastic shear strains.
3. They do not account for the volumetric dilation before the peak stress.
4. They do not account for the nonassociative behavior, that is, the increment of plastic strain is orthogonal to the plastic potential function, Q, and not to yield function, F.

The following hierarchical single surface (HISS) model overcomes the above limitations and is found to provide the general and unified continuous yield plasticity model [1,2,15].

A1.9 HIERARCHICAL SINGLE SURFACE PLASTICITY

The yield function in the HISS δ_0-model for associative behavior is expressed as (Figures A1.14a and A1.14b)

$$F = \bar{J}_{2D} - (-\alpha \bar{J}_1^n + \gamma \bar{J}_1^2)(1 - \beta S_r)^{-0.5} = 0 \qquad \text{(A1.28a)}$$

$$= \bar{J}_{2D} - F_b F_s = 0 \qquad \text{(A1.28b)}$$

where γ and β are material parameters associated with the ultimate (yield) envelope (Figure A1.14a), n is associated with the transition from compactive to dilative volumetric response, $\bar{J}_{2D} = J_{2D}/p_a^2$, $\bar{J}_1 = (J_1 + 3R)/p_a$, R is the term related to the cohesive strength $(= \bar{c}/3\sqrt{\gamma})$ (Figure A1.14a), S_r is the stress ratio $(=(\sqrt{27}/2) \cdot (J_{3D}/J_{2D}^{3/2}))$, J_{3D} is the third invariant of the deviatoric stress S_{ij}, and α is the growth or yield or

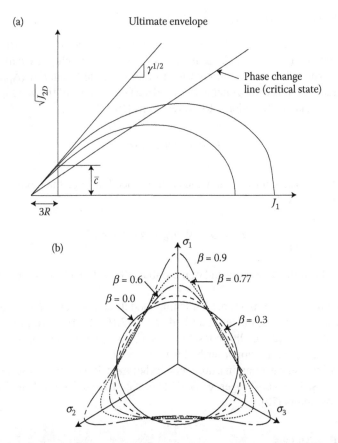

FIGURE A1.14 Yield surfaces in hierarchical single surface (HISS) model. (a) $(J_{2D})^2 - J_1$; (b) octahedral plane ($\beta < 0.756$ for convexity).

hardening function, which in a simple form is expressed in terms of both volumetric and deviatoric plastic strains as

$$\alpha = \frac{a_1}{\xi^{\eta_1}} \tag{A1.29a}$$

where ξ is the accumulated (or trajectory) of plastic strains given by

$$\xi = \xi_D + \xi_v = \int (dE_{ij}^p \, dE_{ij}^p)^{\frac{1}{2}} + \frac{1}{\sqrt{3}} |\varepsilon_{ii}^p| \tag{A1.29b}$$

where E_{ij}^p is the deviatoric plastic strain tensor $= \varepsilon_{ij}^p - (1/3) \, \varepsilon_{ii}^p \, \delta_{ij}$, ε_{ij}^p is the total plastic strain tensor, $\varepsilon_{ii}^p = \varepsilon_v^p$ is the volumetric plastic strain, and δ_{ij} is the Kronecker delta.

A1.9.1   NONASSOCIATED BEHAVIOR (δ_1-MODEL)

For many cohesionless materials, the increment of the plastic strain is orthogonal to the plastic potential function, Q, but not the yield surface, F. They are called nonassociative (δ_1-model) materials in contrast to the associated material (δ_0-model) in which the increment is orthogonal to the yield surface, F. One of the ways to account for the nonassociative behavior is to express Q as [16,17]

$$Q = F + h\,(J_1,\ \alpha) \tag{A1.30a}$$

where h is the correction function, which is introduced through the modified hardening function, α_Q, as

$$\alpha_Q = \alpha + \bar{\kappa}\,(\alpha_0 - \alpha)\,(1 - r_v) \tag{A1.30b}$$

where α_0 is the value of α at the end of the initial (hydrostatic) loading, $\bar{\kappa}$ is a parameter, and $r_v = \xi_v/\xi$.

Nonassociative and anisotropic hardening models can be developed as the extension of the associative δ_0-model described above [2,16–18]. However, the disturbed state concept (DSC) [2,19,20] is considered capable of (partly) accounting for such a behavior. Hence, they are not described here in detail.

The HISS model can contain a number of other models as special cases, for example, von Mises, Drucker–Prager, critical state, cap, Matsuoka and Nakai [21], and Lade and coworkers [22].

A1.9.2   PARAMETERS

Laboratory and/or field testing is essential to define the parameters for various models. We need tests under various factors such as confining stress, stress paths, and temperature. A variety of test devices are available for such testing; this topic is beyond the scope of this book and details are given in Refs. [1,2]. Laboratory tests under hydrostatic, triaxial, and multiaxial conditions are employed in the following description of the determination of parameters.

The HISS-δ_0 (associative) model contains the following parameters.

A1.9.2.1   Elasticity

E and v or K and G. The procedure for finding them has been presented before.

A1.9.2.2   Plasticity

Ultimate yield: γ and β. The ultimate stress condition is obtained (Figure A1.15a), often by assuming that the asymptotic stress is about 5–15% of the maximum stress. Then, the ultimate envelopes for various stress paths are plotted; for example, compression and extension are plotted as in Figure A1.15b. The values of γ and β can be found by using a least squares fit procedure based on the (average) ultimate envelope. The values of the parameters can also be found from the slopes of the ultimate

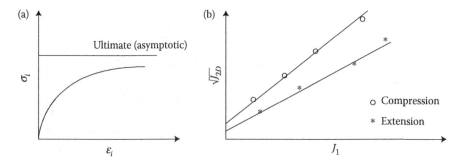

FIGURE A1.15 Parameters γ and β in HISS model. (a) σ_i versus ε_i; (b) $\sqrt{J_{2D}}$ versus J_1.

envelopes for different stress paths in Mohr–Coulomb and in J_1–$\sqrt{J_{2D}}$ spaces (Figure A1.16); details are given in Ref. [2].

A1.9.2.3 Transition Parameter: *n*

Figure A1.17 shows the location of the transition point from compressive to dilative volume changes, which can be considered to relate to the vanishing of the $\delta F/\delta J_1$ (Figure A1.17); the expression for n then can be derived as

$$n = \frac{2}{1 - (J_{2D}/J_1)(1/F_s\gamma)} \quad (\text{at } d\underline{\varepsilon}_v = 0) \tag{A1.31}$$

There are other ways to obtain n [2]; one way is to determine it using the following expression (Figure A1.18):

$$J_{1a}/J_{1m} = (2/n)^{1/(n-2)} \tag{A1.32}$$

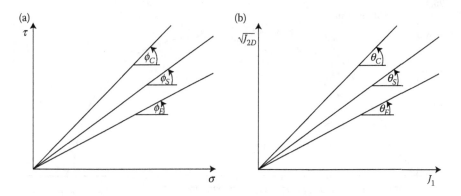

FIGURE A1.16 Ultimate envelopes in (a) Mohr–Coulomb ($\tau - \sigma$); (b) $\sqrt{J_{2D}} - J_1$ spaces. C, compression; S, simple shear; E, extension.

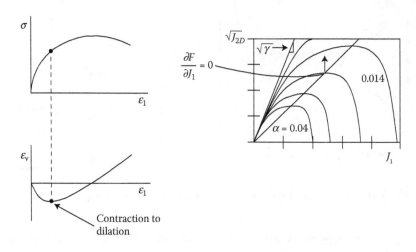

FIGURE A1.17   Phase change parmeter, n: Procedure 1.

A1.9.2.4   Yield Function

Parameters a_1 and η_1 (Figure A1.19): For a given increment of stress on the stress–strain curve (Figure A1.19a), the plastic strain increment is computed using the unloading modulus, which is then used to compute the total plastic strain, ξ (Equation A1.29b). Then, the total stress at the end of the stress increment is found, and knowing γ and β, the value of hardening parameters, α is found from $F = 0$ (Equation A1.28). The values of such ξ and α for various points are plotted as in Figure A1.19b; an average line is usually assumed. Then, the slope of the line gives η_1, and the intercept along $\ln \alpha$ is used to find a_1.

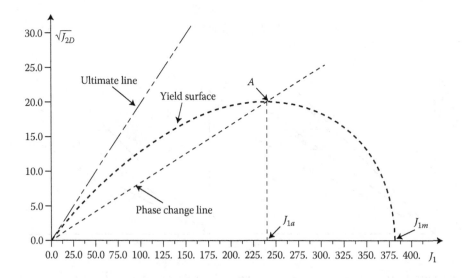

FIGURE A1.18   Phase change parameter, n: Procedure 2.

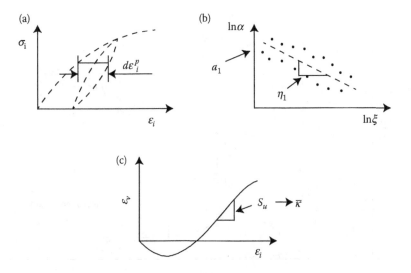

FIGURE A1.19 Determination of hardening parameters, (a) and (b) for a_1 and η_1; and (c) nonassociative parameters, for $\bar{\kappa}$.

A1.9.2.5 Cohesive Intercept

$3R$ or $\bar{c}$ is obtained from the plot in Figure A1.14. Another procedure can be used for the approximate value of R [2,23].

A1.9.2.6 Nonassociative Parameter, κ

Figure A1.19c shows the variation of ε_v versus ε_1. The slope of the curve, S_u, in the final zone is used to find this parameter; details are given in Ref. [2].

Further details of the procedures for the determination of the HISS parameters are given in Ref. [2].

A1.10   CREEP MODELS

A schematic of creep time-dependent behavior is shown in Figure A1.20. Here, we describe mainly the elastoviscoplastic (evp) Perzyna's model [24], which has been often used. In this model, the rate of strain vector, $\dot{\varepsilon}$, is decomposed into two components, elastic ($\dot{\varepsilon}^e$) and viscoplastic ($\dot{\varepsilon}^{vp}$), as follows:

$$\dot{\varepsilon} = \dot{\varepsilon}^e + \dot{\varepsilon}^{vp}$$
(A1.33)

The rate-dependent stress–strain behavior is then expressed as

$$\dot{\sigma} = C^e \dot{\varepsilon}^e$$
(A1.34a)

$$\dot{\varepsilon}^e = C^{(e)-1} \dot{\sigma}$$
(A1.34b)

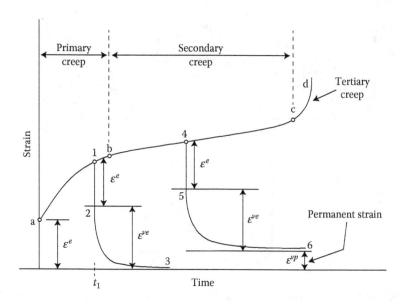

FIGURE A1.20   Schematic of creep behavior.

where σ is the stress vector and $\underline{C}^e$ is the (linear) elastic constitutive matrix, which for isotropic materials contains two constants (E and v or K and G). The viscoplastic strain rate that contains the inelastic or irreversible strain due to both the plastic and viscous effects is expressed as [2,24]

$$\dot{\varepsilon}^{vp} = \Gamma \langle \phi \rangle \frac{\partial Q}{\partial \sigma} \qquad (A1.34c)$$

Here, Q is the plastic potential function. For associative plasticity, the yield function $F \cong Q$; Γ is the fluidity parameter, φ is the flow function, which is expressed in terms of F, the overdot denotes the time rate, and the angle bracket $\langle \rangle$ represents a switch-on–switch-off operator:

$$\left\langle \phi \frac{F}{F_o} \right\rangle = \begin{cases} \left(\dfrac{F}{F_o} \quad \text{if} \quad \dfrac{F}{F_o} > 0 \right) \\ 0 \quad \text{if} \quad \dfrac{F}{F_o} \le 0 \end{cases} \qquad (A1.35)$$

In Equation A1.35, F_o is a reference value of F (e.g., yield stress, σ_y, atmospheric constant, p_a) used to render F dimensionless. The flow function is often used in different forms; two of the common forms are given by [2,24]

$$\phi = \left(\frac{F}{F_o} \right)^N \qquad (A1.36a)$$

$$\phi = \exp\left(\frac{F}{F_o}\right)^{\bar{N}} - 1.0 \qquad\qquad (A1.36b)$$

where N and $\bar{N}$ are material parameters.

A1.10.1 YIELD FUNCTION

The yield function in the Perzyna model [24] can be chosen from various models such as von Mises (Equation A1.16), Drucker–Prager (Equation A1.18), Mohr–Coulomb (Equation A1.20), critical state (Equation A1.24), cap (Equations A1.25 through A1.27), or HISS plasticity (Equation A1.28).

The parameters for elastoviscoplastic (evp) model are

Elastic modulus, E
Poisson's ratio, v
Fluidity, Γ
Flow function, N or $\bar{N}$
Yield function, F, as per the model selected

The determination of parameters for the above items is presented before, except for Γ and N or $\bar{N}$, which are presented next.

From Equation A1.34c, we can derive $\Gamma(\varphi) = \chi$ as

$$\chi = \Gamma(\varphi) = \sqrt{\frac{I_2^{vp}}{(\partial F / \partial Q)^T (\partial F / \partial Q)}} \qquad\qquad (A1.37)$$

where I_{2D} is the second invariant of the deviatoric strain tensor. Now, we can plot χ versus F/F_o as shown in Figure A1.21a. The values of Γ and N can be evaluated from the plot of $\ln \chi$ versus $\ln(F/F_o)$ (Figure A1.21b).

Desai [2] has proposed and described the unified multicomponent DSC (MDSC) procedure from which various types of creep behavior such as viscoelastic (ve),

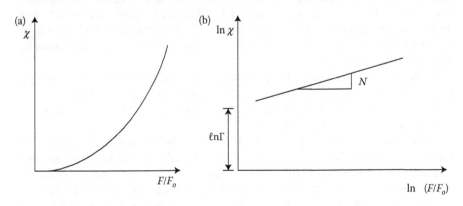

FIGURE A1.21   Determination of creep parameters.

elastoviscoplastic-Perzyna (evp), and viscoelasticviscoplastic (vevp) models can be derived as the special cases. The details of MDSC are given in Ref. [2].

A1.11   DISTURBED STATE CONCEPT MODELS

The previously described models are based on the assumption that the material is continuous. However, when softening or degradation occurs in a deforming material, the continuum approach may not be applicable because the material contains initial and induced discontinuities. The DSC has been developed as a unified and hierarchical approach in which the effect of discontinuities and resulting softening is allowed; the approach contains various continuum models as special cases [2,25]. An advantage of the DSC is that it allows healing and strengthening during deformation.

The DSC is based on the idea that a deforming material, undergoing microstructural modifications, can be assumed to consist of more than one component. For a dry solid, the material is considered to be a mixture of continuum or relative intact (RI) and ultimate or fully adjusted (FA) part. The term "relative" in RI is used because such a state can be defined as being dependent on factors such as confining pressure and temperature. The FA state denotes the asymptotic condition of the material, approached when microcracking grows to (near) disintegration of the material.

Figure A1.22a shows the three typical states in a deforming material. A symbolic representation of the DSC is shown in Figure A1.22b in which the open circle at the top denotes the initial state when no disturbance may exist. As the material deforms, the disturbance grows, denoted by a dark part of the circle. At critical disturbance, D_c, the material may experience the initiation of microstructural instability (failure, liquefaction, etc.), and full failure may take place at D_f. The states D_u and $D = 1$ are not attainable. The schematics of the stress–strain, static, and cyclic behavior are shown in Figures A1.22c and A1.22d, respectively; various states are also marked on these figures.

The idea of the disturbance, D, is introduced, which acts as a coupling function between the RI and FA states. For an initially disturbance-free material, D varies from zero in the beginning and approaches unity as depicted in Figure A1.23a; here, D is plotted versus accumulated plastic (shear) strain, ξ_D, number of cycles (N), or time (t). When a critical disturbance (D_{cm}) is reached, Figure A1.23a, at an intermediate stage, microcracking may initiate. The microcracks grow and coalesce and lead to fracture around the critical disturbance, D_c. The disturbance, D_f, may indicate fracture and failure. The ultimate state, D_u, is asymptotic and usually cannot be achieved in laboratory tests.

In the DSC, we can model softening or degradation, and also stiffening or healing (Figure A1.23b). The idea of DSC is based on a different viewpoint than the classical damage model [26]; the differences and advantages of the DSC compared to the damage model are described in Ref. [2].

In the DSC model, we need to define RI and FA responses and the disturbance. In the intact state, the behavior represents as if the material does not experience any disturbance, and behaves as a continuum. Then, the RI response can be defined by adopting a continuum model such as based on elasticity and plasticity, as described above.

The behavior for each factor such as initial confining pressure can exhibit its own (RI) response and the asymptotic ultimate state. The term "relative" in *relative intact*

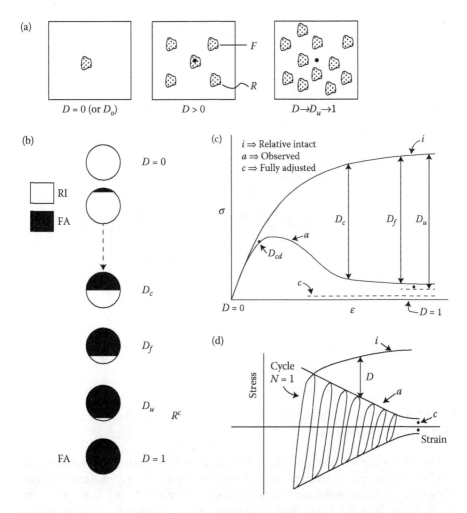

FIGURE A1.22 Representation of disturbed state concept. (a) Clusters of RI and FA parts; (b) symbolic representation of DSC; (c) schematic of static stress–strain response; and (d) cyclic stress–strain response.

refers to such state for each factor. The material parameters for a given model for the RI state are determined on the basis of appropriate tests under various factors; the procedures for the determination of parameters are given before in this appendix.

The FA state, denoted by c, is considered approximately as the asymptotic state (Figures A1.22c and A1.22d). As it is not possible to define the behavior in the asymptotic state, usually, a state such as at D_f or at D_c can be adopted to represent the FA state.

The continuous microstructural changes increase (or decrease in the case of strengthening or healing), and the disturbance passes through various stages or thresholds in the material such as initiation of microcracking, transition from contractive to dilative state, peak, and failure states (Figure A1.22c).

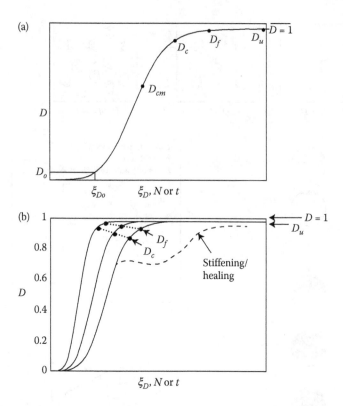

FIGURE A1.23 (a) Disturbance versus ξ_D, N or t; (b) schematic of disturbance during softening and stiffening (healing).

One of the important attributes of the DSC is that the component materials, RI and FA, are coupled together and interact continuously during deformation. The coupling function between RI and FA parts is the disturbance. At any state during the deformation, the disturbance defines the states that lie between the reference states RI (*i*) and FA (*c*) (Figure A1.22c).

A1.11.1 DSC EQUATIONS

The DSC incremental equations are expressed as [2]

$$d\underset{\sim}{\sigma}^a = (1 - D)\underset{\sim}{C}^i d\underset{\sim}{\varepsilon}^i + D\underset{\sim}{C}^c d\underset{\sim}{\varepsilon}^c + dD(\underset{\sim}{\sigma}^c - \underset{\sim}{\sigma}^i) \qquad (A1.38a)$$

or

$$d\underset{\sim}{\sigma}^a = \underset{\sim}{C}^{DSC} d\underset{\sim}{\varepsilon} \qquad (A1.38b)$$

where *a*, *i*, and *c* denote observed, RI, and FA behaviors, respectively (Figures A1.22c and A1.22d), $\underset{\sim}{C}^i$ is the constitutive matrix for the RI state, which can be adopted as a model based on linear or nonlinear elasticity, plasticity, and other continuum models, $\underset{\sim}{C^c}$ is the constitutive matrix for the FA state, and $\underset{\sim}{C^{DSC}}$ is the constitutive matrix for DSC.

For the RI state, the parameters for the matrix $\underset{\sim}{C^i}$ can be based on elasticity, conventional plasticity, continuous yield plasticity or HISS plasticity, and so on. We can characterize the FA state by using various assumptions regarding the strength of the material in the FA state as

1. Zero strength like in the classical Kachanov model [26]. Here, the material in the FA state is assumed to possess no strength at all. In other words, the FA material does not interact with the RI (continuum) part. Such a *local* model is not advisable because it involves certain computational difficulties such as spurious mesh dependence [2,27].
2. Hydrostatic strength or constrained liquid. In this case, the FA part continues to carry only the mean pressure or hydrostatic stress, defined by the bulk modulus, K. Then, the FA and RI parts interact with each other.
3. CS or constrained liquid–solid. When the FA material is assumed to act as a constrained liquid–solid, it can be defined by the CS model [2,12,13] in which the material deforms under shear with invariant volume. The equations to define the CS are given by Equations A1.22a and A1.22b. Other definitions of the FA states are described in Ref. [2].

A1.11.2 DISTURBANCE

Disturbance can be defined in various ways based on the available test data, for example, (1) stress–strain, (2) volumetric or void ratio, (3) pore water pressure, or effective stress, and (4) nondestructive behavior such as shear (S) and volumetric (P) wave velocities [2] (Figure A1.24).

The disturbance, D, based on the stress–strain behavior can be defined as (Figure A1.24a)

$$D = \frac{\sigma^i - \sigma^a}{\sigma^i - \sigma^c} \tag{A1.39a}$$

The stress σ can be in various forms such as $(\sigma_i - \sigma_3)$, $\sqrt{J_{2D}}$, effective stress, void ratio, and pore water pressure. In terms of ultrasonic wave velocity, it is expressed as (Figure A1.24c)

$$D = \frac{V^i - V^a}{V^i - V^c} \tag{A1.39b}$$

where V denotes velocity.

The disturbance, D, can be expressed in terms of internal variables such as accumulated plastic strain or plastic deviatoric strain, or plastic work. In case of the former two, D can be expressed as

$$D = D_u(1 - e^{-A\xi^Z}) \tag{A1.40a}$$

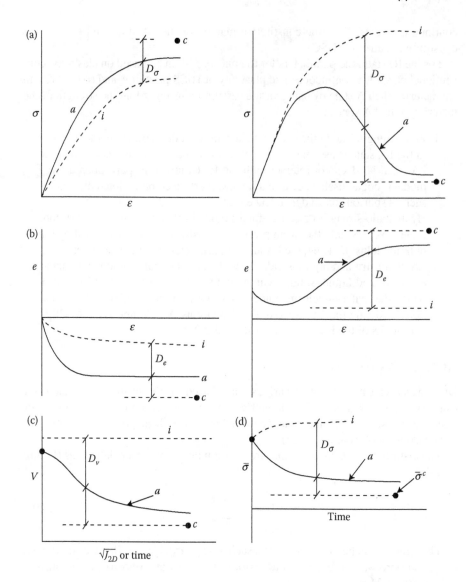

FIGURE A1.24 Disturbance from various test data. (a) Stress-strain; (b) void ratio; (c) non-destructive velocity; and (d) effective stress.

or

$$D = D_u(1 - e^{-A\xi_D^Z})$$ (A1.40b)

where D_u is the ultimate disturbance, which can be sometimes adopted as unity, A and Z are parameters, and ξ and ξ_D are the accumulated plastic strain and deviatoric plastic strains, respectively.

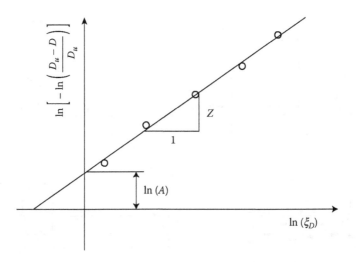

FIGURE A1.25 Determination of disturbance parameter.

The parameters for the RI part are the same as described above for elasticity, plasticity, and evp models. If the FA state is assumed to be the CS, the parameters $\bar{m}$ (M), λ, e_o, and so on can be determined by using Equation A1.22. The determination of the parameters in D, the disturbance function (Equation A1.40), are given below:

D_u can be determined from (approximate) the ultimate state (around c) in Figures A1.22c and A1.22d.

A and Z can be found from the test data involving softening or degradation, by using Equation A1.40 as

$$Z\ell n(\xi_D) + \ell n(A) = \ell n\left[-\ell n\left(\frac{D_u - D}{D_u}\right)\right]$$ (A1.41)

and plotting $\ell n\xi_D$ versus $\ell n[-\ell n((D_u - D)/D_u)]$ (Figure A1.25). The slope of the average line in the plot, (Figure A1.25) gives Z and the intercept along the vertical axis leads to A. Further details of the procedures are given in Ref. [2].

A1.11.3 DSC MODEL FOR INTERFACE OR JOINT

The behavior of an interface or joint can be modeled by using the DSC. As in the case of a solid, the thin-layer interface zone is assumed to contain RI and FA states (Figure A1.5c). The foregoing DSC equations for "solid," soil or rock, can be specialized for the interface. For instance, for the two-dimensional interface (Figure A1.26), the significant stresses that remain are shear stress, τ, and normal stress, σ_n (Figure A1.5b), and the corresponding relative shear (u_r) and normal (v_r) displacements (Figure A1.26d) [2]. The schematics of the interface test behavior in terms of the shear stress versus relative shear displacement (u_r), the normal stress (σ_n) versus relative normal displacement (v_r), and the behavior in terms of u_r vs, and v_r are shown in Figures A1.27a through A1.27c, respectively.

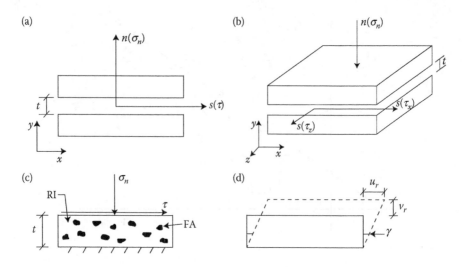

FIGURE A1.26 Interface (joint) idealizations in DSC and relative motions.

The DSC incremental equations for the two-dimensional case can be expressed for $d\tau$ and $d\sigma_n$ as [2,25]

$$\left\{\begin{matrix} d\tau^a \\ d\sigma_n^a \end{matrix}\right\} = (1-D)\,\underset{\sim}{C}_j^i \left\{\begin{matrix} du_r^i \\ dv_r^i \end{matrix}\right\} + D\,\underset{\sim}{C}_j^c \left\{\begin{matrix} du_r^c \\ dv_r^c \end{matrix}\right\} + \mathrm{dD}\left\{\begin{matrix} \tau^c - \tau^i \\ \sigma_n^c - \sigma_n^i \end{matrix}\right\} \qquad (A1.42)$$

where, as before, a, i, and c denote observed, RI, and FA states, respectively. Figure A1.26c shows the interface material element composed of RI and FA states.

The constitutive matrix, $\underset{\sim}{C}_j^i$, is defined according to the assumption for the RI behavior, such as elastic or plastic. For the elastic behavior, the shear stiffness, k_{st}, and the normal stiffness, k_{nt} (Figures A1.27a and A1.27b), define the constitutive matrix. For the conventional plasticity model, such as Mohr–Coulomb, four parameters are needed, elastic: k_{st} and k_{nt}; plasticity: adhesive strength c_a; and interface friction angle, δ.

For the HISS plasticity, the yield function can be specialized from Equation A1.28 for a solid as (Figure A1.28)

$$F = \tau^2 + \alpha\sigma_n^{*n} - \gamma\sigma_n^{*2} = 0 \qquad (A1.43)$$

where $\sigma_n^* = \sigma_n + R$; R is the intercept along the negative σ_n axis, which is used to define the adhesive strength, c_a, and γ and n are related to the ultimate and phase transition (from contractive to dilative behavior) (Figure A1.27c). As in the case of "solid," the continuous yield or hardening function, α, can be defined as

$$\alpha = \frac{a}{\xi^b} \qquad (A1.44a)$$

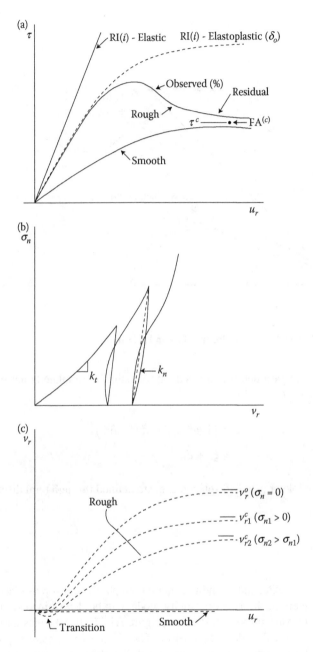

FIGURE A1.27 Schematic of stress-relative displacement for shear, normal and dilative behavior. (a) $\tau - u_r$; (b) σ_n versus v_r; and (c) v_r versus u_r.

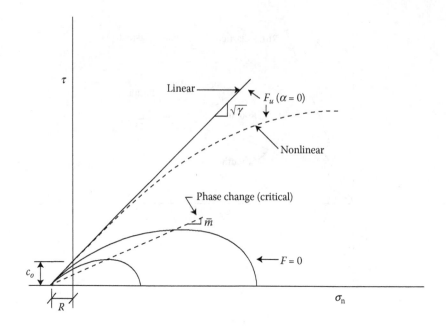

FIGURE A1.28   Yield surfaces for interface in HISS model.

where a and b are parameters and ξ is the irreversible accumulated relative displacements or trajectory given by

$$\xi = \int \left(du_r^p \, dv_r^p + dv_r^p \cdot dv_r^p \right)^{1/2}$$

$$= \xi_v + \xi_D \qquad\qquad\qquad (A1.44b)$$

The irreversible or plastic displacements are defined through total displacements, u_r and v_r, as

$$u_r = u^e + u^p \qquad\qquad\qquad (A1.45a)$$

$$v_r = v^e + v^p \qquad\qquad\qquad (A1.45b)$$

where e denotes elastic, and u^p and v^p contain both plastic and slip displacements [2].

The parameters of the HISS plasticity model can be determined from interface shear tests under various normal stresses (Figure A1.27). For the elasticity behavior, k_{st} and k_{nt} are obtained as described before. The plasticity parameters, γ, n, a, b, are determined by using similar procedures as that for solids, as presented above.

The disturbance D can be defined from test data such as τ versus u_r behavior, and normal effective stress versus u_r. For instance, for the former (Figure A1.29a):

$$D_\tau = \frac{\tau^i - \tau^a}{\tau^i - \tau^c} \qquad\qquad\qquad (A1.46a)$$

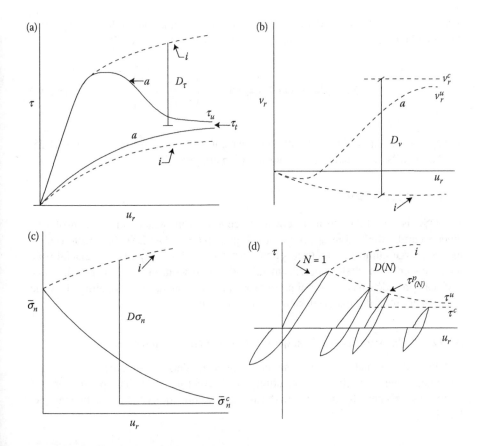

FIGURE A1.29 Disturbance from various test data. (a) τ versus u_r; (b) v_r versus u_r; (c) effective stress $\bar{\sigma}_n$ versus u_r; and (d) cyclic: τ versus u_r.

and for the latter (Figure A1.29c):

$$D_{\sigma_n} = \frac{\bar{\sigma}_n^i - \bar{\sigma}_n^a}{\bar{\sigma}_n^i - \bar{\sigma}_n^c} \tag{A1.46b}$$

The disturbance function can be defined as

$$D = D_{\tau u} \left(1 - e^{-A_\tau \, \xi_D^{Z_\tau}}\right) \tag{A1.47a}$$

and

$$D = D_{nu}\left(1 - e^{-A_n \, \xi_v^{Z_n}}\right) \tag{A1.47b}$$

where D is the disturbance defined from the test data (Equation A1.46a). The parameters can be determined, for example, by expressing Equation A1.47 as

$$Z_\tau \ell n(\xi_D) + \ell n(A_\tau) = \ell n\left[-\ell n\left(\frac{D_{\tau u} - D}{D_{\tau u}}\right)\right] \qquad (A1.48)$$

and by plotting $\ell n\ \xi_D$ versus the term on the right-hand side, similar to Figure A1.25. Further details of the parameter determination are given in Ref. [2].

A1.12   SUMMARY

The DSC is a unified and hierarchical procedure, from which various constitutive models can be derived as special cases (Figure A1.30). The DSC has been used to model a large number of materials and interfaces such as clays, sands, glacial tills, rocks, concrete, asphalt concrete, metals, alloys, silicon, and polymers [2]. Thus, DSC can provide a basic and unique approach for the constitutive modeling of a wide range of engineering materials and interfaces/joints.

A1.12.1   PARAMETERS FOR SOILS, ROCKS, AND INTERFACES/JOINTS

The lists of parameters for various constitutive models for specific materials are presented together with the applications in different chapters. The parameters for various models can also be found in the available publications and texts, for example, Ref. [2].

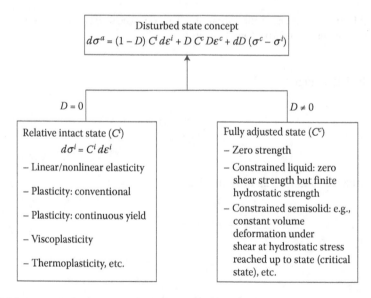

FIGURE A1.30   Hierarchical versions in unified DSC.

REFERENCES

1. Desai, C.S. and Siriwardane, H.J., *Constitutive Laws for Engineering Material*, Prentice-Hall, Englewood Cliffs, NJ, USA, 1984.

2. Desai, C.S., *Mechanics of Materials and Interfaces: The Disturbed State Concept*, CRD Press, Boca Raton, FL, USA, 2001.

3. Chen, W.F. and Han, D.J., *Plasticity for Structural Engineers*, Springer-Verlag, New York, 1988.

4. Kondner, R.L., Hyperbolic stress-strain response: Cohesive soils, *Journal of the Soil Mechanics and Foundations Division, ASCE*, 89(SM1), 1963, 115–163.

5. Duncan, J.M. and Chang, C.Y., Nonlinear analysis of stress and strain in soils, *Journal of the Soil Mechanics and Foundations Division, ASCE*, 96(SM5), 1970, 1629–1653.

6. Kulhawy, F.H., Duncan, J.M., and Seed, H.B., Finite Element Analysis of Stresses and Movements in Embankment During Construction, Report 569-8, U.S. Army Corps of Eng., Waterway Expt. Sta., Vicksburg, MS, USA, 1969.

7. Desai, C.S., Zaman, M.M., Lightner, J.G., and Siriwardane, H.J., Thin-layer element for interfaces and joints, *International Journal for Numerical and Analytical Methods in Geomechanics*, 8(1), 1984, 19–43.

8. Ramberg, W. and Osgood, W.R., Description of Stress-Strain Curves by Three Parameters, Tech. Note 902, National Advisory Committee, Aeronaut., Washington, DC, 1943.

9. Desai, C.S. and Wu, T.H., A general function for stress-strain curves, *Proceedings of the 2nd International Conference on Numerical Methods in Geomechanics*, C.S. Desai (Editor), Blacksburg, VA, ASCE, 1976.

10. Baron, M.L., Nelson, I., and Sandler, I., Influence of Constitutive Models on Ground Motion Predictions, Contract Report S-71-10, No. 2, U.S. Army Corps of Engrs., Waterways Expt. Stn., Vicksburg, MS, USA, 1971.

11. Hill, R., *The Mathematical Theory of Plasticity*, Oxford Univ., Oxford, UK, 1950.

12. Roscoe, K.H., Schofield, A.N., and Wroth, C.P., On yielding of soils, *Geotechnique*, 8, 1958, 22–53.

13. Schofield, A.N. and Wroth, C.P., *Critical State Soil Mechanics*, McGraw-Hill, London, 1968.

14. DiMagio, F.L. and Sandler, I., Material model for granular soils, *Journal of Engineering Mechanics, ASCE*, 19(3), 1971, 935–1950.

15. Desai, C.S., Somasundaram, S., and Frantziskonis, G., A hierarchical approach for constitutive modelling of geologic materials, *International Journal for Numerical and Analytical Methods in Geomechanics*, 10(3), 1986, 225–257.

16. Desai, C.S. and Siriwardane, H.J., A concept of correction functions for non-associative characteristics of geologic me3dfia, *International Journal for Numerical and Analytical Methods in Geomechanics*, 4, 1980, 377–387.

17. Desai, C.S. and Hashmi, Q.S.E., Analysis, evaluation, and implementation of a non-associative model for geologic materials, *International Journal of Plasticity*, 6, 1989, 397–420.

18. Somasundaram, S. and Desai, C.S., Modelling and testing for anisotropic behavior of soils, *Journal of Engineering Mechanics, ASCE*, 114, 1988, 1473–1496.

19. Katti, D.R. and Desai, C.S., Modelling and testing of cohesive soil using the disturbed state concept, *Journal of Engineering Mechanic*, 121(5), 1995, 648–658.

20. Desai, C.S. and Toth, J., Disturbed state constitutive modeling based on stress-strain and nondestructive behavior, *International Journal of Solids Structures*, 33(11), 1996, 1619–1654.

21. Matsuoka, H. and Nakai, T., Stress-deformation and strength characteristics of soil under three different principal stresses, *Proceedings of the Japanese Society of Civil Engineers*, 232, 59–70, 1974.

22. Lade, P.V. and Kim, M.K., Single hardening constitutive model for frictional material—III: Comparisons with experimental data, *Computers and Geotechnics*, 6, 1988, 31–47.

23. Lade, P.V., Three-parameter failure criterion for concrete, *Proceedings of ASCE*, 108(5), 1982, 850–863.

24. Perzyna, P., Fundamental problems in viscoplasticity, *Advances in Applied Mechanics*, 9, 1966, 243–277.

25. Desai, C.S. and Ma, Y., Modelling of joints and interfaces using the disturbed state concept, *International Journal for Numerical and Analytical Methods in Geomechanics*, 16(9), 1992, 623–653.

26. Kachanov, L.M., *Introduction to Continuum Damage Mechanics*, Martinus Nijhoff Publishers, Dondrecht, The Netherlands, 1986.

27. Mühlhaus, H.B. (Ed.), *Continuum Models for Materials with Microstructure*, John Wiley, UK, 1995.

Appendix 2: Computer Software or Codes

A2.1 INTRODUCTION

Solutions to geotechnical problems can be achieved by using software or codes based on various computer-oriented methods such as finite element, finite difference, boundary element, and analytical procedures. A variety of finite element computer codes (List 1 below) developed by the authors are used for the solutions of problems included in various chapters of this book.

Readers can also use other appropriate codes available to them.

Codes acquired from commercial companies (List 2 below) may also be suitable for geotechnical problems.

This appendix contains, in List 1, a number of computer codes developed and used by the authors; for further information, the contact email is csdesai@comcast.net.

Educational software: A number of finite element codes are available for introductory and educational purpose for solutions of one- and two-dimensional problems using linear (elastic) constitutive models. They allow solution of problems such as one-dimensional stress, consolidation, and thermal analysis; two-dimensional stress (plane and plane stress and axisymmetric) analysis; and field problems such as fluid flow (seepage), thermal flow, and torsion. These codes can be downloaded free of cost from the website: http://www.crecpress.com/product/catno/0618 (click "download updates" and 0618.zip) related to the textbook, *Introductory Finite Element Method*, CRC Press, Boca Raton, Florida, USA, 2001.

A2.2 LIST 1: FINITE ELEMENT SOFTWARE SYSTEM: DSC SOFTWARE

The following codes use *disturbed state concept* (DSC) constitutive modeling approach for two- and three-dimensional problems; the continuous yield *hierarchical single surface plasticity (HISS)* model is a part of the DSC; hence, it is referred to as the DSC/HISS approach. The details of various constitutive models are given in Appendix 1.

DSC/HISS is a general approach that contains a number of available models [e.g., linear and nonlinear elastic, conventional plasticity, continuous yield plasticity (e.g., critical state and cap), HISS plasticity, and creep]. DSC/HISS also allows microcracking leading to fracture, degradation or softening, healing or stiffening, and microstructural instability such as failure and liquefaction. At this time, it is perhaps the only unified approach with this unique scope, available for realistic modeling of geomaterials, interfaces, and joints.

For academic and comparative use, some of the following codes contain available models such as linear and nonlinear elastic (hyperbolic, Ramberg–Osgood), von Mises, Drucker–Prager, Mohr–Coulomb, critical state, and cap.

The computer codes developed and used for solutions of problems in various chapters of this book are described first in List 1 below. The chapter number related to the use of a specific code is inserted at the end of the description of each code.

1. *SSTIN-1DFE*: Finite element (FE) code for the solution of problems idealized as one-dimensional, with linear and nonlinear response simulated through p_y–v or p–y curves by using the Ramberg–Osgood model (*Chapter 2*).
2. *DSC-SST2D*: FE code for two-dimensional (plane strain, plain stress, and axisymmetric) static problems with nonlinear analysis using DSC/HISS (*Chapter 3*).
3. a. *STFN-3D Frame*: FE (approximate) code for three-dimensional simulation, including beam-columns, slab or plate, and springs for soil resistance, simulated by using the Ramberg–Osgood model (*Chapter 4*).
   b. *DSC-SST3D*: FE for full three-dimensional analysis for static and dynamic problems using the DSC/HISS models (*Chapter 4*). This is same as 6(b) below.
4. *SEEP-2DFE and SEEP-3DFE*: FE analyses for two- and three-dimensional steady-state and transient free surface seepage problems (*Chapter 5*).
5. *CONS-1DFE*: FE analysis of one-dimensional consolidation (*Chapter 6*).
6. a. *DSC-DYN2D*: FE dynamic nonlinear analysis for coupled (Biot's theory) two-dimensional problems using DSC/HISS models, including liquefaction (*Chapter 7*).
   b. *DSC-SST3D*: FE analysis of static and dynamic nonlinear analysis of coupled (generalized Biot's theory) three-dimensional problems using DSC/HISS models (*Chapter 7*).

All the above codes are written in FORTRAN, except DSC-SS3D, which is written in C++.

A2.3   LIST 2: COMMERCIAL CODES

The software applicable to geotechnical problems can be obtained from various commercial companies. Some of them are described below.

ANSYS, Inc. is one of the world's leading engineering simulation software providers. Its technology has enabled customers to predict with accuracy that their product designs will thrive in the real world. The comprehensive range of engineering tools gives users access to virtually any field of engineering simulation that their design process requires.

The ANSYS® hallmark mechanical product performs nonlinear (finite element) analysis of engineering challenges. Drawing on decades of "firsts" and "bests" in structural simulation technology, the suite offers a wide range of material models, including hyperelastic, perfectly plastic, elastic–plastic hardening, creep, and damage for solids. It includes special models for simulating part contacts or interfaces.

This is accompanied by a comprehensive elements library, including SOLID, SHELL, BEAMS, and so on, which are often based on coupled (u–p) formulation for displacement and hydrostatic pressure for pore water pressure.

The suite of engineering simulation tools from ANSYS is a solution set of unparalleled breadth that goes well beyond finite element analysis (FEA) to include interoperative structural, fluid flow, thermal, electromagnetic, embedded software, and related technologies. These products offer the ability to perform comprehensive multiphysics analysis, critical for high-fidelity simulation of real architecture that integrates components.

A single, unified engineering simulation environment harnesses the core physics and enables their interoperability, which is critical for a quality solution. It also provides common tools for interfacing with CAD, repairing geometry, creating meshes, and postprocessing results.

The above capabilities in ANSYS allow its use for various *geotechnical* applications.

ANSYS, Inc. has operations in more than 60 locations around the world. For more information, contact ANSYS, Inc., Southpointe, 275 Technology Drive, Canonsburg, PA, 15317, USA. Telephone: 1-866-267 9724; fax: 1-704 -514 9494; email: ansysinfo www.ansys.com.

Other commercial companies that provide software for geotechnical applications can be found on the website: *Geotechnical and Geoenvironmental Engineering Software Directory* (www.ggsd.com).

Note: All available software may not contain constitutive models suitable for the realistic behavior of geomaterials, interfaces, and joints. Hence, in using any code, the user should verify whether it allows realistic constitutive models consistent with the observed behavior. The use of an unrealistic constitutive model may lead to unreliable results.

Index

Varved clay foundation consolidation *(Continued)*
  foundation soil details, 530, 532
  initial and applied pressures, 532
  material properties, 530
  parametric study on anisotropy, 533
  permeability coefficient for soil, 533
  varved clay properties, 533
Vertical stress, 149; *see also* Sequential
      construction
vevp models, *see* Viscoelasticviscoplastic models
      (vevp models)
VFM, *see* Viscous flow model (VFM)
VI method, *see* Variational inequality method
      (VI method)
Villarbeney natural slope, 216; *see also* Creeping
      slope analysis
Viscoelasticviscoplastic models (vevp models),
      584
Viscous flow model (VFM), 367, 368
  field analysis of seepage, 383–385
  finite element mesh for, 369
VM method, *see* Variable mesh method
      (VM method)
Volume change behavior of mixture,
      455–456
Volumetric strain
  and axial strain, 207, 209
  in differential equation, 410
  vs. maximum shear strain, 177

  vs. pressure, 414
  total, 576
von Mises model, 568
  compression test, 570
  constants, 569
  stress-paths, 569
  yield function, 568

W

Walls, 60; *see also* Beams; One-dimensional
      simulation; Pile
  depth of embedment of, 60
  p_y–v curve for retaining, 63
  retaining wall, 61
Winkler soil model, *see* Spring soil model

Y

Yield function, 568
  creep models, 582, 583
  in Drucker–Prager models, 570
  HISS, 576–577, 580
  von Mises model, 568
Yield surface, 573
  cap model, 575
  CS model and, 571, 573
  HISS, 577, 592
Young's modulus of soil, 70